FOUNDATIONS OF GAS DYNAMICS

This book covers supersonic and subsonic flow phenomena where compressibility of the fluid cannot be ignored. It finds application in jet and rocket propulsion systems as well as handling industrial gas flow at high speeds. Students and engineers in the mechanical, aerospace, and chemical disciplines should find it useful. It begins with basic concepts such as isentropic flows, shock, and supersonic expansion waves in one dimension. These are followed by one-dimensional flows with friction and heat exchange. Two-dimensional theory with small perturbations is presented, with its applications illustrated by supersonic airfoils. Method of characteristics is used for flows with two independent variables, either with two spatial coordinates or with time variations in one dimension. In later chapters, acoustic wave propagation, supersonic flow combustion, and unsteady shock formation are treated thoroughly. The book ends with a chapter on basic hypersonic flow, with a discussion of similarity rules.

Ruey-Hung Chen is currently the Robert G. Myers Professor and Head of the Mechanical and Aerospace Engineering Department of New Mexico State University. Prior to this appointment, he was a professor at University of Central Florida from 1993 to 2015. Between 2012 and 2015, Dr. Chen served as the Director of the Combustion and Fire Systems Program and also as the Acting Director of the Thermal Transport Processes Program, from 2014 to 2015, of the National Science Foundation. His research interests include combustion, heat and mass transfer, and compressible flow. Dr. Chen received his PhD degree in Aerospace Engineering from the University of Michigan and his BS in Aeronautical Engineering from National Cheng Kung University in Taiwan.

Foundations of Gas Dynamics

Ruey-Hung Chen

New Mexico State University

CAMBRIDGE
UNIVERSITY PRESS

One Liberty Plaza, 20th Floor, New York NY 10006, USA

Cambridge University Press is part of the University of Cambridge.

It furthers the University's mission by disseminating knowledge in the pursuit of education, learning, and research at the highest international levels of excellence.

www.cambridge.org
Information on this title: www.cambridge.org/9781107082700

First published 2017

Printed in the United States of America by Sheridan Books, Inc.

A catalogue record for this publication is available from the British Library.

Library of Congress Cataloging-in-Publication Data
Chen, Ruey-Hung.
Foundations of gas dynamics / Ruey-Hung Chen, University of Central Florida.
New York NY : Cambridge University Press, 2017. | Includes bibliographical references.
LCCN 2016032383 | ISBN 9781107082700
LCSH: Gas dynamics. | Gases – Thermal properties.
LCC QC168 .C487 2017 | DDC 533/.2–dc23
LC record available at https://lccn.loc.gov/2016032383

ISBN 978-1-107-08270-0 Hardback

Contents

Preface

Before embarking on writing this book, I asked myself: With some excellent text-books already covering this subject, why another one? In one aspect, this book is probably not different from the others in the field, in that it results from over thirty years of experience in learning and teaching the subject of compressible flow (or gas dynamics). The present book is written in what I believe to be an efficient way of learning and presenting the subject materials, reflecting my own perspective as a student and, more importantly, with feedback from students I have taught during the last two decades. Examples and exercise problems, which are many for this level of coverage, are chosen similarly.

For a typical mechanical and aerospace engineering curriculum, this book might serve well for senior- or first-year graduate students. The materials are more than sufficient for a one-semester course, allowing the instructor and students to choose materials based on the desired level of coverage. Practicing engineers should also find this book a useful review.

One readily finds that the physics of the subject defines the theme of the book. Results from mathematical derivations are explained so that readers can develop physical understanding and intuition. Therefore, I have made no attempt to incorporate numerical methods for solving governing equations of compressible flow. The approach is intuitive and students should only need a simple scientific calculator to follow and understand the materials. Tables and graphs are included wherever necessary to present key results and solve example and exercise problems. Detailed numerical treatises of gas dynamics can be found in many excellent books on this specialized topic, such as those listed in the Reference section. The example problems are chosen to facilitate understanding essential concepts and they might recur in several chapters, even with the same given Mach numbers, wedge and shock wave angles, and so on, to illustrate the development of theories as one progresses from one chapter to another with increasing depth. No attempt is made to obtain third decimal accuracy in the example problems, whether using a table or graph, because effort should be better focused on understanding the materials.

While striving to be intuitive and emphasizing physical understanding, the rigor of theory derivation is not lost. Detailed derivations are presented when they best illustrate the essence of how commonly used gas-dynamic equations are formulated, while teachers can assign the derivations of other equations as exercise problems. Exercise problems are chosen in a similar manner to the example problems (with the goal of enhancing the grasp of the materials). These exercise problems, although not voluminous, should be most beneficial to help with the student's learning.

The support of my family – my wife, Jessie, and my sons, Dar-Wei and Ryan – has been the greatest asset in all my endeavors. The education I received from National Cheng Kung University in Taiwan and the University of Michigan (Ann Arbor) has helped me prepare for my career as a teacher and researcher. I would like to dedicate this book to my family and these two institutions.

1 Properties and Kinetic Theory of Gases

The properties of gases have direct effects on how gases behave as they undergo mechanical or thermal changes. Thermal changes may result from heating due to heat-releasing chemical reactions, such as those taking place in the chemical propellant in rocket chambers, or from aerodynamic heating as on the external surfaces of high-speed flight vehicles or during reentry to the atmosphere. Mechanical changes are associated with processes such as compression (usually accompanied by deceleration in flowing systems) and expansion (usually accompanied by acceleration in flowing systems). These changes usually occur as a consequence of pressure gradients, changes in volume of a given mass, or the existence of body forces arising from gravitational, electrical, or electromagnetic fields. Mechanical and thermal changes may accompany each other, as in the case of isentropic compression/expansion processes or in a constant-volume heating process. They may also take place simultaneously, as in the case of the chemically reacting flow within a rocket chamber, where heating by heat-releasing chemical reactions occur; these chemical reactions can continue through the converging-diverging nozzle, where the flow undergoes expansion and acceleration, and beyond the nozzle exit. Chemical changes are not within the scope of this book, except where it is noted.

The degree with which thermal changes can be measured depends on the properties of the gas. Gas properties are the average quantities of a large number of gas molecules or atoms, which are in a state of constant motion at temperatures above absolute zero. Under normal conditions (i.e., in the macroscopic or "continuum" limit), gas properties can be easily determined experimentally or from thermodynamic considerations. However, in a rarefied gas, with very low density (as one finds under low-pressure conditions such as in the upper atmosphere), such average properties are not readily defined or obtained. It is therefore desirable to look at the behavior and motion of each individual molecule, that is, at the microscopic level. At the microscopic level, one expects the motion of molecules to be influenced by temperature so as to reflect different thermodynamic properties at the macroscopic level. Gas kinetics (or the kinetic theory of gases) is the study of gas molecules at the microscopic level as influenced by temperature. However, kinetic theory can be applied to both rarefied and dense gases, with their macroscopic properties reflecting the collective behavior of individual molecules.

1.1 Maxwellian Velocity Distribution

Assuming the average spacing between molecules of a given gas species is sufficiently large (larger than the diameter of the molecule so that the gas is "dilute" or the volume taken up by the molecules is negligibly small compared with the volume containing them), molecules do not interact (i.e., do not experience forces from one another) except during collisions. Under such circumstances, the distribution of molecular velocity follows the Maxwellian distribution, for gases in equilibrium states (Vincenti and Kruger, 1975):

$$f(C_1, C_2, C_3) = \left(\frac{m}{2\pi kT}\right)^{3/2} exp\left[-\frac{m}{2kT}\left(C_1^2 + C_2^2 + C_3^2\right)\right] \tag{1.1}$$

where C_1, C_2, and C_3, respectively, are the three velocity components in the Cartesian coordinate system and m, k, and T are the mass of the molecule, Boltzmann constant, and the absolute temperature, respectively. The magnitudes of C_1, C_2, and C_3 range from $-\infty$ to $+\infty$. Equation (1.1) indicates that $f(C_1, C_2, C_3)$ is an even function of C_1, C_2, and C_3. When only the magnitude of molecular speed, $u\ (=\sqrt{C_1^2 + C_2^2 + C_3^2})$, is of interest, the probability of finding molecules with speed between u and $u + du$ (regardless of direction) is

$$\chi(u) = 4\pi\left(\frac{m}{2\pi kT}\right)^{3/2} u^2 exp\left[-\frac{mu^2}{2kT}\right] \tag{1.2}$$

where the magnitude ranges from 0 to $+\infty$. The result of Eqn. (1.2) for $\chi(u)$ vs. u for a family of temperatures T is schematically shown in Fig.1.1. It is seen that for realistic temperatures, the value of $\chi(u)$ is less than 1. This also holds true for $f(C_1, C_2, C_3)$, for the reason provided in the following.

Using the definite integral and by noting that f is an even function of C_1, C_2, and C_3,

$$\int_0^\infty exp\left(-ax^2\right)dx = \frac{1}{2}\left(\frac{\pi}{a}\right)^{1/2}$$

one finds

$$\int_{-\infty}^{+\infty} f(C_1, C_2, C_3)dC_1 dC_2 dC_3$$
$$= 2^3\left(\frac{m}{2\pi kT}\right)^{3/2}\int_0^\infty exp\left[-\frac{mC_1^2}{2kT}\right]dC_1\int_0^\infty exp\left[-\frac{mC_2^2}{2kT}\right]dC_2\int_0^\infty exp\left[-\frac{mC_3^2}{2kT}\right]dC_{13} \tag{1.3}$$
$$= 2^3\left(\frac{m}{2\pi kT}\right)^{3/2}\left[\left(\frac{1}{2}\right)\left(\frac{2\pi kT}{m}\right)^{1/2}\right]^3 = 1$$

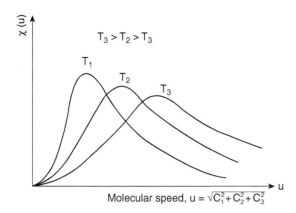

Figure 1.1 Schematic of probability function $\chi(u)$ of finding a molecule having a speed in the range of $(u, u + du)$, as function of absolute temperature; $u = \sqrt{C_1^2 + C_2^2 + C_3^2}$ is the molecular speed. Note that the probability of finding high-speed molecules increases with the temperature.

From Eqn. (1.3) the Maxwellian velocity distribution function, Eqn. (1.1), is thus the probability of finding molecules that have three velocity components between C_1 and $C_1 + dC_1$, C_2 and $C_2 + dC_2$, and C_3 and $C_3 + dC_3$. This probability is also the fraction (which is less than 1) of the molecules that possess the velocity components in the said ranges. The total probability of finding molecules with velocity magnitudes between $-\infty$ and $+\infty$ is 1, as expected, and the probability of finding molecules in any given velocity range is always smaller than 1. The same result of the unity total probability using Eqn. (1.2) can easily be found and is left as an end-of-chapter problem.

1.2 Characteristic Molecular Velocities

The Maxwellian velocity distribution enables the calculation of magnitudes of some characteristic velocities of a large number of molecules. Examples include the following.

1. The most probable speed of the molecules (u_{mp}) is the speed at which $\chi(u)$ has the maximum value. By setting the derivative of $\chi(u)$ to be zero, u_{mp} is found to be

$$u_{mp} = \left(\frac{2kT}{m}\right)^{1/2}. \tag{1.4}$$

2. The average speed ($\bar{u}$) is

$$\bar{u} = \int_0^\infty u\chi(u)du = 4\pi\left(\frac{m}{2\pi kT}\right)^{3/2}\int_0^\infty u^3 exp\left[-\frac{mu^2}{2kT}\right]du. \tag{1.5}$$

By adopting the definite integral

$$\int_0^\infty x^3 exp(-ax^2)dx = \frac{1}{2a^2},$$

one finds

$$\bar{u} = \left(\frac{8kT}{m\pi}\right)^{1/2} \approx 1.13 u_{mp}. \tag{1.6}$$

3. The root-mean-square speed ($u_{rms} = \left(\overline{u^2}\right)^{1/2}$) is related to the kinetic energy of the molecules. $\overline{u^2}$ is obtained as follows:

$$\overline{u^2} = \int_0^\infty u^2 \chi(u) du = 4\pi \left(\frac{m}{2\pi kT}\right)^{3/2} \int_0^\infty u^4 exp\left[-\frac{mu^2}{2kT}\right] du. \tag{1.7}$$

By using the definite integral

$$\int_0^\infty x^4 exp(-ax^2) dx = \frac{3}{8}\left(\frac{\pi}{a^5}\right)^{1/2} = \left(\frac{3kT}{m}\right),$$

the root-mean-square speed is

$$u_{rms} = \left(\frac{3kT}{m}\right)^{1/2} \approx 1.22 u_{mp}. \tag{1.8}$$

The average kinetic energy (or the translational energy) of a molecule is therefore

$$\tilde{e}_{tr} = \frac{1}{2} m u_{rms}^2 = \frac{3}{2} kT \tag{1.9}$$

Finding the values of $u_{mp}, \bar{u}, u_{rms}$, and $\tilde{e}_{tr}$ is straightforward and Problems 1.2 and 1.3 should serve as good examples for such an exercise.

A few observations can be made regarding the Maxwellian velocity distribution and the magnitudes of characteristic molecular velocities shown in Eqn. (1.5) through (1.8).

(i) The probabilities for molecules to move with the same velocity magnitude but in opposite directions are equal, as expected by examining Eqn. (1.1) as $f(C_1, C_2, C_3)$ is an even function of each of the three velocity components. To find the velocity magnitudes, $\chi(u)$ can therefore be used in place of $f(C_1, C_2, C_3)$.

(ii) The result of $u_{rms} > \bar{u} > u_{mp}$ can readily be understood by considering, for example, the distribution shown in Fig. 1.1 and the weighting factors of u^3 and u^4 in Eqns. (1.5) and (1.7), where the increasing exponent favors the contribution from high-speed molecules.

(iii) These characteristic speeds ($u_{rms}, \bar{u}$, and u_{mp}) are of the same order of magnitude. Because the speed of sound (denoted by a) results from molecular motion and collisions, one can expect it to be of the same order of magnitude as these three characteristic molecular speeds. Because pressure results from molecular collisions, it is also expected to be closely related to the magnitude of these characteristic velocities, and specifically to u_{rms}, as shown in the following section.

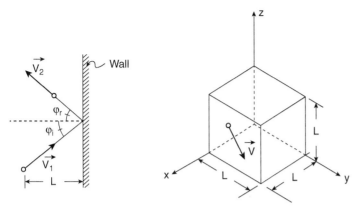

Figure 1.2 Collision of a particle (for example, a gas molecule) with walls (left figure) and a particle having a velocity of $\vec{V}$ in a cubic box with length L on each side (right figure).

1.3 Pressure and the Ideal Gas Law

Pressure (p) is defined as the force normal to the surface divided by the surface area on which it exerts; it is therefore a normal stress. According to Newton's second law of motion, the force exerted by a gas molecule (denoted as the ith molecule) on a surface is the rate of momentum change due to its collision with the surface. Referring to Fig.1.2, assume that the collision is "elastic" (i.e., the angle of reflection equals the angle of incidence, $\varphi_r = \varphi_i$, and the speed does not change as a result of collision, $u_r = |\vec{V}_r| = |\vec{V}_i| = u_i$), that, and that in a small region near the surface, the molecules collide only with the surface and not with any other molecule. Let L denote the average distance the molecule travels during collision with the wall, a box of a volume L^3 can be constructed with each of six surface areas being equal to L^2. Let the magnitude of the velocity normal to the surface be $u_{i,x}$. Then the momentum change throughout the collision process is $2m_i u_x$ and the time to complete the process is $2L/u_{i,x}$. Therefore, the pressure due to this molecular collision using Newton's second law is

$$(2m_i u_{i,x}) \div (2L/u_{i,x}) \div L^2 = \frac{m_i u_{i,x}^2}{L} = \frac{m_i u_{i,x}^2}{V}. \tag{1.10}$$

where $V = L^3$ is the volume of the box. Therefore the pressure on the surface facing the x-direction is

$$P_x = \frac{1}{V} \sum_i m_i u_{i,x}^2 \tag{1.11}$$

In a more general form, the pressure in the volume of L^3 is taken to be the average of the pressures in the three coordinate directions:

$$p = \frac{1}{3}\left(p_x + p_y + p_z\right) = \frac{1}{3V}\sum_i \left(m_i u_{i,x}^2 + m_i u_{i,y}^2 + m_i u_{i,z}^2\right) = \frac{1}{3V}\sum_i \left(m_i u_i^2\right) \tag{1.12}$$

Equation (1.12) can be applied to a mixtures of different species (i.e., $m_i \neq m_j$, where i and j denote different species). For a single gas species, with $m_i = m_j = m$. Comparing Eqns. (1.8) and (1.12) and keeping in mind that u_i represent all possible velocity magnitudes ranging from 0 to $+\infty$ leads to

$$p = \frac{1}{3V}\sum_i \left(m_i u_i^2\right) = \frac{2}{3V}\sum_i \left(\frac{1}{2}m_i u_i^2\right) = \frac{2}{3V}\sum_i \tilde{e}_{tr,i} = \frac{Nmu_{rms}^2}{3V} \tag{1.13}$$

where N is the total number of molecules in the volume. By substituting Eqn. (1.9) into Eqn. (1.13), one finds

$$pV = NkT \tag{1.14}$$

The assumptions of large spacing and non-interaction between molecules except during the collision process are essentially the same as the ideal gas assumption of classical, macroscopic thermodynamics. Therefore Eqn. (1.14) is identical to the equation of state for an ideal gas from classical thermodynamics:

$$pV = n\hat{R}T \tag{1.15}$$

where n and $\hat{R}$ are, respectively, the number of moles of gases contained in volume V and the universal gas constant ($\hat{R} = 8.3143$ kJ/kmol·K). Setting $\hat{N}$ to be the Avogadro's number (6.02252×10^{23} molecules per mole or 6.02252×10^{23}/mole), $n = N/\hat{N}$ and

$$k = \frac{\hat{R}}{\hat{N}} \tag{1.16}$$

which yields the Boltzmann constant $k = 1.3805 \times 10^{-23}$ J/K.

Equations (1.14) and (1.15) are called the ideal gas law, valid for gas in equilibrium. Alternatively, gas that behaves according to these equations are called the ideal gas for the very fact that the size of the molecule and the intermolecular distance do not play a role. Because a molecule has a finite size, the ideal gas law would not accurately describe the gas behavior under the circumstance where the intermolecular distance is comparable to the molecular diameter. The non-ideality typically occurs under high pressures, for which the gas density is high and the intermolecular forces becomes important even between collisions, violating the "*dilute*" requirement for the law's derivation. At low temperatures, non-ideality also occurs because the molecular kinetic energy is small relative to the potential energy between molecules, leading to relatively more important molecular interactions due to potential field.

1.4 Forms of Ideal Gas Law

If a specific gases is of interest, Eqn. (1.14) can be rewritten as

$$PV = \frac{M}{\hat{M}}\hat{R}T \tag{1.17a}$$

where M and $\hat{M}$ are, respectively, the mass of the gas in the volume and the molar mass of the gas. Equation (1.15) then becomes

$$PV = MRT \qquad (1.17b)$$

with $R = \hat{R}/\hat{M}$ being the specific gas constant. Because different gas species have different molar masses, their values of R are accordingly different. A table (Table A.1) listing values of R for commonly encountered gases is provided in Table B.1 in Appendix B. Recognizing that density $\rho = M/V$ and that the specific volume $v = 1/\rho$, the ideal gas law can assume the following forms:

$$p = \rho RT \qquad (1.17c)$$

and

$$pv = RT \qquad (1.17d)$$

For a given n, the mass M in the volume (and the associated quantities such as ρ and v) is gas-specific, leading to the associated use of the specific constant, R, as in Eqn. (1.17b) – (1.17d). If the number of moles (n) or of molecules (N) is independent of the specific type of gas (or they are "universal"), then universal constants k and $\hat{R}$ are used, as in Eqns. (1.14) and (1.15). The two forms of ideal gas law, Eqns. (1.14) and (1.17) are identical for $R = k/m$, thus providing a feasible kinetic explanation of a classical thermodynamic result.

1.5 Temperature and Energy

Equation (1.9) indicates that temperature and the average kinetic energy of molecules are related. Since temperature can only be measured in regions containing a large number of molecules, it is appropriate to consider the total kinetic energy of the molecules in the volume, $\sum_i (\frac{1}{2}m_i u_i^2)$. With $\sum_i (\frac{1}{2}m_i u_i^2) = E_{tr}$, Eqn. (1.13) becomes

$$PV = \frac{2}{3}E_{tr} \qquad (1.18)$$

which relates the macroscopic P and V to E_{tr}, with E_{tr} determined from the microscopic behavior of the molecules described by the Maxwellian velocity distribution. By comparing Eqn. (1.18) with Eqns. (1.15) and (1.17), the temperature, T, is directly proportional to E_{tr}. Several observations can be made from these results:

(i) The term PV represents a form of energy, which is directly proportional to E_{tr}, the total kinetic energy in the volume (or system). A large amount of E_{tr}, or PV in the classical thermodynamics, can be used to perform a large amount of work, such as filling up automobile tires or driving machine tools.

(ii) For a given volume containing a given number of molecules (and a given mass), the system pressure is directly proportional to E_{tr}. This is so because during the collision process, the more energetic molecules (i.e., with high speeds) not only experience larger change in momentum ($2m_i u_{i,x}$ in Eqn. [1.10]) but also strike the surface with larger frequencies (i.e., within shorter times, as given by $2L/u_{i,x}$ in Eqn. [1.10]).

(iii) According to Eqns. (1.9) and (1.18), the kinetic energy of a molecule is also directly proportional to temperature, leading to the conclusion that the system pressure is directly proportional to temperature for a fixed volume. This conclusion was arrived at by empirical deduction in classical thermodynamics without examining the velocity distribution and the collisions or molecules, in the form of Eqn. (1.15) or (1.17).

Since there is no preferred direction of molecular movement and $\tilde{e}_{tr} = \frac{1}{2}mu_{rms}^2 = \frac{3}{2}kT = \frac{1}{2}m[C_1^2 + C_2^2 + C_3^2]$, a case exists where

$$\frac{1}{2}mC_1^2 = \frac{1}{2}mC_2^2 = \frac{1}{2}mC_3^2 = \frac{1}{2}kT. \tag{1.19}$$

This case satisfies the *principle* (or *theorem*) *of equipartition of energy*, which states that for any different mode of the average molecular energy that can be expressed as the sum of quadratic terms, all such terms are equal and each makes a contribution of $\frac{1}{2}kT$.

The principle of equipartition of energy applies to other modes of energy, such as rotation and vibration of molecules, as well. Depending on the structure of the molecule, each of the vibrational and rotational modes has a different number of such terms. The number of these terms of energy is also called the degree of freedom (*dof*). Let *dof* be denoted by the symbol ξ. The degrees of freedom besides translation is called the internal degrees of freedom (ξ_{int}). Because $\xi = \xi_{int} + \xi_{tr} = 2 + 3 = 5$. The energy of a molecule is $\frac{5}{2}kT$. A monatomic gas, such as helium and argon, has three translational terms (i.e., in three independent coordinate directions) and possesses no vibrational energy; its rotational energy is too small compared to its kinetic energy because its negligible moment of inertia. It is therefore "structureless" and has $\xi_{int} = 0$, $\xi = \xi_{tr} = 3$.

A diatomic molecule has a kinetic energy equal to $\frac{3}{2}kT$, as for monatomic molecules, and a vibrational energy due to the relative velocity between the two atoms along their connecting axis, and two terms of rotational energy due to rotation about the axes that are perpendicular to the connecting axis. Rotational and vibrational energies of molecules also contain quadratic terms of the relative velocity among atoms and angular velocity of rotation, respectively and, according to the principle of equipartition of energy, each of the vibrational and rotational mode contributes $\frac{1}{2}kT$ to the total energy of a molecule. At normal temperatures, the vibrational energy (including both stretching and bending modes) is negligible compared to the other two energy modes. So for a diatomic gas molecules, the molecular energy is $\frac{5}{2}kT$ or $\frac{5}{2}RT$. It is noted that for determining pressure, only kinetic energy needs to be considered, although a molecule simultaneously possesses multiple modes of energy.

Knowing the energy content of molecules, two types of specific heats can be defined: the constant-volume (c_v) and constant-pressure specific heats (c_p).

The corresponding specific heat ratio γ is defined as $\gamma \equiv c_p/c_v$. There parameters can be expressed in terms of ξ and R with good accuracy under normal temperatures:

$$c_v = \frac{\xi}{2}R, \quad c_p = \frac{\xi+2}{2}R, \quad \gamma \equiv \frac{c_p}{c_v} = \frac{\xi+2}{\xi} \tag{1.20}$$

It is often of interest to consider kinetic energy on a per unit mass basis. By letting e_{tr} be the kinetic energy per kilogram and noting $\hat{N}k = \hat{R}$ and $R = \hat{R}/\hat{M}$,

$$e_{tr} = \frac{\tilde{e}_{tr}}{\hat{M}/\hat{N}} = \frac{3}{2} \cdot \frac{\hat{N}kT}{\hat{M}} = \frac{3}{2}RT \tag{1.21}$$

For polyatomic molecules, the accounting of the rotational and vibrational terms and the internal degrees of freedom can become involved, although all of them have three kinetic terms. If the number of atoms in a molecule is N ($N > 2$), then there are a total of 3N possible degrees of freedom. A simple principle can be followed to approximately determine ξ_{int} at ordinary temperatures. If the molecule is linear (i.e., all atoms form a straight line, such as in CO_2), then it has $\xi_{int} = \xi_{rot} = (3N-5)$; for a nonlinear molecule (e.g., H_2O), $\xi_{int} = \xi_{rot} = (3N-6)$. For further discussion on internal energies, see Vincent and Kruger (1971),Clarke and McChesnney (1964), and Sonntag and Van Wylen (1991).

EXAMPLE 1.1 Find the values of c_v, c_p, and γ for CO_2 and H_2O vapor molecules at 25°C.

Solution –
At this ordinary temperature, vibrational energy is negligible. The CO_2 molecule ($N = 3$) is a linear molecule, similar to a diatomic molecule. It therefore has three terms of kinetic energy, i.e., $\xi_{tr} = 3$. Therefore, $\xi_{CO_2} = \xi_{tr} + \xi_{int} = 3 + (3 \times 3-5) = 7$. Using Eqn. (1.21),

$$c_{v,CO_2} = \frac{\xi_{CO_2}}{2}R = \frac{\xi_{CO_2}}{2}\frac{\hat{R}}{\hat{M}_{CO_2}} = \frac{7}{2}\frac{8.3143kJ/kmol\cdot K}{44kg/kmol} = 0.6614\frac{kJ}{kg\cdot K}$$

$$c_{v,CO_2} = \frac{\xi_{CO_2}+2}{2}R = \frac{\xi_{CO_2}+2}{2}\frac{\hat{R}}{\hat{M}_{CO_2}} = \frac{9}{2}\frac{8.3143kJ/kmol\cdot K}{44kg/kmol} = 0.8503\frac{kJ}{kg\cdot K}$$

$$\gamma = \frac{\xi+2}{\xi} = \frac{9}{7} = 1.2857$$

For comparison, the published values* of c_v, c_p, and γ for CO_2 at 25°C are, respectively, 0.6529 kJ/kg·K, 0.8418 kJ/kg·K, and 1.2890.

The H_2O vapor molecule ($N = 3$) is not a linear molecule with $\xi_{int} = (3 \times 3 - 6) = 3$ and $\xi_{tr} = 3$. Therefore, $\xi_{H_2O} = 6$. Using Eqn. (1.21) and noting that $\hat{M}_{H_2O} = 18$ kg/kmol,

$$c_{v,H_2O} = 1.3857 \text{ kJ/kg} \cdot \text{K}$$

$$c_{p,H_2O} = 1.8476 \text{ kJ/kg} \cdot \text{K}$$

$$\gamma = 1.333.$$

The published values* of c_v, c_p, and γ for CO_2 at 25°C are, respectively, 1.4108 kJ/kg·K, 1.8723 kJ/kg·K, and 1.3270.

*These values are taken from R.E. Sonntag and G.J. Van Wylen, *Introduction to Thermodynamics, Classical and Statistical*, 3rd ed., John Wiley and Sons, 1991. Also see Table 1.   □

1.6 Characteristic Molecular Length and Knudsen Number

For a dilute gas, the average spacing between molecules is sufficiently large compared to the size of the molecules. A question arises: how dilute is a gas (or air), for example, under standard atmospheric conditions?

Since the molecules are in constant motion, the spacing between them can be chosen to be the average distance traveled by molecules between consecutive collisions. This particular distance is called the *mean free path* and is denoted by λ. The mean free path is expected to decrease with increases in the cross-sectional area of the molecule πd^2 and its number density ρ/m. For collision between like molecules, λ is found to be

$$\lambda = \frac{m}{\sqrt{2}\pi d^2 \rho}. \tag{1.22}$$

There are various ways of determining the effective diameters of molecules. For water and oxygen molecules, the effective diameters are approximately 0.36 nm and 0.4 nm (or 3.6×10^{-10} m and 4.0×10^{-10} m), respectively (CRC Handbook of Chemistry and Physics, 74th ed., 1993). Many other molecules have diameters on the same order of magnitude. For estimation, let the typical diameter of gas molecules be $d \approx 3.8 \times 10^{-10}$ m and consider air as an example with a molar mass of 28.9 grams. At the standard conditions of 1 atm and 0°C, an ideal gas occupies approximately 22.4 liters (22.4×10^{-3} m^3) at the sea level condition. The mean free path is approximately

$$\lambda \approx 5.8 \times 10^{-8} m$$

Another way of answering the same question is to assume that molecules are not moving (i.e., static) and that 6.02252×10^{23} molecules orderly pack the volume of 22.4×10^{-3} m^3. The average spacing between molecules (δ) under this assumption, can be calculated as $\delta \approx (m/\rho)^{1/3}$ and

$$\delta \approx 3.3 \times 10^{-9} \ m$$

Therefore for air at 1 atm and 0°C,

$$\lambda : \delta : d \approx 150 : 10 : 1.$$

Thus the mean free path is much greater than the static molecular spacing, which is in turn much larger than the molecular diameter. The kinetic derivation of the ideal gas law based on the dilute model leads to results consistent with classical thermodynamics, which is based on the so-called continuum medium. So air under standard conditions must be sufficiently dilute, but is still a continuum.

For a continuum assumption (i.e., such as classical thermodynamics) requires the volume of interest to contain a sufficiently large number of molecules for the measurements and definition of temperature, density, and pressure to be meaningful. Such a requirement is apparent as suggested by Eqn. (1.13), as the value of the term $\sum_i \left(m_i u_i^2 \right)/3V$ for a small number of molecules would depend on the few velocities possessed by the few molecules incidentally in the volume at the instant of observation. These few discrete velocities might be found anywhere along the curves in Fig. 1.1 depending on the temperature. In the extreme case when there is no molecule in the volume, *no* temperature can be calculated according to Eqn. (1.14) or (1.15), let alone being observed or measured! Therefore, the ideal gas law is for continuum gases. If the gas is so dilute that the number of molecules in the volume may fluctuate wildly, then a constant density, and therefore a continuum, does not exist in that volume. Under such conditions, the motion of individual gas molecules have to be studied; the field of such study is called *rarefied gas dynamics* and the motion of molecules constitute a rarefied gas flow.

Different flow regimes can be classified by the *Knudsen number, Kn,* defined as $Kn \equiv \lambda/L$, where L is the characteristic length of the system under consideration. For $Kn \ll 1$, the flow is said to be collision-dominated because a molecule traveling over the characteristic length experiences many collisions. With many collisions in a given volume (e.g., a small region), the gas tends to achieve uniform properties, equal to the average property values in the volume. The flow regime of $Kn \ll 1$ constitutes continuum mechanics. Because of the domination of molecular collisions, the viscous no-slip condition holds at fluid-solid boundaries. For $Kn \geq 1$, the flow is called *free-molecule flow* as the number of collisions experienced by a molecule is no greater than one while traversing the length. In this case the velocities of individual molecules need to be considered during their interaction with the solid boundaries and among themselves. As a consequence, the no-slip viscous boundary condition is not valid. The flow regime separating $Kn \ll 1$ and $Kn \geq 1$ is the transitional regime, which exhibits a hybrid of continuum and free-molecular flows.

EXAMPLE 1.2 Find the values of the Knudsen number for air at sea level, at 100 km, and at typical low earth orbit (LEO; ≈ 322 km for the space shuttle) above sea level for a small flight vehicle that has a characteristic dimension of 1 m.

Solution –

As mentioned in the above discussion, $\lambda \approx 5.8 \times 10^{-8} m$ at 1 atmosphere and 0°C (273 K). At sea level $T \approx 15°C$ (288 K) and $\rho_{SL} = 1.2250$ kg/m³, then

$$\lambda = \frac{m}{\sqrt{2}\pi d^2 \rho} = \frac{0.0288 \frac{kg}{6.02252 \times 10^{23}}}{\sqrt{2}\pi \left(3.8 \times 10^{-10} m\right)^2 \times 1.2250 \text{ kg}/m^3} = 6.085 \times 10^{-8} \ m$$

$$Kn = \frac{\lambda}{L} = \frac{6.085 \times 10^{-8} m}{1m} = 6.085 \times 10^{-8} \ll 1$$

At any altitude, H, the density (ρ_H) can be sufficiently approximated as

$$\frac{\rho_H}{\rho_{SL}} = e^{-\alpha H} \ (\alpha = 1.394 \ \times \ 10^{-4}/m)$$

Therefore at $H = 120$ km, $\rho_H = 5.43 \times 10^{-8} \rho_{SL} = 6.657 \times 10^{-8}$ kg/m³.

$$\lambda = \frac{m}{\sqrt{2}\pi d^2 \rho} = \frac{0.0288 \frac{kg}{6.02252 \times 10^{23}}}{\sqrt{2}\pi (3.8 \times 10^{-10} m)^2 \times \left(6.657 \times 10^{-8} \frac{kg}{m^3}\right)} = 1.120 \ m$$

$$Kn = \frac{\lambda}{L} = \frac{1.120m}{1m} = 1.120 \approx 1$$

At the shuttle altitude $H = 322$ km, $\rho_H = 3.21 \times 10^{-21} \rho_{SL} = 3.93 \times 10^{-20}$ kg/m³

$$\lambda = \frac{m}{\sqrt{2}\pi d^2 \rho} = \frac{0.0288 \frac{kg}{6.02252 \times 10^{23}}}{\sqrt{2}\pi (3.8 \times 10^{-10} \ m)^2 \times \left(3.93 \times 10^{-20} \frac{kg}{m^3}\right)} = 1.897 \times 10^{12} \ m$$

$$Kn = \frac{\lambda}{L} = \frac{1.897 \times 10^{12} m}{1m} = 1.897 \times 10^{12} \gg 1$$

Comments –

One may therefore expect that a lunch vehicle will go through continuum and free-molecule flow regimes as it ascends to the low earth orbit.   □

It is also of interest to find out the collision frequency (Θ) under various conditions. For molecules having the most probable speed, the value of Θ can be found by using Eqn. (1.4) and (1.22),

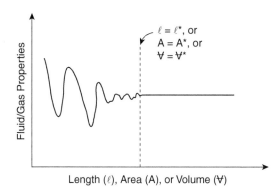

Figure 1.3 Gas properties as a function of scales (*l* for length, *A* for surface area, *V* for volume) of interest; the asterisks (*) denotes the scales above which the gas behaves as a continuum medium.

$$\Theta = \frac{u_{mp}}{\lambda} = \left(\frac{2kT}{m}\right)^{1/2} \bigg/ \frac{m}{\sqrt{2}\pi d^2 \rho} = 2\pi d^2 \rho \sqrt{\frac{kT}{m^3}}. \tag{1.23}$$

The estimation is straightforward and is assigned as Problem 1.4.

1.7 The Continuum Gas

For *Kn* « 1, not only the flow is collision-dominated but also that the number of molecules in the volume defined by the characteristic length is so numerous that the gas density remains constant all the time and throughout the flow field. Such a scale in length/surface area/volume with various properties, such as density, temperature, and pressure, is illustrated in Fig. 1.3. In the continuum approach of compressible flow, a fluid "particle" can be as small as the mass contained in this minimal volume (V^*) with the corresponding minimal area and length equal to A^* and L^*, respectively. The continuum gas particle thus consists of many molecules. Thus for the continuum gas,

$$\rho \equiv \lim_{\Delta V \to V^*} \frac{\Delta M}{\Delta V} \quad \text{and} \tag{1.24}$$

$$p \equiv \lim_{\Delta A \to A^*} \frac{\Delta F_{normal}}{\Delta A}. \tag{1.25}$$

Except when specifically dealing with rarefied gas dynamics (Chapter 12), the governing equations are derived and solved for the continuum gas throughout this book.

Problems

Problem 1.1

(a) Show that the total probability of finding molecules with speed between *u* and *u* + *du* regardless of direction is unity by using the integral

$$\int x^2 \exp(-ax^2)\,dx = \frac{1}{4}\left(\frac{\pi}{a^3}\right)^{1/2}\mathrm{erf}(x\sqrt{a}) - \frac{x}{2a}\exp(-ax^2)$$

(b) Consider nitrogen gas (molar mass = 28 kg/kmol) in equilibrium at 1 atmosphere and 300 K. Find the probability of nitrogen molecules having the speed of $0 - 347$ m/s, by using Eqn. (1.2)

(c) Following Part (b), find u_{mp}.

(d) Following Part (b), find the fraction of molecules that possess speeds in the range of $0 - 2u_{mp}$.

(e) Repeat Part (d) for CO_2, which has a molar mass of 44 kg/kmol.

Problem 1.2 Find the values of $u_{mp}, \bar{u}$, and u_{rms} for air at 1 atm and 0°C. The average mass of an air molecule is approximately $m = 4.782 \times 10^{-26}$ kg.

Problem 1.3 Find the values of e_{tr} and $\tilde{e}_{tr}$ of nitrogen molecules at room temperature (25°C).

Problem 1.4 Take the sea level condition as an example to estimate the collision frequency of the air molecule with the most probably velocity. Answer: $\Theta \approx 7 \times 10^9 s^{-1}$.

Problem 1.5 Find the values of specific heats for helium (4.003 kg/kmol) and argon (39.945 kg/kmol) on both a per mole and per unit mass bases.

2 Basic Equations and Thermodynamics of Compressible Flow

Fluids consist of two categories: gases and liquids. For the subject of compressible fluid dynamics in the continuum regime (i.e., $Kn \ll 1$), only the gas medium is of interest, as liquids can be treated as incompressible in most circumstances and applications. This and following chapters are devoted to compressible flows in the continuum regime. Governing equations and thermodynamic relationships useful for compressible fluid flow are derived in the following sections.

2.1 Introduction

A fluid is defined as a medium that cannot sustain a shear stress without continuously deforming. The continuous deformation results in a shear strain rate. For a Newtonian fluid, the shear stress is linearly proportional to the shear strain rate, with the viscosity (μ) as the proportionality constant. In the presence of a solid boundary, the fluid sticks to the solid surface because of fluid viscosity (the "no-slip" condition), resulting in a zero relative velocity at the surface. At a distance immediately away from the surface, the relative velocity is not zero.

Let τ_{yx} denote the shear stress pointing in the x direction on a surface (which can be a solid or an imaginary surface within the fluid) facing the y direction. Consider, for example, the surface of a flat plate shown in Fig. 2.1). The associated shear strain rate arises because layers of fluid near and parallel to the surface have different velocity magnitudes (u) in the x direction, causing a non-zero velocity gradient, du/dy (which has the unit of 1/time and is the strain rate). Then for Newtonian fluids, for which the shear stress is directly proportional to the shear strain rate, the relationship between stress and strain rate is expressed as

$$\tau_{yx} = \mu \frac{du}{dy} \tag{2.1}$$

(A non-Newtonian fluid is one for which the relationship between shear stress and strain rate is nonlinear; if τ_{yx} is to be expressed only in terms of du/dy, then μ would appears as if it is a function of du/dy as well.) Shear stresses can be found in flows in the vicinity of an arbitrarily shaped surface (such as that shown in Fig. 2.1b with

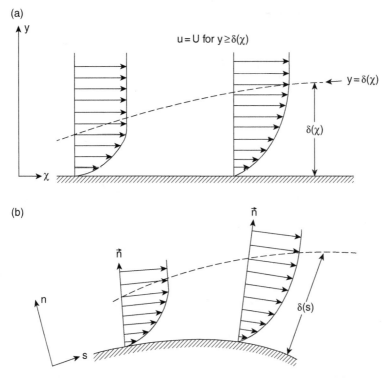

Figure 2.1 Velocity profiles over a viscous flow over a flat plate (a) and over an arbitrarily curved surface (b). In each figure, the dashed line denotes the edge of the boundary layer within which a velocity gradient exists because the gas feels the presence of the surface and outside of which the velocity is uniform.

a curvature). Because the shear stress is associated with a shear strain rate, it exists where there is a non-zero velocity gradient, whether at, near, or away from a surface, as shown in Fig. 2.1.

In the case of a uniform flow over a flat surface (Fig. 2.1a), a velocity gradient exists in the near-wall region. As the distance increases, the flow velocity increases and the effect of the surface diminishes. At a sufficiently large distance the fluid velocity assumes the free stream value and does not feel the existence and the effect of the surface. The near-wall region, where the presence of the wall is felt due to the effect of viscosity, is called the *viscous boundary layer*. Alternatively, the boundary layer is defined as where the velocity gradient is not zero, as qualitatively depicted in Fig. 2.1a. Outside the boundary layer $\partial u/\partial y = 0$, nullifying the effect of viscosity as if viscosity were zero. In the absence of a velocity gradient or the effect of viscosity, the flow (not the fluid itself) is *inviscid*. The boundary layer thickness decreases with flow speed (and, for the high speeds associated with compressible flows, it is particularly small). Due to the near-wall negligible velocity in the normal, y-direction, $\partial p/\partial y = 0$. As a consequence, the external flow "impresses" on the boundary layer and the boundary layer possesses the same pressure in the x-direction as the outer flow. Solving the inviscid flow equations in the region outside the boundary layer thus also yields the pressure field inside within the boundary, with the pressure varying only in the *x*-direction. For flow over an arbitrarily shaped surface (e.g., Fig. 2.1b) the local curvature causes a centripetal force and the pressure field near the surface becomes multidimensional.

2.2 Control-Volume Approach of Governing Equations

Fluid motion are governed by laws of conservation of mass, momentum, and energy. The gas particle mentioned at the end of Chapter 1 moves with motions resembling those of a tennis ball: translation, rotation, and deformation. While it is easy to follow a ball along its trajectory, tracking or "tagging" gas particles is no trivial matter, because a system of gas usually comprises many identical particles. The formulation of governing equations can based on a system (or *Lagrangian*) approach, by following a fixed mass, which is akin to tracking all gas molecules in kinetics (Chapter 1). It can be done, on the other hand, by using the control-volume (or *Eulerian*) approach that considers a fixed volume in the flow field through which the gas flows. The control volume is chosen such that it contains the fluid at a location of interest. In many applications, such as analyzing the thrust of an aircraft engine, the states of fluid at the inlet and exhaust of the engine and within the engine are of importance, rather than those of the gas that has exited the engine and is already some distance downstream. For these reasons, the control-volume approach is adopted.

The graphical representations of a system and a control volume are shown in Fig. 2.2 and Fig. 2.3, respectively. Because conservation laws for mass, momentum, and energy are for a system, a relationship is needed between the system and the control-volume approaches. Such a relationship is the *Reynolds transport theorem* (RTT), which relates at any instant the rate of change in a property of the system to its rate of change inside the control volume (C∀) and the flux of the property across the control surface (CS):

$$\left.\frac{D\Phi}{Dt}\right)_{system} = \frac{\partial}{\partial t}\int_{C\forall}\phi\rho dV + \int_{CS}\phi(\rho\vec{V}\cdot d\vec{A}) \tag{2.2}$$

where Φ, and ϕ are the extensive and intensive properties, respectively (and $\forall$ denotes volume, different from V used for velocity), related by

$$\phi = \frac{\Phi}{M}, \tag{2.3}$$

and M is the mass of the system. In Eqn. (2.2) ρdV is the infinitesimal mass (dm) inside C∀ that carries property ϕ, and $\rho\vec{V}\cdot d\vec{A}$ is the mass flow rate carrying property ϕ across CS over an infinitesimal area of $d\vec{A}$. Convenience may dictate the choice of the C∀ (i.e., the CS) in ways to simplify the integration of the term $\rho\vec{V}\cdot d\vec{A}$. For example, because $\vec{V}\cdot d\vec{A} = |\vec{V}| \times |d\vec{A}| \cos\theta$, with θ being the angle between the two vectors, it is convenient to have the value of θ that leads to a readily known value of $\cos\theta$ (one possible and obvious choice of θ is 0°). Figure 2.4 illustrates the choice of

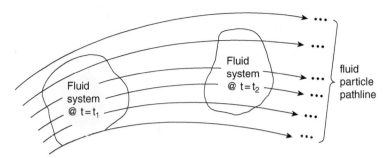

Figure 2.2 A system consists of identical fluid particles, although its boundary might deform as it travels throughout the space, as indicated by the solid boundaries at $t = t_1$ and at a later time t_2.

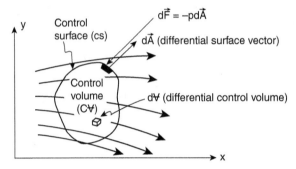

Figure 2.3 A control volume (C⊽) is an arbitrarily chosen volume in space through which fluid particles pass through (for the interest of problem solving, the C⊽ is usually selected to facilitate the solution). The boundary of C⊽ is the control surface (CS). The differential C⊽ and CS are also shown.

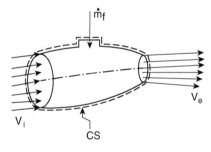

Figure 2.4 The choice of C⊽ and CS to facilitate problem solving for a schematic jet engine is illustrated (and used for Example 2.1).

C⊽ and CS for analyzing the flow through a schematic jet engine. If $\int_{CS} \left(\rho \vec{V} \cdot d\vec{A} \right) > 0$, then there is a net rate of mass flowing out of C⊽, resulting in positive rate of mass loss in C⊽ while keeping the mass of the system unchanged all the time.

As shown in Fig. 2.3, if the angle between the two vectors $\vec{V}$ and $d\vec{A}$ is smaller than $90°$, then $\vec{V} \cdot d\vec{A} > 0$ and $\rho \vec{V} \cdot d\vec{A}$ represents mass flow out of C⊽; if the angle is greater than, then an influx occurs. Thus the integration over CS sums up

Table 2.1 *Summary of conservation equations*

	Φ	ϕ	$\frac{D\Phi}{Dt}\big)_{system}$	Comments
Mass	M	1	0	The mass of the system cannot be generated or destroyed (i.e., fixed)
Linear momentum	$M\vec{V}$	$\vec{V}$	$\sum_i \vec{F}_i\big)_{system}$	The change in the system's linear momentum results from the forces (including body and surface forces) acting on the system
Angular momentum	$M\vec{r} \times \vec{V}$	$\vec{r} \times \vec{V}$	$\sum_i \vec{T}_i\big)_{system}$	The change in the system's angular momentum results from the torques acting on the system
Energy	E	e	0	The mechanical energy of the system cannot be generated or destroyed
Entropy	S	s	$\int_{cs} \frac{\delta Q}{T} + I$	$I\,(>0)$ is the entropy increase due to irreversibility and δQ is the heat transfer across system boundaries

contributions from all the inflows and outflows. For conservation of mass, momentum, and energy, the properties and the results of the rate of change of the system properties are summarized in Table 2.1. Entropy generation is also described in Table 2.1.

2.3 Conservation of Mass (Continuity Equation)

Based on the results of the previous section, the conservation of mass (or the *continuity*) equation becomes

$$0 = \frac{\partial}{\partial t}\int_{CV} \rho dV + \int_{CS}(\rho\vec{V}\cdot d\vec{A}) \ or \ \frac{\partial}{\partial t}\int_{CV}\rho dV = -\int_{CS}(\rho\vec{V}\cdot d\vec{A}) \qquad (2.4)$$

Under the steady state condition,

$$\frac{\partial}{\partial t}\int_{CV}\rho dV = -\int_{CS}\left(\rho\vec{V}\cdot d\vec{A}\right) = 0$$

requiring that the net mass accumulation is zero within the CV all the time and that whatever the mass flowing into CV must flow out at the same rate. For an unsteady flow,

$$\frac{\partial}{\partial t}\int_{CV}\rho dV \neq 0 \ and \ \int_{CS}\left(\rho\vec{V}\cdot d\vec{A}\right) \neq 0$$

However, the mass of the system still remains unchanged at all times and thus Eqn. (2.4). For an incompressible flow, density is not a function of location or time and therefore

$$\frac{\partial}{\partial t}\int_{CV}\rho dV = \rho\frac{\partial}{\partial t}\int_{CV}dV$$

Incompressibility implies $\int_{CV} dV = $ constant and, therefore, $\int_{CS} \left(\vec{V} \cdot d\vec{A} \right) = 0.$

EXAMPLE 2.1 The preliminary design for a jet engine is based on a steady state operation at an altitude of 12,000 m, where the air density is 0.312 kg/m³. Assume that both the inlet and exit cross-sections are circular in shape; the inlet and exit diameters are 0.8 m and 0.39 m, respectively. The air speed is 240 m/s and the engine exit speed is 765 m/s (due mainly to thermal expansion by burning the fuel). Also assume that the velocity at both the inlet and the exit is uniform and that the exit gas density at the exit is 0.434 kg/m³.and pressure is same as the ambient. How much fuel in (kg/s) is needed per second?

Solution –
Mass conservation for this steady state engine (refer to schematic shown in Fig. 2.4) is

$$0 = \frac{\partial}{\partial t} \int_{CV} \rho dV + \int_{CS} \left(\rho \vec{V} \cdot d\vec{A} \right) = -\rho_i V_i A_i - \dot{m}_{F,i} + \rho_e V_e A_e$$

For steady state operation,

$$\frac{\partial}{\partial t} \int_{CV} \rho dV = 0$$

Therefore,

$$\dot{m}_{F,i} = -\rho_i V_i A_i + \rho_e V_e A_e = -0.312 \frac{kg}{s} \times 240 \frac{m}{s} \times \frac{\pi}{4} (0.8 \ m)^2$$

$$+ 0.434 \frac{kg}{s} \times 765 \frac{m}{s} \times \frac{\pi}{4} (0.39 \ m)^2$$

$\dot{m}_{F,i} = 2.02 \ kg/s$ (Also $\dot{m}_e = 39.66 \ kg/s$ and $\dot{m}_i = 37.64 \ kg/s$.)   □

2.4 Conservation of Momentum: Newton's Second Law for Fluid Flow

The mass (gas particles) flowing through CV carries both momentum and energy with them. With $\phi = \vec{V}$ substituted into Eqn. (2.2), the conservation of linear momentum is

$$\left. \frac{D\Phi}{Dt} \right)_{system} = \sum_i \vec{F}_i \Big)_{system} = \frac{\partial}{\partial t} \int_{CV} \rho \vec{V} dV + \int_{CS} \vec{V} \left(\rho \vec{V} \cdot d\vec{A} \right) \qquad (2.5)$$

Note that in Eqn. (2.5) $\rho \vec{V} dV = \vec{V} dm$ is the momentum of an infinitesimal mass inside CV. Equation (2.5) states that all the forces as a whole exerted on the system (the fluid) would cause the time rate of change of momentum of the mass within the CV plus the net rate of momentum flux out of the CS. In a steady flow, $\frac{\partial}{\partial t} (\cdot) = 0$ and the forces on the system cause its momentum flux to change between the entrance

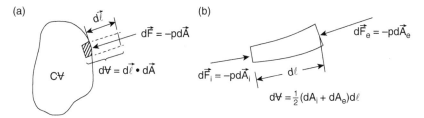

Figure 2.5 Mass flow through and forces acting on a differential control volume ($d\mathcal{V}$) and surface (dS), with the differential length being dl.

and the exit of the C$\mathcal{V}$. Therefore in a steady state flow, the system of gas can be accelerated or decelerated, while the C$\mathcal{V}$ itself is not. For example, the aircraft jet engine produces the net effect of accelerating the gas from its inlet to exit, which generates forces on the fluid within the engine and the reaction force on the engine (the control volume) is the thrust. In the steady state operation during cruise, the thrust is equal to and is used to overcome the aerodynamic drag, and the C$\mathcal{V}$ (the engine and, by extension, the aircraft) does not accelerate.

The forces acting on the system include all surface and body forces (denoted by $\vec{F}_S$ and $\vec{F}_B$, respectively) acting on the system

$$\sum_i \vec{F}_i \bigg)_{system} = \sum_i \vec{F}_{Bi}\bigg)_{system} + \sum_i \vec{F}_{Si}\bigg)_{system} \tag{2.6}$$

In the absence of electrical or electromagnetic fields and charged gas particles, the other body force of interest is due to gravity. The gravitational force can usually be neglected unless the elevation or gas density differs significantly from one point to the other in the control volume. For $\vec{F}_S$, all forces acting on the fluid due to contact should be included, including the pressure and viscous shear forces. It is known that among forces due to pressure only the gauge pressure ($p - p_{amb}$, where p_{amb} is the ambient pressure) on the surface areas at inlets and outlets contributes to $\vec{F}_S$. Viscous shear forces and those by mechanical devices such as compressors in a jet engine can be grouped together and designated by $\vec{R}_S$. Therefore,

$$\sum_i \vec{F}_{Si}\bigg)_{system} = -\int_{CS}(p - p_{amb}) \cdot d\vec{A} + \vec{R}_{S,system} \tag{2.7}$$

where the negative sign for the pressure term is due to the fact that pressure force acts in the opposite direction of the local surface vector ($d\vec{A}$, as shown in Fig. 2.5). Equation (2.5) then becomes

$$-\int_{CS}(p - p_{amb}) \cdot d\vec{A} + \vec{R}_{S,system} = \frac{\partial}{\partial t}\int_{C\mathcal{V}}\rho\vec{V}d\mathcal{V} + \int_{CS}\vec{V}\left(\rho\vec{V}\cdot d\vec{A}\right) \tag{2.8}$$

When momentum ($M\vec{V}$) and force ($\sum_i\vec{F}_i)_{system}$ are replaced in Eqn. (2.5) by angular moment ($M\vec{r} \times \vec{V}$) and torque ($\sum_i\vec{T}_i)_{system}$, respectively, the conservation of angular moment (or the angular principle) is established.

EXAMPLE 2.2 All conditions are the same as those in Example 2.1 with the pressures of gas equal to 19.40 kPa and 304.89 kPa, respectively at the inlet and the exit. What is the thrust force generated by the engine?

Solution –
The momentum equation, Eqn. (2.8)

$$-\int_{CS}(p - p_{amb})\cdot d\vec{A} + \vec{R}_{S,system} = \frac{\partial}{\partial t}\int_{CV}\rho\vec{V}\,dV + \int_{CS}\vec{V}\left(\rho\vec{V}\cdot d\vec{A}\right)$$

with the steady state condition, leads to

$$R_{S,system} - [(p - p_{amb})A]_e + [(p - p_{amb})A]_i = \left(\rho V^2 A\right)_e - \left(\rho V^2 A\right)_i$$

Note that at the inlet, $p = p_{amb}$. Therefore,

$$R_{S,system} = [(p - p_{amb})A]_e + \left(\rho V^2 A\right)_e - \left(\rho V^2 A\right)_i$$

or

$$R_{S,system} = (p_e - p_{amb})A_e + \dot{m}_e V_e - \dot{m}_i V_i$$
$$= (304.89 - 19.40) \times 10^3 \frac{N}{m^2} \times \frac{\pi}{4}(0.39\ m)^2 + 39.66\frac{kg}{s} \times 765\frac{m}{s}$$
$$- 37.64\frac{kg}{s} \times 240\frac{m}{s}$$
$$R_{S,system} = 64,446.7\ N$$

Since this is the force acting on the system (the exhaust gas), the reaction on the engine is in the opposite direction, in which the aircraft is propelled. Therefore, the thrust force is

$$T = -64,446.7\ N\ \text{(i.e., in the negative direction)}$$

Note – the thrust due to the fuel at the inlet to the engine is assumed to be small. If the fuel is injected in the direction normal to the x-axis, it does not contribute to the momentum in the x-direction and does not affect the thrust produced.   □

2.5 Conservation of Energy: First Law of Thermodynamics

The principle of conservation of energy when applied to a system leads to the first law of thermodynamics. When it is applied to the CV, it becomes

$$\left.\frac{DE}{Dt}\right)_{system} = \frac{\partial}{\partial t}\int_{CV}e\rho\,dV + \int_{CS}e\left(\rho\vec{V}\cdot d\vec{A}\right) \tag{2.9}$$

Similar to mass, energy is a scalar that cannot be generated or destroyed in a system. In general, $\frac{\partial}{\partial t}\int_{CV}e\rho\,dV = -\int_{CS}e\left(\rho\vec{V}\cdot d\vec{A}\right)$. The steady state would require that $\int_{CS}e\left(\rho\vec{V}\cdot d\vec{A}\right) = 0$, that is, the net energy accumulation is zero within the CV at all time and that whatever the amount of energy flowing into CΨ must flow out at the same rate. For an unsteady flow, $\frac{\partial}{\partial t}\int_{C\Psi} e\rho d\Psi = -\int_{CS} e\left(\rho\vec{V}\cdot d\vec{A}\right) \neq 0$ and the influx and outflux of energy are not equal, resulting in accumulation of energy within the CΨ.

Two forms of energy are of interest: heat and work. For the system, the first law of thermodynamics states that the net balance of heat transfer and work performed across the system boundaries is the amount of energy to be stored in the system. When the time rates of these processes are of interest, the first law is expressed as

$$\dot{Q} - \dot{W} = \left.\frac{DE}{Dt}\right)_{system} \qquad \text{or} \qquad (2.10a)$$

$$\delta Q - \delta W = dE \qquad (2.10b)$$

where $\dot{Q}$ and $\dot{W}$ are, respectively, the rates of heat transfer and work across the system boundaries and Q and W are the amount of heat transfer and work, respectively. The sign conventions for heat and work are: $\dot{Q}$ and Q are positive/negative when the heat transfer is to/from the system (i.e., heat addition/removal is positive/negative) and $\dot{W}$ and W are positive/negative when it is done by/on the system. Like heat, work is a form of energy and can be envisioned to be taken away from the system when it is expended to perform work (hence the negative sign in Eqn. [2.10]).

The energy E consists of internal, kinetic, and potential energies (denoted by U, KE, and PE, respectively) and the energy associated with the fluid having to "push" its way across the control surface (some call this form of energy is sometimes called "flow work"). The fluid gains energy when flowing from a low pressure inlet to an outlet at a higher pressure, like undergoing compression and receiving work, and loses energy when flowing from high to low pressures. As shown in Fig. 2.5a, the pressure force working against the motion of a differential fluid particle (with a volume of $d\vec{l}\cdot d\vec{A}$) moving out of CΨ is $d\vec{F} = -pd\vec{A}$, which is opposite in direction to the pressure force. The work expanded (i.e., done) by the fluid particle when it moves the distance of $d\vec{l}$ out of the control volume is $dW_p = -d\vec{F}\cdot d\vec{l} = pd\vec{A}\cdot d\vec{l}$, which is positive; it is negative for the inflow. When the fluid particle is tracked from the inlet to the outlet (Fig. 2.5b), this is the amount of energy change in the fluid due to the work done by pressure on the boundaries. Because $d\vec{A}\cdot d\vec{l} = d\Psi$, on a per mass basis (using $dm = \rho d\Psi$),

$$w_p = \frac{dW_p}{dm} = \frac{p}{\rho} = p\Psi \qquad (2.11)$$

where $\Psi = 1/\rho$ is the specific volume (volume per unit mass). This amount of pressure work is done on the mass fluxes across the control surface. Summing over the entire control surface and taking the rate of change $d\vec{l}/dt = \vec{V}$:

$$\dot{W}_p = \int_{CS} w_p \left(\rho \vec{V} \cdot d\vec{A} \right) = \int_{CS} \frac{p}{\rho} \left(\rho \vec{V} \cdot d\vec{A} \right) \text{ or } \int_{CS} pv \left(\rho \vec{V} \cdot d\vec{A} \right) \qquad (2.12)$$

The total amount of work done by pressure can be separated from other forms of work:

$$\dot{W} = \dot{W}_p + \dot{W}_{shear} + \dot{W}_{others} \qquad (2.13)$$

where $\dot{W}_{shear}$ denotes energy dissipation rate due to shear stresses (pressure is the normal stress) and $\dot{W}_{others}$ is work done by other mechanical mechanisms such as a shaft. Then Eqn. (2.9) becomes

$$\dot{Q} - \dot{W}_p - \dot{W}_{shear} - \dot{W}_{others} = \frac{\partial}{\partial t} \int_{CV} e\rho dV + \int_{CS} e \left(\rho \vec{V} \cdot d\vec{A} \right) \qquad (2.14)$$

Substituting Eqn. (2.12) into Eqn. (2.14) yields

$$\dot{Q} - \dot{W}_{shear} - \dot{W}_{others} = \frac{\partial}{\partial t} \int_{CV} e\rho dV + \int_{CS} (e + pv) \left(\rho \vec{V} \cdot d\vec{A} \right) \qquad (2.15)$$

where, on a per unit mass basis,

$$e = u + \frac{1}{2} V^2 + gz \qquad (2.16)$$

with $u, \frac{1}{2} V^2$, and gz being, respectively, the internal, kinetic, and potential energies per unit mass, g being the magnitude of gravitational acceleration $\vec{g}$ and z, the elevation in the direction of $\vec{g}$. The internal energy u is the molecular energy discussed in Chapter 1, where knowledge of the bulk gas velocity ($\vec{V}$) and the bulk kinetic energy $\frac{1}{2} V^2$ is not required. Inserting Eqn. (2.16) into Eqn. (2.12) leads to

$$\dot{Q} - \dot{W}_{shear} - \dot{W}_{others} = \frac{\partial}{\partial t} \int_{CV} e\rho dV + \int_{CS} \left(u + pv + \frac{1}{2} V^2 + gz \right) \left(\rho \vec{V} \cdot d\vec{A} \right) \quad (2.17)$$

The thermodynamic property enthalpy h is defined as

$$h \equiv u + pv \qquad (2.18)$$

so that Eqn. (2.17) is also customarily written as

$$\dot{Q} - \dot{W}_{shear} - \dot{W}_{others} = \frac{\partial}{\partial t} \int_{CV} e\rho dV + \int_{CS} \left(h + \frac{1}{2} V^2 + gz \right) \left(\rho \vec{V} \cdot d\vec{A} \right) \qquad (2.19)$$

EXAMPLE 2.3 Consider the same conditions of Example 2.2 with the inlet and exit gas temperature equal to 216 K and 1,010 K, respectively. Find the energy content of the fuel (in kJ/kg). Assume that enthalpy is only a function of temperature $h = c_p T$ (which will be discussed in more detail later in this chapter) and that values of c_p for air and the exhaust gas in the temperature range of interest are approximately 1.0 $kJ/kg \cdot K$ and 1.1 $kJ/kg \cdot K$, respectively.

Solution –

The energy added to the system by burning the fuel can be found using the steady-state energy equation. Assume that the velocity and temperature are uniform at both the inlet and exit planes. Then

$$\left.\frac{DE}{Dt}\right)_{system} = \int_{CS}\left(h+\frac{1}{2}V^2+gz\right)\left(\rho\vec{V}\cdot d\vec{A}\right) = \left[\dot{m}\left(c_pT+\frac{1}{2}V^2+gz\right)\right]_e$$
$$-\left[\dot{m}\left(c_pT+\frac{1}{2}V^2+gz\right)\right]_i$$

It is reasonable to assume $z_e \approx z_i$. Then

$$\left.\frac{DE}{Dt}\right)_{system} = \left[\left(39.66\frac{kg}{s}\right)\left(1.1\frac{kJ}{kg\cdot K}\times 1,010\ K\right)+\frac{1}{2}\left(39.66\frac{kg}{s}\right)\times\left(756\frac{m}{s}\right)^2\right]$$
$$-\left[\left(37.64\frac{kg}{s}\right)\left(1.0\frac{kJ}{kg\cdot K}\times 216\ K\right)+\frac{1}{2}\left(37.64\frac{kg}{s}\right)\times\left(240\frac{m}{s}\right)^2\right]$$
$$= 22,864\ kJ/s$$

The energy content of the fuel (H) can be estimated to be

$$H = \left.\frac{DE}{Dt}\right)_{system}\div \dot{m}_{F,i} = \frac{46,181.5\ kJ/s}{2.02\ kg/s} = 22,862.0\frac{kJ}{kg}$$

Comments –

(1) A significant portion of the fuel energy is used to overcome irreversibilities such as friction, heat loss from the engine, sudden expansion, and rapid processes such as heat-releasing combustion. The energy content of the fuel thus should be larger than the calculated value. Most hydrocarbon fuels, including aviation fuels, have energy contents approximately 45,000 kJ/kg.

(2) The effect of irreversibility will be discussed in more detail in Section 2.7. □

2.6 Special Cases of Conservation Equations

For a steady flow with uniform properties at any flow location, the mass conservation equation is reduced to

$$\int_{CS}\left(\rho\vec{V}\cdot d\vec{A}\right) = 0 \text{ and } (\rho VA)_e = (\rho VA)_i \tag{2.20}$$

where the subscripts i and e denote values determined at the inlet and exit surface, respectively. Much more simplified forms of the momentum equation, Eqn. (2.8), and the energy equation, Eqn. (2.19) can be obtained by considering the following conditions: (1) steady flow, (2) the control volume is a thin stream tube having a differential cross-sectional area (dA), as shown in Fig. 2.6, (3) all properties are uniform at the inlet and exit plane, and (4) heat transfer across the CS (i.e., adiabatic

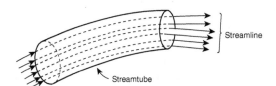

Figure 2.6 Schematic of a control volume bounded by streamlines with uniform properties.

surface), shear work, and other mechanical work are either zero or negligibly small. Under these conditions, Eqn. (2.8) becomes

$$0 = \int_{CS} \left(h + \frac{1}{2}V^2 + gz \right) \left(\rho \vec{V} \cdot d\vec{A} \right) \tag{2.21}$$

Because the only inlet and outlet of the flow are located at the ends of the tube, the continuity, Eqn. (2.5), requires that $\rho \vec{V} \cdot d\vec{A} = dm =$ constant at both ends. Then,

$$\left(h + \frac{1}{2}V^2 + gz \right)_e = \left(h + \frac{1}{2}V^2 + gz \right)_i = \text{constant} \tag{2.22}$$

In the limit where the stream tube consists of only one streamline, Eqn. (2.22) becomes the well-known *Bernoulli equation along a streamline*.

EXAMPLE 2.4

Consider a compressible air flow in the diverging section of a converging-diverging nozzle (the flow schematic is shown in Fig. E2.4) with adiabatic frictionless wall. At location 1, the gas temperature and pressure are 500 K and 300 kPa, respectively, and the velocity is 539 m/s. The gas temperature and pressure measured at a downstream location, 2, are 358 K and 93 kPa, respectively. What is the gas velocity at location 2? What is the ratio of the cross-sectional areas of location 2 to location 1? Assume that the properties are uniform at any location in the nozzle and the value of c_p for air is $1.004 \, kJ/kg \cdot K$.

Solution –

Because there is no mechanical and shear work (due to the absence of shaft and walls being frictionless), and no heat transfer (adiabatic walls) and the change in elevation can be neglected, Eqn. (2.21) provides a relationship between the gas velocity and temperature:

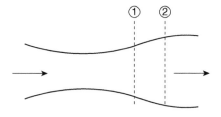

Figure E2.4 Flow schematic.

$$c_p T_2 + \frac{1}{2}V_2^2 = c_p T_1 + \frac{1}{2}V_1^2$$

$$V_2 = \sqrt{2 \times c_p(T_1 - T_2) + V_1^2} = \sqrt{2 \times 1004 \; \frac{J}{kg \cdot K} \times (500 - 358)K + \left(539\frac{m}{s}\right)^2}$$

$$= 758 \text{ m/s}$$

The area ratio according to the continuity equation, Eqn. (2.20), is

$$\frac{A_2}{A_1} = \frac{\rho_1}{\rho_2}\frac{V_1}{V_2}$$

The velocity ratio has been known and the density ratio can be determined by again using the ideal gas law

$$\frac{\rho_1}{\rho_2} = \frac{p_1}{p_2}\frac{T_2}{T_1} = \frac{300}{93} \times \frac{358}{500} = 2.310$$

Therefore,

$$\frac{A_2}{A_1} = \frac{\rho_1}{\rho_2}\frac{V_1}{V_2} = 2.310 \times \frac{539}{758} = 1.64$$

Comments –

It is noted that the increase in velocity is associated with an increase in area, which is not possible in incompressible flows. However, in compressible flows, such a phenomenon is expected in the supersonic flow regime. As will be discussed in Chapter 3, the flow in this example is supersonic. The local acoustic speed is $a = \sqrt{\gamma R T}$. So

$$a_1 = \sqrt{\gamma R T_1} = \sqrt{1.4 \times \frac{8314 \; J/kmol \cdot K}{28.8 \; kg/kmol} \times (500 \; K)} \approx 450\frac{m}{s}$$

$$a_2 = \sqrt{\gamma R T_2} = \sqrt{1.4 \times \frac{8314 \; J/kmol \cdot K}{28.8 \; kg/kmol} \times (358 \; K)} \approx 380\frac{m}{s}$$

The Mach number is defined as $M \equiv V/a$. Therefore,

$$M_1 = \frac{539}{450} = 1.20 \text{ and } M_2 = \frac{758}{380} \approx 2.0$$

At both locations, the Mach number is greater than 1 and the flow is supersonic and accelerates from location 1 to location 2. Note that $A_2/A_1 > 1$; it will be seen in Chapter 4 that an increase in the cross-sectional area for a supersonic flow leads to acceleration, contrary to a subsonic flow. $\square$

With the same assumptions as for the above energy equation (frictionless and adiabatic) and requiring that the streamline is a straight line in the x-direction, the momentum equation, Eqn. (2.8), is simplified to

$$R_{S,system} - \int_{CS}(p - p_{amb})dA = \int_{CS}V\left(\rho\vec{V}\cdot d\vec{A}\right) \qquad (2.23a)$$

and

$$R_{S,system} - [(p - p_{amb})A]_e + [(p - p_{amb})A]_i = \left(\rho V^2 A\right)_e - \left(\rho V^2 A\right)_i \qquad (2.23b)$$

Since at any instant the system consists of the fluid moving through the CV, $R_{S,system}$ is also the external force on the fluid at a given instant in the control volume and the reaction force is the "propulsive" force on the CV and the structure (e.g., the aircraft, rocket, etc.), as already demonstrated in Example 2.2 (where $T = -R_{S,system}$).

For a constant-area ($A_i = A_e$) duct flow with uniform properties and without external forces ($R_{S,system} = 0$) at any location,

$$p + \rho V^2 = \text{constant}. \qquad (2.24)$$

This form of momentum equation is useful when one-dimensional flows through a normal shock wave and within a constant-area duct with wall friction and heat transfer, as will be shown in Chapters 6 and 7.

EXAMPLE 2.5 A supersonic air flow at velocity 760 m/s and a pressure and temperature of 93 kPa and 358 K, respectively, passes through a plane shock wave in a perpendicular direction. As will be discussed in Chapter 4, a shock wave is a strong compression wave, whose thickness is very small (usually on the order of the mean free path). The air pressure and temperature downstream of the shock wave are 418.5 kPa and 603.4 K, respectively. What is the velocity downstream of the shock wave?

Solution –

Because the shock wave is thin and is normal to the flow, the constant-area assumption is good. The control volume can be drawn as shown Fig. E2.5. Because the control volume is thin and the flow is one-dimensional, the heat transfer and frictional forces through the edge of the CV are negligible ($R_{S,system} = 0$). The heat transfer and the work terms are absent in Eqn. (2.19) and therefore

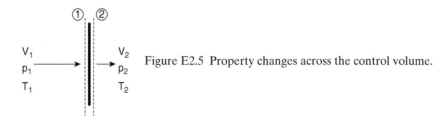

Figure E2.5 Property changes across the control volume.

$$p_1 + \rho_1 V_1^2 = p_2 + \rho_2 V_2^2$$

$$V_2 = \sqrt{\frac{1}{\rho_2}(p_1 - p_2) + \frac{\rho_1}{\rho_2} V_1^2} = \sqrt{\frac{RT_2}{p_2}(p_1 - p_2) + \frac{p_1 T_1}{p_2 T_2} V_1^2}$$

where the ideal gas law for downstream gas is used and $R = (8314 \ J/kmol \cdot K)/(28.8 \ kg/kmol) = 288.7 \ kJ/kg \cdot K$. Thus

$$V_2 = \sqrt{\frac{288.7 \frac{J}{kg \cdot K} \times 603.4 \ K}{418.5 \times 10^3 \ N/m^2} (93 \times 10^3 - 418.5 \times 10^3) \ N/m^2 + \frac{93}{418.5} \times \frac{358}{603.4} \times \left(760 \frac{m}{s}\right)^2}$$

$$\approx 276 \ m/s$$

Comments –

Since the gas is compressed through a shock wave (one of the subjects of Chapter 4), one expects a smaller speed downstream of the shock wave in the steady state flow. Following the comments of Example 2.4, here $M_2 = V_2/\sqrt{\gamma R T_2} = 276/\sqrt{1.4 \times 288 \times 603.4} = 0.56$ and $M_1 = V_1/\sqrt{\gamma R T_1} = 760/\sqrt{1.4 \times 288 \times 358} = 2.0$, that is a supersonic flow decelerates to be subsonic across a normal shock (a topic of Chapter 4). $\square$

2.7 Second Law of Thermodynamics and Entropy

The second law of thermodynamics addresses whether a prescribed process undergone by the system can possibly take place, even if it satisfies the first law of thermodynamics. It is known that not all processes are possible; for example, a baseball lying on the floor would not jump to the top of a desk without work done by an external source. All processes can be classified into two categories: reversible and irreversible processes. A reversible process is an idealized one in which all details can be reversed and the system restored to its initial state without leaving any effect (or change) on the system or its surroundings. Irreversible processes are those in which some finite changes occur that cannot be removed without external effect or work. Commonly encountered irreversibilities include friction (for example, due to fluid viscosity), heat transfer due to finite temperature differences, mixing of different liquids, combustion reactions, and so on. It takes some effort to separate oxygen and nitrogen in the air, which is the mixture of the two pure gases; energy expended to overcome friction dissipates into the surroundings is practically impossible to recovered. The reversible heat transfer only takes place with zero temperature difference, and thus takes an infinitely long time. Other finite-rate processes are similarly irreversible: the heat-releasing combustion process and the pressure jump across a shock wave within a very short distance and time (as illustrated in Example 2.5).

Perhaps naturally occurring events best illustrate irreversibility. For example, water flows from high to lower grounds. To restore it to the high ground, external work, such as by pumping, is needed that constitutes an effect on and by the surroundings. Therefore, the second law dictates in what direction a process can or cannot proceed. In terms of entropy, the second law states that the net effect of a thermodynamic process can only increase the combined entropy of the system and its surroundings. The thermodynamic property entropy (S, in J/kg) is defined for a system undergoing a reversible process, as

$$dS \equiv \left(\frac{\delta Q}{T}\right)_{rev} \tag{2.25}$$

Therefore,

$$(S_2 - S_1)_{sys} = \int_1^2 \left(\frac{\delta Q}{T}\right)_{rev} \tag{2.26}$$

where δQ is the heat exchange between the system and its surroundings, positive when the heat is transferred to the system and negative out of the system, and the subscripts 1 and 2 denote the initial and the final states, respectively. Keeping in mind that there exists irreversibility in any realistic process, a fixed mass of fluid flowing through the control volume with heat transfer would experience a change in entropy, greater than that due to a reversible process with the same initial and final states:

$$(S_2 - S_1)_{sys} = \int_{CS} \frac{\delta Q}{T} + I > \int_1^2 \left(\frac{\delta Q}{T}\right)_{rev} \tag{2.27}$$

where I is the entropy increase due to irreversibility and is always positive in value. The exact value of I is dependent on the path or the process taken (a reversible process is associated with $I = 0$). For example, to deliver the same amount of heat transfer, the larger temperature difference between the system and its surroundings, the greater the value is of I. Consequently, the assumption of reversibility of any realistic process is only an idealization with its quality depending on how sudden or finite the changes are.

A process that involves no change in entropy is an *isentropic* process. From Eqns. (2.26) and (2.27), an isentropic process must possess two features: it is both *adiabatic* ($\delta Q = 0$) and *reversible* ($I = 0$). A reversible process with heat transfer leads to $(S_2 - S_1)_{sys} > 0$ (or < 0) if $\delta Q > 0$ (or < 0). An irreversible process without heat transfer still leads to an increase in the system entropy, as $I > 0$. For an irreversible process with heat loss, whether an increase or decrease in S occurs would then depend on the relative magnitudes of $\int_1^2 \frac{\delta Q}{T}$ and I, as $\int_1^2 \frac{\delta Q}{T} < 0$ and as $I > 0$.

It is therefore possible that the system entropy decreases if there is heat loss from the system to the surroundings. This, however, causes the entropy in the surrounding to increase. On the other hand, heat addition to the system leads to an increase and a decrease in entropy of the system and the surroundings, respectively. The *principle of the entropy increase* states that the sum of these two changes in entropy is greater than zero if the heat transfer process is not reversible. The mathematical proof of this principle is shown in the following.

Let ΔS_{sys} and ΔS_{surr} denote the entropy changes of the system and its surroundings, respectively, due to the process. Thus

$$\Delta S_{sys} + \Delta S_{surr} = \frac{\delta Q}{T_{sys}} - \frac{\delta Q}{T_{surr}} + I = \delta Q \left(\frac{1}{T_{sys}} - \frac{1}{T_{surr}} \right) + I \qquad (2.28)$$

where δI is the entropy increase due to irreversibility and is ≥ 0. Two, and only two, scenarios of heat transfer are possible:

1. For heat transfer from the system to its surroundings ($\delta Q < 0$) to take place, $T_{sys} > T_{surr}$.
2. For heat transfer from the surroundings to the system ($\delta Q > 0$) to take place, $T_{surr} < T_{sys}$.

In either scenario,

$$\Delta S_{sys} + \Delta S_{surr} \geq 0 \qquad (2.29)$$

with the equal sign for the reversible process ($T_{sys} = T_{surr}$ and $I = 0$). The total entropy of the universe, consisting of the system and its surroundings, can only increase, although local (i.e., system) entropy may increase or decrease. The universe is a special case of an isolated system; an isolated system is one through whose boundaries there is no mass or energy flow. For an isolated system,

$$\Delta S_{Isolated} = I \geq 0 \qquad (2.30)$$

The entropy balance for the control volume approach is

$$\left. \frac{DS}{Dt} \right)_{system} = \frac{\partial}{\partial t} \int_{CV} s\rho dV + \int_{CS} s \left(\rho \vec{V} \cdot d\vec{A} \right) = \int_{CS} \frac{\delta \dot{Q}}{T} + \dot{I} = \int_{CS} \frac{\dot{q}}{T} dA + \dot{I} \qquad (2.31)$$

where the rate form of Eqn. (2.27) is used and $\dot{q} = \dot{Q}/A$ is the heat transfer per unit area per unit time.

2.8 Isentropic Process

For a simple compressible substance such as air and most gases, the only reversible work performed on or by the system is in the form of expansion or compression; that is, $\delta W = pdV$ (Sonntag and Van Wylen, 1993). By combining the definition of entropy, Eqn. (2.25), and the energy conservation, Eqn. (2.10b), without kinetic and potential energies,

$$TdS - pdV = dE = dU \quad \text{or} \qquad (2.32a)$$

In terms of intensive properties, Eqn. (2.23a) is rewritten as

$$Tds - pdv = de = du \tag{2.32b}$$

Then

$$Tds = pdv + du = d(u + pv) - vdp = dh - vdp \tag{2.33}$$

For ideal gases enthalpy (h) and internal energy (u) are both functions only of temperature and their changes with temperature are expressed as

$$dh = c_p dT \quad \text{and} \quad du = c_v dT \tag{2.34}$$

where c_p and c_v are the constant-pressure and constant-volume specific heats. Alternatively,

$$c_p \equiv \left(\frac{\partial h}{\partial T}\right)_p \quad \text{and} \quad c_v \equiv \left(\frac{\partial u}{\partial T}\right)_v \tag{2.35}$$

In general, both c_p and c_v are also functions of temperature. Equation (2.33) can be written as

$$Tds = c_p dT - vdp \quad \text{or} \quad Tds = c_v dT + pdv \tag{2.36}$$

By using the ideal gas law, Eqn. (2.36) becomes

$$ds = c_p \frac{dT}{T} - R\frac{dp}{p} \quad \text{or} \quad ds = c_v \frac{dT}{T} + Rdv \tag{2.37}$$

If the values of c_p and c_v are constant over the temperature range under consideration, then

$$s_2 - s_1 = c_p \ln\frac{T_2}{T_1} - R \ln\frac{p_2}{p_1} \quad \text{or} \quad s_2 - s_1 = c_v \ln\frac{T_2}{T_1} + R \ln\frac{v_2}{v_1} \tag{2.38}$$

Equation (2.38) can be understood in the following manner. From the viewpoint of gas kinetics, entropy is related to the degree of randomness of the molecule in a system. Fast-moving molecules and large volume contribute to higher degrees of randomness as they both make molecules more difficult to locate. As shown in Chapter 1, the molecular speed increases with $T^{1/2}$. Thus, keeping all other properties constant, an increase in temperature leads to an increase in entropy and so does an increase in volume for a given system of molecules. For the same reason, an increase in pressure while keeping temperature constant leads to a decrease in volume and entropy.

For an ideal gas undergoing an isentropic process, the changes in p, T, and v occur in such a way that $s_2 - s_1 = 0$. Therefore, Eqn. (2.38) leads to

$$\frac{p_2}{p_1} = \left(\frac{T_2}{T_1}\right)^{c_p/R} \quad \text{and} \quad \frac{v_2}{v_1} = \left(\frac{T_2}{T_1}\right)^{-c_v/R}$$

Noting that

$$\gamma = \frac{c_p}{c_v} \text{ and } R = c_p - c_v \tag{2.39}$$

The following relationships are obtained

$$\frac{c_p}{R} = \frac{\gamma}{\gamma - 1} \text{ and } \frac{c_v}{R} = \frac{1}{\gamma - 1} \tag{2.40}$$

Therefore for *isentropic relationships* one finds

$$\frac{p_2}{p_1} = \left(\frac{T_2}{T_1}\right)^{\gamma/(\gamma-1)}; \quad \frac{v_2}{v_1} = \frac{\rho_1}{\rho_2} = \left(\frac{T_2}{T_1}\right)^{-1/(\gamma-1)}; \quad \frac{p_2}{p_1} = \left(\frac{\rho_2}{\rho_1}\right)^{\gamma} \tag{2.41}$$

Very often the isentropic nature of the process can also be expressed by

$$p v^{\gamma} = \text{constant} \tag{2.42}$$

Those processes that deviate from the idealization can be described by the following expression for easy comparison,

$$p v^{n} = \text{constant} \tag{2.43}$$

where $n \neq \gamma$ and $n \neq 1$; these processes are called *polytropic* processes (when $n = 1$ the process is called the isothermal process as the ideal gas law $pv = RT$ states). The value of n is dependent on the details of process (and path) and is not unique even for a given system of gas. On the other hand, γ is not process-dependent; it is the ratio of gas properties c_p and c_v.

An calorically or thermally perfect gas is one for which c_p and c_v (and thus γ) are constant. An ideal gas that is calorically perfect is also called a *perfect gas. Except where noted throughout this book, gasses are assumed to be perfect.* One notable case of imperfect gas behavior is that across very strong shock waves in hypersonic flow, where the extremely large temperature rise may produce dissociation species causing values of c_p and c_v to change across the shock.

EXAMPLE 2.6 A cylinder-piston device containing 1 kg of air is undergoing an expansion process. The initial and the final pressures are 500 kPa and 200 kPa, respectively, and the initial temperature is 327 °C. For air $c_p = 1.004$ kJ/kg· K.

(a) What is the amount of work performed through the isentropic process? What are the final temperature and volume?
(b) What is the amount of work for the same pressure condition if the expansion is done through a polytropic process with $n = 1.33$? What is the heat transfer? What is the final volume?

Solution –

(a) <u>Isentropic process</u>

By using the extensive form of the isentropic relationships and the ideal gas law,

$$pV^\gamma = \text{constant} = C \text{ and } PV = MRT$$

the expansion work is

$$W_{1-2} = \int_1^2 pdV = \int_1^2 \frac{C}{V^\gamma}dV - \frac{C}{1-\gamma}\left(V_2^{1-\gamma} - V_1^{1-\gamma}\right) = \frac{p_2 V_2^\gamma V_2^{1-\gamma} - p_1 V_1^\gamma V_1^{1-\gamma}}{1-\gamma}$$

$$= \frac{p_2 V_2 - p_1 V_1}{1-\gamma} = \frac{MR(T_2 - T_1)}{1-\gamma}$$

$$\frac{T_2}{T_1} = \left(\frac{P_2}{P_1}\right)^{(\gamma-1)/\gamma} = \left(\frac{200\ kPa}{500\ kPa}\right)^{0.4/1.4} = 0.77$$

$T_2 = 462\ K$

$$W_{1-2} = \frac{1\ kg\ \times 0.288\ kJ/kg{\cdot}K\ \times (462\ K - 600\ K)}{1 - 1.4} = 99.4\ kJ$$

To find the final volume, Eqn. (2.41) can be rewritten as

$$\frac{V_2}{V_1} = \frac{v_2}{v_1} = \left(\frac{p_2}{p_1}\right)^{-1/\gamma} = \left(\frac{200\ kPa}{500\ kPa}\right)^{-1/1.4} = 1.924$$

$$V_1 = \frac{MRT_1}{p} = \frac{1\ kg \times 0.288\ kJ/kg{\cdot}K \times 600\ K}{500 \times kJ/m^3} = 0.3456 m^3$$

$$V_2 = 0.6649 m^3$$

(b) <u>Polytropic process</u>

Replacing γ with n,

$$W_{1-2} = \frac{MR(T_2 - T_1)}{1-n} \text{ and } \frac{T_2}{T_1} = \left(\frac{P_2}{P_1}\right)^{(n-1)/n}$$

$$\frac{T_2}{T_1} = \left(\frac{P_2}{P_1}\right)^{(n-1)/n} = \left(\frac{200\ kPa}{500\ kPa}\right)^{0.33/1.33} = 0.80$$

$T_2 = 480\ K$

$$W_{1-2} = \frac{1\ kg\ \times 0.288\ kJ/kg{\cdot}K\ \times (480\ K - 600\ K)}{1 - 1.33} = 103.8\ kJ$$

The first law of thermodynamics for the system is

$$Q_{1-2} - W_{1-2} = E_2 - E_1 = Mc_v(T_2 - T_1) = M\frac{c_p}{\gamma}(T_2 - T_1)$$

$$Q_{1-2} = W_{1-2} + M\frac{c_p}{\gamma}(T_2 - T_1) = 103.8 \ kJ + 1 \ kg \times \frac{1.004 \ kJ/kg\cdot K}{1.4}(480 \ K - 600 \ K)$$

$$= 17.7 \ kJ$$

$$\frac{V_2}{V_1} = \frac{v_2}{v_1} = \left(\frac{p_2}{p_1}\right)^{-1/n} = \left(\frac{200 \ kPa}{500 \ kPa}\right)^{-1/1.33} = 1.992$$

$$V_2 = 0.6883 \ m^3$$

Comments –
(1) The polytropic process performs slightly more work, 4.4 kJ, than the isentropic process for the same expansion ratio. However, it requires four times the amount of work in the form of heat transfer to the system. In contrast the isentropic process does not require heat transfer to perform work. Because of the heat transfer to the system and the higher final temperature, the polytropic process results in a larger expanded final volume for the same final pressure. (2) for the isentropic process,

$$Q_{1-2}W_{1-2} = M\frac{c_p}{\gamma}(T_2 - T_1) = 99.4 \ kJ + \frac{1 \ kg\times0.288 \ kJ/kg\cdot K}{1.4}(462 \ K - 600 \ K) = 0,$$

which satisfies the adiabatic requirement for an isentropic process. □

Problems

Problem 2.1 Consider air (assumed an ideal and perfect gas) in a container undergoing thermodynamic changes.

(a) If the volume is reduced by a factor of 3 due through an isentropic process, what is the change (increase or decrease) in entropy?
(b) What is the pressure change?
(c) If the above volume change takes place under isothermal conditions, what is the entropy change and what is the pressure change?
(d) If the volume change takes place under the constant-pressure condition, what is the final temperature? How can you achieve this change?

Problem 2.2 A perfectly insulated box was initially divided into two chambers by a diaphragm. One of the chambers was one-third of the total volume and was filled with a perfect gas with specific gas constant R, while the other was a vacuum. The divider was then removed and the system was allowed time to reach an equilibrium state. For the system of perfect gas in the box, what is the change in entropy?

Problem 2.3 A perfectly insulated box was initially divided by a diaphragm into two chambers containing same gas species. In the initial separated state (state A), one chamber, chamber 1, has volume, pressure, temperature, and mass equal to $V_1, p_1, T_1,$ and m_1 respectively, while the other has $V_2, p_2, T_2,$ and m_2. The diaphragm is then removed so that the mixing of gas takes place in a reversible manner. The state after the mixing is completed is state B.

(a) Find the final pressure, p_B, and entropy difference between the two states $(S_B - S_A)$.
(b) For $T_1 = T_2 = T, m_1 = m_2 = m$ (so that $m_B = 2\ m$), find p_B and $(S_B - S_A)/m$ and show that $(S_B - S_A) > 0$.
(c) Show that the result in (a) can be reduced to be identical to that in Problem 2.2 for $m_1 = 0$.

Problem 2.4 Within a perfectly insulated box, a football-shaped object was initially suspended from its upper surface by a string. The string was then cut and the object dropped. After the possible bouncing motion had subsided, the center of mass of the object is lowered by a height h. (a) Find the entropy change from entropy change in the gas. (b) Does the inelastic collisions between the object (for example, effects of friction and acoustic waves due to collisions) play a role in the entropy change?

Problem 2.5 An air container is initially at a pressure of 1MPa and a temperature of 600 K. A nozzle attached to it discharges its content into the surroundings until the mass in the container is one third of its original value. What are the final temperature and pressure if the process is isentropic?

Problem 2.6 A perfect gas in a piston-cylinder device (initial pressure and temperature are p_1 and T_1, respectively) undergoes volume change isothermally and reversibly.

(a) Find the heat transfer necessary (q_{1-2}) for the specific volume to change from v_1 to v_2 and the entropy change, $(s_2 - s_1)$, as a function of p_2/p_1.
(b) The above process is followed by heat transfer that brings the system pressure back to p_1, i.e., $p_3 = p_1$, while keeping $v_3 = v_2$. Find the heat transfer $q_{2-3}, T_3,$ and $(s_3 - s_2)$ in terms of p_2/p_1.
(c) What is the entropy change, $(s_3 - s_1)$, expressed in c_p, after completing the above two processes?

Problem 2.7 A piston-cylinder system, as shown in the figure, contains air $(R = 0.287\,\text{kJ/kg} \cdot \text{K}$ and $c_p = 1.005\,\text{kJ/kg} \cdot \text{K})$ is initially at a temperature of 300 K. The cylinder has a cross-sectional area of 0.1 m^2 and an initial volume of 0.02 m^3 and the piston mass M = 200 kg. The ambient pressure is 1 atmosphere. The cylinder is then submerged in a thermal bath at 500 K. Time is allowed for the equilibrium to be reached with the thermal bath. Assume no friction between the piston and the cylinder wall. Calculate the following: the heat transfer to, the work done by, and the entropy change in, the air in cylinder. Comment on the findings.

Problem 2.8 Consider a quasi-one-dimensional gas flow in a variable-area channel with frictionless walls. At two known locations, designated by subscripts 1 and 2, the cross-sectional areas and pressures are $A_1, A_2, p_1,$ and p_2. The gas is considered to be ideal with constant c_p and c_v (and therefore a constant γ), that is, it is a perfect gas. If the flow between locations 1 and 2 is to be isothermal, find (a) V_2/V_1 and (b) heat transfer from location 1 to location 2, q_{12}.

3 Acoustic Wave and Flow Regime

This chapter describes the mechanism and the speed with which small disturbances propagate in continuum fluids, especially in gases, where compressible and supersonic flows are likely to occur. Small disturbances, such as human voices, are known to propagate with the speed of sound waves. The mechanism of propagation determines how the disturbance is felt, thus constituting the signaling mechanism. The speed of sound will be determined in terms of macroscopic fluids properties, although the signaling mechanism has a microscopic, or molecular, root.

3.1 Speed of Sound in Compressible Media

Chapter 1 discusses some characteristic molecular velocities of gases based on the kinetic theory without needing to know the continuum behavior of gases. Consider molecules adjacent to the surface. These molecules have a preferred direction of motion (i.e., away from the surface) because they rebound from the surface. As a consequence, the molecules present next to this immediate layer also experience more collisions with molecules coming from the direction of the surface than they would if the surface is not present. Such a preferred direction of molecular motion continues into the fluid far away from the surface, with diminishingly smaller effect in that direction. The frequency and intensity of such collisions must be related to some characteristic speed of the molecules, which is a function of the molecular mass and temperature. The fact (or the message) of the presence of the surface propagates in a preferred direction, that is, away from the surface, with that characteristic speed.

Recall from Chapter 1 the three characteristic speeds: the most probable (u_{mp}), the average ($\bar{u}$), and the root-mean-square (u_{rms}), as Eqns. (1.4), (1.5), and (1.8) show:

$$u_{mp} = \left(\frac{2\,kT}{m} \right)^{1/2}$$

$$\bar{u} = \left(\frac{8\,kT}{\pi m}\right)^{1/2}$$

$$u_{rms} = \left(\frac{3\,kT}{m}\right)^{1/2}$$

The values of these three velocities depend on the molecular mass (m) and temperature (T). As speculated in Chapter 1, the speed of sound is on the same order of magnitude of these three velocities. However, it is desirable to express the speed of sound in terms of macroscopic (or continuum) properties of the fluids, such as density, temperature, pressure, the degree of compressibility of the medium, and other properties specific to the particular gas (note that k is a universal constant and is not specific to any gas). Recall that to arrive at the ideal gas law Eqn. (1.14) using the kinetic approach, it is implicit that a sufficient number of molecules are contained in a volume. The ideal gas law can be written as, by combining Eqn. (1.14), which is based on molecular velocity distribution, and Eqn. (1.17b), based on macroscopic observations:

$$pV = NkT \tag{1.14}$$

$$pV = MRT \text{ or } p = \rho RT \tag{1.17}$$

As discussed in Section 1.3 these two forms of ideal gas law are identical for $R = k/m$. Equation (1.17) includes density, the potential compressibility (p vs. ρ, depending on the nature of the compression process), and the specific gas constant (R instead of k).

The gas kinetics theory that leads to the ideal gas law assumes no intermolecular interactions except during collisions and a negligible volume taken by molecules. Therefore, forces such as gravitational and electromagnetic forces are not of concern. At high pressures and low temperatures, when molecules could come into each other's potential fields, the intermolecular force can be appreciable, constituting the real, non-ideal gas effects. One of the occasions, where the non-ideal gas effect plays a role in the speed of sound in these media, will be seen in the next section.

To obtain the speed of sound in terms of macroscopic properties, consider how the signal of an infinitesimal disturbance propagates in a long tube filled with a gas with a piston at one end (shown in Fig. 3.1a). The piston is initially at rest and is given a sudden motion of a constant but infinitesimal constant velocity dV toward the gas. The gas molecules in the immediate vicinity experience not only the presence of the piston surface but also the added velocity dV; the gas thus experiences compression because the collision with molecules freshly rebounding from the surface with a larger velocity leads to a larger pressure. This immediate layer of molecules relays this message of piston motion to the next layer and into the bulk of the gas. Because of the infinitesimal piston motion, this signal of compression (the compression wave) travels

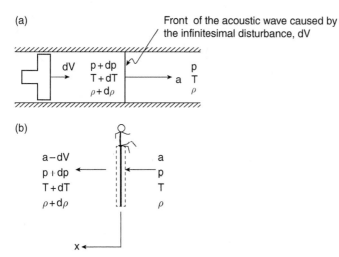

(a)

Front of the acoustic wave caused by
the infinitesimal disturbance, dV

dV $p+dp$ p
       $T+dT$ a T
       $\rho+d\rho$ ρ

(b)

$a-dV$ a
$p+dp$ ←   ← p
$T+dT$ T
$\rho+d\rho$ ρ

x ←

Figure 3.1 (a) Acoustic (or sound)
wave generated by the motion or
disturbance of a piston in a piston-
cylinder device; (b) flow and fluid
properties seen by an observer
traveling with (or "riding on") the
sound wave. The inherently
unsteady wave propagation
problem of (a) is transformed into
a steady state flow problem in (b).

with the acoustic speed, denoted by a. As illustrated in Fig. 3.1a, the bulk gas behind the wave has infinitesimal increases in pressure (dp), temperature (dT), density $(d\rho)$, and velocity $(dV$, from the initial value of zero). The wave propagation is unsteady in nature. A gas particle initially at rest ahead of the wave will at a later time experience these changes in properties; that is, at a given location, gas properties vary with time.

The unsteady problem is inherently more complicated as it involves one more independent variable, that is, time. It can be made tractable by imagining that an observer riding on the sound wave will observe the bulk gas ahead of the wave flow toward the wave with a constant velocity equal to the speed of sound a and leave with a constant speed $a - dV$, as shown in Fig. 3.1b. By fixing the coordinate system on the traveling wave, the observation becomes steady. That is, besides the velocity, the properties are p, ρ, T, and so on, ahead of the wave and are $p + dp, \rho + d\rho, T + dT$, etc. downstream of the wave, again as shown in Fig. 3.1b. Such a coordinate transformation is called *Galilean transformation* and is frequently used when dealing with unsteady problem with a constant reference velocity. For a one-dimensional sound wave (in that its thickness is much smaller than the diameter of the tube, or if the portion far away from the wall is examined, where frictional force and heat transfer are negligible) then the Reynolds transport theorem for mass and momentum (Eqns. (2.5) and (2.8)) become, respectively:

$$0 = (\rho + d\rho)(a - dV)A - \rho a A \tag{3.1}$$

$$(-p - dp)A + pdA = (a - dV)(\rho + d\rho)(a - dV)A - (a\rho aA) \tag{3.2}$$

Note that the surface force the control volume shown in Fig. 3.1b experiences is only due to pressure. By neglecting the second-order term, Eqn. (3.1) is reduced to

$$ad\rho - \rho dV = 0 \tag{3.3}$$

By substituting $(\rho + d\rho)(a - dV)A = \rho a A$ into Eqn. (3.2), the momentum equation becomes

$$dp = \rho a \, dV \tag{3.4}$$

Combining Eqns. (3.3) and (3.4) yields

$$a^2 = \frac{dp}{d\rho} \tag{3.5}$$

Apparently, the value of a described by Eqn. (3.5) depends on the process of generating the disturbance. For example, an isothermal process (where $p = \rho RT =$ constant would lead to $a^2 = RT$) and an isentropic process (where $p\nu^\gamma =$ constant) would lead to different results. Due to the vanishingly small perturbation, heat transfer, and frictional force, the process leading to Eqn. (3.5) is isentropic. Therefore, it is appropriate to express a in terms of a partial derivative while keeping s constant:

$$a^2 = \left(\frac{\partial p}{\partial \rho}\right)_s \tag{3.6}$$

Because for a perfect gas undergoing isentropic processes, $p\nu^\gamma =$ constant, $p =$ constant $\times \rho^\gamma$, and $p = \rho RT$. Therefore,

$$\left(\frac{\partial p}{\partial \rho}\right)_s = \text{constant} \times \gamma \times \rho^{\gamma-1} = p\rho^\gamma \times \gamma \times \rho^{-1} = \frac{\gamma p}{\rho} = \gamma RT \tag{3.7}$$

So the speed of sound is

$$a = \sqrt{\gamma RT} \tag{3.8}$$

Equation (3.4) indicates that a compression wave ($dp > 0$) causes $dV > 0$, which is depicted in Fig. 3.1a; that is, the wave carries the gas behind it along with it. On the other hand, an expansion wave ($dp < 0$) is generated by withdrawing the piston ($dV < 0$) away from the gas. While the wave propagates into the bulk gas at rest, it pushes the gas away from it and the gas behind the expansion wave moves along with the piston. For either the compression or expansion wave, the gas between the piston and the wave moves with the cylinder, while that ahead of the wave remains at rest. As long as the amplitude of either wave is small, the propagation is isentropic and the speed of sound is the same for compression and expansion waves; this is left as an end-of-chapter problem (Problem 3.1).

The *speed of sound* is also called the *acoustic speed* or *acoustic velocity*. The wave generated by a small disturbance is called a sound wave or acoustic wave. From Eqn. (3.8), the following observations can be made for a sound wave.

(1) The speed of sound is specific for a gas at a given temperature. Recall that $R = \hat{R}/\hat{M}$ or k/m, which is specific to a gas as shown in Chapter 1 and that $\gamma = c_p/c_v$, with both c_p and c_v also specific to the gas.

(2) For a given gas, the speed of sound is proportional to $\sqrt{T}$ with T being the temperature ahead of the sound wave.

(3) Equations (1.4), (1.5), and (1.8) can be rewritten as $u_{mp} = \sqrt{2\,RT}$, $\bar{u} = \sqrt{8\,RT/\pi}$, and $u_{rms} = \sqrt{3\,RT}$. For air, $\gamma = 1.4$ and $a < u_{mp} < \bar{u} < u_{rms}$. For many common gases, $\gamma < 2$, such an ordering of molecular and acoustic speeds are also true.

(4) The motion of $dV < 0$ in Fig. 3.2 will also propagate away from the piston, sending out a signal of expansion. In the isentropic limit, the signal also travels with a speed $a = \sqrt{(\partial p/\partial \rho)_s} = \sqrt{\gamma RT}$. The derivation for this result can be done following similar procedures leading to Eqn. (3.6) and is left to the reader as an end-of-chapter problem.

In a medium that is not a perfect gas, the speed of sound can be derived using its *isentropic compressibility*.

$$k_s \equiv \frac{1}{\rho}\left(\frac{\partial \rho}{\partial p}\right)_s \qquad (3.9)$$

Therefore,

$$a = \sqrt{\left(\frac{\partial p}{\partial \rho}\right)_s} = \sqrt{\frac{1}{\rho k_s}} \qquad (3.10)$$

For liquids and solids the value of the bulk modulus (E_v) are often reported and it is related to k_s by

$$E_v = \frac{1}{k_s} \qquad (3.11)$$

Therefore, for solids and liquids,

$$a = \sqrt{\frac{E_v}{\rho}} \qquad (3.12)$$

Because it takes a very large pressure change (or compressive stress, Δp) to cause a small density change ($\Delta \rho$) in liquids and solids, much larger speeds of sound can be expected than in gases. If the gas in Fig. 3.1a is replaced by a perfectly rigid solid, for which no deformation or change in ρ is appreciable and $k_s = 0$, the speed of sound is infinity, according to Eqn. (3.10).

> **EXAMPLE 3.1** Find the speed of sound in air, water, and stainless steel at 20 °C. The values of E_v for water and stainless steel are $2.24 \times 10^9\,N/m^2$ and $160 \times 10^9\,N/m^2$, respectively. The water density is 1,000 kg/m^3 and the steel density is approximately 8,000 kg/m^3.
>
> **Solution –**
> Assuming air is a perfect gas,

$$a_{air} = \sqrt{\gamma RT} = \sqrt{1.4 \times (288 \ N \cdot m/kg \cdot K) \times (293 \ K)} = 343.7 \ m/s$$

$$a_{water} = \sqrt{\frac{E_v}{\rho}} = \sqrt{\frac{2.24 \times 10^9 \ N/m^2}{1,000 \ kg/m^3}} = 1,496.7 \ m/s$$

$$a_{ss} = \sqrt{\frac{E_v}{\rho}} \approx \sqrt{\frac{160 \times 10^9 \ N/m^2}{8,000 \ kg/m^3}} = 4,472.1 \ m/s$$

$\square$

3.2 Speed of Sound in Non-Ideal Gases

In Section 3.2 the speed of sound in perfect gases, $a = \sqrt{\gamma RT}$, is based on the ideal gas law (or the equation of state) $p = \rho RT$ and constant specific heats. Behavior of gases at high pressures and low temperatures tend to deviate from the ideal gas law, as the intermolecular forces and molecular volume cannot be neglected. Under these conditions, the gases are "real" in that the ideal gas law becomes an inadequate description of the gas behavior. A measure of the deviation from the ideal state is the compressibility factor, Z:

$$Z \equiv \frac{p v}{RT} \tag{3.13}$$

According to this definition, the closer Z is to unity, the closer the gas behaves like an ideal gas. So Z is a measure and says nothing about the "law" a real gas should follow. A number of equations of state have been proposed for the real gas. Two of them are here described. The first is *Van der Waals's equation of state* that takes into account effects of intermolecular forces and finite, non-negligible volumes taken by the molecules (Sonntag and Van Wylen, 1993):

$$\left(p + \frac{\bar{a}}{v^2} \right) (v - \bar{b}) = RT \tag{3.14}$$

where $\bar{a}$ and $\bar{b}$ are positive constants specific to each gas. For $p \gg \bar{a}/v^2$ and $v \gg \bar{b}$, Eqn. (3.14) reduces to the ideal gas law. Replacing v with $1/\rho$ and after some algebraic manipulations, Eqn. (3.14) becomes

$$p = \frac{\rho RT}{1 - \bar{b}\rho} - \bar{a}\rho^2 \tag{3.15}$$

This expression again reduces to $p = \rho RT$ for $\bar{a} \to 0$ and $\bar{b} \to 0$. Equation (3.6) or (3.10) will still be used for calculating speed of sound, except that k_s in Eqn. (3.10) is a species-specific physical property and has to be predetermined. A more general expression is derived in the following.

For a given function $z = z(x, y)$,

$$\left(\frac{\partial y}{\partial x}\right)_z = -\frac{(\partial z/\partial x)_y}{(\partial z/\partial y)_x}$$ (3.16)

Similar for

$$s = s(p, \rho)$$

$$\left(\frac{\partial p}{\partial \rho}\right)_s = -\frac{(\partial s/\partial \rho)_p}{(\partial s/\partial T)_\rho} = -\left[\frac{(\partial s/\partial T)_p}{(\partial s/\partial p)_\rho}\right]\left[-\frac{(\partial T/\partial \rho)_p}{(\partial T/\partial p)_\rho}\right]$$

By using Eqn. (3.16) once again for the term in the second bracket where $T = T(p, \rho)$,

$$\left(\frac{\partial p}{\partial \rho}\right)_s = \frac{(\partial s/\partial T)_p}{(\partial s/\partial p)_\rho}\left(\frac{\partial p}{\partial \rho}\right)_T$$ (3.17)

By taking partial derivatives of s in Eqn. (2.37) with respective to T and p while keeping p and ρ fixed respectively, Eqn. (3.17) becomes

$$\left(\frac{\partial p}{\partial \rho}\right)_s = \frac{(c_p/T)}{(c_v/T)}\left(\frac{\partial p}{\partial \rho}\right)_T = \gamma\left(\frac{\partial p}{\partial \rho}\right)_T$$ (3.18)

Therefore,

$$a = \sqrt{\left(\frac{\partial p}{\partial \rho}\right)_s} = \sqrt{\gamma\left(\frac{\partial p}{\partial \rho}\right)_T}$$ (3.19)

Substituting the Van der Waals's equation into Eqn. (3.18) yields the speed of sound in a non-ideal gas:

$$a = \sqrt{\frac{\gamma RT}{(1 - \bar{b}\rho)^2} - 2\gamma\bar{a}\rho}$$ (3.20)

An improvement in predicting the speed of sound, particularly for low temperature, is achieved by modifying Van der Waal's equation as follows:

$$p = \frac{\rho RT}{1 - \bar{b}\rho} - \frac{\bar{a}\rho^2}{T}$$ (3.21)

which is Bethelot's equation of state, so that

$$a = \sqrt{\frac{\gamma RT}{(1 - \bar{b}\rho)^2} - \frac{2\gamma\bar{a}\rho}{T}}$$ (3.22)

3.3 Signaling Mechanism

Because small disturbances send out signals traveling at the speed of sound (a), one natural question is how the fluid reacts to the approaching object with $V > a$. In this scenario, the fluid cannot sense the approaching of the object until being "hit" by the object. The object in this case is said to be traveling with a supersonic speed. Likewise, a flow approaching an object (stationary or moving) with a relative speed that is greater than a is called a supersonic flow; it does not sense and therefore cannot adjust to the presence of the object until it hits the object, when very abrupt changes need to be made (in velocity, density, temperature, and pressure through a shock wave, as will be discussed in details in Chapter 4).

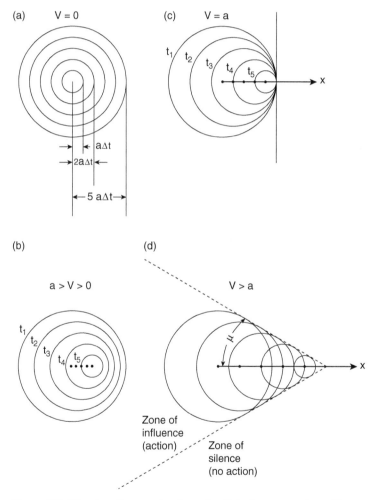

Figure 3.2 The signaling mechanism of a disturbance moving with a speed V: (a) $V = 0$, (b) $0 < V < a$ (i.e., subsonic) (c) the sonic speed, $V = a$ (i.e., sonic) and (d) $V > a$, (i.e., supersonic), where $a = \sqrt{\gamma R T}$ is the speed of sound. Note that only under supersonic conditions does the distinction between zones of silence (no action) and of influence (action) exists.

It is instructive to examine how acoustic waves generated by moving sources propagate in a quiescent fluid. The results for various speeds ranging from stationary source to sources moving with supersonic speeds are shown in Fig. 3.2. In these figures, the circle with the label 1 and radius 4 $a\Delta t$ was generated when the object was at location 1, the circle with the label 2 and radius 3 $a\Delta t$ was generated when the object was at location 2, and so on, while the current location is denoted by 5. For $V = 0$ (a stationary source), the acoustic wave fronts generated at consecutive time intervals of multiples of an arbitrary length Δt form concentric circles, as shown in Fig. 3.2a. If the object moves with a speed $V < a$ (as illustrated in Fig. 3.2b), then the acoustic wave generated at earlier instant still enclose those generated at later moments, like the case of $V = 0$. This is because the sound wave travels ahead of the object. However, the wave front spacing is compressed in the direction of the source motion. With $V = a$, the moving object is catching up with the acoustic wave generated by itself (the wave front spacing is zero in the direction of the source motion), as illustrated in Fig. 3.2c. For $V > a$, the more recently generated wave fronts intersect the earlier ones, as shown in Fig. 3.2d. In this case, the observer moving with the source would not even hear his/her own voice!

For $V > a$ (Fig. 3.2d), lines tangent to the spherical wave fronts can be drawn to form a cone, called Mach cone. This cone has a cone angle, called Mach angle, μ:

$$\sin\mu = \frac{at}{Vt} = \frac{1}{M} \tag{3.23}$$

where M is the Mach number defined as

$$M \equiv \frac{V}{a} \tag{3.24}$$

For $M > 1$ (Fig. 3.2d), the source is said to be moving with a *supersonic* speed. The region inside the cone is called the *zone of action* or the *zone of influence*. When located outside the cone, the *zone of silence*, an observer would not feel the disturbance from the source until sometime after the cone moves past the observer. The cone expands with time with a speed of sound in the direction normal to its surface, because the moving source sends out the signal of itself at the speed of sound, even when it is moving with a supersonic speed. It can be said that a concoction such as a supersonic ambulance might be of little use!

By using the Galilean coordination transformation for the situation shown in Fig. 3.1b, or in the case of a supersonic flow approaching a stationary object or with a relative speed that is supersonic, the fluid ahead of the object would not sense the existence of the object until the moment it hits the object. Therefore, in flow regime where the relative speed is supersonic, the flow cannot adjust to the object well ahead of the object.

If $M < 1$, the flow is said to be *subsonic*. In the case of subsonic flow regime, neither a cone nor a cone angle exists, as suggested by Eqn. (3.23) and demonstrated in Fig. 3.2a and 3.2b. A typical subsonic laminar flow over an airfoil is illustrated in

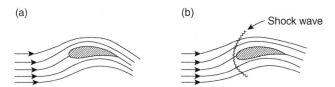

Figure 3.3 Streamline patterns over an airfoil of (a) a subsonic flow and (b) a supersonic flow. Note that the streamlines adjust in the upstream region of the airfoil under subsonic conditions while, under supersonic conditions, they are straight until being deflected by the shock wave that forms near the airfoil surface.

Fig. 3.3, where the streamline curvature suggests that the fluid adjusts its path well ahead before reaching the airfoil (the "up-wash" for a lifting airfoil). In other words, the adjustment is made gradually, with changes in the streamline curvature upstream of the airfoil.

The gradual adjustment of the fluid in the subsonic regime is not possible in the supersonic regime, as the fluid does not sense that it is approaching the airfoil. Some finite adjustment has to be made at the instant when the fluid "hits" the airfoil. Such a finite and abrupt adjustment (with finite changes in density, pressure, and temperature) is not an isentropic process will be further discussed in the next chapter.

Problems

Problem 3.1 Referring to Fig. 3.1b, show for a perfect gas that $a = \sqrt{(\partial p/\partial \rho)_s} = \sqrt{\gamma R T}$ if $dV < 0$ (associated with property changes to $p - dp, \rho - d\rho, T - dT$, etc. downstream of the wave; i.e., the expansion wave still propagates to the right).

Problem 3.2 Find the speed of sound using Van der Waals's equation of state for nitrogen at (a) 101 kPa and 233 K, and (b) 60 atmospheres and 1,200 K. The Van der Waals constants are $\bar{a} = 182.0$ N$\cdot$m^4/kg^2 and $\bar{b} = 1.40 \times 10^{-3}$m^3/kg.

Problem 3.3 Find the speed of sound using Van der Waals's equation of state for oxygen at 101.3 kPa, and 2,000 K. The Van der Waals constants are $\bar{a} = 136.0$ N$\cdot$m^4/kg^2 and $\bar{b} = 0.995 \times 10^{-3}$m^3/kg.

Problem 3.4 A one-dimensional, small-amplitude pressure pulse (having a magnitude dp) propagates into a quiescent perfect gas at a pressure and temperature of p_o and T_o, respectively. Show that the gas velocity and temperature changes are

$$dV = \frac{(\gamma R T_o)^{1/2}}{\gamma p_o} dp$$

and

$$dT = \frac{(\gamma - 1)T_o}{\gamma p_o} dp$$

Problem 3.5 A weak incident pressure wave propagates in quiescent air toward a flat wall with the wave front parallel to the wall (i.e., one-dimensional wave propagation). After reflecting from the wall, the wave travels back into the air, leaving the air behind it quiescent again due to the non-moving boundary condition of the wall. Using Eqn. (3.4) to show that compared to the initially quiescent air, the pressure rise of the air behind the reflected wave is twice that of the air in front of it.

4 One-Dimensional Isentropic Flow, Shock, and Expansion Waves

Thermodynamic relationships for isentropic processes derived in Chapter 2 are useful simplifications for solving a variety of flow problems. In an isentropic flow, the fluid and flow properties at two locations along a stream line are related by these isentropic relations. In this chapter they are used for flow regimes ranging from subsonic to supersonic. Flow transition from subsonic to supersonic conditions are specifically described in ducts with variable cross-sectional area, with applications to rocket nozzle or other internal flows. Once the possibility for supersonic flow is established, the formation of shock waves becomes a concern. As described in Chapter 3, shock waves are compression waves where the compression process is accomplished over a distance of a few mean free paths. They are therefore very abrupt and thus highly non-isentropic; the stronger the shock wave is (i.e., the larger the compression ratio is), the further the compression process deviates from the ideal isentropic process. Although the flow regimes both upstream and downstream may still be approximated as isentropic, the flow properties across the shock can no longer be related by isentropic thermodynamic relationships. Under isentropic conditions, the energy equation (or energy conservation law) can be used for solution over the entire flow field. With shock waves, changes in flow and fluid properties across them can only be related by solving the three conservation laws (mass, momentum, and energy). To focus on the changes in isentropic flow and due to shock waves, one-dimensional flows of perfect gas are assumed throughout this chapter for simplicity.

4.1 One-Dimensional Isentropic Flow of a Perfect Gas

A steady, one-dimensional isentropic flow is a uniform flow with identical constant for all streamlines, for which the Bernoulli equation states

$$h + \frac{1}{2}V^2 + gz \left(= u + \frac{p}{\rho} + \frac{1}{2}V^2 + gz \right) = \text{constant} \tag{2.22}$$

Often in the case for compressible flows, changes in enthalpy and the kinetic energy are much larger than that in elevation. Thus between two arbitrary locations on a streamline

$$h_1 + \frac{1}{2}V_1^2 = h_2 + \frac{1}{2}V_2^2$$

or simply

$$h + \frac{1}{2}V^2 = \text{constant} \tag{4.1}$$

Although Eqn. (4.1) is derived for an isentropic flow, only the adiabatic condition needs to be satisfied. This is so because both $V^2/2$ and h are energies possessed by a fluid particle and are thus not path-dependent. Therefore, it is valid across one-dimensional shock waves (further discussed in Section 4.3). Equation (4.1) requires that a change in velocity would lead to changes in h and temperature. In high-speed compressible flow, such changes can be significant. It is desirable to relate these changes to the change in Mach number. Recall that

$$h = c_p T \quad \text{and}$$

$$\frac{c_p}{R} = \frac{\gamma}{\gamma - 1}$$

Therefore, by assuming a constant c_p,

$$T + \frac{1}{2}\frac{V^2}{c_p} = T\left(1 + \frac{\gamma - 1}{2}\frac{V^2}{\gamma RT}\right) = T\left(1 + \frac{\gamma - 1}{2}M^2\right) = \text{constant} \tag{4.2}$$

and

$$\frac{T_2}{T_1} = \frac{1 + \frac{\gamma-1}{2}M_1^2}{1 + \frac{\gamma-1}{2}M_2^2} \tag{4.3}$$

By recalling other isentropic relationships,

$$\frac{p_2}{p_1} = \left(\frac{T_2}{T_1}\right)^{\gamma/(\gamma-1)} = \left(\frac{1 + \frac{\gamma-1}{2}M_1^2}{1 + \frac{\gamma-1}{2}M_2^2}\right)^{\gamma/(\gamma-1)} \tag{4.4}$$

$$\frac{v_2}{v_1} = \frac{\rho_1}{\rho_2} = \left(\frac{1 + \frac{\gamma-1}{2}M_1^2}{1 + \frac{\gamma-1}{2}M_2^2}\right)^{-1/(\gamma-1)} \tag{4.5}$$

4.2 Stagnation Properties in an Isentropic Flow

In the above discussion, if at one of the locations (for example location 1) the flow is brought to *adiabatically* rest (i.e., $V = 0$), the enthalpy at this location is the *stagnation* (or *total*) *enthalpy*. Because there is no heat transfer, the value of enthalpy remains constant on any given stream line. Letting the subscript "t" denote the stagnation condition, Eqns. (4.1) and (4.2) become

$$h_t = c_p T_t = h + \frac{1}{2} V^2 \tag{4.6a}$$

and

$$T_t = T + \frac{1}{2} \frac{V^2}{c_p} = T\left(1 + \frac{\gamma - 1}{2} \frac{V^2}{\gamma RT}\right) = T\left(1 + \frac{\gamma - 1}{2} M^2\right) \tag{4.6b}$$

Therefore

$$\frac{T_t}{T} = 1 + \frac{\gamma - 1}{2} M^2 \tag{4.6c}$$

Equation (4.6c) is valid as long as the flow is adiabatic (even if not isentropic) and the total enthalpy is fixed. For an isentropic flow $p \gamma^\gamma = $ constant and Eqn. (2.41) becomes

$$\frac{p_t}{p} = \left(\frac{T_t}{T}\right)^{\gamma/(\gamma-1)} = \left(1 + \frac{\gamma - 1}{2} M^2\right)^{\gamma/(\gamma-1)} \tag{4.7}$$

$$\frac{\rho_t}{\rho} = \left(\frac{T_t}{T}\right)^{1/(\gamma-1)} = \left(1 + \frac{\gamma - 1}{2} M^2\right)^{1/(\gamma-1)} \tag{4.8}$$

Equations (4.6c) through (4.8) suggest that stagnation properties remain constant on a given stream line and throughout the one-dimensional flow, as the stagnation state can be achieved by bringing the flow from any point to a rest (in an adiabatic manner for T_t and in an isentropic manner for p_t and ρ_t). Thus they can serve as "*reference*" states for a stream line. If the flow is uniform and isentropic, then they serve as the reference states throughout the entire flow. Using Eqns. (4.6c) through (4.8), thermodynamic properties (T, p, and ρ) can be determined knowing the value of Mach number M, or vice versa. Numerical results of equations are tabulated in Appendix C as functions of the Mach number for a number of values of γ. For $\gamma = 1.4$, these results are plotted in Fig. 4.1, which also contains the Prandtl-Meyer function, ν, and the area ratio A/A^* for isentropic channel flows. Both ν and A/A^* are discussed in details in Chapter 5.

EXAMPLE 4.1 Consider a uniform steady flow of air in an adiabatic nozzle that has the following properties at location 1: $T = 400$ K, $p = 200$ kPa, and $V = 400$ m/s.

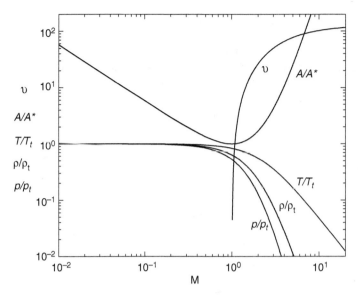

Figure 4.1 Graphical representation of isentropic relations among thermodynamic properties (p, ρ, and T) and channel cross-sectional area (A) with Mach number (M). Note that all properties are referred to the reference/stagnation value (with the subscript t), while the reference area is that of the throat of a channel (A^*). Note that ν is the Prandtl-Meyer function, to be discussed in Chapter 5.

At a downstream location (location 2), the pressure decreases to 150 kPa. What are the velocity and temperature at this location?

Solution –

Assume that the flow is steady, that there is no heat transfer due to adiabatic nozzle walls, and that air is an ideal gas with constant specific heats over the temperature range of the problem. Because the flow is uniform, there is no shear work and the flow is thus isentropic. Therefore stagnation properties can be used as references throughout the flow. First of all, the Mach number is needed for finding out these reference properties.

$$M_1 = \frac{V_1}{a_1} = \frac{V_1}{\sqrt{\gamma R T_1}} = \frac{400 \ m/s}{\sqrt{1.4(288 \ J/kg \cdot K) \times 400 \ K}} = 0.996$$

$$\frac{p_{t1}}{p_1} = \left(1 + \frac{\gamma - 1}{2} M_1^2\right)^{\gamma/(\gamma-1)} = \left(1 + \frac{1.4 - 1}{2}[0.996]^2\right)^{1.4/(1.4-1)} = 1.884$$

or

$$p_{t1} = 376.8 \ kPa$$

$$\frac{T_{t1}}{T_1} = 1 + \frac{\gamma - 1}{2} M_1^2 = 1 + \frac{1.4 - 1}{2} [0.996]^2 = 1.198 \quad \text{or} \quad T_{t1} = 479.2 \ K$$

For the isentropic flow, $p_{t2} = p_{t1}$ and $T_{t2} = T_{t1}$. Therefore,

$$\frac{p_{t2}}{p_2} = \left(1 + \frac{\gamma - 1}{2} M_2^2\right)^{\gamma/(\gamma-1)} = \left(1 + \frac{1.4 - 1}{2} M_2^2\right)^{1.4/(1.4-1)} = \frac{p_{t2}}{p_{t1}} \frac{p_{t1}}{p_1} \frac{p_1}{p_2}$$

$$= (1)(1.884)\left(\frac{200 \ kPa}{150 \ kPa}\right) = 2.508$$

$$M_2 = 1.226$$

Therefore,

$$T_2 = \frac{T_{t2}}{1 + \frac{\gamma-1}{2} M_2^2} = \frac{479.2 \ K}{1 + \frac{1.4-1}{2}[1.226]^2} = 368.4 \ K$$

$$V_2 = M_2 \sqrt{\gamma R T_2} = 1.226 \times \sqrt{1.4(288 \ J/kg \cdot K)(368 \ K)} = 472.3 \ m/s$$

Notes – A pressure decrease is accompanied by an increase in velocity, which is consistent with the results expected from the Bernoulli principle. The value of T_2 can also be found by directly using Eqn. (4.3); this is left as an exercise. The ratios are often expressed as static to stagnation values, as shown in Appendix C. □

EXAMPLE 4.2 At low Mach numbers, flow may be assumed to be incompressible. Under such conditions, the difference between stagnation and static values of a given property is small. If a 5% difference is the allowable limit for such an assumption, what would be the Mach number for that limit for air? Is the flow around a car moving at 65-mph (miles per hour) incompressible?

Solution –
An examination of Eqn. (4.6c) through (4.8) reveals that the pressure ratio possesses the largest exponent. Therefore, setting

$$\frac{p_t}{p} = \left(1 + \frac{\gamma - 1}{2} M^2\right)^{\gamma/(\gamma-1)} = 1.05$$

$$M = 0.26$$

An air speed of 65 mph is equal to $V = 28.9 \ m/s$. Under the normal atmospheric condition, $T = 293$ K and

$$M = \frac{V}{a} = \frac{28.9 \ m/s}{\sqrt{1.4(288 \ J/kg \cdot K)(293 \ K)}} = 0.084 < 0.26$$

The flow around a 65-mph car can be assumed to be incompressible. □

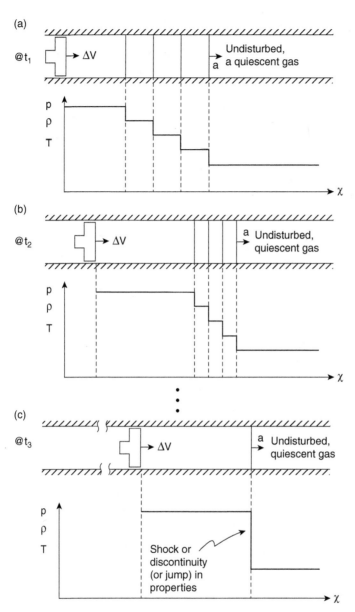

Figure 4.2 (a) Compression waves generated by an infinitesimal velocity of the piston in a piston-cylinder device, at an arbitrary time, $t = t_1$; (b) at a later time $t = t_2 > t_1$, the waves generated at later times catch up with those generated earlier; (c) at $t = t_s > t_2$, all the compression waves collapse to form a shock wave. Note that the wave fronts depicted are representative, as there are an infinite number of waves across which the changes in gas properties are infinitesimally small. The piling up of compression waves is due to rises in temperature and the speed of sound caused by compression waves. Across the thin shock wave, the changes in gas properties are abrupt with large gradients, mathematically approximated as infinite.

4.3 Formation of Normal Shock Waves and Expansion Waves

If the piston in the piston-tube device, shown in Fig. 3.1a, is given another velocity increment dV to the right, a second sound wave is generated and it propagates to the right following the first sound wave. Because the second wave propagates in the gas that was slightly compressed and heated by the first wave, it has a slightly higher speed than the first wave, as given by $a = \sqrt{\gamma RT}$, with higher T ahead of the second wave. Given time, the second sound wave will catch up with the first one. One can imagine that a third, a fourth, and more sound waves are generated sequentially, each by increasing the velocity by additional dV to the right over the previous one.

If the piston is accelerated from rest to a finite velocity increment (ΔV, which can be viewed as the accumulation of a large number of the infinitesimal increment dV) moving to the right, as sketched in Fig. 4.2a, the change in gas properties is expected to be finite in magnitude (now Δp, ΔT, and $\Delta \rho$ instead of their infinitesimal counterparts across an acoustic wave: dp, dT, and $d\rho$). As the later sound waves eventually overtake those generated earlier, the waves collapse into a very thin region across which finite changes can be observed. The progression toward this collapse of acoustic, infinitesimal acoustic waves are shown in Figs. 4.2b and 4.2c. The collapse of a very large number of infinitesimal compression waves (compression because the piston motion is into the gas) causes a finite increase in pressure across a thin wave. Similar sharp gradients in temperature and density also take place. Since the changes are abrupt, the thin region where they occur constitutes a shock wave. The sharp gradient causes irreversibility and an increase in entropy, as discussed in Chapter 2. Considering the signaling mechanism, the shock wave represents the inability of the gas at rest to sense the piston motion or the arrival of the piston itself; thus the gas could not make infinitesimal adjustment. The time and location of shock formation will be discussed with more details in Chapter 11.

If the piston is pulled away from the gas with a constant infinitesimal velocity $-dV$ (to the left), as shown in Fig. 4.3, a signal of expansion (i.e., expansion wave) is sent into the quiescent gas to the right of the piston. This expansion wave travels with the acoustic speed $a = \sqrt{\gamma RT}$, with T being the temperature ahead of the wave as in compression wave propagation. If the acceleration of the piston to the left is from rest to a finite value of $-\Delta V$ that consists of a large number of infinitesimal magnitude of dV, then the series of expansion waves traveling to the right into the gas at rest will travel with decreasing speeds. This is because a later wave propagates into the gas that now has a lower temperature, and thus a low speed of sound, due to the preceding expansion wave. As a consequence, the spacing between wave fronts increases with time and no "expansion shock" is possible, as shown in Fig. 4.3. For an expansion wave consisting of a series of infinitesimal expansion waves, due to the increasing spacing between the waves fronts, the gradients become increasing smaller with time. The expansion wave can be assumed to be close to being an isentropic process.

From the results shown in Fig. 4.2 for the compression wave, the gas at rest assumes a positive absolute velocity once it is downstream of the compression wave,

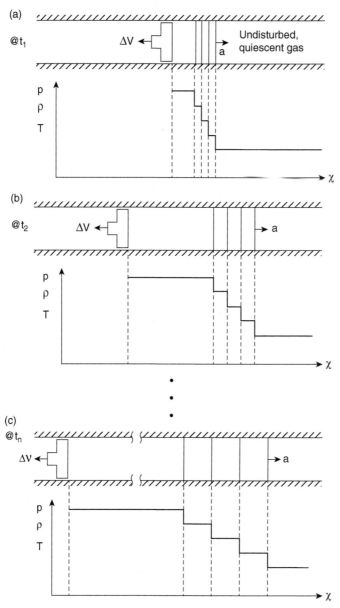

Figure 4.3 (a) Expansion waves generated by an infinitesimal velocity of the piston in a piston-cylinder device, at an arbitrary time, $t = t_1$; (b) at a later time $t = t_2 > t_1$, the waves generated at later times lack behind those generated earlier; (c) at $t = t_3 > t_2$, all the expansion waves are pulling further away from each other. Note that the wave fronts depicted are representative, as there are an infinite number of waves across which the changes in gas properties are infinitesimally small. The increasing separation between waves is due to the decrease in temperature and the speed of sound caused by expansion waves, so that each later wave propagates with a smaller speed than the previous one and the changes in gas properties are spread over increasing larger distance, resulting in ever decreasing gradients with time. Thus, no shock wave forms, due to expansion.

as seen by an observer in the laboratory. This observation can be intuitively understood, as the piston motion push the gas to the right. In other words, the induced gas motion is in the same direction as the compression wave propagation. It can also be stated that because the piston motion is to the right the gas after being passed by the compression wave (i.e., after receiving the signal of compression action) will have no choice but to move along with the wave to the right due to the motion of the impermeable piston surface. In what direction will be the gas motion downstream of an expansion wave? As illustrated in Fig. 4.3, the expansion wave propagates to the right, even though the piston motion is to the left. To satisfy continuity (i.e., the piston cannot leave a vacuum behind), the gas has to move, after sensing the signal for expansion, to the left. One may then conclude that the gas motion will be in the same direction as the compression wave and the moving normal shock wave, and in the opposite direction to the moving expansion wave.

The effect of piston motion described in Fig. 4.3 can be replaced by a diaphragm separating a high-pressure gas and a low-pressure gas, as illustrated in Fig. 4.4a. Assume that the pressure difference is sufficient to cause a shock wave. The diaphragm is ruptured at $t = 0^+$. At some time later ($t > 0$), the pressure and the temperature distributions throughout the flow are illustrated in Figs. 4.4b and 4.4c, respectively. The entire flow field now is divided into four regions, with regions 1 and 4 remaining undisturbed at this moment and having pressure and temperature equal to their values at $t = 0$. The interface between the high- and low-pressure gases, originally represented by the diaphragm before the rupture, acts in similar manner to that of the piston in Fig. 4.3. Upon the diaphragm's rupture, it simultaneously creates a compression wave propagating in the $+x$ direction and an expansion wave in the $-x$ direction. Because of the need to equalize the pressure across the interface, the action due to compression and expansion is such that $p_4 > p_3 = p_2 > p_1$. The change from p_1 to p_2 is abrupt, characteristic of a shock wave, while that from p_3 to p_4 takes place over a finite distance, as the expansion waves spread out over time.

The formation of shock waves is an unsteady process and the shock waves after formation continue to move. Analysis of moving shock waves is presented in Chapter 11. This section is focused on one-dimensional stationary shock waves. The analyses of stationary and moving normal shock waves are similar, except that Galileo coordinate transformation is needed for analyzing the latter, which will be briefly discussed later in this chapter.

Examples of normal shock wave include that in a supersonic flow approaching a blunt body, where a normal shock wave forms due to the inability of the flow to sense the presence of the body (as that demonstrated in Fig. 3.3b). Figure 4.5 shows the visualization of the shock wave system formed around a blunt body in a supersonic stream. The portion of the shock wave immediately in front of the nose of the blunt body can be approximately as a normal shock wave, with the flow direction normal to the wave. Away from the nose region, the shock wave assumes the shape of a curve (or a bow). In the bow shock region, the incoming flow makes an angle with the shock wave, thus called oblique shock. A bow shock therefore consists

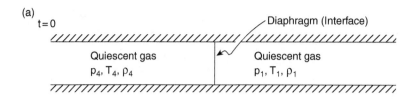

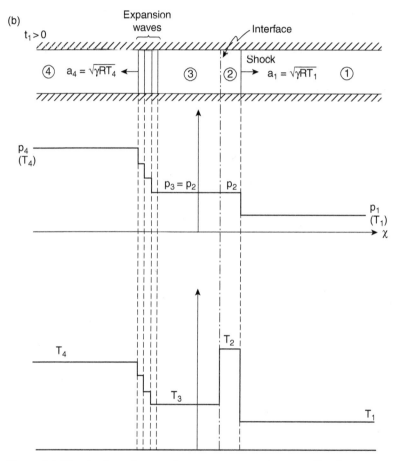

Figure 4.4 The shock- and expansion wave system, at $t = t_1 > 0$ and at $t = t_2 > t_1 > 0$ (Figures [b] and [c]), established by rupturing the diaphragm that initially, at $t = 0$ (Figure [a]), separates a high-pressure gas (region 4) from a low-pressure gas (region 1) in a long cylinder. As depicted, the gases in both regions are initially quiescent. Note that the compression propagates into the low-pressure gas, while the expansion wave, into the high-pressure region, and that the gradients in gas properties decrease with time due to the passing of expansion waves.

of a series of oblique shock waves with continuously varying flow angles. The analysis of oblique shock waves is presented later in Section 4.5. This blunt body could be the Pitot-static tube on an aircraft. As an approximation, a portion of shock wave is normal to the flow, as indicated in Fig. 4.5.

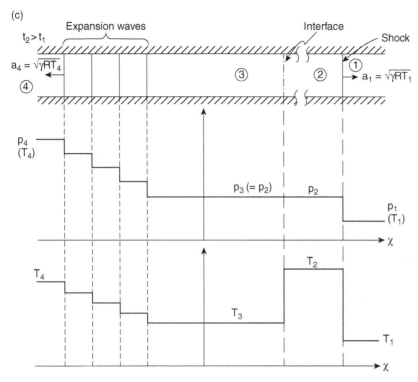

Figure 4.4 (cont.)

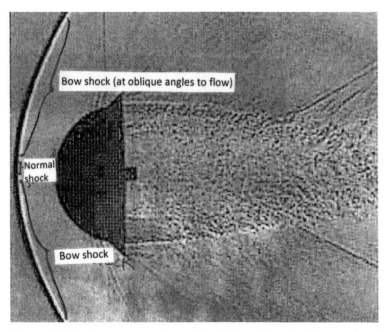

Figure 4.5 Shock wave system formed by placing a blunt body in a supersonic stream. (NASA - http://www.hq.nasa.gov/pao/History/sp-440/ch6.2hmt)

The steady state one-dimensional momentum equation, Eqn. (2.24), can be used to help understand the aforementioned normal shock phenomena:

$$p + \rho V^2 = \text{constant} \tag{2.24}$$

One-dimensional continuity requires that $\rho V = \text{constant}$, and the differential form of Eqn. (2.24), called *Euler's equation*, is

$$dp = -\rho V dV \tag{4.9}$$

That is, a positive pressure rise ($dp > 0$ for a compression or normal shock wave) results in a decrease in the velocity magnitude, or causes the flow to decelerate across the wave. For a traveling compression wave like those depicted in Fig. 4.3, this means that the relative gas velocity relative to the wave decreases after the passage of the compression wave, consistent with the results of Fig. 3.1b. In other words, the gas behind the wave is not leaving the wave with the same speed as its approaching speed, because the gas is also not being pushed to the right by the piston, causing a reduction of speed across a compression/shock wave. To a laboratory observer, the wave thus *carries along* with it the gas it passes. These results of stationary vs. traveling compression/shock wave are illustrated in Fig. 4.6.

For a stationary expansion wave, $dp < 0$ results in acceleration of the gas flow across it. For a traveling expansion wave, $dp < 0$ causes an increase in the downstream gas velocity relative to the wave after the passage of the expansion wave. In both cases, the expansion wave "*repels*" the gas. Figure 4.7 illustrates the differences between observations of a stationary and a traveling expansion waves.

The traffic resulting from the change in traffic signals provides good analogies for compression and expansion waves. Consider the steady stream of vehicles approaching a red light at an intersection. The first to sense the signal will have to stop, which in turn sends a stop signal to the immediately following vehicle that will also stop after the first and decrease the space between them. This wave of the stop signal propagates away from the intersection into the still moving vehicles, causing them to decelerate and eventually stop, resulting in an increase in vehicle density, like a compression wave. Some moments later when the red light turns into a green

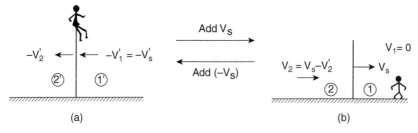

Figure 4.6 Schematic demonstrating how an inherently unsteady shock wave propagation problem (part b) can be transformed and treated as a steady state problem by Galileo coordinate transformation (part a). Such a coordinate transformation can be achieved by making observations and measurements by *riding on the wave*. (Note: $V'_2 < V_s$; $V_2 = V_s - V'_2 > V_s - V_s = 0$.)

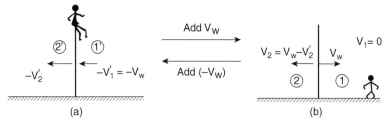

Figure 4.7 Schematic demonstrating how an inherently unsteady expansion wave propagation problem (part b) can be transformed and treated as a steady state problem by Galileo coordinate transformation (part a). Such a coordinate transformation can be achieved by making observations and measurements by *riding on the wave*. (Note: $V'_2 > V'_1$; $V_2 = V_w - V'_2 < V_w - V'_1 = 0$.)

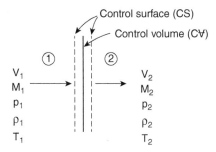

Figure 4.8 Schematic of a stationary normal shock wave.

light, the first vehicle will accelerate passing the intersection and away from it. The second and then the third vehicles follow, and so forth, resulting in a wave that sends the signal for motorists to accelerate. In the process of acceleration, the density of vehicles decreases, similar to an expansion wave.

4.4 Governing Equations of Stationary Normal Shock Waves

It is desirable to be able to relate fluid properties downstream of the normal shock wave to those in the upstream region. Consider a stationary one-dimensional normal shock wave depicted in Fig. 4.8. Flows both down- and upstream are isentropic and once the relationship between their Mach numbers is found, the isentropic relationships, Eqns. (4.6) through (4.8) can be used. The steady state continuity, Eqn. (2.4), can be rewritten as

$$\rho_1 V_1 = \rho_2 V_2 \tag{4.10}$$

where subscripts 1 and 2 denote the regions upstream and downstream of the shock wave, respectively. Similarly, the steady state momentum equation, Eqn. (2.7), becomes identical to Eqn. (2.24):

$$p_1 + \rho_1 V_1^2 = p_2 + \rho_1 V_2^2 \tag{4.11a}$$

For an ideal gas, $\rho = p/RT = p\gamma/\gamma RT$. Therefore, $\rho V^2 = p\gamma M^2$ and the momentum equation, Eqn. (2.8), becomes

$$p_1(1 + \gamma M_1^2) = p_2(1 + \gamma M_2^2) \tag{4.11b}$$

For constant values of c_p, γ, and R (*i.e.*, perfect gases), $h = c_p T$ and $c_p = \gamma R/(\gamma - 1)$. The energy equation, Eqn. (4.22), becomes

$$h_1 + \frac{1}{2}V_1^2 = h_2 + \frac{1}{2}V_2^2 \tag{4.12a}$$

and can be rewritten as

$$T_1\left(1 + \frac{\gamma - 1}{2}M_1^2\right) = T_2\left(1 + \frac{\gamma - 1}{2}M_2^2\right) \tag{4.12b}$$

Combining the ideal gas law and the definitions of Mach number and speed of sound for perfect gases, the continuity equation, Eqn. (4.10), can be rewritten as

$$\frac{p_1}{RT_1}M_1\sqrt{\gamma RT_1} = \frac{p_2}{RT_2}M_2\sqrt{\gamma RT_2} \tag{4.13}$$

Substituting Eqns. (4.11b) and (4.12b) into Eqn. (4.13) yields

$$\frac{M_1^2}{1 + \gamma M_1^2}\sqrt{1 + \frac{\gamma - 1}{2}M_1^2} = \frac{M_2^2}{1 + \gamma M_2^2}\sqrt{1 + \frac{\gamma - 1}{2}M_2^2} \tag{4.14a}$$

Squaring both sides of Eqn. (4.14a) produces

$$\frac{M_1^2\left(1 + \frac{\gamma - 1}{2}M_1^2\right)}{\left(1 + \gamma M_1^2\right)^2} = \frac{M_2^2\left(1 + \frac{\gamma - 1}{2}M_2^2\right)}{\left(1 + \gamma M_2^2\right)^2} \tag{4.14b}$$

This equation relates the downstream Mach number M_2 to the upstream Mach number M_1 by collecting terms containing M_2

$$M_2^4\left(\frac{\gamma - 1}{2} - \gamma^2 Z\right) + M_2^4(1 - 2\gamma Z) - Z = 0$$

where

$$Z = \frac{M_1^2\left(1 + \frac{\gamma - 1}{2}M_1^2\right)}{\left(1 + \gamma M_1^2\right)^2}$$

Solving the quadratic equation for M_2^2 yields

$$M_2^2 = \frac{M_1^2 + \frac{2}{\gamma - 1}}{\frac{2\gamma}{\gamma - 1}M_1^2 - 1} \tag{4.15}$$

The plot of M_2 vs. M_1 for $\gamma = 1.4$ is shown in Fig. 4.9.

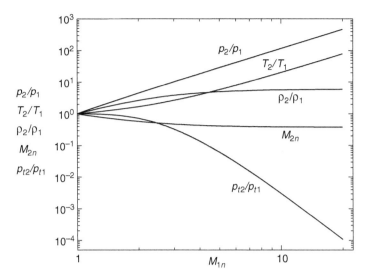

Figure 4.9 Property changes across a stationary normal shock wave, as functions of upstream Mach number, M_1 for $\gamma = 1.4$.

For $M_1 > 1$, $M_2 < 1$ and the flow decelerates across the wave and, according to Eqn. (4.11b), $p_2/p_1 = (1 + \gamma M_1^2)/(1 + \gamma M_2^2) > 1$. Therefore the wave is a compression shock. Equation (4.14b) also has a trivial solution $M_1 = M_2 = 1$, which is the limiting case where no flow changes occur across the wave.

Conversely for $M_1 < 1$, $M_2 > 1$ and the flow accelerates across the wave and, according to Eqn. (4.11b), $p_2/p_1 = (1 + \gamma M_1^2)/(1 + \gamma M_2^2) < 1$. Therefore the wave is an expansion shock, which is not possible because the expansion leads to decreasing gradients of properties, as discussed in Section 4.3. This impossible scenario for the one-dimensional flow will be later explained further in terms of the entropy change. Note that the above derivation does not assume the existence of a shock wave. An expansion wave is also a possible solution, although an expansion shock is not. As will be demonstrated in Section 4.10, a supersonic flow can continue to accelerate (i.e., expand) by "turning" and, in Section 5.1, by an increase in cross-sectional area in a channel.

Now that M_2 can be related to M_1, ρ_2, V_2, T_2, and p_2 can also be related to ρ_1, V_1, T_1, and p_1, using Eqns. (4.10) through (4.12) and Eqn. (4.15). The following relationships across a stationary normal shock wave can easily be derived. Equation (4.11) is rewritten as

$$\frac{p_2}{p_1} = \frac{1 + \gamma M_1^2}{1 + \gamma M_2^2} \tag{4.16a}$$

Substituting Eqn. (4.15) into this equation yields

$$\frac{p_2}{p_1} = \frac{2\gamma M_1^2}{\gamma + 1} - \frac{\gamma - 1}{\gamma + 1} \tag{4.16b}$$

Similarly, combining Eqns. (4.12b) and (4.15), the temperature ratio is

$$\frac{T_2}{T_1} = \frac{\left(1 + \frac{\gamma-1}{2} M_1^2\right)\left(\frac{2\gamma}{\gamma-1} M_1^2 - 1\right)}{\left[\frac{(\gamma+1)^2}{2(\gamma-1)}\right] M_1^2} \tag{4.16c}$$

By using the continuity and substituting Eqns. (4.15) for M_2 and (4.16c) for the temperature ratio, the density ratio is

$$\frac{\rho_2}{\rho_1} = \frac{V_1}{V_2} = \frac{M_1\sqrt{\gamma R T_1}}{M_2\sqrt{\gamma R T_2}} = \frac{(\gamma+1)M_1^2}{(\gamma-1)M_1^2 + 2} \tag{4.16d}$$

Equation (4.16a) shows that the pressure jump across the normal shock wave increases with increasing upstream Mach number M_1, accompanied by a correspondingly smaller downstream Mach number M_2, as seen in Fig. 4.9 for $\gamma = 1.4$.

For convenience, it is desirable to define the *shock strength β*, as $(p_2 - p_1)/p_1$:

$$\beta \equiv \frac{p_2 - p_1}{p_1} = \frac{2\gamma}{\gamma+1}\left(M_1^2 - 1\right) \tag{4.16e}$$

Equation (4.16e) provides a useful relation for choosing M_1 for the desired shock strength:

$$M_1^2 = 1 + \left(\frac{\gamma+1}{2\gamma}\right)\beta \tag{4.16f}$$

Because of the one-dimensional nature of the normal shock wave, there is no heat loss. Therefore,

$$T_{t2} = T_{t1} \tag{4.17a}$$

However, the irreversibility due to the finite compression process must lead to changes in some thermodynamic properties such as entropy and stagnation pressure. By using Eqn. (4.12) for static temperatures and the isentropic relationships, Eqns. (4.7) and (4.8), and after performing some algebraic manipulations, Eqn. (2.38) becomes

$$\frac{s_2 - s_1}{R} = \frac{c_p}{R}\ln\left(\frac{T_2}{T_1}\right) - \ln\left(\frac{p_2}{p_1}\right) = \frac{c_p}{R}\ln\frac{1 + \frac{\gamma-1}{2} M_1^2}{1 + \frac{\gamma-1}{2} M_2^2} - \ln\left\{\frac{p_{t2}}{p_{t1}}\left[\frac{1 + \frac{\gamma-1}{2} M_1^2}{1 + \frac{\gamma-1}{2} M_2^2}\right]^{\gamma/(\gamma-1)}\right\} \tag{4.17b}$$

Because $c_p/R = \gamma/(\gamma - 1)$, this expression becomes

$$\frac{s_2 - s_1}{R} = -\ln\frac{p_{t2}}{p_{t1}} \tag{4.17c}$$

It is noted from Eqns. (4.15), (4.16a), and (4.17c) that for the trivial case of $M_2 = M_1 = 1$, $s_2 = s_1$ (i.e., the flow is isentropic) and $p_{t2} = p_{t1}$. Equation (4.16f) requires that for one-dimensional shock (finite pressure jump and $\beta > 0$, unlike an isentropic acoustic wave) to occur, it is necessary that $M_1 > 1$. For an oblique shock (Section 4.6) to occur, it is necessary that $M_1 > 1$ (and $M_{1n} > 1$ for oblique shocks). In fact M_1 in Eqns. (4.16) and (4.17) can simply be replaced by M_{1n} for oblique shocks.

For a normal shock, being an irreversible process with large gradients, $s_2 - s_1 > 0$. The ratio of the stagnation pressure is

$$\frac{p_{t2}}{p_{t1}} = e^{-(s_2-s_1)/R} = \left[\frac{\frac{\gamma+1}{2} M_1^2}{1 + \frac{\gamma-1}{2} M_1^2} \right]^{\gamma/(\gamma-1)} \left[\frac{1}{\frac{2\gamma}{\gamma+1} M_1^2 - \frac{\gamma-1}{\gamma+1}} \right]^{1/(\gamma-1)} < 1 \qquad (4.17d)$$

Since $(s_2 - s_1)/R > 0$, Eqn. (4.17d) requires that $p_{t2} < p_{t1}$. Therefore for a stationary normal shock, the irreversibility leads to a *loss* in the stagnation (or total) pressure. On the other hand, Eqn. (4.16) requires that $p_2 > p_1$ (because it is necessary that $M_1 > 1$), due to the compression effect of the normal shock. The loss in the stagnation pressure means that the ability of the gas to expand and perform the pdV work is reduced. This has important implications when designing supersonic inlets, where normal shock waves are to be avoided. The reason for this is so that the stagnation pressure remains high entering the jet engine, and the capability is preserved for the gas to expand through the turbine stages and the nozzle exit (to perform a large amount of work in driving the compressor and the high-speed exit flow for propulsion).

The left-hand side results of Eqns. (4.15) – (4.17) are all functions only of the upstream Mach number M_1 and the specific heat ratio γ. They can be conveniently tabulated, in Appendix D, where M_{1n} is adopted to indicate the usefulness of the table for oblique shocks; for normal shocks, $M_{1n} = M_1$. It is also possible to express the ratios of different properties as functions of each other, rather than as functions of M_1. For example, expressing M_1 in terms of the density ratio in Eqn. (4.16c) and substituting it into Eqn. (4.16a) yields

$$\frac{p_2}{p_1} = \frac{\left(\frac{\gamma+1}{\gamma-1}\right)\left(\frac{\rho_2}{\rho_1}\right) - 1}{\left(\frac{\gamma+1}{\gamma-1}\right) - \left(\frac{\rho_2}{\rho_1}\right)} \qquad (4.18a)$$

Alternatively, the density ratio can be expressed as the pressure ratio, as

$$\frac{\rho_2}{\rho_1} = \frac{\left(\frac{\gamma+1}{\gamma-1}\right)\frac{p_2}{p_1} + 1}{\left(\frac{\gamma+1}{\gamma-1}\right) + \frac{p_2}{p_1}} \qquad (4.18b)$$

The result of the *Rankine-Hugoniot relation* will be discussed in further detail in Section 4.18.

EXAMPLE 4.3 An air flow with Mach number 2.0 passes through a normal shock wave immediately ahead of a jet engine inlet. The static air temperature and pressure are, respectively, 216 K and 19.4 kPa. What is the air velocity downstream of the shock? What is the pressure loss? What are the downstream pressure, temperature, and density? Also use the Rankine-Hugoniot relation to find downstream density.

Solution –
First of all the Mach number and the sonic speed are needed downstream of the shock wave. For $M_1 = 2.0$, Appendix A provides $T_1/T_{t1} = 0.5556$ while the shock table (Appendix D) provides $M_2 = 0.5774$ and $T_2/T_1 = 1.6875$, $p_{t2}/p_{t1} = 0.7209$, $\rho_2/\rho_1 = 2.667$, and $p_2/p_1 = 4.50$. Therefore,

$$V_2 = M_2\sqrt{\gamma R T_2} = (0.5774) \times \sqrt{(1.4)(288\ J/kg \cdot K)(1.6875 \times 216\ K)} = 221.4\ m/s.$$

$$p_{t1} = p_1\left(1 + \frac{\gamma - 1}{2}M_1^2\right)^{\gamma/(\gamma-1)} = (19.4\ kPa)(7.8247) = 151.80\ kPa;$$

$$p_{t2} = 0.7209 p_{t1} = 109.43\ kPa$$

$$\Delta p_t = p_{t2} - p_{t1} = 42.47\ kPa$$

The pressure loss is approximately 28%.

$$p_2 = 87.3\ kPa$$

$$T_2 = 364.5\ K$$

$$T_{t2} = T_{t1} = (T_{t1}/T_1) \times T_1 = 216\ K/0.5556 = 388.8\ K$$

$$\rho_2 = 2.667 \times \frac{p_1}{RT_1} = 2.667 \times \frac{19.4 \times 10^3\ kPa}{(288\ J/kg \cdot K) \times (216\ K)} = 0.832\ kg/m^3$$

By using the Rankine-Hogoniot relation with $\gamma = 1.4$

$$\frac{\rho_2}{\rho_1} = \frac{\left(\frac{\gamma+1}{\gamma-1}\right)\frac{p_2}{p_1} + 1}{\left(\frac{\gamma+1}{\gamma-1}\right) + \frac{p_2}{p_1}} = \frac{6 \times 4.5 + 1}{6 + 4.5} = 2.667$$

which is same as using the tabulated value for normal shock.   □

4.5 Strong and Weak Normal Shock Waves

For strong shock waves, the shock strength β is large and, according to Eqn. (4.17e), M_1 must be large. On the other hand, M_1 must be very close to unity for weak shock waves. For strong shock waves, Eqn. (4.15) takes the following form:

$$M_2^2 = \lim_{M_1 \to \infty} \left[\frac{M_1^2 + \frac{2}{\gamma-1}}{\frac{2\gamma}{\gamma-1} M_1^2 - 1} \right] \tag{4.19a}$$

Therefore,

$$M_2 = \sqrt{\frac{\gamma - 1}{2\gamma}} \tag{4.19b}$$

Therefore, for $\gamma = 1.4$, $M_2 \approx 0.378$ is the minimum value possible if M_1 is infinity. As a comparison, use $M_1 = 10$, $M_2 = 0.388$, as seen in Appendix D. After substituting Eqn. (4.19b) into Eqn. (4.16), ratios across the normal shock of other fluid properties can be similarly written as

$$\frac{T_2}{T_1} \approx M_1^2 \frac{2\gamma(\gamma - 1)}{(\gamma + 1)^2} \approx \frac{\gamma - 1}{\gamma + 1} \beta \tag{4.19c}$$

$$\frac{p_2}{p_1} \approx \frac{2\gamma M_1^2}{\gamma + 1} \approx \beta \tag{4.19d}$$

$$\frac{\rho_2}{\rho_1} = \frac{V_1}{V_2} \approx \frac{(\gamma + 1)}{(\gamma - 1)} \tag{4.19e}$$

$$\frac{p_{t2}}{p_{t1}} = e^{-(s_2 - s_1)/R} \approx \left[\frac{\gamma + 1}{\gamma - 1} \right]^{\gamma/(\gamma-1)} \left[\frac{\gamma + 1}{2\gamma M_1^2} \right]^{1/(\gamma-1)} = \left[\frac{1}{\beta} \right]^{1/(\gamma-1)} \tag{4.19f}$$

Therefore, as $M_1 \to \infty$, (T_2/T_1) and (p_2/p_1) are linearly related rather than related by the nonlinear relations of Eqns. (4.16b) and (4.16c). Furthermore, ρ_2/ρ_1 becomes independent of M_1 and $p_{t2}/p_{t1} \sim M_1^{-2/(\gamma-1)}$, which is $\sim M_1^{-5}$ for $\gamma = 1.4$.

EXAMPLE 4.4 It is not possible for M_1 to be infinity, for which the downstream pressure and temperature will also be infinity while the pressure loss is 100%. Above what value of M_1 are Eqns. (4.19a) through (4.19f) good approximations?

Solution –
One estimation can be obtained by assuming M_2 to be 110% of $\sqrt{(\gamma - 1)/2\gamma}$.

In this case for $\gamma = 1.4$, $M_2 \approx 0.416$ and the normal shock table (Appendix D) gives $M_1 \approx 4.94$. The results with and without the strong shock assumption for fluid properties are

$\frac{T_2}{T_1} = 5.684$ 4.745 (strong shock approximation)

$\frac{p_2}{p_1} = 28.304$ 28.47 (strong shock approximation)

$\frac{\rho_2}{\rho_1} = 4.980$ 6.0 (strong shock approximation)

$\frac{p_{t2}}{p_{t1}} = 0.065$ 0.122 (strong shock assumption)

If M_2 is 105% of $\sqrt{(\gamma-1)/2\gamma}$, then $M_2 \approx 0.397$ and $M_1 \approx 7.0$. Therefore,

$\frac{T_2}{T_1} = 10.469$ 9.528 (strong shock approximation)

$\frac{p_2}{p_1} = 57.0$ 57.167 (strong shock approximation)

$\frac{\rho_2}{\rho_1} = 5.444$ 6.0 (strong shock approximation)

$\frac{p_{t2}}{p_{t1}} = 0.015$ 0.021 (strong shock assumption)

Comments –

The strong-shock approximation leads to the stagnation pressure having the largest deviation among all fluid properties, while the static pressure shows the best approximation. Overall, the approximation yields much better results for $M_1 \approx 7.0$ than for $M_1 \approx 5.0$. Other criteria, such as p_2/p_1, can be used, but the ratio p_{t2}/p_{t1} is the most stringent, consistent with the β–dependence shown in Eqns. (4.19c) through (4.19f). □

For small values of β (weak shock approximation), let $M_1^2 = 1 + \epsilon$, where $\epsilon \ll 1$ and after substituting this expression into Eqn. (4.15), M_2 becomes (John and Keith, 2006)

$$M_2^2 = 1 - \epsilon = 2 - M_1^2 \tag{4.20a}$$

The derivation is straightforward and is left as a problem at the end of the chapter.

For the weak normal shock wave, the fluid properties are found by substituting $M_1^2 = 1 + \epsilon$ into Eqn. (4.16):

$$\frac{T_2}{T_1} \approx 1 + \frac{2\gamma(\gamma-1)}{(\gamma+1)^2}(M_1^2 - 1) = 1 + \frac{\gamma-1}{\gamma+1}\beta \tag{4.20b}$$

$$\frac{p_2}{p_1} \approx 1 + \frac{2\gamma}{\gamma+1}(M_1^2 - 1) = 1 + \beta \tag{4.20c}$$

$$\frac{V_2}{V_1} \approx 1 - \frac{2}{\gamma+1}(M_1^2 - 1) = 1 - \frac{\beta}{\gamma} \tag{4.20d}$$

$$\frac{\rho_2}{\rho_1} \approx \frac{V_1}{V_2} = 1 + \frac{2}{\gamma+1}(M_1^2 - 1) = 1 + \frac{\beta}{\gamma} \tag{4.20e}$$

$$\frac{s_2 - s_1}{R} \approx \frac{2\gamma}{3(\gamma+1)^2}\left(M_1^2 - 1\right)^3 = \frac{\gamma+1}{12\gamma^2}\beta^3 \tag{4.20f}$$

In Eqn. (4.20f), $\left(M_1^2 - 1\right)$ was replaced with $\left(\frac{p_2}{p_1} - 1\right)\left(\frac{\gamma+1}{2\gamma}\right)$ from (4.20c).

EXAMPLE 4.5 Find the upstream Mach number for a good weak shock approximation.

Solution –
For a weak normal shock, assume the criterion to be $p_2/p_1 = 1.05$. Then for $\gamma = 1.4$, $M_1^2 - 1 = \epsilon = 0.043$, for which $M_1 \approx 1.02$.

$\dfrac{p_2}{p_1} = 1 + \dfrac{2\gamma}{\gamma+1}\left(M_1^2 - 1\right) = 1.050$ 1.050 (given)

$\dfrac{T_2}{T_1} \approx 1 + 2\left(\dfrac{\gamma-1}{\gamma+1}\right)\left(M_1^2 - 1\right) = 1.014$ 1.013 (normal shock relation)

$\dfrac{\rho_2}{\rho_1} \approx \dfrac{V_1}{V_2} = 1 + \dfrac{2}{\gamma+1}\left(M_1^2 - 1\right) = 1.037$ 1.033 (normal shock relation)

$\dfrac{p_{t2}}{p_{t1}} = e^{-(s_2-s_1)/R} \approx e^{\frac{-2\gamma}{3(\gamma+1)^2}\left(M_1^2 - 1\right)^3} = 0.9999 \approx 1$ ≈ 1 (normal shock relation)

If $p_2/p_1 = 1.1$. Then for $\gamma = 1.4$, $M_1 \approx 1.04$ and $\left(M_1^2 - 1\right) = \epsilon = 0.086$.

$\dfrac{p_2}{p_1} = 1 + \dfrac{2\gamma}{\gamma+1}\left(M_1^2 - 1\right) = 1.10$ 1.095 (given)

$\dfrac{T_2}{T_1} \approx 1 + 2\left(\dfrac{\gamma-1}{\gamma+1}\right)\left(M_1^2 - 1\right) = 1.029$ 1.026 (normal shock relation)

$\dfrac{\rho_2}{\rho_1} \approx \dfrac{V_1}{V_2} = 1 + \dfrac{2}{\gamma+1}\left(M_1^2 - 1\right) = 1.072$ 1.067 (normal shock relation)

$\dfrac{p_{t2}}{p_{t1}} = e^{-(s_2-s_1)/R} \approx e^{\frac{-2\gamma}{3(\gamma+1)^2}\left(M_1^2 - 1\right)^3} = 0.9998 \approx 1$ 0.9999 (normal shock relation)

If $p_2/p_1 = 1.2$. Then for $\gamma = 1.4$, $M_1 \approx 1.08$ and $\left(M_1^2 - 1\right) = \epsilon = 0.166$.

$\dfrac{p_2}{p_1} = 1 + \dfrac{2\gamma}{\gamma+1}\left(M_1^2 - 1\right) = 1.19$ 1.1941 (given)

$\dfrac{T_2}{T_1} \approx 1 + 2\left(\dfrac{\gamma-1}{\gamma+1}\right)\left(M_1^2 - 1\right) = 1.055$ 1.0522 (normal shock relation)

$\dfrac{\rho_2}{\rho_1} \approx \dfrac{V_1}{V_2} = 1 + \dfrac{2}{\gamma+1}\left(M_1^2 - 1\right) = 1.139$ 1.1349 (normal shock relation)

$\dfrac{p_{t2}}{p_{t1}} = e^{-(s_2-s_1)/R} \approx e^{\frac{-2\gamma}{3(\gamma+1)^2}\left(M_1^2 - 1\right)^3} = 0.9998$ 0.9994 (normal shock relation)

Comments –
The accuracy of the weak normal shock approximation appears to be more limited by the density, as values of other fluid properties deviate less from the predictions using the full normal shock relation.   □

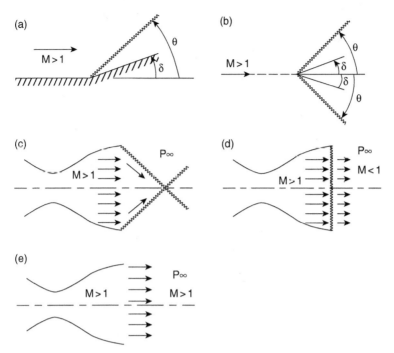

Figure 4.10 Attached oblique shock waves formed (a) over a two-dimensional inclined surface, (b) over a two-dimensional wedge, and (c) downstream of a supersonic nozzle exit when the exit flow pressure is less than that in the ambient; (d) Normal shock forms at the supersonic nozzle exit plane when the ambient pressure is sufficiently higher than that of the exit flow ($p_e < p_\infty = p_b$); (e) when exit flow pressure matches or is greater than the ambient pressure, no shock wave forms and the nozzle (of the flow). For $p_e = p_\infty = p_b$ the nozzle flow is *perfectly expanded* and for $p_e > p_\infty = p_b$, under-expanded.

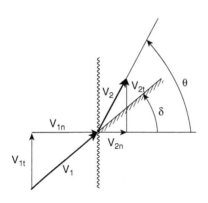

Figure 4.11 Velocity components and inclination angles across an oblique shock wave.

4.6 Oblique Shock Waves

Oblique shock waves are commonly seen in a supersonic flow deflected by a surface that is at an inclined angle to the flow. Consider the planar oblique shock wave formed by a two-dimensional wedge and by a two-dimensional corner with a deflection (or turning) angle δ, which cause the shock wave to be inclined at an angle of θ to the flow, as shown in Figs. 4.10a and 4.10b, respectively. Such a flow can also be constructed by superimposing a uniform velocity component parallel to the normal shock ($V_{1t} = V_{2t} = V_t$) as shown in Fig. 4.11 (the validity of this superimposition is shown by Eqn. (4.21d)). Because the magnitude of the normal velocity component decreases across the shock wave ($V_{1n} > V_{2n}$) the resultant velocity downstream of the shock is turned through an angle δ (here $\delta > 0$ for turning into the flow), which varies with the magnitude of the superimposed tangential component. In this case, the surrounding fluid, being at a higher pressure, pushes toward the exit flow with a tangential velocity component. Here $\delta > 0$ if the turning is made into the flow and thus causes compression. The case for $\delta < 0$ will be seen in Prandtl-Meyer expansion, the subject of Section 4.9.

An oblique shock can thus be formed without a physical wedge, an example of which is the oblique shock wave formed at the exit of a nozzle, as shown in Fig. 4.10c if the pressure of the surrounding is sufficiently high but still lower than that required for a normal shock wave to form. When the surrounding pressure (p_∞, or interchangeably, the back pressure, p_b) is sufficiently high, a normal shock wave will form at the exit plane of the nozzle (Fig. 4.10d), and even within the diverging section of it. Only with $p_b \leq p_e$ would there be no shock wave of any type formed for the nozzle flow under this condition; the exit flow is supersonic (Fig. 4.10e). The location of the normal shock wave in the nozzle and the conditions for formation of the normal and oblique shocks will be discussed in Chapter 5.

Consider a uniform flow having negligible frictional and body forces passing across an oblique shock. By placing the control volume and control surface shown in Fig. 4.11, the governing equations for the normal component velocity are the same as those for the normal shock wave:

$$\rho_1 V_{1n} = \rho_2 V_{2n} \tag{4.21a}$$

$$p_1 + \rho_1 V_{1n}^2 = p_2 + \rho_1 V_{2n}^2 \tag{4.21b}$$

$$h_1 + \frac{1}{2} V_{1n}^2 = h_2 + \frac{1}{2} V_{2n}^2 \tag{4.21c}$$

The momentum equation for the tangential velocity component becomes

$$\int_{CS} V_t \left(\rho \vec{V} \cdot d\vec{A} \right) = 0$$

Therefore,

$$V_{1t}(\rho_1 V_{1n}) - V_{2t}(\rho_2 V_{2n}) = 0$$

Because $\rho_1 V_{1n} = \rho_2 V_{2n}$, this expression becomes

$$V_{1t} = V_{2t} \tag{4.21d}$$

Due to the similarity Eqns. (4.10) through (4.12) bear with Eqn. (4.21), one finds that the shock relations from Eqn. (4.15) to (4.18) for the normal shock are applicable to the oblique shock for the normal velocity component. This means that, M_{2n} and all the ratios of temperature, density, and pressure can be determined using the normal shock relations with M_1 replaced by M_{1n} in these relations; similarly, Table B.2 can be used as long as M_{1n} is known. Therefore the oblique shock relations are

$$M_{2n}^2 = \frac{M_{1n}^2 + \frac{2}{\gamma - 1}}{\frac{2\gamma}{\gamma - 1} M_{1n}^2 - 1} \tag{4.22a}$$

$$\frac{p_{t2}}{p_{t1}} = e^{-(s_2 - s_1)/R} = \left[\frac{\frac{\gamma+1}{2} M_{1n}^2}{1 + \frac{\gamma-1}{2} M_{1n}^2} \right]^{\gamma/(\gamma-1)} \left[\frac{1}{\frac{2\gamma}{\gamma+1} M_{1n}^2 - \frac{\gamma-1}{\gamma+1}} \right]^{1/(\gamma-1)} \tag{4.22b}$$

$$\frac{p_2}{p_1} = \frac{2\gamma M_{1n}^2}{\gamma + 1} - \frac{\gamma - 1}{\gamma + 1} \tag{4.22c}$$

$$\frac{T_2}{T_1} = \frac{\left(1 + \frac{\gamma-1}{2} M_{1n}^2\right)\left(\frac{2\gamma}{\gamma-1} M_{1n}^2 - 1\right)}{\left[\frac{(\gamma+1)^2}{2(\gamma-1)}\right] M_{1n}^2} \tag{4.22d}$$

$$\frac{\rho_2}{\rho_1} = \frac{V_{1n}}{V_{2n}} = \frac{M_{1n}\sqrt{\gamma R T_1}}{M_{2n}\sqrt{\gamma R T_2}} = \frac{(\gamma+1)M_{1n}^2}{(\gamma-1)M_{1n}^2 + 2} \tag{4.22e}$$

$$\frac{T_{t2}}{T_{t1}} = 1 \tag{4.22f}$$

The absolute Mach number M_2 is the vector sum of the normal and tangential components. By using $V_{1t} = V_{2t}$,

$$M_{1t}\sqrt{\gamma R T_1} = M_{2t}\sqrt{\gamma R T_2}$$

M_{2t} is determined as

$$M_{2t} = M_{1t}\sqrt{\frac{T_1}{T_2}} \tag{4.23}$$

Therefore,

$$M_2 = \sqrt{M_{2n}^2 + M_{2t}^2} \tag{4.24}$$

It is necessary to first determine M_{1n} (and knowing M_{1n}, T_2/T_1 can be found) and M_{1t}. From the velocity diagrams shown in Fig. 4.6,

$$M_{1t} = M_1 \cos\theta; \quad V_{1t} = V_1 \cos\theta \tag{4.25a}$$

$$M_{1n} = M_1 \sin\theta; \quad V_{1n} = V_1 \sin\theta \tag{4.25b}$$

$$M_{2t} = M_2 \cos(\theta - \delta); \quad V_{2t} = V_2 \cos(\theta - \delta) \tag{4.25c}$$

$$M_{2n} = M_2 \sin(\theta - \delta); \quad V_{2n} = V_2 \sin(\theta - \delta) \tag{4.25d}$$

It is interesting to note that the downstream Mach number of the oblique shock does not have to be less than unity, as required for the normal shock. However, its normal component must be subsonic, i.e., $M_2 \sin(\theta - \delta) < 1$, although it is possible that $M_2 > 1$. The value of M_2 depends on the value of $(\theta - \delta)$, which in turn depends on the combination of M_1 and δ, as shown in Example 4.6 below.

EXAMPLE 4.6 Air at 30°C and 200 kPa travels at Mach 4.0. (a) Find the speed, temperature, pressure, shock strength, entropy increase, and pressure loss after the air passing through a normal shock. (b) For comparison, if the air flow experiences an oblique shock at an angle $\theta = 30°$, find the same parameters downstream of this oblique shock. (c) In case of oblique shock, also find the angle through which the air flow is deflected. Assume for air $c_p = 1.004 \ kJ/kg \cdot K$.

Solution –
(a) For the normal shock,

$$M_2^2 = \frac{M_1^2 + \frac{2}{\gamma-1}}{\frac{2\gamma}{\gamma-1}M_1^2 - 1} = \frac{16 + \frac{2}{0.4}}{\frac{2.8}{0.4} \times 16 - 1} = 0.1892; \quad M_2 = 0.4350$$

$$\frac{T_2}{T_1} = \frac{\left(1 + \frac{\gamma-1}{2}M_1^2\right)\left(\frac{2\gamma}{\gamma-1}M_1^2 - 1\right)}{\left[\frac{(\gamma+1)^2}{2(\gamma-1)}\right]M_1^2} = \frac{\left(1 + \frac{0.4}{2} \times 16\right)\left(\frac{2\times1.4}{0.4} \times 16 - 1\right)}{\left[\frac{2.4^2}{2(1.4-1)}\right] \times 16} = 4.0469;$$

$$T_2 = 1,226.2 \ K$$

$$V_2 = M_2\sqrt{\gamma R T_2} = 0.4350 \times \sqrt{1.4 \times 287 \times 1,226.2} = 305.3 \ m/s$$

$$\frac{p_2}{p_1} = \frac{2\gamma M_1^2}{\gamma+1} - \frac{\gamma-1}{\gamma+1} = \frac{2 \times 1.4 \times 16}{1.4+1} - \frac{1.4-1}{1.4+1} = 18.50; \quad p_2 = 3,700 \ kPa$$

$$\frac{\Delta p}{p_1} = \frac{p_2}{p_1} - 1 = 17.50 \quad \text{(shock strength)}$$

$$s_2 - s_1 = c_p \ln \frac{T_2}{T_1} - R\ln\frac{p_2}{p_1}$$
$$= (1.004 \ kJ/kg\cdot K) \ \ln(4.0469) - (0.287 \ kJ/kg\cdot K)\ln(18.5) = 0.5661 \ kJ/kg\cdot K$$

$$\frac{p_{t2}}{p_{t1}} = \left[\frac{\frac{\gamma+1}{2}M_1^2}{1+\frac{\gamma-1}{2}M_1^2}\right]^{\gamma/(\gamma-1)} \left[\frac{1}{\frac{2\gamma}{\gamma+1}M_1^2 - \frac{\gamma-1}{\gamma+1}}\right]^{1/(\gamma-1)}$$

$$= \left[\frac{\frac{1.4+1}{2}\times 16}{1+\frac{1.4-1}{2}\times 16}\right]^{1.4/0.4} \left[\frac{1}{\frac{2\times 1.4}{1.4+1}\times 16 - \frac{1.4-1}{1.4+1}}\right]^{1/0.4} = 0.1388$$

$$\frac{p_{t1}}{p_1} = \left(1+\frac{\gamma-1}{2}M_1^2\right)^{\gamma/(\gamma-1)} = \left(1+\frac{1.4-1}{2}\times 16\right)^{1.4/0.4} = 151.52$$

$$\Delta p_t = \left(1 - \frac{p_{t2}}{p_{t1}}\right)\frac{p_{t1}}{p_1}p_1 = (1-0.1388)\times 151.52 \times 200 \ kPa = 26.098 \ MPa$$

(b) <u>For the oblique shock, the first parameter to know is M_{1n}.</u>

$$M_{1n} = M_1 \sin\theta = 4 \sin 30^\circ = 2$$

$$M_{2n}^2 = \frac{M_{1n}^2 + \frac{2}{\gamma-1}}{\frac{2\gamma}{\gamma-1}M_{1n}^2 - 1} = \frac{4+\frac{2}{0.4}}{\frac{2.8}{0.4}\times 4 - 1} = 0.3333; \ M_{2n} = 0.5774$$

$$M_{2t} = M_{1t}\sqrt{\frac{T_1}{T_2}}; \ \frac{T_2}{T_1} = \frac{\left(1+\frac{\gamma-1}{2}M_{1n}^2\right)\left(\frac{2\gamma}{\gamma-1}M_{1n}^2 - 1\right)}{\left[\frac{(\gamma+1)^2}{2(\gamma-1)}\right]M_{1n}^2}$$

$$= \frac{\left(1+\frac{0.4}{2}\times 4\right)\left(\frac{2\times 1.4}{0.4}\times 4 - 1\right)}{\left[\frac{2.4^2}{2(1.4-1)}\right]\times 4} = 1.6875; \ T_2 = 511.3 \ K$$

$$M_{2t} = M_{1t}\sqrt{\frac{T_1}{T_2}} = M_1 \cos 30^\circ \sqrt{\frac{T_1}{T_2}} = 4\cos 30^\circ \sqrt{\frac{1}{1.6875}} = 2.6667$$

$$M_2 = \sqrt{M_{2n}^2 + M_{2t}^2} = \sqrt{(0.5774)^2 + (2.6667)^2} = 2.7285$$

It is noted that in this case the contribution of M_{2n} to M_2 is relatively small.

$$V_2 = M_2\sqrt{\gamma R T_2} = 2.7285 \times \sqrt{1.4 \times 287 \times 511.3} = 1,236.7 \ m/s$$

$$\frac{p_2}{p_1} = \frac{2\gamma M_{1n}^2}{\gamma + 1} - \frac{\gamma - 1}{\gamma + 1} = \frac{2 \times 1.4 \times 4}{1.4 + 1} - \frac{1.4 - 1}{1.4 + 1} = 4.50; \ p_2 = 900 \ kPa$$

$$\frac{\Delta p}{p_1} = \frac{p_2}{p_1} - 1 = 3.50$$

(This shock strength is five times smaller than that across a normal shock with the same upstream Mach number.)

$$s_2 - s_1 = c_p \ln\frac{T_2}{T_1} - R\ln\frac{p_2}{p_1} =$$
$$(1.004 \ kJ/kg \cdot K) \ \ln(1.6875) - (0.287 \ kJ/kg \cdot K)\ln(4.5) = 0.0936 \ kJ/kg \cdot K$$

$$\frac{p_{t2}}{p_{t1}} = \left[\frac{\frac{\gamma + 1}{2}M_{1n}^2}{1 + \frac{\gamma - 1}{2}M_{1n}^2}\right]^{\gamma/(\gamma-1)} \left[\frac{1}{\frac{2\gamma}{\gamma + 1}M_{1n}^2 - \frac{\gamma - 1}{\gamma + 1}}\right]^{1/(\gamma-1)}$$

$$= \left[\frac{\frac{1.4 + 1}{2} \times 4}{1 + \frac{1.4 - 1}{2} \times 4}\right]^{1.4/0.4} \left[\frac{1}{\frac{2 \times 1.4}{1.4 + 1} \times 4 - \frac{1.4 - 1}{1.4 + 1}}\right]^{1/0.4} = 0.7209$$

$$\Delta p_t = \left(1 - \frac{p_{t2}}{p_{t1}}\right)\frac{p_{t1}}{p_1}p_1 = (1 - 0.7209) \times 151.52 \times 200 \ kPa = 8.458 \ MPa$$

(c) The flow deflection angle can be found using either Eqn. (4.24c) or Eqn. (4.24d).

$$\frac{M_{2n}}{M_2} = \frac{0.5774}{2.7285} = \sin(\theta - \delta) = \sin(30° - \delta); \ \delta = 17.78°$$

Comments –

As discussed in the text, when the back pressure is insufficient to cause a normal shock (900 kPa < 3,700 kPa needed for a normal shock), an oblique shock may result. Both the pressure and temperature rises across an oblique shock are smaller than those across a normal shock, as an oblique shock is a weaker compression wave than a normal shock; it also causes a smaller entropy increase and thus a smaller pressure loss. The flow deflection angle can also be a result of a two-dimensional wedge (or surface) at an angle of 17.78° measured from the flow direction.

The downstream properties can also be found after M_2 is calculated. As an illustration, for $M_2 = 2.7285$, $\frac{p_{t2}}{p_2} = \left(1 + \frac{\gamma-1}{2}M_2^2\right)^{\gamma/(\gamma-1)} = \left(1 + \frac{1.4-1}{2} \times 7.4445\right)^{1.4/0.4} = 24.32$

$$p_2 = \frac{p_{t2}}{24.32} = \frac{0.7209 p_{t1}}{24.32} = \frac{0.7209 \times 151.52 \times 200 \; kPa}{24.32} = 898.1 \; kPa \approx 900 \; kPa$$

$\square$

Example 4.6 illustrates that the flow deflection angle (δ) and the downstream Mach number M_2 can be calculated by knowing the incoming flow Mach number (M_1) and the inclination of the shock wave angle (i.e., θ). It is then desirable to obtain an expression of θ in terms of δ and M_1. The ratio of normal velocity components (Eqn. [4.22e]) is:

$$\frac{V_{1n}}{V_{2n}} = \frac{V_{1t}\tan\theta}{V_{2t}\tan(\theta-\delta)} = \frac{\tan\theta}{\tan(\theta-\delta)} = \frac{p_2}{p_1} = \frac{(\gamma+1)M_{1n}^2}{(\gamma-1)M_{1n}^2+2} = \frac{(\gamma+1)M_1^2\sin^2\theta}{(\gamma-1)M_1^2\sin^2\theta+2}$$

$$(4.26)$$

After some trigonometric manipulations,

$$\tan\delta = 2\cot\theta \frac{M_1^2\sin^2\theta - 1}{M_1^2(\gamma+\cos2\theta)+2} \tag{4.27}$$

For $\delta = 0$, $M_1^2\sin^2\theta = 1$ and this occurs in two scenarios: (i) for $\theta = \pi/2$ (i.e., the normal shock, for which no flow redirection occurs immediately downstream of the shock) and (ii) $\theta = \sin^{-1}(1/M_1)$; in this case θ is also the Mach wave angle μ as expressed by Eqn. (3.12). Therefore for $\delta > 0$, θ is expected to fall in the range given by

$$\sin^{-1}\frac{1}{M_1} \leq \theta \leq \frac{\pi}{2} \tag{4.28}$$

The relation between θ and δ, Eqn. (4.27), is graphically shown in Fig. 4.12 for $\gamma = 1.4$. A few observations and comments can be made:

(1) As shown in Fig. 4.12, for a given pair of M_1 and δ, there are in general two possible oblique shock wave angles – one for a weak shock (smaller θ value associated with smaller M_{1n}) and the other for a stronger shock (larger θ value associated with larger M_{1n}). Usually, the weak oblique shock appears and, in the case of a two-dimensional wedge as the flow turning device, the shock wave attaches to the vertex. For a given M_1, if the deflection angle δ is too large, there is no solution as represented in Fig. 4.11. For example, for $M_1 = 3$, there is no attached shock solution (i.e., no value of θ exists) for δ greater than approximately $34°$. The flow is forced to adjust to it through a detached normal shock instead of an oblique shock, as illustrated in Fig. 4.13.

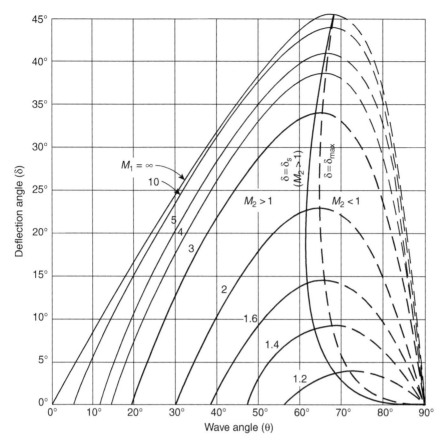

Figure 4.12 Oblique shock solutions, expressed by Eqn. (4.27) for $\gamma = 1.4$ (adapted from Liepmann, H. W., and Roshko, A., *Elements of Gasdynamics*, John Wiley and Sons, New York, 1957).

(2) The supersonic nozzle exit flow discussed earlier may enter the surroundings without encountering physical deflecting object (or obstruction). Depending on the pressure of the surroundings (or rather the pressure ratio of the surroundings to the exit gas), an oblique shock may form with the flow turning angle and the shock wave angle given by Eqn. (4.27). In the case of $\delta = 0$, either a normal shock ($\theta = \pi/2$, Fig. 4.10d) due to sufficiently high pressure ratio p_2/p_1 given by Eqn. (4.16a) or no shock of either type (normal and oblique, Fig. 4.10e) is formed. In the latter case, the angle $\theta = \sin^{-1}(1/M_1)$ is the Mach cone angle (or the Mach wave angle) described in Chapter 3 implying that the pressure gradient is zero. In this case of a nozzle exit flow, the flow is said to be *perfectly expanded* (to the ambient pressure, p_b, as shown in Fig. 4.10e). In contrast, the flow that forms either a normal or an oblique shock at the exit is said to be *overexpanded* because its pressure falls below that of the ambient (i.e., $p_e < p_b$) and, due to its supersonic speed, it adjusts to the ambient condition through a shock wave.

(3) In this $\theta - \delta$ plot, two contours for the particular flow deflection angles are also shown: (i) δ_{max} for causing the maximum shock wave angle θ_{max} and (ii) δ_s for causing sonic downstream flow (($M_2 = 1$). The value of θ_{max} represents the maximum shock wave angle with which the oblique shock remains attached. They were determined from the following expressions (Ferri, 2005):

$$\sin^2\theta_{max} = \frac{1}{\gamma M_1^2}\left\{\frac{\gamma+1}{4}M_1^2 - 1 + \left[(\gamma+1)\left(1 + \frac{\gamma-1}{2}M_1^2 + \frac{\gamma+1}{16}M_1^4\right)\right]^{1/2}\right\} \quad (4.29)$$

$$\sin^2\theta_s = \frac{1}{\gamma M_1^2}\left\{\frac{\gamma+1}{4}M_1^2 - \frac{3-\gamma}{4} + \left[(\gamma+1)\left(\frac{\gamma+9}{16} - \frac{3-\gamma}{8}M_1^2 + \frac{\gamma+1}{16}M_1^4\right)\right]^{1/2}\right\}$$
$$(4.30)$$

For a given M_1 there are corresponding values of δ that would cause θ_{max} and $M_2 = 1$. As an example, for $M_1 = 3.0$, $\delta_s \approx \delta_{max} \approx 34°$ with corresponding $\theta_{max} \approx 65°$. For $M_1 = 1.4$ and $\delta_s = 8.5°$ ($\theta_s = 63.5°$) and $\delta_{max} \approx 9.25°$ and $\theta_{max} \approx 67.5°$. As a consequence, for $M_1 = 1.4$ the shock wave becomes detached for $\delta > 9.25°$. Therefore for $M_1 = 1.4$ and $9.25° > \delta > 9°$ the oblique shock is attached with $M_2 < 1$.

(4) There exist subsonic values of M_2 for a narrow range of δ for a given M_1. For $M_1 = 1.6$, for example, $M_2 < 1$ for $\delta \approx 14.3° - 14.7°$, as shown by the solid line segment between the $\delta = \delta_s$ and $\delta = \delta_{max}$ curve. For most values of δ, the weak solution results in $M_2 > 1$. Strong solutions require the shock to detach from the wedge and $M_2 < 1$.

Figure 4.12 can be used to illustrate how the shock wave angle may evolve as M_1 increases. Consider a two-dimensional wedge with a half wedge angle $\delta = 15°$ symmetrically aligned with the flow. As M_1 increases from subsonic condition to approximately $M_1 = 1.6$, there is no oblique solution – instead there is either no shock (subsonic flow) or detached normal shock as the horizontal line of $\delta = 15°$ and the solution curves for these Mach numbers do not intersect. As M_1 increases to approximately 1.62, the oblique shock forms with a wave angle $\theta \approx 63°$. When M_1 reaches 2, 3, 4, 5, and so on, and toward infinity, the oblique shock wave angle $\theta \approx 45°, 32°, 27°, 23°$ and toward $18°$, respectively. The Mach number that divides the regimes of attached and detached shocks is called the *detachment Mach number*. The result of Fig 4.12 indicates that this number is dependent on the deflection angle. For $\delta = 15°$, it is 1.62, as discussed, and for $\delta = 10°$, it is approximately equal to 1.43. Oblique shock solutions similar to Fig. 4.12 with higher resolution are found in Appendix E.

The results of Fig. 4.12 suggest that for a given δ, the difference between θ and δ decreases as M_1 increases. High Mach number flows are thus more able to "press" onto the wedge. For $M_1 \to \infty$, as shown in Fig. 4.12, θ is larger than δ by a decreasingly smaller fraction. For example for $\delta = 10°$, $\theta \approx 12°$ for $M_1 \to \infty$ and, for

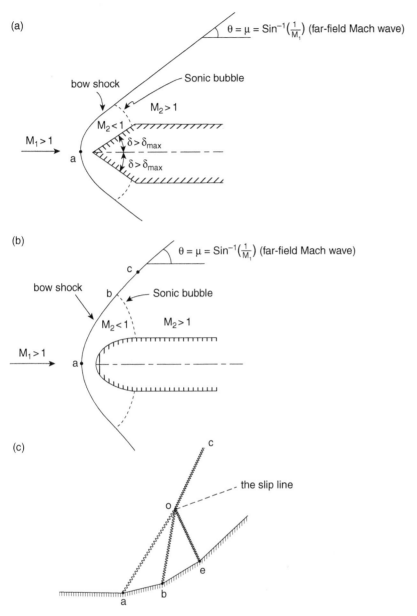

Figure 4.13 (a) Schematic of (a) the detached shock wave system on a wedge with an afterbody, (b) the detached shock wave system on a blunt-nose body, and (c) merging of oblique shocks to form a stronger shock over a concave corner. Note in (a) and (b) that the far-field shock wave becomes the Mach wave, while in (c) the entropy change through the three oblique shocks is different from that through the merged shock, leading to the formation of the slip line.

comparison, $\theta \approx 51°$ for $M_1 = 1.6$. As will be seen in Chapter 12 on hypersonic flows, for Eqn. (4.33c), $\theta \to (\gamma + 1)\theta/2$ as $M_1 \to \infty$.

Figure 4.13a qualitatively illustrates a detached shock wave system on a wedge with an afterbody. At point a along the shock, the wave angle θ is equal to $\pi/2$. In this region, the shock wave is a normal shock and the downstream Mach number $M_2 < 1$. Moving away from point a, the shock wave becomes weaker and beyond point b (where $M_2 = 1$), the downstream flow is supersonic. Between points a and b, where the flow is subsonic, the effect of the shoulder can be felt by the flow, helping to explain the curvature of the shock wave. Moving further away from point b, the shock becomes so weak that its inclination (θ) asymptotes to the Mach angle, as if not affected by the wedge or the afterbody. As an example, for $M_1 = 2.0$, the inclination is $30° \left(= \mu = \sin^{-1}[1/M_1]\right)$, the same as the Mach (or wave) angle for $\delta = 0°$ as can be found on Fig. 4.12. Therefore, conditions along a detached shock wave correspond to the entire range of the curve for a given M_1. Different values of M_1 not only give different asymptotic values of $\theta(= \mu)$, but also different values of θ_{max} ($\approx 67.5°$ for $M_1 = 1.4$ and $\approx 64.4°$ for $M_1 = 2.0$).

Referring to Fig. 4.13, one can expect that the shape of the detached shock will depend on M_1 for a given geometry. As can be seen in Fig. 4.11, one would also expect that flows having smaller M_1 detach at smaller δ. It is also natural to expect it to depend on the geometry for a given M_1. Figure 4.13b depicts the detached shock wave system on a blunt-nose body, which can be thought of as having $\delta > \delta_{max}$ at the nose. The wave near point a has a larger segment for $\theta = \pi/2$ and $M_2 < 1$ than the corresponding point a in Fig. 4.13a. Because of the curvature, the shock strength gradually diminishes along a detached shock away from point a in both Figs. 4.12a and 4.12b. The resultant gradient of entropy would cause generation of vorticity, even if the incoming flow is uniform and without vorticity – this will be shown in Section 8.7. The associated entropy rise also decreases along the shock wave in the direction toward the far field.

Oblique shocks may merge to form a stronger shock wave, as shown in Fig. 4.13c. One expects the entropy increase behind segment oc to be larger than that behind segment oe. A dividing line might be identified (the "slip" line as shown in Fig. 4.13c), across which entropies do no match. Along a gradually curved shock wave, such as those shown in Figs. 4.12a and 4.12b, the slip line region is more diffuse, and thus forms an *entropy layer*.

For attached shocks, once the corresponding values of θ and δ are found, the downstream Mach number can be calculated by combining Eqns. (4.22a), (4.25b), and (4.25d), so that M_2 is a function of M_1 and δ

$$M_2^2 \sin^2(\theta - \delta) = \frac{M_1^2 \sin^2\theta + \frac{2}{\gamma-1}}{\frac{2\gamma}{\gamma-1}M_1^2 \sin^2\theta - 1} \tag{4.31}$$

where it is understood that θ is also a function of M_1 and δ.

By referring to Fig. 4.13 and Eqn. (4.22), all shock waves, including bow shocks, can be treated as normal shocks, as long as the local normal component of the Mach number of the approaching flow is used (Thompson, 1972). For this reason, the solution process in Example 4.6 can be followed along with using the table in Appendix D.

EXAMPLE 4.7 For the same oblique shock wave in Example 4.6, use Fig. 4.12 to find the downstream Mach number. Also compare the shock strengths for $M_1 = 4$ and 2 for the same wedge angle.

Solution –
For $M_1 = 4$ and $\theta = 30°$, Fig. 4.12 indicates that $\delta \approx 17.75°$.
Eqn. (4.31) becomes

$$M_2^2 \sin^2(30° - 17.5°) = \frac{M_1^2 \sin^2 30° + \frac{2}{\gamma - 1}}{\frac{2\gamma}{\gamma - 1} M_1^2 \sin^2 30° - 1}$$

$$M_2 \approx 2.67$$

This value is slightly different from that calculated in Example 4.4, due to the different values of δ. For $M_1 = 2$ and $\delta \approx 17.75°$, Fig. 4.7 gives $\theta \approx 48.5°$. Thus $M_{1n} = M_1 \sin\theta = 2 \times \sin 48.5° \approx 1.50$. Form the normal shock table, $\frac{p_2}{p_1} = 2.458$ and $\beta = \frac{p_2}{p_1} - 1 = 1.458$.
    $M_1 = 4$, $M_{1n} = M_1 \sin\theta = 4 \times \sin 30° = 2$.   Therefore, $\frac{p_2}{p_1} = 4.50$ and $\beta = \frac{p_2}{p_1} - 1 = 3.50$.

Comments –
As M_1 is increased for a given wedge angle, the shock strength increases even though the shock wave angle decreases. This is because $M_{1n} = M_1 \sin\theta$ still increases as long as the shock remains attached, further illustrating that higher Mach number flows are more able to "press" on the wedge, with higher pressure on the surface and more inclined shock wave.   □

A useful alternative to Eqn. (4.27) for relating δ and θ can be derived by rearranging Eqn. (4.26) so that

$$\frac{1}{M_1^2 \sin^2\theta} = \frac{\gamma + 1}{2} \frac{\tan(\theta - \delta)}{\tan\theta} - \frac{\gamma - 1}{2} \tag{4.32}$$

Applying necessary trigonometric identities, this expression is reduced to

$$M_1^2 \sin^2\theta - 1 = \frac{\gamma + 1}{2} M_1^2 \frac{\sin\theta \sin\delta}{\cos(\theta - \delta)} \tag{4.33a}$$

The usefulness of Eqn. (4.33a) can be seen in large-Mach number flows, where small turning angles are desirable, i.e., $\delta \ll 1$, so that $\sin\delta \approx \delta$ and $\cos(\theta - \delta) \approx \cos\theta$. In these flows,

$$M_1^2 \sin^2\theta - 1 = \left(\frac{\gamma + 1}{2} M_1^2 \tan\theta\right) \cdot \delta \tag{4.33b}$$

For extremely large Mach numbers ($M_1 \rightarrow \infty$), results from Fig. 4.12 suggest $\theta \ll 1$. Using small-angle approximation for Eqn. (4.33b), $\sin\theta \approx \tan\theta \approx \theta$, and $M_1^2 \sin^2\theta \gg 1$, one finds

$$\theta = \frac{\gamma + 1}{2}\delta \tag{4.33c}$$

Equation (4.33c) appears to be an excellent approximation even for moderately large deflection angles. For and $M_1 \rightarrow \infty$ and $M_1 = 10$, the $\theta - \delta$ plot preserves the slope of $(\gamma + 1)/2 \approx 1.2$ for δ up to and beyond 25° (i.e., $\delta = 0.436$). The flow with $M_1 \rightarrow \infty$ is called the hypersonic flow. More details related to hypersonic flows are presented in Chapter 12.

EXAMPLE 4.8 For $M_1 \rightarrow \infty$, find the attached oblique shock solution for the deflection angle, shock wave angle, downstream Mach number. Consider $M_1 = 10$ as approaching ∞. Find ratios of pressure and temperature across the shock wave for $\gamma = 1.4$.

Solution –
The maximum shock wave angle θ_{max} can be directly determined without knowing δ, according to Eqn. (4.29):

$$\sin^2\theta_{max} = \lim_{M_1 \rightarrow \infty} \frac{1}{\gamma M_1^2} \left\{ \frac{\gamma + 1}{4}M_1^2 - 1 + \left[(\gamma + 1)\left(1 + \frac{\gamma - 1}{2}M_1^2 + \frac{\gamma + 1}{16}M_1^4\right)\right]^{1/2}\right\}$$

$$= \frac{\gamma + 1}{2\gamma}$$

$$\theta_{max} = \sin^{-1}\sqrt{\frac{\gamma + 1}{2\gamma}}$$

Then

$$\tan\delta_{max} = \lim_{M_1 \rightarrow \infty} 2\cot\theta \frac{M_1^2 \sin^2\theta - 1}{M_1^2(\gamma + \cos 2\theta) + 2} = 2\frac{\cos\theta_{max}}{\sin\theta_{max}}\frac{\sin^2\theta_{max}}{\gamma + \cos2\theta_{max}}$$

$$= \frac{\sin2\theta_{max}}{\gamma + \cos2\theta_{max}}$$

For $\gamma = 1.4$,

$$\theta_{max} = 67.792°$$

$$\delta_{max} = 45.579°$$

These values are consistent with those than can be read from Fig. 4.12.
$$M_2^2 \sin^2(\theta - \delta) = \frac{M_1^2\sin^2\theta + \frac{2}{\gamma - 1}}{\frac{2\gamma}{\gamma - 1}M_1^2\sin^2\theta - 1} = \frac{\gamma - 1}{2\gamma} \text{ in the limit of } M_1 \rightarrow \infty.$$

$$M_2 = \sqrt{\frac{\gamma - 1}{2\gamma \sin^2(\theta - \delta)}} = \sqrt{\frac{0.4}{2 \times 1.4 \times \sin^2(67.792^\circ - 45.579^\circ)}} = 0.9998 \approx 1$$

For $M_1 = 10$,

$$M_{1n} = M_1 \sin\theta \approx M_1 \sqrt{\frac{\gamma + 1}{2\gamma}} = 9.26$$

$$\frac{p_2}{p_1} = \frac{2\gamma M_{1n}^2}{\gamma + 1} - \frac{\gamma - 1}{\gamma + 1} \approx M_1^2 = 100(\approx 100 \text{ from Appendix D})$$

$$\frac{T_2}{T_1} = \frac{2\gamma(\gamma - 1)M_{1n}^2}{(\gamma + 1)^2} = \frac{(\gamma - 1)}{(\gamma + 1)}M_1^2 \approx 16.7(\approx 17.5 \text{ from Appendix D})$$

Comments –

This near-unity value of the downstream Mach number is to be expected, as Eqns. (4.29) and (4.30) for θ_{max} and θ_s (shock wave angle for $M_2 = 1$) are the same as $M_1 \to \infty$. This result can be seen in Fig. 4.12. From Appendix D, the pressure and temperature ratios are approximately 100 and 17.5, respectively. $\square$

4.7 Weak Oblique Shock

As discussed in Section 4.6, shock waves formed with moderate values of M_1 are prone to detachment. Small turning (or deflection) angles are desirable to avoid detachment for this range of Mach numbers. For the limiting case where $\delta \to 0$, either the right-hand side of Eqn. (4.33b) is finite and $\tan\theta \to \infty$ (*i.e.*, $\theta = \pi/2$) or the left-hand side approaches zero (i.e., $\theta = \sin^{-1}(1/M_1) = \mu$, the Mach angle), for attached oblique shock waves $\theta \leq \theta_{max}$ (as shown in Fig. 4.12). As a consequence, $\tan\theta$ has to be finite and therefore

$$\tan\theta = \tan\mu = \frac{1}{\sqrt{M_1^2 - 1}}$$

Equation (4.33b) now becomes

$$M_1^2 \sin^2\theta - 1 = \left(\frac{\gamma + 1}{2} \frac{M_1^2}{\sqrt{M_1^2 - 1}}\right) \cdot \delta \tag{4.34}$$

The shock strength depends on $M_{1n} = M_1 \sin\theta$, and Eqn. (4.16e) is rewritten for the weak oblique shock (with small δ) as

$$\beta = \frac{p_2 - p_1}{p_1} = \frac{2\gamma}{\gamma + 1}\left(M_1^2 \sin^2\theta - 1\right) = \left(\frac{\gamma M_1^2}{\sqrt{M_1^2 - 1}}\right) \cdot \delta \qquad (4.35)$$

Therefore for weak oblique shocks, the shock strength is proportional to the turning angle, δ.

4.8 Compression in Supersonic Flow by Turning

Similar to the weak normal shock wave, the weak oblique shock causes an entropy increase proportional to the third power of the shock strength. Eqn. (4.22f) for the normal shock can be written in terms of the normal component of the Mach number M_{1n}:

$$\frac{s_2 - s_1}{R} \approx \frac{2\gamma}{3(\gamma + 1)^2}\left(M_{1n}^2 - 1\right)^3 = \frac{\gamma + 1}{12\gamma^2}\beta^3 \propto \delta^3 \qquad (4.36)$$

For small values of δ, the entropy increase becomes vanishingly smaller and so does the loss in stagnation pressure, as

$$\frac{p_{t1} - p_{t2}}{p_{t1}} = 1 - e^{-(s_2 - s_1)/R} = O(\delta^3) \qquad (4.37)$$

where binomial expansion was used for the exponential term for small $(s_2 - s_1)/R$. This result is similar to those shown in Example 4.5 for weak normal shocks, where the loss in the stagnation pressure is vanishingly negligible with a small supersonic upstream Mach number. For weak oblique shock waves, the entropy increase and pressure loss are now related to the turning angle, as shown in Eqns. (4.36) and (4.37). It is thus desirable to achieve a turning angle (i.e., compression) through a series of weak oblique shocks using smaller turning angles than by one single oblique shock through a finite turning angle.

Figure 4.14a shows an oblique shock formed at a small finite turning angle δ. To further minimize entropy increase due to turning, let $n\Delta\delta = \delta$, where n is the number of much smaller equal turning angles $\Delta\delta$ that comprise the finite turning angle δ, as shown in Figs. 4.14a and 4.14b. Equations (4.36) and (4.37) can be written for each $\Delta\delta$ as

$$\Delta\beta \sim \Delta\delta$$

$$\Delta s \sim (\Delta\delta)^3$$

The overall changes after completing the turn are

$$\beta = \frac{p_2 - p_1}{p_1} \sim \delta \qquad (4.38)$$

(a)

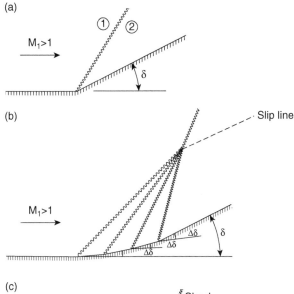

(b)

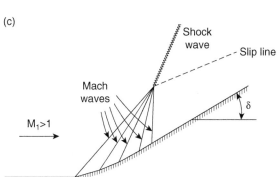

Figure 4.14 To turn a supersonic flow through an angle δ (part a) with decreasing amounts entropy increase, the turn can be achieved by gradual turning through a series of n turns, each with $\Delta\delta$ so that $n\Delta\delta = \delta$, allowing the flow to pass a series of weaker shock waves (part b). As $n \rightarrow \infty$, the shock waves become Mach waves. Similar to Fig. 4.12, slip lines form in parts (b) and (c). No slip line exists as the flow experiences uniform entropy increase through the single oblique shock.

$$\frac{\Delta s}{R} = \frac{s_2 - s_1}{R} = n(\Delta\delta)^3 \sim n\Delta\delta(\Delta\delta)^2 \sim \delta(\Delta\delta)^2 \sim \frac{\delta^3}{n^2} \tag{4.39a}$$

$$\frac{p_{t1} - p_{t2}}{p_{t1}} = O\left(\frac{\delta^3}{n^2}\right) \tag{4.39b}$$

Recall that for compression (i.e., wedge turning into the flow), $\delta > 0$ and $\Delta\delta > 0$. Comparing with compression by one single oblique shock, the same degree of compression (β) can be achieved with a much reduced entropy increase and a much reduced pressure loss by a factor of $1/n^2$. In the limiting case of $\Delta\delta \rightarrow 0^+$ and $n \rightarrow \infty$ the turning is a continuous procedure and the oblique shock waves becomes Mach waves, as illustrated in Fig. 4.14c where some representative Mach lines are shown. In this limiting procedure, the compression is isentropic. The flow between the two successive Mach lines is uniform. However, in the limiting compression process, the region between the representative Mach lines also becomes vanishingly small and the change in flow becomes continuous. The following numerical example illustrates the benefits of such successive turning through smaller angles.

EXAMPLE 4.9 An air flow at Mach 4.0 and 216 K is being turned by (1) a single turning angle of 10° and (2) two successive turning angles, both of 5°. Assume $c_p = 1.0$ kJ/kg·K. Compare pressure increases, the stagnation pressure change, entropy change, and the final flow speeds of the two scenarios.

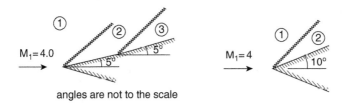

angles are not to the scale

Solution –
From Fig. 4.12, for $M_1 = 4.0$ and $\delta = 10°$, $\theta \approx 23°$. Therefore, $M_{1n} = M_1 \sin 23° \approx 1.56$. Therefore, using shock table or Eqns. (4.16) and (4.17d) yields

$$\frac{p_2}{p_1} = 2.673 \text{ and } \beta = \frac{p_2 - p_1}{p_1} = 1.673$$

$$\frac{p_{t2}}{p_{t1}} = 0.910 \quad \frac{p_{t2} - p_{t1}}{p_{t1}} = 9\%$$

$$\frac{T_2}{T_1} = 1.361$$

$$M_{2n} = 0.681$$

$$M_{2t} = M_{1t}\sqrt{\frac{T_1}{T_2}} = M_1 \cos\theta_1 \sqrt{\frac{T_1}{T_2}} = 4 \ \cos 23°\sqrt{\frac{1}{1.361}} = 3.156$$

Therefore $M_2 = 3.229$ and $V_2 = M_2\sqrt{\gamma R T_2} = 3.229 \times \sqrt{1.4 \times 288 \times 1.361 \times 216} = 1,111.7$ m/s

$$\Delta s = s_2 - s_1 = c_p \ln\frac{T_2}{T_1} - R \ln\frac{p_2}{p_1} = 1.0 \times \ln(1.361) - 0.287 \times \ln(2.673)$$

$$= 0.026 \ kJ/kg \cdot K$$

With successive turning, the first turning results in (with $M_1 = 4.0$ and $\delta_1 = \Delta\delta = 5°$) $\theta_1 \approx 18°$ and $M_{1n} = M_1 \sin 18° \approx 1.24$. Therefore, using shock table or Eqns. (4.16) and (4.17d) again yields

$$\frac{p_2}{p_1} = 1.627$$

$$\frac{p_{t2}}{p_{t1}} = 0.988 \text{ or } \frac{p_{t2} - p_{t1}}{p_{t1}} = 1.2\%$$

$$\frac{T_2}{T_1} = 1.153$$

$$M_{2n} = 0.818$$

$$M_{2t} = M_{1t}\sqrt{\frac{T_1}{T_2}} = M_1\cos\theta_1\sqrt{\frac{T_1}{T_2}} = 4\cos 18^\circ\sqrt{\frac{1}{1.153}} = 3.543$$

Therefore $M_2 \approx 3.64$

Now with $\delta_2 = \Delta\delta = 5^\circ, \theta_2 \approx 19^\circ$. Therefore $M_{2n} = M_2\sin 19^\circ \approx 1.19$. Thus

$$\frac{p_3}{p_2} = 1.486$$

$$\frac{p_{t3}}{p_{t2}} = 0.993$$

$$\frac{T_3}{T_2} = 1.120$$

$$M_{n3} = 0.848$$

$$M_{t3} = M_{t2}\sqrt{\frac{T_1}{T_2}} = M_2\cos\theta_2\sqrt{\frac{T_1}{T_2}} = 3.64\cos 19^\circ\sqrt{\frac{1}{1.120}} = 3.253$$

$$M_3 = 3.360 \text{ and } V_3 = M_3\sqrt{\gamma R T_3} = 3.36 \times \sqrt{1.4 \times 288 \times 1.293 \times 216}$$
$$= 1,127.5\text{m/s}$$

$$\frac{p_{t3}}{p_{t1}} = \frac{p_{t3}}{p_{t2}}\frac{p_{t2}}{p_{t1}} = 0.993 \times 0.988 = 0.981 \text{ and } \frac{p_{t3} - p_{t1}}{p_{t1}} \approx 2\%$$

$$\frac{p_3}{p_1} = \frac{p_3}{p_2}\frac{p_2}{p_1} = 1.486 \times 1.627 = 2.418 \text{ and } \beta = \frac{p_3 - p_1}{p_1} = 1.418$$

$$\frac{T_3}{T_1} = \frac{T_3}{T_2}\frac{T_2}{T_1} = 1.121 \times 1.153 = 1.293$$

$$\Delta s = s_3 - s_1 = c_p\ln\frac{T_3}{T_1} - R\ln\frac{p_3}{p_1} = 1.0 \times \ln(1.293) - 0.287 \times \ln(2.418)$$
$$= 0.004 \ kJ/kg\cdot K$$

Comments –

It can be seen that indeed $\Delta s \sim 1/n^2$. In this example, $n = 2$ and the values of Δs differ by a factor of 4.5, which is close to $n^2 = 4$. Furthermore, the pressure loss given by Eqn. (4.37), $\frac{p_{t1}-p_{t2}}{p_{t1}} = 1 - e^{-(s_2-s_1)/R} = O(\delta^3)$ suggests that by halving the angle, the pressure loss should decrease by a factor of 8. The first turning by results in $\frac{p_{t2}-p_{t1}}{p_{t1}} = 1.2\%$ close to being one-eighth ($= 1/n^3$, with $n = 2$) of loss due to the $10°$ turning. If p_1 is known, both the static and stagnation pressures before and after the shock waves can be calculated. The static pressure contributes to the force acting on a wing surface and its value is therefore necessary when aerodynamic lift and drag are of interest. For the same total turning angle, a series of smaller turning angle results in smaller β.   □

A series of oblique shock waves can be used for pressure recovery in supersonic inlets. The following example illustrates such an application.

EXAMPLE 4.10 Consider two different two-dimensional supersonic inlets for a free stream Mach number $M_1 = 4$. One operates with a standing normal shock wave at the inlet as the way to slowing down the flow, the other uses the scheme similar to that described in Example 4.9, with wedge-shaped diffuser, which generate two oblique shocks through two $5°$ turning angles for deceleration, followed by further deceleration through a normal shock. Compare the stagnation pressure losses.

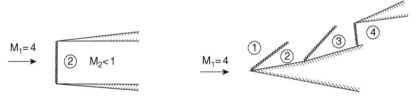

Solution –

For the one single normal shock for $M_1 = 4$, $M_2 = 0.435$ and $p_{t2}/p_{t1} = 0.1388$ (a loss of approximately 86%).

For the second case, $M_3 = 3.36$ from Example 4.9. Therefore through the normal shock, $M_4 = 0.457$ and $p_{t4}/p_{t3} = 0.2404$. The $p_{t4}/p_{t1} = (p_{t4}/p_{t3})(p_{t3}/p_{t1}) = 0.2404 \times 0.981 = 0.236$ and the stagnation pressure loss is approximately 76%. Therefore there is an advantage by adopting the second flow deceleration scheme for supersonic diffuser design, even though the final compression may still be done through a normal shock.   □

For small but non-vanishing values of $\Delta\delta > 0$, the shock wave angle θ deviates from the Mach angle μ slightly. Let the deviation be ε, with $\varepsilon \ll \mu$, such that

$$\theta = \mu + \varepsilon \qquad (4.40)$$

and

$$\sin\theta = \sin(\mu + \varepsilon) = \sin\mu\,\cos\varepsilon + \sin\varepsilon\,\cos\mu \approx \sin\mu + \varepsilon\,\cos\mu \qquad (4.41)$$

Since $\sin\mu = 1/M_1$, $\cot\mu = (M_1^2 - 1)^{1/2}$. Dividing the above expression through by $\sin\mu$ yields

$$M_1\,\sin\theta \approx 1 + \varepsilon\cot\mu = 1 + \varepsilon(M_1^2 - 1)^{1/2} \qquad (4.42a)$$

A series expansion for $M_1^2\sin^2\theta$ and neglecting terms containing ε^2 leads to

$$M_1^2\,\sin^2\theta \approx 1 + 2\varepsilon(M_1^2 - 1)^{1/2} \qquad (4.43b)$$

Comparing with Eqn. (4.34) leads to the following expression:

$$\varepsilon = \left(\frac{\gamma+1}{4}\frac{M_1^2}{M_1^2 - 1}\right)\cdot\Delta\delta \qquad (4.44)$$

It is also of interest to find the velocity change across the wave. By referring to Fig. 4.11, with $V_{2t} = V_{1t} = V_t$ one arrives at the following:

$$\frac{V_2^2}{V_1^2} = \frac{V_{2n}^2 + V_{2t}^2}{V_{1n}^2 + V_{1t}^2} = \frac{(V_{2n}/V_t)^2 + 1}{(V_{1n}/V_t)^2 + 1} = \frac{\tan^2(\theta - \delta) + 1}{\tan^2\theta + 1} = \frac{\cos^2\theta}{\cos^2(\theta - \Delta\delta)}$$

This expression can be rewritten as

$$\frac{V_2}{V_1} = \frac{\cos\theta}{\cos(\theta - \Delta\delta)} = \frac{\cos\theta}{\cos\theta\,\cos\Delta\delta + \sin\theta\,\sin\Delta\delta} \approx \frac{\cos\theta}{\cos\theta + \sin\theta\cdot\Delta\delta} = \frac{1}{1 + \tan\theta\cdot\Delta\delta}$$

where small angle approximations, $\sin\Delta\delta \approx \Delta\delta$ and $\cos\Delta\delta \approx 1$, are used. Equation (4.42a) can be manipulated to provide information for $\tan\theta$ by assuming $\varepsilon \ll 1$ (as $\varepsilon \ll \mu$ and μ is of the order of unity or smaller), resulting in

$$\tan\theta = \frac{1}{\sqrt{M_1^2 - 1}}$$

Substituting this expression for $\tan\theta$ back to the expression for V_2/V_1, and noting that for a small value x, $1/(1 + x) \approx (1 - x)$ and letting $x = \tan\theta\cdot\Delta\delta$ yields

$$\frac{V_2}{V_1} = 1 - \frac{\Delta\delta}{\sqrt{M_1^2 - 1}} \qquad (4.45a)$$

$$\frac{\Delta V}{V_1} = \frac{V_2 - V_1}{V_1} = -\frac{\Delta\delta}{\sqrt{M_1^2 - 1}} \qquad (4.45b)$$

In the differential expression, Eqn. (4.45b) becomes

$$\frac{dV}{V} = -\frac{d\delta}{\sqrt{M^2 - 1}} \qquad (4.46)$$

Because compression makes M and μ decrease while making ε increase, according to Eqns. (4.40) and (4.44), Mach wave and shock wave angles shown in Figs. 4.14a and 4.14c become increasingly steeper. At a sufficient distance away from the turning corner, depending how gradual the turn is, these Mach lines converge to form a shock wave, as shown in Fig. 4.14c.

Because there is no length scale to characterize the corner region, the flow and fluid parameters in the near-wall region must only be dependent upon the angle through which the flow is turned. There is a Mach wave angle corresponding to each turning angle and thus the flow and fluid properties must be uniform at locations along the Mach lines from the corner. The flow is uniform between two Mach lines. Since a continuous turn creates an infinite number of Mach lines with infinitesimal spacing, it generates a continuous compression process near the turning corner.

Derivations leading to Eqns. (4.15) and (4.16) relating the downstream and upstream Mach numbers do not involve the assumption of a shock wave. In fact a supersonic flow may expand and accelerate to be more supersonic, *i.e.*, $M_2 > M_1 > 1$ and $p_2 < p_1$ as allowed by

$$\frac{p_2}{p_1} = \frac{1 + \gamma M_1^2}{1 + \gamma M_2^2} \tag{4.16a}$$

Because the flow turning associated with weak oblique shocks can be accomplished by small turning angles, the compression is nearly isentropic. Would supersonic

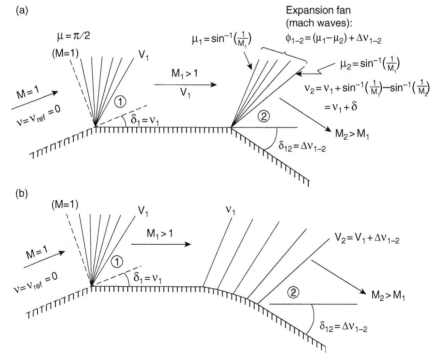

Figure 4.15 Schematic of Prandtl-Meyer expansion from region 1 to region 2 through (a) a sharp turning angle and (b) through a smooth but equal turning angle.

expansion be accomplished with $d\delta < 0$ (i.e., deflection is away from the flow and the flow turns around a *convex* angle)? The case of $d\delta < 0$ $(\delta_2 < \delta_1 < 0)$ $\varepsilon < 0$ constitutes the *Prandtl-Meyer expansion* discussed in the following section.

4.9 The Prandtl-Meyer Function

Imagine a sharp (or "centered") turn away from the flow ($d\delta < 0$ and $\delta_2 < \delta_1 < 0$, as shown in Fig. 4.15a) as consisting of a series of much more gradual turns through much smaller turning angles that make up the same total turning angle (shown in Fig. 4.15b). Due to the small turning angles, the flow near the corner is isentropic as in the case of isentropic compression. As a result, the fluid and flow properties are constant along each of these Mach waves and the flow is considered uniform in the near-corner region between two successive Mach waves. Eqn. (4.46) is applicable for its analysis and is rewritten as

$$\pm d\delta = \sqrt{M^2 - 1}\,\frac{dV}{V} \tag{4.47a}$$

where the $+$ and $-$ signs denote turning away from and into the flow, respectively. Integration of this expression for Prandtl-Meyer expansion yields

$$+\delta + \text{constant} = \int \sqrt{M^2 - 1}\,\frac{dV}{V} = \nu(M) \tag{4.47b}$$

where $\nu(M)$ is called the *Prandtl-Meyer function*. It is desirable to express V in terms of M to obtain an explicit expression only in M. The following relations can be used:

$$V = Ma$$

and

$$\frac{a_t^2}{a^2} = \frac{\gamma R T_t}{\gamma R T} = 1 + \frac{\gamma - 1}{2} M^2 \tag{4.47c}$$

where the subscript t once again denotes the stagnation state. For isentropic expansion, $T_t = \text{constant}$. It follows that

$$\frac{dV}{V} = \frac{dM}{M} + \frac{da}{a} \tag{4.47d}$$

Because $T_t = \text{constant}$ (i.e., $a_t^2 = \gamma R T^t$). After some algebraic manipulations,

$$\frac{da}{a} = \frac{(\gamma - 1) M dM}{2\left(1 + \frac{\gamma-1}{2} M^2\right)}$$

and

$$\frac{dV}{V} = \frac{dM}{M}\left(\frac{1}{1+\frac{\gamma-1}{2}M^2}\right) \tag{4.47e}$$

By substituting Eqn. (4.47d) into Eqn. (4.47a) the Prandtl-Meyer function becomes

$$\nu(M) = \int \frac{\sqrt{M^2-1}}{1+\frac{\gamma-1}{2}M^2}\frac{dM}{M} = \sqrt{\frac{\gamma+1}{\gamma-1}}\tan^{-1}\sqrt{\frac{\gamma-1}{\gamma+1}(M^2-1)} - \tan^{-1}\sqrt{M^2-1}$$

$$\tag{4.47f}$$

Equation (4.47e) suggests that constant $= 0$ in Eqn. (4.47a) and that ν is the total turning angle necessary for accelerating $M = 1$ to a supersonic Mach number. Therefore, ν is also called the *Prandtl-Meyer angle*. The numerical value of $\nu(M)$ is tabulated in Appendix C. For $\gamma = 1.4$, this result is shown graphically in Fig. 4.1.

The change in Mach number through an infinitesimal turning angle is found by combining Eqns. (4.46) and (4.47b):

$$\frac{dM}{M} = \left(1+\frac{\gamma-1}{2}M^2\right)\frac{d\delta}{\sqrt{M^2-1}} \tag{4.48a}$$

This is then used to obtain

$$d\delta = \frac{-\sqrt{M^2-1}}{1+\frac{\gamma-1}{2}M^2}\frac{dM}{M} = d\nu$$

followed by integration, which yields

$$\nu - \delta = R \tag{4.48b}$$

where R is a constant. Therefore $\nu_2 = \nu_1 + (\delta_2 - \delta_1)$ and $\delta_2 > \delta_1 > 0$ suggest that $\nu_2 > \nu_1$ and $M_2 > M_1$. Let $\Delta\nu_{1-2} \equiv (\delta_2 - \delta_1)$ denote the turning angle. Then

$$\nu_2 = \nu_1 + \Delta\nu_{1-2} \tag{4.48c}$$

where $\Delta\nu_{1-2} > 0$ results in Prandtl-Meyer expansion and $\Delta\nu_{1-2} < 0$, oblique shock, consistent with the sign convention and results for oblique shock waves in Section 4.6. The use of $\Delta\nu_{1-2}$ in Eqn. (4.48c) indicates that the turning angle $\Delta\nu_{1-2} > 0$ directly contributes to the Prandtl-Meyer angle for isentropic expansion. $\Delta\nu_{1-2} < 0$ may be used to cause isentropic compression in the near-wall region of a gradually concaved surface such as the one depicted in Fig. 4.14c.

4.10 Expansion in Supersonic Flow by Turning

The Prandtl-Meyer function shown in Eqn. (4.47f) has a zero value $\nu = 0$ for $M = 1$. It cannot be defined for subsonic flows, and expansion by turning in subsonic flows is not possible. It can be seen from the table in Appendix C that as the turning angle away from the oncoming flow increases leads to $dV > 0$, which in turn leads to $dM > 0$ (i.e., $M_2 > M_1$). Such isentropic acceleration results in a decrease in pressure and thus flow expansion ($p_2 < p_1$, according to Eqn. (4.16a)). Due to the isentropic nature of the process, both p_t and T_t remain unchanged through the turning process:

$$p_{t2} = p_{t1}$$

$$T_{t2} = T_{t1}$$

Due to the increase in supersonic value of M through expansion, μ decreases correspondingly and causes the Mach line to tilt, in addition to turning, in the downstream direction. Therefore, the waves diverge in the direction away from the turning corner, as shown in Fig. 4.15, decreasing possible gradients and further ensuring that the expansion is isentropic.

When dealing with supersonic expansion, the flow can be viewed as if it were an $M = 1$ flow that has undergone the Prandtl-Meyer expansion through an angle ν corresponding to the Mach number. Then the angle of the corner $\Delta\nu$ contributes to further turning. Therefore, by treating $\nu + \Delta\nu$ as the total turning from the imaginary $M = 1$ flow, conditions after the expansion can easily be found. Figure 4.15b graphically demonstrates such an idea. The following example illustrates the calculation procedure.

EXAMPLE 4.11 An air flow at $M = 4$, $T = 300\ K$, and $p = 101\ kPa$ undergoes Prandtl-Meyer expansion by turning around a $10°$ convex corner. Find the flow speed and pressure both before and after the expansion.

Solution –
Because $\nu = 0$ for $M = 1$, the flow with $M_1 = 4$ can be imagined as having turned through an angle $\nu_1 = 65.785°$ (the value can be found in Appendix C). Now $\nu_2 = \nu_1 + \Delta\nu = 65.785° + 10° = 75.785°$. Using the table in Appendix C again,

$$M_2 \approx 4.88.$$

(For comparison, Example 4.9 illustrates that an $M_1 = 4$ flow decelerates to $M_2 = 3.233$ by turning around a $10°$ concave corner. Isentropic relations are used to find pressures and temperatures.)

$$\frac{p_2}{p_1} = \frac{p_2}{p_{t2}}\frac{p_{t2}}{p_{t1}}\frac{p_{t1}}{p_1} = 0.0022 \times 1 \times \frac{1}{0.0066} = \frac{1}{3}$$

$p_2 = 33.367\ kPa$ ($p_2 < p_1$ – expansion results in a lower pressure)

$$\frac{T_2}{T_1} = \frac{T_2}{T_{t2}}\frac{T_{t2}}{T_{t1}}\frac{T_{t1}}{T_1} = 0.1735 \times 1 \times \frac{1}{0.2381} = 0.729;\ T_2 = 218.61°$$

$$V_1 = M_1\sqrt{\gamma R T_1} = 4.0 \times \sqrt{1.4 \times 287 \times 300} = 1{,}388.8\ m/s$$

$$V_2 = M_2\sqrt{\gamma R T_2} = 4.88 \times \sqrt{1.4 \times 287 \times 218.61} = 1,446.3 \ m/s$$

$(V_2 > V_1 -$ flow acceleration due to supersonic expansion) $\qquad\qquad \square$

Similar to isentropic compression by turning, the lack of a length scale in the region near the corner requires that fluid and flow properties must be constant along the Mach line. Since these Mach lines diverge from the corner while undergoing supersonic expansion, the first and the last Mach lines define the extent of the *expansion fan*, wherein the flow and fluid properties undergo changes from uniform flow in region 1 to region 2. As depicted in Fig. 4.15a, the fan angle is

$$\phi_{1-2} = \Delta\nu_{1-2} + \sin^{-1}\frac{1}{M_1} - \sin^{-1}\frac{1}{M_2} = \Delta\nu_{1-2} + \mu_1 - \mu_2 \qquad (4.49)$$

where the subscript "1.2" denotes from regions 1 to 2 and $\Delta\nu_{1-2}$ is the corresponding turning angle.

EXAMPLE 4.12 Find the expansion fan angle for the flow configuration in Example 4.11.

Solution –

$$\phi_{1-2} = \Delta\nu_{1-2} + \sin^{-1}\frac{1}{M_1} - \sin^{-1}\frac{1}{M_2} = 10^\circ + 14.478^\circ - 11.825^\circ = 12.653^\circ$$

Comments –

The fan angle is always greater than the turning angle because in addition to the turning, the Mach line is further inclined in the flow direction due to expansion and acceleration. $\qquad\qquad \square$

Although the sharp turn and expansion depicted in Fig. 4.15a are not thermo-dynamically possible, their representation by a series of gradual turns is not unique. The development of the Prandtl-Meyer function does not assume a known degree of gradualness. Therefore, when dealing with supersonic expansion, one may treat the total turning angle as the only relevant parameter, as established by Eqn. (4.47f), while ignoring the effect of gradualness. The following example illustrates that the effect of one larger turning angle is the same as that achieved by two (which can be extended to many) small turnings with the same total turning.

EXAMPLE 4.13 For the flow in Example 4.11, instead of expanding through a single 10° convex corner, allow the expansion to occur at first around a 6° convex corner, followed by a 4° convex corner. Find the final Mach number and the expansion fan angles emanating from these corners.

Solution –

Again $M_1 = 4$ and $\nu_1 = 65.785°$. $\nu_2 = \nu_1 + \Delta\nu_{1-2} = 65.785° + 6° = 71.785°$. Therefore, $M_2 \approx 4.50$. $\nu_3 = \nu_2 + \Delta\nu_{2-3} = 71.785° + 4° = 75.785$. Therefore, $M_3 \approx 4.88$.

$$\phi_{1-2} = \Delta\nu_{1-2} + \sin^{-1}\frac{1}{M_1} - \sin^{-1}\frac{1}{M_2} = 6° + 14.478° - 12.840° = 7.638°$$

$$\phi_{2-3} = \Delta\nu_{2-3} + \sin^{-1}\frac{1}{M_2} - \sin^{-1}\frac{1}{M_3} = 4° + 12.840° - 11.825° = 5.015°$$

Comments –

(1) Similar results for the Mach number cannot be obtained for turning around concave angles, as demonstrated by Example 4.9, where oblique shocks are not isentropic processes. Even with a very gradually curved concave corner, the Mach lines converge in the region away from the corner and the resultant shock wave causes a non-uniform flow downstream of the last Mach line (Section 8.7 describes how a non-uniform entropy field generates vorticity even though the incoming flow is uniform and free of vorticity). Prandtl-Meyer expansion is isentropic throughout the whole flow field. While the flow experiences changes in the fan region, the entropy remains the same everywhere in the flow, in both near and far fields. The total increase in Mach numbers is the sum of the increases due to expansion around two consecutive corners with smaller angles with the same total angle of turning. (2) The fan angles add up to be the same as that for a single $10°$ turn, although they do not have the same corner as their vertices. □

A curious question is: How large a turning angle of a convex corner is needed to accelerate the flow from $M = 1$ to ∞? Applying these two Mach numbers as the lower and the upper limits to Eqn. (4.47c) yields the turning angle needed, ν_{max}:

$$\nu_{max} = \frac{\pi}{2}\left(\sqrt{\frac{\gamma+1}{\gamma-1}} - 1\right) \tag{4.50}$$

For $\gamma = 1.4, \nu_{max} = 134.454°$. Since the Prandtl-Meyer expansion is isentropic, the isentropic relation, Eqn. (4.50), suggests that a supersonic flow into a vacuum (as in the case of a rocket exhaust in the outer space) would lead to a maximum flow turning for a rocket nozzle exit (use Fig. 4.9e as an example), Mach number $M_e = 4$ (i.e., $\nu_e = 65.785°$), $\Delta\nu = (\nu_{max} - \nu_e) = 68.669°$; that is, the exit flow turns through an angle $68.669°$ away from the nozzle axis.

4.11 Simple and Non-Simple Regions

The Prandtl-Meyer function for isentropic compression and expansion waves discussed so far can be summarized as

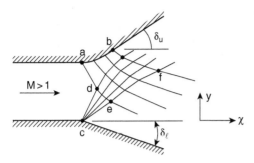

Figure 4.16 Schematics illustrating characteristics and wave regions. (a) Simple wave regions are: a-d-f-g, c-e-d, downstream region of b-g-f, and downstream region of c-e-f. The non-simple wave region is where the two different families of characteristics intersect.

Non-simple wave region: d–e–f–g

Simple wave regions: d–e–f–g, c–e–d
downstream of b–g–f
downstream of c–e–f

$$\nu = \nu_1 \pm \delta \tag{4.51}$$

where the $+$ sign indicates the turning is *away* from the flow and thus an expansion, whereas the $-$ sign indicates the turning is *into* the flow and thus a compression. Each of these waves (compression and expansion) are called *simple waves*, characterized by straight Mach lines with constant fluid and flow properties and by the simple relation, Eqn. (4.51), between δ and ν.

Waves generated by adjacent walls may intersect (or interact), as illustrated in Fig. 4.16. These simple waves belong to one of the two families ($+$ or $-$) depending on whether they emanate from walls that lie to the left (for the "$+$" family) or right (for the "$-$" family) of the flow. In literature, these waves are also called the characteristic lines or simply characteristics. The $-/+$ characteristic is also called the η/ξ characteristic or the left-/right-running characteristic. The angle between the characteristic and the flow (i.e., streamline) is equal to the Mach angle, $\mu = \sin^{-1}(1/M)$, where M is the local Mach number. For the right-running characteristics, emanating from the upper wall in Fig. 4.16, the turning angle (away from the flow and thus resulting in expansion) and the Prandtl-Meyer angle are related by Eqn. (4.48b):

$$\nu + \delta_u = Q \text{ (right-running or } + \text{ characteristics)} \tag{4.52a}$$

Similarly along a left-running characteristic, a similar turning angle (away from the flow) results in compression

$$\nu + \delta_l = R \text{ (left-running or } - \text{ characteristics)} \tag{4.52b}$$

where, in general $\delta_u \neq \delta_l$ and therefore $Q \neq R$. In Eqn. (4.52) R and Q are different constants. For the supersonic flow over a thin airfoil shown in Fig 4.17, only left-running characteristics exist as the airfoil surface itself is a streamline. In the region where *simple waves* of the two opposite families interact, the flow is not simple and the relation between δ and ν given by Eqn. (4.51) does not hold. As illustrated in Fig. 4.16, the characteristics in the non-simple region are curved and the local values

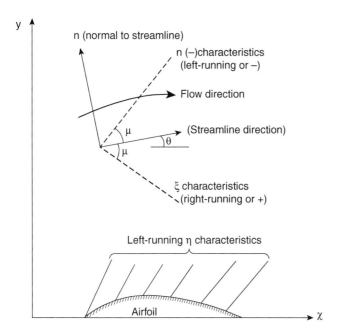

Figure 4.17 Flow region bounded by one wall surface generates only simple wave regions.

of ν and δ (hence the Mach number and velocity) at intersections of the left- and right-running characteristics can be determined by solving the simultaneous equations (4.52a) and (4.52b). The flow field is thus determined by the method of characteristics (a similar method for one-dimensional unsteady flow will be discussed in Chapter 10).

EXAMPLE 4.14 Use Eqn. (4.48) to calculate the Mach numbers after the turning around the corner described in Example 4.9.

Solution –
Equation (4.48) is rewritten for approximation as

$$\frac{\Delta M}{M} = -\left(1 + \frac{\gamma - 1}{2} M^2\right) \frac{\Delta \delta}{\sqrt{M^2 - 1}}$$

For $M_1 = 4$ and a single turn with $\Delta \delta = 10° = 0.1745$, $\Delta M_1 = -0.757$ and $M_2 = M_1 + \Delta M_1 = 3.243$ (compared with the value in Example 4.9, $M_2 = 3.229$, which can be considered as the "exact" value for no approximation was made.)

With two turns, with each of $\Delta \delta = 5° = 0.0873$, $\Delta M_1 = -0.379$, and $M_2 = M_1 + \Delta M_1 = 3.621$.

$\Delta M_2 = -0.329$ and $M_3 = M_2 + \Delta M_2 = 3.292$ (compared with $M_3 \approx 3.36$ in Example 4.9)   □

EXAMPLE 4.15 Repeat finding Mach numbers after the expansion processes in Examples 4.11 and 4.13 using Prandtl-Meyer relations.

Solution –

Again, use

$$\frac{\Delta M}{M} = +\left(1 + \frac{\gamma - 1}{2} M^2\right) \frac{\Delta \delta}{\sqrt{M^2 - 1}}$$

$\Delta \delta = +10° = +0.1745$, $\Delta M_1 = 0.757$ and $M_2 = M_1 + \Delta M_1 = 4.757$ ($M_2 \approx 4.88$ in Example 4.11)

For expansion by turning $\Delta \delta = 6° = 0.105$ followed by $\Delta \delta = 4° = 0.070$, $\Delta M_1 = 0.455$ and $M_2 = M_1 + \Delta M_1 = 4.455$ ($M_2 \approx 4.50$ in Example 4.12)

Following the turning with $\Delta \delta = 4° = 0.070$, $\Delta M_2 = 0.357$ and $M_3 = M_2 + \Delta M_2 = 4.812$ ($M_2 \approx 4.88$ in Examples 4.11 and 4.13).   □

4.12 Reflected Shock and Expansion Waves

So far the deflection of supersonic flow by a wall/surface is either accompanied by an oblique shock wave emanating from a concave corner or through an expansion wave emanating from a convex corner. In this section, reflection of waves from the wall and interaction of these waves are of concern.

Figure 4.18a illustrates the phenomenon of reflection of an oblique shock wave, created by a frictionless wedge surface AB, which impinges on another surface CDE. The surface CDE is parallel to the free-stream flow (Region 1, with Mach number equal to M_1). The wedge in inclined at an angle δ_1 to the incident flow and an oblique shock is, for the present discussion, attached to the vertex of the wedge. The oblique shock wave, as shown by the line AD, reaches the inner surface of the supersonic inlet at point D. Downstream of the oblique shock wave AD (Region 2), the Mach number is M_2. It is assumed that $M_2 > 1$ for this discussion ($M_2 < 1$ is possible as Fig. 4.12 shows). The flow in Region 2 makes an angle $\delta_2 (= \delta_1)$ with the surface CDE and, upon its impingement on CDE, the flow has to turn to be parallel to the surface. This means that the surface CDE acts as a wedge to the incident flow from Region 2 and an oblique shock wave (again, assumed to be attached) is formed emanating from point D. One expects that $M_1 > M_2 > M_3$, $p_{t1} > p_{t2} > p_{t3}$, and $p_1 < p_2 < p_3$ as a consequence of non-isentropic compression by the original and the reflected oblique shocks. When the incident Mach number and the deflection angle permit formation of an attached reflected shock, such a reflection is called a *regular reflection*. The *strength of reflection* is defined by $p_3/p_1 = (p_3/p_2)(p_2/p_1)$, which is calculated once M_{1n} and M_{2n} become known.

The reflected shock can be cancelled if it is not desirable, by turning the portion of the wall DE, shown in Fig. 4.13a, by an angle δ_1 away from the incident flow in Region 2. By doing this as shown in Fig. 4.18b, *wave cancellation* or *neutralization* is achieved.

If the combination of the incident oblique shock wave and the associated deflected flow angle (from Region 2 in Fig. 4.18a, for example) does not permit an attached oblique shock wave to form at the surface, the shock wave becomes normal

(a)

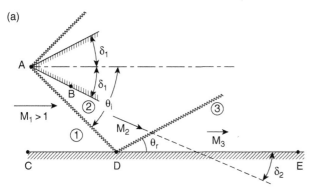

(b)

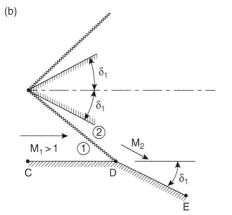

Figure 4.18 Reflection and cancellation of oblique shocks (parts a and b) and Mach reflection (part c, where the shock DG is the Mach shock wave).

(c)

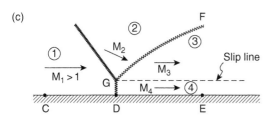

to the wall at the wall and is curved away from the surface to be tangent to the incident shock wave, as illustrated in Fig. 4.18c. Such reflections are called *Mach reflections* and the shock wave DG is called a *Mach shock wave*. The Mach shock wave occurs because the combination of the incident Mach and the wedge angle does not allow an attached oblique shock wave at point D in Fig. 4.18c. In this case, the surface CDE acts as a blunt-nose body to the incident flow with M_2. The shock wave GF resembles that shown in Fig. 4.5, a bow shock. The shock strengths of DG and GF are therefore different, with $M_3 > 1$ (downstream of the Mach shock wave DG) while $M_4 < 1$ (behind the bow shock GF), and the discontinuities of flow properties exist across the interface of Regions 3 and 4. The interface is thus a *slip line*. In many occasions, such sharp discontinuities do not exist because of the varying curvature of the shock wave. Instead, a layer of finite thickness results across which some characteristic change in properties occurs. Because entropy is one of the properties that changes over the thickness, the layer is sometimes called the *entropy layer*. Because the Mach shock wave is not exactly normal to the wall in the region near point G, the flow in Regions 3 and 4 (in the entropy layer) in Fig. 4.18c is more complicated than that in regular reflection (e.g., Region 3 in Fig. 4.18c). The following example illustrates the simpler analysis that can be done for a regular reflection.

EXAMPLE 4.16 Consider the flow depicted in Fig. 4.18a with for $M_1 = 4.0$ with $T_1 = 216.7\ K$ and $p_1 = 19.1\ kPa$ and $\delta_1 = 10^\circ$. Find the air speed, loss in stagnation pressure, shock strength of the reflection, and the reflected shock angle.

Solution –
From Fig. 4.12, for $M_1 = 4.0$ with $T_1 = 216.7\ K$ and $p_1 = 19.1\ kPa$ and $\delta_1 = 10^\circ$, $\theta_i \approx 23^\circ$. Therefore, $M_{1n} = M_1 \sin 23^\circ \approx 1.56$. Therefore, using shock table or Eqn. (4.18) yields

$$\frac{p_2}{p_1} = 2.673 \text{ and } \beta = \frac{p_2 - p_1}{p_1} = 1.673$$

$$\frac{p_{t2}}{p_{t1}} = 0.910 \text{ or } \frac{p_{t2} - p_{t1}}{p_{t1}} = 9\%$$

$$\frac{T_2}{T_1} = 1.361$$

$$M_{2n} = 0.681$$

$$M_{2t} = M_{1t}\sqrt{\frac{T_1}{T_2}} = M_1 \cos\theta_1 \sqrt{\frac{T_1}{T_2}} = 4 \cos 23^\circ \sqrt{\frac{1}{1.361}} = 3.156$$

Therefore $M_2 = \sqrt{M_{2t}^2 + M_{2n}^2} = 3.229$

So far the solution is the same as that shown in part (1) of Example 4.9.

For the flow to be parallel to the lower wall, it must be turned through an angle of $\delta_2 = 10^\circ$. By again using Fig 4.12 for $M_2 = 3.229$ and $\delta_2 = 10^\circ$, $\theta_r \approx 25.5^\circ$. Therefore, $M_{2n} = M_2 \sin 25.5^\circ \approx 1.39$. Using the shock table or Eqn. (4.18), one finds

$$\frac{p_3}{p_2} \approx 2.088 \text{ and } \beta = \frac{p_3 - p_2}{p_2} \approx 1.088$$

$$\frac{p_{t3}}{p_{t2}} \approx 0.961 \text{ or } \frac{p_{t3} - p_{t2}}{p_{t2}} \approx 3.9\%$$

$$\frac{T_3}{T_2} \approx 1.248$$

$$M_{3n} \approx 0.681$$

$$M_{3t} = M_{2t}\sqrt{\frac{T_2}{T_3}} = M_2\cos\theta_r\sqrt{\frac{T_2}{T_3}} = 3.229 \times \cos 25.5°\sqrt{\frac{1}{1.248}} \approx 2.609$$

Therefore $M_3 = \sqrt{M_{3t}^2 + M_{3n}^2} \approx 2.696$

$$\frac{p_{t3}}{p_{t1}} = \frac{p_{t3}}{p_{t2}} \times \frac{p_{t2}}{p_{t1}} \approx 0.961 \times 0.910 = 0.875$$

The stagnation pressure loss due to the incident and the reflected shocks is 12.5%, which translates into

$$\Delta p_t = p_{t3} - p_{t1} = p_{t1} \times \left(\frac{p_{t3}}{p_{t1}} - 1\right) = p_1 \times \frac{p_{t1}}{p_1} \times \left(\frac{p_{t3}}{p_{t1}} - 1\right)$$

$$= 19.1 \times \frac{1}{0.0066} \times (0.875 - 1) - 361.7 \ kPa$$

$$\Delta p_t = -361.7 \ kPa$$

$$T_3 = \frac{T_3}{T_2} \times \frac{T_2}{T_1} \times T_1 = 1.248 \times 1.361 \times 216.7 \ K = 368.1 \ K$$

The speed in Region 3 is $V_3 = M_3\sqrt{\gamma R T_3} = 2.696 \times \sqrt{1.4 \times 287 \times 368.1} = 1,036.8 \ m/s$

The overall shock strength due to the reflection is $p_3/p_1 = (p_3/p_2)(p_2/p_1) = 2.088 \times 2.673 = 5.581$

The reflected shock angle is $\theta_r - \delta_2 \approx 15.5°$.

Comments –
The reflection is not specular, i.e., $(\theta_r - \delta_2) \neq \theta_i$; here $\theta_i > (\theta_r - \delta_2)$. This is mainly because $M_2 \neq M_1$ and that the wave angle depends on the Mach number and the deflection angle. □

4.13 Intersection of Shock Waves

Figure 4.19a shows the intersection (or interaction) between two oblique shocks of equal strength (that is, caused by the same wedge angle). Because of the symmetry, the central streamline plays the identical role as the flat wall CDE in Fig. 4.18a.

(a)

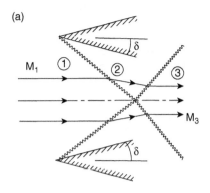

(b)

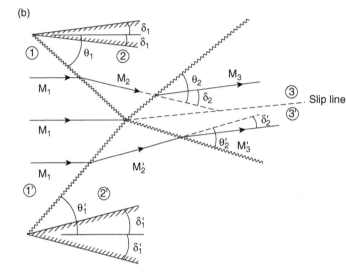

Figure 4.19 Interactions of two oblique shocks (a) with equal strength and (b) with unequal strengths. Note the slip line in the case of unequal strengths.

The analysis for flows in Regions 1 through 3 is the same as that for the flow in Fig. 4.18a.

The intersection of two shock waves of different strengths is shown in Fig. 4.19b. The deflection angles are $\delta_1 \neq \delta_1'$ and $M_2 \neq M_2'$. The final results, downstream of the reflected shocks, require that in Regions 3 and 3' (1) the pressures are equal, $p_3 = p_3'$, and (2) the flow directions are the same. Other properties (Mach number, temperature, speed, and entropy) are not expected to be the same in Regions 3 and 3', which are separated by the slip line (or the shear layer due to different flow speeds, or the entropy layer due to different values in shock strength). With the requirement of $p_3 = p_3'$, the final flow direction δ and p_3 and other properties can then be determined, as demonstrated in the following example.

EXAMPLE 4.17 Referring to Fig. 4.19b, consider air flow at $M_1 = 4$ and $\delta_1 = 5°$ and $\delta_1' = 10°$. Find $\theta_1, \theta_1', \theta_2, \theta_2', \delta$ (the final resultant turning angle) and all other properties in Regions 3 and 3'.

Solution –

For $M_1 = 4$ and $\delta_1 = 5°$, $\theta_1 \approx 18°$ from Fig. 4.12. Therefore, $M_{1n} = M_1 \sin 18°$ ≈ 1.24. Therefore, using shock table or Eqn. (4.20) yields

$$\frac{p_2}{p_1} = 1.6272 \text{ and } \beta = \frac{p_2 - p_1}{p_1} = 0.6272$$

$$\frac{p_{t2}}{p_{t1}} = 0.9884 \quad \frac{p_{t2} - p_{t1}}{p_{t1}} = 1.2\%$$

$$\frac{T_2}{T_1} = 1.1531$$

$$M_{2n} = 0.8183$$

$$M_{2t} = M_{1t}\sqrt{\frac{T_1}{T_2}} = M_1 \cos\theta_1 \sqrt{\frac{T_1}{T_2}} = 4 \cos 18°\sqrt{\frac{1}{1.1531}} = 3.543$$

Therefore, $M_2 = 3.64$.

$\delta_1' = 10°$ leads to $\theta_1' = 23°$, also from Fig. 4.12 (or Appendix E). Therefore, $M_{1n} = M_1 \sin 23° \approx 1.56$. Therefore, using the shock table or Eqn. (4.20) yields

$$\frac{p_2'}{p_1} = 2.673 \text{ and } \beta = \frac{p_2' - p_1}{p_1} = 1.673$$

$$\frac{p_{t2}'}{p_{t1}} = 0.910 \quad \frac{p_{t2}' - p_{t1}}{p_{t1}} = 9\%$$

$$\frac{T_2'}{T_1} = 1.361$$

$$M_{2n}' = 0.681$$

$$M_{2t}' = M_{1t}\sqrt{\frac{T_1}{T_2}} = M_1 \cos\theta_1' \sqrt{\frac{T_1}{T_2}} = 4 \cos 23°\sqrt{\frac{1}{1.361}} = 3.156$$

Therefore, $M_2' = 3.229$.

Because the flow in Regions $2'$ is turning through a larger angle upward than that in Regions 2 downward, it is reasonable to expect that the final flow is turned upward from the incident flow. By guessing the value of δ, p_3, and p_3' can be found. If their values are equal, then the guessed value of δ is the final flow deflection angle and other properties in Regions 3 and $3'$ can be determined.

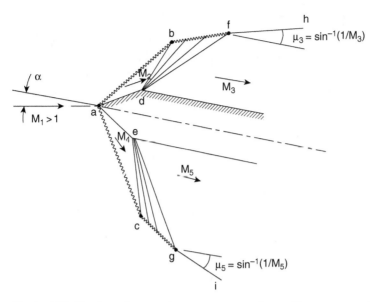

Figure 4.20 Shock- and expansion wave systems formed by a supersonic flow approaching a *finite* wedge followed by an afterbody.

First guess $\delta = 6°$, then $\delta_2 = \delta_1 + \delta = 5° + 6° = 11°$ for $M_2 = 3.64$, $\theta_2 \approx 23.0°$. Therefore, $M_{2n} = M_2 \sin 23.0° \approx 1.42$ and $\frac{p_3}{p_2} \approx 2.186$ and $\frac{p_3}{p_1} = \frac{p_3}{p_2} \times \frac{p_2}{p_1} \approx 2.186 \times 1.627 \approx 3.557$.

Then $\delta_2' = \delta_1' + \delta = 10° - 6° = 4°$ for $M_2' = 3.229$, $\theta_1' \approx 20.0°$. Therefore, $M_{2n}' = M_2' \sin 20° \approx 1.10$ and $\frac{p_3}{p_2} \approx 1.245$ and $\frac{p_3}{p_1} = \frac{p_3}{p_2} \times \frac{p_2}{p_1} \approx 1.245 \times 2.673 \approx 3.352$.

Therefore for $\delta = 6°$, $\frac{p_3}{p_1} > \frac{p_3}{p_1}$.

Guess $\delta = 5°$, then $\delta_2 = \delta_1 + \delta = 5° + 5° = 10°$ for $M_2 = 3.64$, $\theta_2 \approx 22.0°$. Therefore, $M_{2n} = M_2 \sin 22.0° \approx 1.36$ and $\frac{p_3}{p_2} \approx 1.991$ and $\frac{p_3}{p_1} = \frac{p_3}{p_2} \times \frac{p_2}{p_1} \approx 1.991 \times 1.627 \approx 3.239$.

For $\delta = 5°$, then $\delta_2' = \delta_1' + \delta = 10° - 5° = 5°$ for $M_2' = 3.229$, $\theta_2' \approx 20.0°$. Therefore, $M_{2n}' = M_2' \sin 21° \approx 1.16$ and $\frac{p_3}{p_2} \approx 1.403$ and $\frac{p_3}{p_1} = \frac{p_3}{p_2} \times \frac{p_2}{p_1} \approx 1.403 \times 2.673 \approx 3.750$.

Therefore for $\delta = 5°$, $\frac{p_3}{p_1} < \frac{p_3}{p_1}$. The solution must be $5° < \delta < 6°$.

One may assume the final solution is $\delta = 5.5°$. The best final value can be found by further iterations. Once the value of δ is found, the answers for M_3, M_3' and all other properties can then be found following similar procedures to those in Examples 4.9 and 4.16. □

4.14 Interaction of Shock- and Expansion Waves

In Section 4.6, the weakening of bow shock in the far field from the nose region of the wedge was discussed. The reason for weakening can be understood by considering a simpler attached oblique shock formed by a supersonic flow approaching a *finite* wedge followed by an afterbody, sketched in Fig. 4.20. Assume that the finite wedge was symmetrical in shape and is place at an angle of attachment, α, to the free stream. As illustrated in Fig. 4.20, two attached oblique shock waves emanate from the vertex of the wedge, resulting in uniform flows in Regions 2 and 3. Due to different deflection angles, the shock strengths are different and $M_2 \neq M_3$. If both M_2 and M_3 are > 1, the Prandtl-Meyer expansion will occur at corners c and d (i.e., the wedge-plate junctions).

Because of the qualitative similarity between the flow over the upper- and lower surfaces, it suffices to discuss the effect of the upper surface of the afterbody. These expansion fans emanating from point d accelerate the flow velocity and decrease the pressure in the flow region bdf. As a consequence, the shock wave inclines gradually from point b to point f and approaches the Mach wave angle $\mu_3 = \sin^{-1}(1/M_3)$. The curved shock wave ceg can be explained in a similar manner (Thompson, 1972).

4.15 Applications of Shock- and Expansion Waves

For a given free-stream Mach number $M_1 > 1$, the larger the deflection (or wedge) angle is, the larger the shock wave angle (as Fig. 4.12 or Eqn. (4.27) shows), accompanied by the higher shock strength β, which is given by

$$\beta = \frac{p_2 - p_1}{p_1} = \frac{2\gamma M_{1n}^2}{\gamma + 1} - \frac{\gamma - 1}{\gamma + 1} = \frac{2\gamma M_{1n}^2}{\gamma + 1} - \frac{2\gamma}{\gamma + 1} \tag{4.53}$$

where subscripts 1 and 2 designate the free-stream and the post-shock conditions. The drag on the wedge due to $(p_2 - p_1)$ is called *shock drag* or *supersonic wave drag*, in addition to and different from skin/viscous drag and flow separation drag. Therefore, to minimize shock drag, it is desirable to reduce the shock wave angle. For a given free-stream Mach number, this can be achieved by reducing wedge angle and/or by avoiding a blunt nose. As shown in Example 4.9, turning through a series of small deflection angles results in a smaller β than a single large deflection with the same total turning angle. From the thermodynamic point of view, as little entropy increase as possible is desirable, which is suggested by Eqn. (4.36).

Some examples illustrating supersonic wave drag are shown in Fig. 4.21. A diamond-shaped airfoil (symmetric with respect to its midsection as well as the chord line) at zero angle of attack, along with the surface pressure, is shown in Fig. 4.21a. Such an airfoil experiences drag force because the pressure in Region 2 is greater than that in Region 3, as the pressure distribution on the upper (and same on the lower) surface indicates that $p_2 > p_1$ due to the oblique shock and $p_3 < p_2$ due to the Prandtl-Meyer expansion around the convex corner at the midsection of the airfoil. In this case, the drag force per unit wing span on *either* the upper *or* the lower surface is

$$D = (p_2 - p_3)\left(\frac{c/2}{\cos\delta}\right)\sin\delta = (p_2 - p_3)\frac{c}{2}\tan\delta$$

(a)

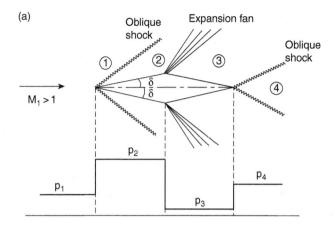

(b)

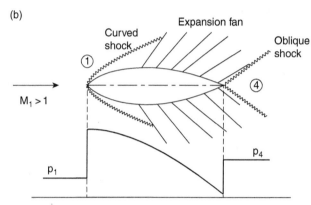

Figure 4.21 (a) and (b): Pressure distributions over symmetric airfoils in supersonic flows at zero angle of attack, explaining the non-zero supersonic wave drag. (c) Pressure distribution over a thin flat-plate at an angle of attack in a supersonic flow.

(c)

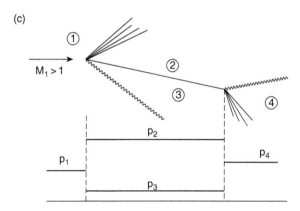

and the total drag is

$$D = (p_2 - p_3)c\tan\delta \qquad (4.54a)$$

or alternatively, when expressed in terms of the airfoil thickness (t)

$$D = (p_2 - p_3)t \qquad (4.54b)$$

where c is the chord length and t is the thickness at the midsection. This drag force exists in the absence of lift force, is independent of the viscous force, and is proportional to the midsection thickness, t. Equation (4.54) suggests that thin airfoils are desirable for minimizing drag in supersonic flows, the reasons being: (1) they result in small values of $(p_2 - p_3)$ because of smaller values of β, and (2) since $D \propto \tan \delta \approx \delta$, the thinner the airfoil the smaller δ and t, and thus D.

A symmetric and curved airfoil at zero angle of attack is shown in Fig. 4.21b. The supersonic wave drag on this airfoil would require integration of the surface pressures on both upper and lower surfaces, whose geometry affects local slopes (and turning angles) on the surfaces and thus the surface pressure. For the shock wave to be attached, the half angle at the leading and trailing edges should be less than θ_{max} (determined by Eqn. [4.27] or graphically by Fig. 4.12).

An extremely thin airfoil may be approximated as a thin flat-plate, as shown in Fig. 4.20c. It has neither lift nor supersonic wave drag at zero angle of attack ($\alpha = 0$). For $\alpha \neq 0$, it is clear that the lift and drag forces per unit wing span are, respectively, L and D:

$$L = (p_L - p_U)c \cos \alpha \tag{4.55a}$$

$$D = (p_L - p_U)c \sin \alpha \tag{4.55b}$$

where the subscripts L and U denote the lower and the upper surfaces, respectively. Here the angle of attack serves as the wedge angle (δ) for the flow along the lower surface and as the turning angle for Prandtl-Meyer expansion (ν_{1-U}) on the upper surface. Unlike the symmetry seen in Figs. 4.21a and 4.21b, a slip line emanates from the trailing of the flat-plate with $\alpha \neq 0$. Slip lines (or more diffused entropy layers) are also expected for symmetric airfoils with $\alpha \neq 0$.

For thin airfoils at a small angle of attack, isentropic compression and expansion theory can be used to relate lift and drag forces to the incoming Mach number and the angle of turning. For the flat-plate depicted in Fig. 4.21c, Eqn. (4.35) can be rewritten as

$$\frac{p - p_1}{p_1} = \pm \left(\frac{\gamma M_1^2}{\sqrt{M_1^2 - 1}} \right) \cdot \alpha \tag{4.56}$$

Where the + sign is associated with the lower surface that causes compression while the − sign, with the upper wall that causes expansion. It is convenient to express pressure in terms of the pressure coefficient (C_p):

$$C_p \equiv \frac{p - p_1}{\frac{1}{2}\rho_1 V_1^2} = \frac{p - p_1}{\frac{1}{2}\frac{\gamma p_1}{\gamma R T_1} V_1^2} = \frac{2}{\gamma M_1^2} \frac{p - p_1}{p_1} \tag{4.57}$$

where the positive and negative signs are, respectively for the lower and the upper surfaces. It is similarly convenient to obtain the lift and the drag coefficients:

$$C_L \equiv \frac{(p_L - p_U)c \cos \alpha}{\frac{1}{2}\rho_1 V_1^2 c} = (C_{pL} - C_{pU}) \cos \alpha \tag{4.58a}$$

$$C_D \equiv \frac{(p_L - p_U)c \sin \alpha}{\frac{1}{2}\rho_1 V_1^2 c} = (C_{pL} - C_{pU}) \sin \alpha \tag{4.58b}$$

Substituting Eqn. (4.57) yields for small angle α ($\cos \alpha \approx 1; \sin \alpha \approx \alpha$):

$$C_L = \frac{4\alpha}{\sqrt{M_1^2 - 1}} \tag{4.59a}$$

$$C_D = \frac{4\alpha^2}{\sqrt{M_1^2 - 1}} \tag{4.59b}$$

Returning to the diamond-shaped airfoil at $\alpha = 0$ in Fig. 4.21a, the pressure coefficients on the surfaces in Regions 2 and 3 are, respectively,

$$C_{p_{2,3}} = \frac{\pm 2\delta}{\sqrt{M_1^2 - 1}} \tag{4.59c}$$

where the "+" sign is for Region 2 (compression) and the "−" sign is for Region 3 (expansion). Therefore, with $t = c \tan \delta$

$$D = 2(p_2 - p_3)t = (p_2 - p_3)c \tan \delta = (C_{p2} - C_{p3}) \left(\frac{1}{2}\rho_1 V_1^2\right) c \tan \delta$$

$$\approx \frac{4\delta^2}{\sqrt{M_1^2 - 1}} \left(\frac{1}{2}\rho_1 V_1^2\right) c$$

Thus

$$C_D = \frac{D}{(\frac{1}{2}\rho_1 V_1^2)c} = \frac{4\delta^2}{\sqrt{M_1^2 - 1}} \approx \frac{4}{\sqrt{M_1^2 - 1}} \left(\frac{t}{c}\right)^2 \tag{4.60}$$

4.16 Thin Airfoil Theory

Combining Eqn. (4.59b) for a flat-plate and Eqn. (4.60) for a diamond-shaped symmetric airfoil, one obtains the drag coefficient for this airfoil with thickness-to-chord ratio of t/c placed at an angle of attack of α to the free stream. There exists one more contribution to drag if the airfoil is not symmetric with respective to its chord, that is, if the camber is not zero, as shown in Fig. 4.22. The thin airfoil theory is developed in the following to take care of the effect of camber. Assuming that the upper and the lower surfaces of a thin airfoil is described by $y_U = y_U(x)$ and $y_L = y_L(x)$, respectively. Then

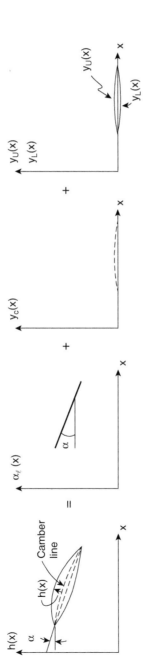

Figure 4.22 Decomposition of a thin, asymmetric airfoil geometry into a combination of three simpler geometric components: angle of attack, camber, and thickness of a symmetric airfoil. Because of the linear approximation used in the thin airfoil theory, the coefficients of pressure/lift/drag of the airfoil is the sum of the respective coefficients of the three components.

$$C_{pU} = \frac{2}{\sqrt{M_1^2 - 1}} \frac{dy_U}{dx} \tag{4.61a}$$

$$C_{pL} = \frac{2}{\sqrt{M_1^2 - 1}} \left(-\frac{dy_L}{dx} \right) \tag{4.61b}$$

Because the camber line consists of the midpoints between y_U and y_L, the profile can be decomposed into the camber line $y_c(x)$ and the symmetric thickness $t(x)$. Thus

$$\frac{dy_U}{dx} = \frac{d}{dx} [y_c + h(x)] = -\alpha_l(x) + \frac{dh}{dx} \tag{4.62a}$$

$$\frac{dy_L}{dx} = \frac{d}{dx} [y_c - h(x)] = -\alpha_l(x) - \frac{dh}{dx} \tag{4.62b}$$

where $\alpha_l(x) = \alpha + \alpha_c(x)$ and $\alpha_c(x)$ is the local angle of attack of the camber line. For lift and drag coefficients, the lift and drag forces need to be calculated first. Then

$$L = \frac{1}{2} \rho_1 V_1^2 \int_0^c (C_{pL} - C_{pU}) dx$$

$$D = \frac{1}{2} \rho_1 V_1^2 \int_0^c \left[C_{pL} \left(-\frac{dy_L}{dx} \right) + C_{pU} \left(\frac{dy_U}{dx} \right) \right] dx$$

Combining Eqns. (4.61) and (4.62) with these expressions yields

$$L = \frac{2 \left(\frac{1}{2} \rho_1 V_1^2 \right)}{\sqrt{M_1^2 - 1}} \int_0^c \left(-2 \frac{dy_c}{dx} \right) dx = \frac{4 \left(\frac{1}{2} \rho_1 V_1^2 \right)}{\sqrt{M_1^2 - 1}} \int_0^c \alpha_l(x) dx$$

$$D = \frac{2 \left(\frac{1}{2} \rho_1 V_1^2 \right)}{\sqrt{M_1^2 - 1}} \int_0^c \left[\left(\frac{dy_L}{dx} \right)^2 + \left(\frac{dy_U}{dx} \right)^2 \right] dx = \frac{4 \left(\frac{1}{2} \rho_1 V_1^2 \right)}{\sqrt{M_1^2 - 1}} \int_0^c \left[(\alpha_l(x))^2 + \left(\frac{dh}{dx} \right)^2 \right] dx$$

where the factor 4 accounts for the symmetry of the upper and lower surfaces. The average value over the chord length is defined as

$$\overline{(\cdot)} = \frac{1}{c} \int_0^c (\cdot) dx$$

Thus $\overline{(\alpha_c)} = 0$ by its definition and

$$\overline{(\alpha_l)} = \overline{(\alpha + \alpha_c)} = \overline{(\alpha)} + \overline{(\alpha_c)} = \alpha$$

$$\overline{(\alpha_l)^2} = \overline{(\alpha + \alpha_c)^2} = \overline{\alpha^2} + 2\overline{\alpha\alpha_c} + \overline{\alpha_c^2} = \alpha^2 + \overline{\alpha_c^2}$$

where $\overline{\alpha\alpha_c} = 0$ because α is given and $\overline{\alpha\alpha_c} = \alpha\overline{\alpha_c} = 0$. It follows that

$$C_L = \frac{L}{\frac{1}{2}\rho_1 V_1^2 c} = \frac{4}{\sqrt{M_1^2 - 1}} \frac{1}{c} \int_0^c \alpha_l(x)dx = \frac{4\alpha}{\sqrt{M_1^2 - 1}} \qquad (4.63\text{a})$$

$$C_D = \frac{D}{\frac{1}{2}\rho_1 V_1^2 c} = \frac{4}{\sqrt{M_1^2 - 1}} \frac{1}{c} \int_0^c \left[\left(\alpha_l(x)\right)^2 + \left(\frac{dh}{dx}\right)^2 \right] dx$$

$$= \frac{4}{\sqrt{M_1^2 - 1}} \left[\alpha^2 + \overline{\alpha_c^2} + \overline{\left(\frac{dh}{dx}\right)^2} \right] \qquad (4.63\text{b})$$

The following observations are made from Eqn. (4.63) for the thin airfoil theory.

1. The lift coefficient and the lift force are proportional to the angle of attack and do not depend on the camber and the thickness of the airfoil.
2. The drag coefficient and drag force have three contributing factors: First, the drag arises due to the lift, which is proportional to α^2. Second, camber contributes in proportion to $\overline{\alpha_c^2}$, which is always positive regardless of the sign of α_c. Third, the thickness also makes a positive contribution to the overall drag, with $\overline{\left(\frac{dh}{dx}\right)^2}$ being analogous to $\left(\frac{t}{c}\right)^2$ for the symmetric diamond-shaped airfoil. These three drag forces are in addition to frictional or shear drag forces due to fluid viscosity.
3. Both large camber and large thickness are to be avoided for drag reduction.

EXAMPLE 4.18 Consider the symmetric diamond airfoil depicted in Fig. 4.21a, with its upper front surface alignment parallel to the free stream and its half angle equal to $\delta = 5°$, as shown in the figure. Assume the free stream Mach number $M_1 = 4$ and $P_1 = 20\ kPa$ and a chord length of 0.6 m. Find C_L and C_D.

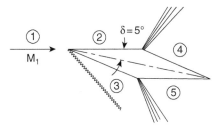

Solution –
One way of finding C_L and C_D is to first find L and D as described by Eqn. (4.57). There are now, however, four surfaces to consider. Region 2 in the figure is not disturbed and $p_2 = p_1$.

Region 3:

$M_1 = 4$ and $\delta_{13} = 10^\circ$ leads to $\theta_{13} \approx 23^\circ$ and $M_{1n} = M_1 \sin 23^\circ \approx 1.56$. Therefore, using the shock table or Eqn. (4.20) yields

$$\frac{p_3}{p_1} = 2.673$$

$$\frac{p_{t3}}{p_{t1}} = 0.910$$

$$\frac{T_3}{T_1} = 1.361$$

$$M_{3n} = 0.681$$

$$M_{3t} = M_{1t}\sqrt{\frac{T_1}{T_3}} = M_1 \cos\theta_{13}\sqrt{\frac{T_1}{T_3}} = 4\cos 23^\circ\sqrt{\frac{1}{1.361}} = 3.156$$

Therefore $M_3 = 3.229$ and the corresponding Prandtl-Meyer angle is $\nu_3 \approx 54^\circ$.

Region 4:

$$\nu_2 \approx 65.78^\circ \quad \nu_4 = \nu_2 + \Delta\nu_{24} = 65.78^\circ + 10^\circ = 75.78^\circ \text{ and } M_4 \approx 4.88,$$

$$\frac{p_4}{p_{t4}} = 0.0022$$

$$\frac{p_4}{p_1} = \frac{p_4}{p_{t4}} \times \frac{p_{t4}}{p_{t2}} \times \frac{p_{t2}}{p_{t1}} \times \frac{p_{t1}}{p_1} = (0.0022)(1)(1)\left(\frac{1}{0.0066}\right) = 0.333$$

Region 5:

$$\nu_5 = \nu_3 + \Delta\nu_{35} = 54^\circ + 10^\circ = 64^\circ. \quad \text{Thus} \quad M_5 \approx 3.87 \quad \text{and} \quad \frac{p_5}{p_{t5}} = 0.0079.$$

$$\frac{p_5}{p_1} = \frac{p_5}{p_{t5}} \times \frac{p_{t5}}{p_{t3}} \times \frac{p_{t3}}{p_{t1}} \times \frac{p_{t1}}{p_1} = (0.0079)(1)(0.910)\left(\frac{1}{0.0066}\right) = 1.089$$

Eqn. (4.57) states

$$C_p \equiv \frac{p - p_1}{\frac{1}{2}\rho_1 V_1^2} = \frac{p - p_1}{\frac{1}{2}\frac{\gamma p_1}{\gamma R T_1}V_1^2} = \frac{2}{\gamma M_1^2}\frac{p - p_1}{p_1}$$

Using Eqn. (4.58) by incorporating the angle of the four surfaces, each of which has a length of $c/2\cos\delta$, leads to

$$C_L = \frac{2}{\gamma M_1^2} \left[\left(\frac{p_3 - p_1}{p_1} \right) \frac{1}{2 \cos\delta} \cos2\delta + \left(\frac{p_5 - p_1}{p_1} \right) \frac{1}{2 \cos\delta} \cos0 \right.$$
$$\left. - \left(\frac{p_2 - p_1}{p_1} \right) \frac{1}{2 \cos\delta} \cos0 - \left(\frac{p_4 - p_1}{p_1} \right) \frac{1}{2 \cos\delta} \cos2\delta \right]$$

$$C_D = \frac{2}{\gamma M_1^2} \left[\left(\frac{p_3 - p_1}{p_1} \right) \frac{1}{2 \cos\delta} \sin2\delta + \left(\frac{p_5 - p_1}{p_1} \right) \frac{1}{2 \cos\delta} \sin0 \right.$$
$$\left. - \left(\frac{p_2 - p_1}{p_1} \right) \frac{1}{2 \cos\delta} \sin0 - \left(\frac{p_4 - p_1}{p_1} \right) \frac{1}{2 \cos\delta} \sin2\delta \right]$$

Because $2\delta = 10°$, $= 1.4$, $p_2 = p_1$, $\frac{p_3 - p_1}{p_1} = 1.673$, $\frac{p_5 - p_1}{p_1} = 0.089$, $\frac{p_4 - p_1}{p_1} = -0.667$ and $M_1 = 4$, $C_L = 0.1072$ and $C_D = 0.018$.

Now if the thin-airfoil theory is used, $\alpha = \delta = 5° = 0.0873$(radians), $\frac{dh}{dx} = \tan\alpha \approx \alpha$, $\overline{\alpha_c} = 0$ because of symmetry, and also $\overline{\alpha_c^2} = 0$ along the whole chord length.

$$C_L = \frac{4\alpha}{\sqrt{M_1^2 - 1}} = \frac{4 \times 0.0873}{\sqrt{4^2 - 1}} = 0.090$$

$$C_D = \frac{4}{\sqrt{M_1^2 - 1}} \left[\alpha^2 + \overline{\alpha_c^2} + \overline{\left(\frac{dh}{dx}\right)^2} \right] = \frac{4}{\sqrt{4^2 - 1}} [0.0873^2 + 0 + 0.0873^2] = 0.0158$$

$\square$

EXAMPLE 4.19 Consider a flat-plate that is at an angle of attack of $\alpha = 10°$ to a free stream of $M_1 = 4$. (a) Find C_L and C_D considering the shock- and expansion waves. (b) Repeat for $\alpha = 5°$. (c) For both angle attacks also use the thin airfoil theory.

Solution –

At the lower surface:

(a) $M_1 = 4$ and $\delta_{1L} = 10°$ leads to $\theta_L \approx 23°$ and $M_{1n} = M_1 \sin 23° \approx 1.56$. Therefore, using the shock table or Eqn. (4.20) yields

$$\frac{p_L}{p_1} = 2.673$$

At the upper surface:

$\nu_1 \approx 65.78° \nu_U = \nu_1 + \Delta\nu_{1U} = 65.78° + 10° = 75.78°$ and $M_U \approx 4.88 \frac{p_U}{p_{tU}} = 0.0022$

$$\frac{p_U}{p_1} = \frac{p_U}{p_{tU}} \times \frac{p_{tU}}{p_{t1}} \times \frac{p_{t1}}{p_1} = (0.0022)(1)\left(\frac{1}{0.0066}\right) = 0.333$$

Therefore,

$$C_L = \frac{2}{\gamma M_1^2}\left[\left(\frac{p_L - p_1}{p_1}\right)\cos\alpha - \left(\frac{p_U - p_1}{p_1}\right)\cos\alpha\right] = 0.206$$

$$C_D = \frac{2}{\gamma M_1^2}\left[\left(\frac{p_L - p_1}{p_1}\right)\sin\alpha - \left(\frac{p_U - p_1}{p_1}\right)\sin\alpha\right] = 0.036$$

Simple geometric considerations also yield $C_D = C_L \tan\alpha$.

(b) $M_1 = 4$ and $\delta_{1L} = 5°$ leads to $\theta_L \approx 18°$ and $M_{1n} = M_1\sin 18° \approx 1.24$. Therefore, using the shock table or Eqn. (4.20) yields

$$\frac{p_L}{p_1} = 1.6272 \text{ and } \beta = \frac{p_L - p_1}{p_1} = 0.6272$$

At the upper surface:

$$\nu_1 \approx 65.78° \quad \nu_U = \nu_1 + \Delta\nu_{1U} = 65.78° + 5° = 70.78° \text{ and } M_U \approx 4.41 \frac{p_U}{p_{tU}} = 0.0039$$

$$\frac{p_U}{p_1} = \frac{p_U}{p_{tU}}\times\frac{p_{tU}}{p_{t1}}\times\frac{p_{t1}}{p_1} = (0.0039)(1)\left(\frac{1}{0.0066}\right) = 0.591$$

Therefore,

$$C_L = \frac{2}{\gamma M_1^2}\left[\left(\frac{p_L - p_1}{p_1}\right)\cos\alpha - \left(\frac{p_U - p_1}{p_1}\right)\cos\alpha\right] = 0.092$$

$$C_D = \frac{2}{\gamma M_1^2}\left[\left(\frac{p_L - p_1}{p_1}\right)\sin\alpha - \left(\frac{p_U - p_1}{p_1}\right)\sin\alpha\right] = 0.008$$

The thin-airfoil theory yields

$$\delta = 10° = 0.1754:$$

$$C_L = \frac{4\alpha}{\sqrt{M_1^2 - 1}} = \frac{4\times 0.1754}{\sqrt{4^2 - 1}} = 0.180$$

$$C_D = \frac{4\alpha^2}{\sqrt{M_1^2 - 1}} = \frac{4\times 0.1745^2}{\sqrt{4^2 - 1}} = 0.0314$$

$$\delta = 5° = 0.0873:$$

$$C_L = \frac{4\alpha}{\sqrt{M_1^2 - 1}} = \frac{4 \times 0.0873}{\sqrt{4^2 - 1}} = 0.090$$

$$C_D = \frac{4\alpha^2}{\sqrt{M_1^2 - 1}} = \frac{4 \times 0.0873^2}{\sqrt{4^2 - 1}} = 0.0078$$

Comments –
Based on the results of Examples 4.18 and 4.19, the thin-airfoil appears to under-predict both C_L and C_D compared with the shock-expansion theory. As the angle of attack is reduced, the agreements between thin-airfoil predictions and shock-expansion results improve. □

4.17 Prandtl's Relation

A useful *Prandtl relation* is now derived to directly relate velocities across normal or oblique shock waves without needing to know the Mach numbers. Equations (4.10) through (4.11) constitute the governing equations for a normal shock wave. With some algebraic manipulation for a normal shock, one obtains

$$V_1 V_2 = a^{*2} \tag{4.64}$$

where the superscript * denotes the sonic condition. For an oblique shock wave the Prandtl relation becomes

$$V_{1n} V_{2n} = a^{*2} - \frac{\gamma - 1}{\gamma + 1} V_t^2 \tag{4.65}$$

It is noted that across an oblique shock the tangential velocity component does not change, that is, $V_{1t} = V_{2t} = V_t$. The derivation of Eqns. (4.64) and (4.65) is left as an exercise problem. When deriving Eqn. (6.45), attention should be paid to the fact that it is the normal component of the velocity that contributes the changes across the wave. The usefulness of Eqns. (4.64) and (4.65) can be seen in the following examples, and in a later chapter (Chapter 11) on traveling shock waves.

EXAMPLE 4.20 For the same condition given in Example 4.3, use the Prandtl relation to calculate the air velocity downstream of the normal shock.

Solution –
For $M_1 = 2.0$ $T_1 = 216$ K, $V_1 = \sqrt{\gamma R T_1} = 590.2$ m/s and

$$T^* = T_t / (\gamma + 1) = 2\, T\left(1 + \frac{\gamma - 1}{2} M_1^2\right)\left(\frac{2}{\gamma + 1}\right) = 324 \text{ K}.$$

$$a^* = 361.4 \text{ m/s}$$

$$V_2 = \frac{a^{*2}}{V_1} = 221.3 \text{ m/s}$$

Comments –

The result of V_2 is the same as in Example 4.3, where the approach is intuitive. □

EXAMPLE 4.21 Applying the Prandtl relation for the oblique shock problem of Example 4.6, find the downstream velocity and Mach number.

Solution –

In Example 4.6, the given conditions are: air at 30 °C and traveling at Mach 4.0, experiencing an oblique shock at an angle $\theta = 30°$.

Following results of Example 4.6, $a_1 = 349.53$ m/s

$M_{1n} = M_1 \sin\theta = 4 \sin30° = 2$, $V_{1n} = M_{1n}a_1 = 699.1$m/s, and $T_2/T_1 = 1.6875$

$M_{1t} = M_1 \cos\theta = 4 \cos30° = 2.667$, $V_t = V_{1t} = V_{2t} = M_{1t}a_1 = 1210.8$ m/s

$$a^* = \sqrt{\gamma R T^*} = \sqrt{\gamma R T_1 \left(1 + \frac{\gamma - 1}{2} M_1^2\right) \Big/ \left(\frac{\gamma + 1}{2}\right)} = 653.9 \text{ m/s}$$

Therefore, Eqn. (4.65) gives

$$(699.1\text{m/s}) V_{2n} = (653.9\text{m/s})^2 - \frac{1.4 - 1}{1.4 + 1}(1210.8 \text{ m/s})^2,$$

$V_{2n} = 262.1$ m/s, and

$$V_2 = \sqrt{V_{2n}^2 + V_{2t}^2} = 1,238.9 \text{ m/s}$$

$$M_2 = V_2/a_2 = V_2/\sqrt{\gamma \times R \times 1.6875 \times (T_2/T_1)}$$

$$= v_2/\sqrt{1.4 \times 288 \times 1.6875 \times 303} = 2.73$$

These values of V_2 and M_2 are the same as for those obtained in Example 4.6. However, the approach used in Example 4.6 is intuitive, using shock relations and without the need for manipulating conservation laws for the Prandtl relation. However, the Prandtl relation is easy to use, as this example demonstrates. □

4.18 The Rankine-Hugoniot Equation

The Prandtl relation directly relates gas velocities across shock waves without needing to know the Mach number. It is also desirable and advantageous to relate thermodynamic quantities across the shock bypassing the need for Mach number.

The following derivation leads to the useful Rankine-Hugoniot relation that directly relates thermodynamic quantities across the shock.

Recalling the conservation equations for mass and momentum in one dimension, one can rewrite them respectively as

$$\rho_1 V_1 = \rho_2 V_2 \tag{4.66}$$

and

$$\rho_1 V_1^2 + p_1 = \rho_2 V_2^2 + p_2 \tag{4.67}$$

The energy equation is

$$h_1 + \frac{V_1^2}{2} = h_2 + \frac{V_2^2}{2} \tag{4.69}$$

Noting that $h = c_p T$, Eqn. (4.68) becomes

$$V_1^2 - V_2^2 = 2c_p(T_2 - T_1) = 2c_p \left(\frac{p_2}{\rho_2 R} - \frac{p_1}{\rho_1 R} \right) = \frac{2\gamma}{\gamma - 1} \left(\frac{p_2}{\rho_2} - \frac{p_1}{\rho_1} \right) \tag{4.70}$$

By combining Eqn. (4.66) and Eqn. (4.67), one finds

$$V_1 - V_2 = \left(\frac{p_2}{\rho_2 V_2} - \frac{p_1}{\rho_1 V_1} \right)$$

Multiplying this expression by $(V_1 + V_2)$ yields

$$V_1^2 - V_2^2 = (p_2 - p_1) \left(\frac{1}{\rho_1} + \frac{1}{\rho_2} \right) \tag{4.71}$$

Equating the right-hand sides of Eqns. (4.70) and (4.71) leads to the useful *Rankine-Hugoniot relation* (also called the equation for a shock-adiabat curve for a perfect gas)

$$\frac{p_2}{p_1} = \frac{\left(\frac{\gamma+1}{\gamma-1} \right) - \frac{v_2}{v_1}}{\left(\frac{\gamma+1}{\gamma-1} \right) \frac{v_2}{v_1} - 1} \tag{4.72}$$

or

$$\frac{\rho_2}{\rho_1} = \frac{\left(\frac{\gamma+1}{\gamma-1} \right) \frac{p_2}{p_1} + 1}{\left(\frac{\gamma+1}{\gamma-1} \right) + \frac{p_2}{p_1}} \tag{4.73}$$

The derivation of Eqn. (4.72) is left as an exercise problem. Equation (4.73) is qualitatively sketched in Fig. 4.23. The upper branch of the Hugoniot curve (to the upper left of Point 1, the condition downstream of the shock wave) is represented by a solid line. For comparison, the isentropic $p - v$ relation is also presented, by a dashed line, due to the known (i.e., isentropic) thermodynamic

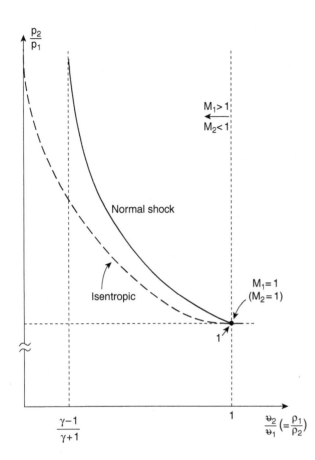

Figure 4.23 The Hugoniot curve (the shock adiabat) vs. the isentropic solution. The Hugoniot curve results from having to satisfy three conservation equations (mass, momentum, and energy), while the isentropic curve requires only conservation of energy and the isentropic condition.

path. It is noted that the key difference between the two curves in Fig. 4.23 is that the Hugoniot curve is the result of three conservation equations (mass, momentum, and energy), while the isentropic curve is obtained using only conservation of energy and the requirement that the process be isentropic. Because Eqns. (4.72) and (4.73) contain only thermodynamic quantities and they reduce to a function $p_2 = p_2(\textit{v}_2)$ or $p_2 = p_2(\rho_2)$, the function is called the *shock adiabat* (Shivamoggi, 1998; Thompson, 1972).

Note that the Rankine-Hugoniot relation is valid for expansion as well as for compression. For very strong normal shocks ($p_2/p_1 \to \infty$),

$$\lim_{p_2/p_1 \to \infty} \left(\frac{\rho_2}{\rho_1} \right) = \frac{\gamma + 1}{\gamma - 1}$$

EXAMPLE 4.22 For a pressure ratio of 4.5 across a normal shock, find the temperature ratio using the Rankine-Hugoniot relation. Assume $\gamma = 1.4$. Compare the result using the normal shock relationships or table.

Solution –

It is convenient to use Eqn. (4.72b) to find that $\frac{v_1}{v_2} = 2.667$. By using the ideal gas law,

$$\frac{T_2}{T_1} = \frac{p_2 \, v_2}{p_1 \, v_1} = 4.5 \times \frac{1}{2.667} = 1.6875$$

Alternatively,

$$\frac{T_2}{T_1} = \frac{p_2 \, v_2}{p_1 \, v_1} = \frac{p_2}{p_1} \frac{\left(\frac{\gamma+1}{\gamma-1}\right) + \frac{p_2}{p_1}}{\left(\frac{\gamma+1}{\gamma-1}\right)\frac{p_2}{p_1} + 1} = \frac{\left(\frac{\gamma+1}{\gamma-1}\right)\frac{p_2}{p_1} + \left(\frac{p_2}{p_1}\right)^2}{\left(\frac{\gamma+1}{\gamma-1}\right)\frac{p_2}{p_1} + 1} = \frac{6 \times 4.5 + 4.5^2}{6 \times 4.5 + 1} = 1.6875$$

By using the following normal shock relation (or table)

$$\frac{p_2}{p_1} = \frac{2\gamma M_1^2}{\gamma + 1} - \frac{\gamma - 1}{\gamma + 1} = 4.5$$

one finds

$$M_1 = 2.0$$

Then either by using the normal shock relation (or table in Appendix D), the temperature ratio across the shock is

$$\frac{T_2}{T_1} = 1.6875$$

□

EXAMPLE 4.23 Use Eqn. (4.72a) to find a Hugoniot relationship for pressure ratio (p_2/p_1) in terms of density ratio (ρ_2/ρ_1). Use the result to find the percentage increase in pressure for a 2% increase in density.

Solution –

Either by replacing v_2/v_1 with ρ_1/ρ_2 in Eqn. (4.72a) or rearranging Eqn. (4.72b) leads to

$$\frac{p_2}{p_1} = \frac{\left(\frac{\gamma+1}{\gamma-1}\right)\frac{\rho_2}{\rho_1} - 1}{\left(\frac{\gamma+1}{\gamma-1}\right) - \frac{\rho_2}{\rho_1}}$$

The percentage changes in density and pressure are, respectively, $(\rho_2 - \rho_1)/\rho_1$ and $(p_2 - p_1)/p_1$. Therefore,

$$\frac{p_2 - p_1}{p_1} = \frac{p_2}{p_1} - 1 = \frac{\left(\frac{\gamma+1}{\gamma-1}\right)\frac{\rho_2}{\rho_1} - 1}{\left(\frac{\gamma+1}{\gamma-1}\right) - \frac{\rho_2}{\rho_1}} - 1 = \frac{\left(\frac{\gamma+1}{\gamma-1}\right)\frac{\rho_2-\rho_1}{\rho_1} - 1 + \frac{\gamma+1}{\gamma-1}}{\left(\frac{\gamma+1}{\gamma-1}\right) - \frac{\rho_2-\rho_1}{\rho_1} - 1} - 1$$

For an x change in density,

$$\frac{p_2 - p_1}{p_1} = \frac{(1 + x)\left(\frac{\gamma+1}{\gamma-1}\right) - 1}{\left(\frac{2}{\gamma-1}\right) - x}$$

In air ($\gamma = 1.4$) and $x = 0.002$, $(p_2 - p_1)/p_1 = 0.00280 = 2.8\%$.   □

4.19 Jump Conditions

Because of the sharp gradients, the gas and flow properties are said to "jump" across the shock. It is useful to directly specify these jump conditions, by using $[\Psi] \equiv \Psi_2 - \Psi_1$ to denote the difference between the downstream (subscript 2) and upstream (subscript 1) values of the parameter Ψ; the jump conditions are frequently used in literature. The one-dimensional continuity equation, for example, in terms of jump conditions is $[\rho V] = 0$, while for a one-dimensional nozzle flow where the flow cross-sectional area varies (to be discussed in detail in the next chapter) the continuity is $[\rho V A] = 0$. It is left as an end-of-chapter exercise to rearrange Eqns. (4.22a), (4.22c), (4.22e) to arrive at the following jump conditions:

$$M_{2n}^2 = \frac{M_{1n}^2 + \frac{2}{\gamma-1}}{\frac{2\gamma}{\gamma-1}M_{1n}^2 - 1} \tag{4.74a}$$

$$\frac{[p]}{p_1} = \frac{[p_2 - p_1]}{p_1} = \beta = \frac{2\gamma M_{1n}^2}{\gamma + 1} - \frac{\gamma - 1}{\gamma + 1} \tag{4.74b}$$

$$\frac{[V]}{a_1} = \frac{V_{2n} - V_{1n}}{a_1} = -\frac{2}{\gamma + 1}\left(M_{1n} - \frac{1}{M_{1n}}\right) \tag{4.74c}$$

$$\frac{[v]}{v_1} = \frac{v_2 - v_1}{v_1} = -\frac{2}{\gamma + 1}\left(1 - \frac{1}{M_{1n}^2}\right) \tag{4.74d}$$

The Rankine-Hugoniot relation can also be rearranged in terms of jump conditions as

$$\frac{[v]}{v_1} = -\frac{2\frac{[p]}{p_1}}{2\gamma + (\gamma + 1)\frac{[p]}{p_1}} \tag{4.75}$$

One useful relationship relating jump conditions of pressure and velocity is found by combining Eqns. (4.74b) through (4.74d) and Eqn. (4.75); it is left as an exercise at the end of the chapter:

$$\frac{[p]}{p_1} = \frac{\gamma(\gamma+1)}{4}\left(\frac{[V]}{a_1}\right)^2\left[1+\sqrt{1+\left(\frac{4}{\gamma+1}\frac{a_1}{[V]}\right)^2}\right] \tag{4.76}$$

For normal shocks, results of Eqn. (4.74a) through (4.74d) are directly applicable by simply dropping the subscript n that denotes the normal component in case of oblique shocks.

4.20 Criterion for Strong Shocks

So far a strong shock is associated with large values of M_1 or M_{1n} or $\beta \gg 1$. Jump conditions should provide a more quantitative criterion. The criterion for strong and weak shocks in this section is adopted from Thompson (1972). Following the results of Problem 4.29, the non-dimensional pressure jump, denoted by Π, is defined as

$$\Pi \equiv \frac{[p]}{\rho_1 a_1^2} = \frac{[p]}{\gamma p_1} \tag{4.77}$$

which is equal to β/γ. Thus, the weak and strong shock approximation can be adopted, respectively, for $\Pi \ll 1$ and $\Pi \gg 1$.

4.21 Hodograph Diagrams for Shock- and Prandtl-Meyer Expansion Waves

For one-dimensional oblique shock waves, Fig. 4.12 provides graphical relations among the free stream Mach number (M_1), the turning angle (δ), and the shock wave angle (θ), which are visually easy to grasp. The graphical relations between upstream and downstream velocities can be achieved by plotting them on the *hodograph plane*, where the horizontal and vertical axes represent, respectively, the

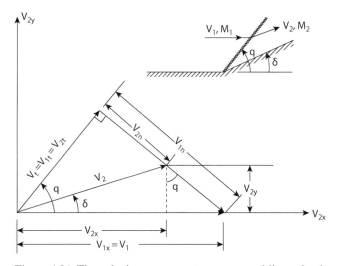

Figure 4.24 The velocity components across an oblique shock on the hodograph plane.

downstream x- and y-velocity components. To this end, an oblique-shock velocity diagram can be constructed and is shown in Fig. 4.24. Note that in Fig. 4.23, $V_t = V_{1t} = V_{2t}$, and $V_t \perp V_{1n}$ and V_{2n}. The geometry indicates that

$$\frac{V_t}{V_{1n}} = \frac{V_{2y}}{V_1 - V_{2x}} \tag{4.78a}$$

and

$$\frac{V_t}{V_1} = \frac{V_{2y}}{V_{1n} - V_{2n}} \tag{4.78b}$$

Combining with the energy equation, Eqn. (4.1), for a perfect gas and $(h_2 - h_1) = c_p(T_2 - T_1) = c_p(\gamma R T_2 - \gamma R T_1)/\gamma R = (a_2^2 - a_1^2)/(\gamma - 1)$, one arrives for normal shock waves

$$V_1^2 + \frac{a_1^2}{\gamma - 1} = V_2^2 + \frac{a_2^2}{\gamma - 1} = \frac{\gamma + 1}{\gamma - 1} a^{*2} \tag{4.79a}$$

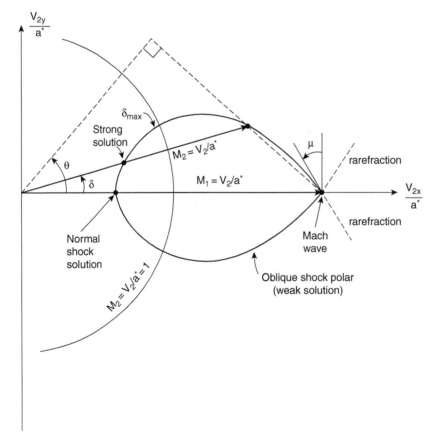

Figure 4.25 Equation (4.80) is plotted as shock polar (solid lines) that represents weak solutions of the oblique shock equation. The dashed lines represent the rarefaction branches of the shock polar. The rarefaction branch is shown in further details in Fig. 4.26.

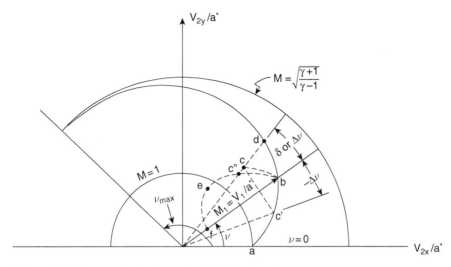

Figure 4.26 The rarefaction branch of the shock polar.

and when modified for oblique shocks

$$V_{1n}^2 + \frac{a_1^2}{\gamma - 1} = V_{2n}^2 + \frac{a_2^2}{\gamma - 1} = \frac{\gamma + 1}{\gamma - 1} a^{*2} - V_t^2 \tag{4.79b}$$

After combining Eqns. (4.78a), (4.78b) and (4.79b) and some manipulations, the equation of the shock polar is obtained as

$$\left(\frac{V_{2y}}{a^*}\right)^2 = \frac{\left(\frac{V_1}{a^*} - \frac{V_{2x}}{a^*}\right)^2 \left(\frac{V_1}{a^*}\frac{V_{2x}}{a^*} - 1\right)}{\frac{2}{\gamma + 1}\left(\frac{V_1}{a^*}\right)^2 - \left(\frac{V_1}{a^*}\frac{V_{2x}}{a^*} - 1\right)} \tag{4.80}$$

which relates V_{2y}/a^* to V_{2x}/a^* for the know value of V_1/a^*. Equation (4.80) is plotted and is qualitatively shown in Fig. 4.24.

For given M_2 and δ, there are two possible solutions: the weak solution and the strong solution. In Fig. 4.25, the shock polar represents weak solutions of the oblique shock equation, corresponding to the solid-line solutions. The Mach wave occurs when the supersonic flow encounters no obstacle (due to the absence of a two-dimensional wedge in this case). The symmetry about the horizontal axis simply means that the deflection could be either upward or downward turning.

Oblique shock results are represented by the curve to the left of the point of the Mach wave in Fig. 4.25. Recall that there are two different types of results downstream of oblique shock, represented by the strong and the weak solutions. The strong shock solution indicates subsonic downstream velocity ($M_2 = V_2/a^* < 1$), recalling that it is represented by dashed lines in Fig. 4.12. The weak solutions lead to mostly supersonic downstream conditions except for a narrow range of δ between the $M_2 = 1$ and the δ_{max} lines in Fig. 4.12, corresponding to the similarly portion of the shock polar between the $M_2 = 1$ circle and the point denoted by δ_{max}. The maximum turning angle (δ_{max}) for a given M_1 results in an M_2 value only slightly smaller than one, consistent with findings in Fig. 4.12. The normal shock solution is a special (and also the extreme) case on the shock polar, and is at the intersection of the shock polar and the V_{2x} axis, with no y-velocity component as expected.

The curves (represented by the dashed lines) to the right of the Mach wave are the rarefaction branches, due to Prandtl-Meyer expansion. While the shock polar correctly indicates that downstream velocity magnitude decreases as δ increases, the Prandtl-expansion branch indicates increasing velocity as δ increases. The details of the rarefaction branch is plotted in Fig. 4.26 using Eqn. (4.47c), given again below for quick reference:

$$\nu(M) = \int \frac{\sqrt{M^2 - 1}}{1 + \frac{\gamma - 1}{2} M^2} \frac{dM}{M} = \sqrt{\frac{\gamma + 1}{\gamma - 1}} \tan^{-1} \sqrt{\frac{\gamma - 1}{\gamma + 1}(M^2 - 1)} - \tan^{-1}\sqrt{M^2 - 1}$$

Figure 4.26 represents Mach number (M) as a function of the flow turning angle (δ and $\delta = \nu$) and demonstrates that the solutions for the rarefaction branch of the shock polar only exists for $M > 1$, as the solution lies between the concentric circles with radius equal to 1 and $\sqrt{(\gamma + 1)/(\gamma - 1)}$, the latter being the maximum Mach number achievable by Prandtl-Meyer expansion, and for $\nu \leq \nu_{max}$. The horizontal axis is for $\nu = 0$, on which the hodographic solution is $M = 1$ and the freestream velocity $V_1/a^* > 1$ shown in Fig. 4.25 is the point at the rarefaction branch with a corresponding Prandtl-Meyer angle ν. The dashed curve $b - c'' - e - f$ is the shock polar, same as that depicted in Fig. 4.25.

Consider a flow with $M_1 > 1$, denoted by point b on the rarefaction branch in Fig. 4.26. The flow slows down when ν is decreased, moving along the rarefaction branch toward the horizontal axis ($\nu = 0$) and toward $M = 1$ (point a). By a decrease equal to $-\Delta\nu$, the flow decelerates to point c', with a smaller M. Such deceleration is isentropic, implied by the assumption for Eqn. (4.47c) and is indeed supersonic compression by isentropic turning discussed in Section 4.8. Similarly, increasing ν from point b by $\Delta\nu$ results in the flow represented by point c, corresponding to Prandtl-Meyer expansion to a higher Mach number. The symmetry of the rarefaction branches in Fig. 4.24 makes the conditions of points c' and c identical. Turning the flow by a wedge angle of $\delta = \Delta\nu$ from point b, however, leads to the oblique shock solution represented by point c'' on the shock branch (curve $b - c'' - e - f$) in Fig. 4.26. Points c'', c, and d in Fig. 4.26 thus represent results of turning the a supersonic flow through an angle of $\Delta\nu$ by a two-dimensional wedge, by isentropic compression, and Prandtl-Meyer expansion, with the resultant Mach numbers of $M_{c''}$, M_c, and M_d. By now it is understood that $M_{c''} < M_c < M_d$ and this result is qualitatively shown in Fig. 4.26.

Problems

Problem 4.1 Derive Eqn. (4.15).

Problem 4.2 Derive Eqns. (4.16a), (4.16b), (4.16c), and (4.17b).

Problem 4.3 Derive Eqns. (4.18a) and (4.18b).

Problem 4.4 A converging-diverging nozzle made of unspecified materials and having unknown quality of surface finishing is attached to a reservoir of pressurized air. In the flow direction, three locations were selected for temperature and pressure measurements. The results are

Location 1 – $M_1 = 0.2$, $T_1 = 50°C$, and $p_1 = 3.00$ MPa

Location 2 – $M_2 = 0.7$, $T_2 = 20°C$, and $p_2 = 2.13$ MPa

Location 3 – $M_3 = 1.2$, $T_3 = 5°C$, and $p_3 = 1.20$ MPa

The ambient air has a temperature of $25°C$. For air, $c_p = 1.004$ kJ/kg·K and $R = 0.287$kJ/kg·K. Find T_{t1}, T_{t2}, p_{t1}, p_{t2}, the amounts of heat transfer to the air and entropy changes between locations 1 and 2 and between locations 1 and 3. Is the flow isentropic? Is the flow at all possible (or, in other words, could the measurement results be wrong)?

Problem 4.5 Show that for a very weak shock wave

$$M_{1n}^2 - 1 = 1 - M_{2n}^2$$

Consult the normal shock table to determine in what range of M_{1n} this approximation is satisfactory.

Problem 4.6 An air flow with a static temperature and pressure of 25 °C and 101.3 kPa, respectively, is moving at Mach 2.5 before passing a normal shock. Let subscripts 1 and 2 denote the states of air upstream and downstream of the shock. Find: (1) the air velocities upstream and downstream of the shock wave; (2) the pressure rise across the shock; (3) the pressure of air if it is slowed down isentropically to the same speed as with the shock; (4) the loss in stagnation pressure due to the shock.

Problem 4.7 A supersonic flow at Mach 2.5 and with a static temperature and pressure of 25 °C and 101.3 kPa, respectively, is to be slowed down to achieve a static pressure of 1,450 kPa using a supersonic diffuser. Determine the diffuser geometry (e.g., the proportions of inlet, throat, and the exit cross-sectional areas).

Problem 4.8 Use the Rankine-Hugoniot relation to show that $p_2/p_1 > 1$ for a shock wave.

Problem 4.9 Derive (4.22a)-(4.22f) for weak normal shock waves.

Problem 4.10 Derive Eqns. (4.33a-c) from Eqn. (4.32)

Problem 4.11 Designate the sonic condition, with the superscript *, as the reference condition for flows both down and upstream of a normal shock wave. By using the energy equation, Eqn. (4.6), show that

(1) $T_1^* = T_2^* = \frac{2T_t}{\gamma+1}$

(2) $\frac{V_1^2}{2} + \frac{a_1^2}{\gamma-1} = \frac{V_2^2}{2} + \frac{a_2^2}{\gamma-1} = \text{constant} = \frac{1}{2}\frac{\gamma+1}{\gamma-1}a^{*2}$ or $a^{*2} = \frac{\gamma-1}{\gamma+1}V_1^2 + \frac{2}{\gamma+1}a_1^2$

(3) For an oblique shock: $V_{1n}^2 + \frac{a_1^2}{\gamma-1} = V_{2n}^2 + \frac{a_2^2}{\gamma-1} = \frac{\gamma+1}{\gamma-1}a^{*2} - V_t^2$

Problem 4.12 (a) Using the energy equation, Eqn. (4.6), for a perfect gas flow across a normal shock, show that

$$V_1 V_2 = \frac{2}{\gamma+1}a_t^2$$

or

$$V_1 V_2 = a^{*2}$$

where subscripts 1 and 2 designate the upstream and the downstream regions, respectively. The above two relations are the *Prandtl relation*. This result makes readily clear that for $M_1^* = v_1/a^* > 1$ (corresponding to $M_1 > 1$, because $T^* > T_1$ and $a^* > a_1$), $M_2^* < 1$ (corresponding to $M_2 < 1$), as explained in Section 4.4. Then use the limiting case where $M_1 = M_2 = 1$ to show that

$$a^* = \sqrt{\frac{2}{\gamma+1}}a_t$$

which can also be alternatively shown in Problem 5.13.

   (b) Then show that for an oblique shock wave

$$V_{1n} V_{2n} = a^{*2} - \frac{\gamma-1}{\gamma+1}V_t^2$$

Problem 4.13 For the flow depicted in Fig. 4.17 find the reflected shock angle for $M_1 = 2.0$ and 1.6, with $\delta_1 = 10^\circ$ in both cases. Compare with the results of Example 4.16.

Problem 4.14 Show that for an isentropic flow of perfect gas

$$\frac{dp/p}{dT/T} = \frac{\gamma}{\gamma-1}.$$

That is, the rate of change in pressure is larger than that in temperature for a given Mach number as the flow accelerates or decelerates.

Problem 4.15 Use energy conservation to show that the sonic (or critical) velocity, $V_c (= a^*)$, and the maximum (or limiting) velocity, V_l, to obtain for an ideal gas in tank are, respectively,

$$V_c = a^* = \sqrt{\frac{2\gamma}{\gamma+1}\frac{p_t}{\rho_t}} = \sqrt{\frac{2}{\gamma+1}}a_t$$

and

$$V_l = \sqrt{\frac{2\gamma}{\gamma-1}\frac{p_t}{\rho_t}} = \sqrt{\frac{2}{\gamma-1}}a_t$$

Also show that at any point in an isentropic flow, the velocity V is

$$V = \sqrt{\frac{2\gamma}{\gamma-1}\frac{p_t}{\rho_t}\left[1-\left(\frac{p}{p_t}\right)^{\frac{\gamma-1}{\gamma}}\right]}$$

Problem 4.16 Show that across a stationary normal shock wave (with subscripts 1 and 2 denoting upstream and downstream states of the shock, respectively)

(1) $V_{1n}V_{2n} = \frac{p_2-p_1}{\rho_2-\rho_1}$

(2) $J^2 = -\frac{p_2-p_1}{v_2-v_1}$

(3) $\frac{p_2-p_1}{\rho_1 a_1^2} = -M_1\frac{V_2-V_1}{a_1}\left(\text{or} = -M_{1n}\frac{V_{2n}-V_{1n}}{a_1} = -M_{1n}^2\frac{v_2-v_1}{v_1}\right)$

(4) $(V_{2n} - V_{1n})^2 = -(p_2 - p_1)(v_2 - v_1)$

where $J = \rho_1 V_1 = \rho_2 V_2$ is the mass flux across the shock wave and M_{1n}, V_{1n}, and V_{2n} are used for an oblique shock; drop the subscript n for a normal shock.

Problem 4.17 Use the isentropic relationship

$$\frac{p_t}{p} = \left[1+\frac{\gamma-1}{2}M^2\right]^{\frac{\gamma}{\gamma-1}}$$

to show that

(1) At low Mach numbers the local static pressure can be expressed as

$$p = p_t - \frac{\rho V^2}{2}\left[1+\frac{1}{4}M^2+\frac{2-\gamma}{24}M^4+...\right]$$

(2) The velocity determined by Pitot measurement by neglecting the Mach number, or compressibility, effect, has an error of

$$\frac{1}{4(\gamma - 1)}\left[\frac{p_t}{p_\infty} - 1\right]^{\frac{\gamma - 1}{\gamma}}.$$

where p_∞ is the free-stream static pressure.

Problem 4.18 For a weak shock wave $(M_{1n}^2 - 1 = \epsilon \ll 1)$, find $(p_{t2} - p_{t1})/p_{t1}$ in terms of $(M_{1n}^2 - 1)$ and M_{2n} in terms of M_{1n}. Assume a constant specific heat ratio, γ.

Problem 4.19 Can the entropy change across a shock wave be written as the following expression?

$$\frac{s_2 - s_1}{R} = \frac{c_p}{R}\ln\frac{T_2}{T_1} - \ln\frac{p_2}{p_1} = \frac{c_p}{R}\ln\frac{T_{t2}}{T_{t1}} - \ln\frac{p_{t2}}{p_{t1}}$$

where subscripts 1 and 2 denote, respectively, upstream and downstream of the shock wave. Hint: Entropy is a thermodynamic property and there exists a stagnation entropy.

Problem 4.20 Combine continuity, momentum, and energy considerations to show that for a normal shock in a general fluid that is not an ideal gas:

$$h_2 - h_1 = \frac{1}{2}\frac{\rho_1 + \rho_2}{\rho_1\rho_2}(p_2 - p_1)$$

where subscripts 1 and 2 denote, respectively, upstream and downstream of the shock wave.

Problem 4.21 Find p_2/p_{t2} downstream of a normal shock wave in terms of M_1, the upstream Mach number. Also show that

$$\frac{p_1}{p_{t2}} = \frac{\left[\frac{2\gamma}{\gamma+1}M_1^2 - \frac{\gamma-1}{\gamma+1}\right]^{\frac{1}{\gamma-1}}}{\left[\frac{\gamma+1}{2}M_1^2\right]^{\frac{\gamma}{\gamma-1}}}$$

which is the *Rayleigh supersonic Pitot formula* that allows determining M_1 by measuring p_{t2} that is the reading by the Pitot probe in supersonic flow, unlike p_{t1} read by the subsonic Pitot probe.

Problem 4.22 A normal shock occurs standing immediately in front of a pitot-static probe placed facing and parallel to a supersonic air flow. Upstream of the shock the static pressure and temperature are measured to be $p_1 = 11.0$ kPa and $T_1 = 200$ K. Downstream of the shock the measured values are $p_2 = 54.8$ kPa and $p_{t2} = 68.1$ kPa. Find V_1. Demonstrate that $p_{t2} = 68.1$ kPa is a reasonable reading.

Problem 4.23 Derive the Rankine-Hugoniot relationship between pressure and specific volume, Eqn. (4.72a), and then show that

(1) it can be rewritten as Eqn. (4.72b), and

(2) for a very weak shock wave, it is reduced to the isentropic relationship

$$\frac{dp}{p} = \gamma\frac{d\rho}{\rho} \text{ or } \frac{p_2}{p_1} = \left(\frac{\rho_2}{\rho_1}\right)^{\gamma}$$

(3) For a weak shock in air that causes a 0.2% increase in density, what is the corresponding increase in pressure? Compare this result with that in Example 4.23.

Problem 4.24 A supersonic air flow ($M = 3.0$) is going through a centered expansion fan through an 8° turning and then returns to the original flow direction through an oblique shock, as here shown.

(1) Find the Mach number and pressure in each flow region shown in the figure using the shock/expansion theory, and

(2) Use the Prandtl-Meyer isentropic turning (for both expansion and compression) to find the Mach number and pressure in region 3; compare these results with those of (1).

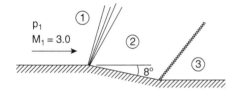

Problem 4.25 The same supersonic flow is going through the expansion and oblique shock in the reverse order as those in Problem 4.24, as here shown. Find the Mach number and pressure in regions 2 and 3, and compare with the results of Problem 4.24.

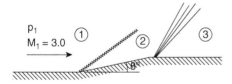

Problem 4.26 A thin flat-plate airfoil with a trailing-edge flap as shown is placed in a uniform supersonic air stream. Both angles of attack and the flap deflection are small (i.e., $\alpha \ll 1$ and $\varphi \ll 1$). Find the drag coefficient using the thin-airfoil theory.

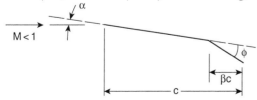

Problem 4.27 A symmetric diamond-shaped airfoil has a chord length c and a midsection thickness t. It is placed in a uniform supersonic air stream at Mach 3.0 and zero angle of attack. Find the drag for $t/c = 0.05$ using the thin-airfoil theory; this is the wave drag that does not exist in subsonic inviscid flows with a symmetric airfoil at zero angle of attack.

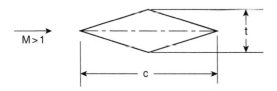

Problem 4.28 A thin airfoil with flat surfaces is placed in a uniform supersonic flow with a Mach number M, as shown in the figure. Assuming $2t/c \ll 1$, find the drag coefficient.

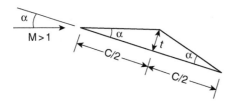

Problem 4.29 A thin flat-plate is placed in an air stream with $M = 1.6$, at an angle of attack of $8°$. Find the lift and drag coefficients.

Problem 4.30

A supersonic air flow ($p = 100$ kPa and $M = 2.0$) is deflected by a two-dimensional wedge through an angle of $10°$. Find the pressure and entropy change due to the deflection using (1) oblique shock theory and (2) isentropic turning. Compare the results.

Problem 4.31 Derive Eqn. (4.74).

Problem 4.32 Use the Rankine-Hugoniot equation to show the following jump conditions without invoking the Mach number:

$$\frac{[\rho]}{\rho_1} = \frac{2\frac{[p]}{p_1}}{2\gamma + (\gamma - 1)\frac{[p]}{p_1}}$$

$$\frac{[v]}{v_1} = -\frac{2\frac{[p]}{p_1}}{2\gamma + (\gamma + 1)\frac{[p]}{p_1}}$$

Problem 4.33 Combine Eqns. (4.74b) through (4.74d) and Eqn. (4.75) to show that

$$\frac{[p]}{p_1} = \frac{\gamma(\gamma+1)}{4}\left(\frac{[V]}{a_1}\right)^2\left[1 + \sqrt{1 + \left(\frac{4}{\gamma+1}\frac{a_1}{[V]}\right)^2}\right]$$

Problem 4.34 Use the jump conditions, Eqn. (4.74) to show the following:

(1) $V_1 V_2 = \frac{[p]}{[\rho]}$

(2) $J^2 = -\frac{[p]}{[\nu]}$, where $J = \rho_1 V_1 = \rho_2 V_2$ is the mass flux across the shock wave

(3) $\frac{[p]}{\rho_1 a_1^2} = -M_{1n}\frac{[V]}{a_1} = M_{1n}^2\frac{[\nu]}{\nu_1}$

(4) $[V]^2 = -[p][\nu]$

Problem 4.35 A shock wave system resulting from supersonic flow and wave inter-action is described in the figure below. Find out the values of the following para-meters – M_3, V_3, p, T_3, and p_{t3}.

Problem 4.36 On a hodograph plane qualitative locate and explain the results of a supersonic flow undergo turning by a two-dimensional wedge through an angle of 2δ in two scenario: (i) by one single turning of 2δ and (ii) by two successive wedges each having a turning angle of δ. Also locate the result with a normal shock. Assume attached oblique shocks. For illustration, use the freestream Mach number $M_1 = 4$ and $\delta = 5°$ on the hodograph plane (the numerical results are readily available from Examples 4.9, 4.10, and 4.11).

5 Steady One-Dimensional Flows in Channels

Chapter 4 is mainly concerned with shock and expansion waves in external flows; it has also discussed the concept of supersonic inlet flows that can be slowed down either by a normal shock, accompanied by a loss in stagnation pressure, or a series of weaker, oblique shocks with a smaller loss in stagnation pressure. Flow in nozzles with a varying cross-sectional area was also mentioned in Section 4.19. In fact a supersonic flow can be slowed down without any shock waves by using a properly designed diffuser, which has a decreasing cross-sectional area in the flow direction. A diffuser is a nozzle in reverse, with a nozzle having an increasing cross-sectional area in the flow direction. One might therefore expect a nozzle to be used to accelerate supersonic flows, in applications such as rockets where exit flows should be as supersonic as allowed by the rocket chamber design. Both diffuser and nozzle flows are channel flows. For best performance, it is a general rule to avoid shock waves in the channel. This chapter addresses flows in channels, including supersonic wind tunnel operation and design. Conditions under which shock wave might occur in channels are of practical interest to thrust generation in propulsion applications.

5.1 Isentropic Flow of an Perfect Gas in Channels of Variable Cross-Sectional Area

Consider a steady one-dimensional flow in a channel of variable cross-sectional area, schematically shown in Fig. 5.1. To focus on the effect of area change, assume (1) the flow is one-dimensional, (2) the flow is steady, (3) effects of viscosity and gravity are negligible, (4) there is no heat transfer through the channel wall (i.e., the wall is adiabatic). As a result of these assumptions, the flow is everywhere isentropic except where shock waves occur.

The conservation of mass requires that

$$\rho V A = \dot{m} = \text{constant} \tag{5.1}$$

Differentiating both sides of this equation and dividing through by $\rho V A$ yields

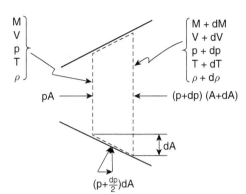

Figure 5.1 Schematic of a steady one-dimensional flow in a channel of variable cross-sectional area.

$$\frac{d\rho}{\rho} + \frac{dV}{V} + \frac{dA}{A} = 0 \tag{5.2}$$

The momentum conservation is governed by Euler's equation, Eqn. (4.11):

$$dp + \rho V dV = 0 \tag{5.3}$$

The energy conservation is given by Eqn. (4.6)

$$h_t = c_p T_t = h + \frac{1}{2} V^2 = \text{constant}$$

In differential form, it becomes

$$d\left(h + \frac{1}{2}V^2\right) = dh + d\left(\frac{V^2}{2}\right) = 0 \tag{5.4}$$

Note that Reynolds transport theorem applied to the control volume shown in Fig. 5.1 will also lead to the same differential forms of the governing equations (5.2) through (5.4). This is left as a problem at the end of this chapter.

Isentropic flows are reversible, and therefore

$$T ds = dh - \frac{dp}{\rho} \tag{2.36} \tag{5.5}$$

and because $ds = 0$

$$dh = \frac{dp}{\rho}$$

Combining this expression with Eqn. (5.4) leads to an equation identical with Eqn. (5.3); that is, for an isentropic flow, the momentum and energy considerations results in the same equation. It is rewritten as

$$dp = -\rho V dV \text{ or } \frac{dV}{V} = -\frac{dp}{\rho V^2} \tag{5.6}$$

Substituting this expression into the mass conservation relation, Eqn. (5.2), yields

$$dp + \rho V^2 \left[-\frac{d\rho}{\rho} - \frac{dA}{A} \right] = 0 \qquad (5.7)$$

Because $a^2 = (\partial p / \partial \rho)_s$, $d\rho = dp / a^2$, which upon substituting into Eqn. (5.7) yields

$$dp + \rho V^2 \left[-\frac{dp}{\rho a^2} - \frac{dA}{A} \right] = 0$$

$$dp(1 - M^2) = \rho V^2 \frac{dA}{A} \qquad (5.8a)$$

For a one-dimensional flow, A is a function only of the flow direction x, and Eqn. (5.8a) can be written as

$$\frac{dp}{dx} = \frac{\rho V^2}{A(1 - M^2)} \frac{dA}{dx} \qquad (5.8b)$$

Again by using $dp = -\rho V dV$ for dp, one obtains

$$(M^2 - 1) \frac{dV}{V} = \frac{dA}{A} \qquad (5.9)$$

For supersonic flows, $(M^2 - 1) > 0$ and, therefore, an increase in the cross-sectional area (i.e., $dA > 0$) leads to flow acceleration ($dV > 0$). According to Eqn. (5.6), an increase in the cross-sectional area also causes $dp < 0$, a decrease in pressure. As the flow accelerates, $(d[V^2/2] > 0)$, and conservation of energy requires that dT (or dh) < 0 according to Eqn. (5.4), which causes a to also decrease. These results for supersonic flows are in contrast to those in subsonic channel flows. In subsonic flows, $(M^2 - 1) < 0$, $dA > 0$ causes $dV < 0$ (i.e., deceleration), $dp > 0$ and $dT > 0$; for $dA < 0$, $dV > 0$ (i.e., acceleration), $dp < 0$. The pressure variation with velocity discussed is thus consistent with the Bernoulli principle, whether $M < 1$ or $M > 1$. These results are summarized in Fig. 5.2.

The results in Fig. 5.2 indicate that a subsonic flow cannot be accelerated to be supersonic in a converging nozzle and that a supersonic cannot be decelerated to be

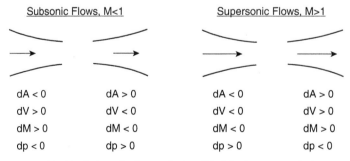

Figure 5.2 Variations in flow velocity (V), Mach number (M), and pressure (p) in a steady one-dimensional flow in a channel of variable cross-sectional area.

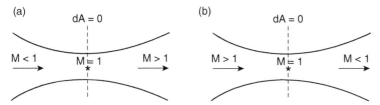

Figure 5.3 Variations in Mach number (M) in a steady one-dimensional flow in (a) a converging-diverging nozzle and (b) supersonic inlet.

subsonic in a converging nozzle, either. Equation (5.9) points to a possibility that $M = 1$ can be achieved at a location where $dA = 0$, that is, at a location where the cross-sectional area is the minimum (i.e., the *throat*). Thus with a sufficiently large pressure ratio of the flow to that in the ambient, the combination of a converging section and a diverging section (i.e., a converging-diverging channel) would allow the acceleration of a subsonic flow in the converging section to achieve the sonic condition at the throat and to continue to become supersonic in the diverging section. Such a converging-diverging channel serves as a *nozzle*, for flow acceleration from subsonic to supersonic velocities, as shown in Fig. 5.3a. The reverse scenario, shown in Fig. 5.3b, constitutes the essence of a *supersonic diffuser* or inlet.

5.2 Reference Parameters in Nozzle and Diffuser Flows

In the previous section, the cross-sectional area of the throat (A_{th}) takes on significance as it is associated with sonic conditions. As such it can serve as a useful reference parameter. For isentropic flows, P_t and T_t have earlier been shown to be useful reference parameters that remain constant throughout the flow. This section examines the connection between all the reference parameters and the reference cross-sectional area. The reference cross-sectional area might not physically exist in a given nozzle or diffuser; nonetheless, it is a reference based on which useful calculations and results can be obtained. Relations between the ratio of P_t/p and T_t/T and M are given by Eqns. (4.6c) and (4.7):

$$\frac{T_t}{T} = 1 + \frac{\gamma - 1}{2}M^2 \qquad \text{(4.6c) (5.10)}$$

$$\frac{p_t}{p} = \left(\frac{T_t}{T}\right)^{\gamma/(\gamma-1)} = \left(1 + \frac{\gamma - 1}{2}M^2\right)^{\gamma/(\gamma-1)} \qquad \text{(4.7) (5.11)}$$

The following derivation shows that the ratio of A/A_{th} can be related to M when $M_{th} = 1$, under which condition the throat area is called the critical throat area and A_{th} is replaced by A^*. For the rest of the book, the superscript * denotes sonic, or critical, conditions.

For steady flows, $\rho VA = \dot{m} = $ constant. By making use of $\rho = p/RT$ and $V = M\sqrt{\gamma RT}$ and then substituting Eqns. (4.6c) and (4.7) for p and T, one arrives at

$$\dot{m} = \frac{P_t A}{\sqrt{RT_t}} \sqrt{\gamma} M \left(1 + \frac{\gamma - 1}{2} M^2 \right)^{(\gamma+1)/2(1-\gamma)} \tag{5.12}$$

If sonic conditions exist at the throat location ($A = A_{th} = A^*$ and $M = M_{th} = 1$), Eqn. (5.12) becomes

$$\dot{m} = \frac{P_t A}{\sqrt{RT_t}} \sqrt{\gamma} M \left(1 + \frac{\gamma - 1}{2} M^2 \right)^{(\gamma+1)/2(1-\gamma)} = f(M, \gamma) \tag{5.13a}$$

For steady flows,

$$\dot{m} = f(1, \gamma) = \frac{P_t A^*}{\sqrt{RT_t}} \sqrt{\gamma} \left(\frac{\gamma + 1}{2} \right)^{(\gamma+1)/2(1-\gamma)} \tag{5.13b}$$

Thus

$$\frac{A}{A^*} = \frac{1}{M} \left[\left(\frac{2}{\gamma + 1} \right) \left(1 + \frac{\gamma - 1}{2} M^2 \right) \right]^{(\gamma+1)/2(\gamma-1)} \tag{5.14}$$

The relation between A/A^* and M is established and is only a function of M and γ, just like p_t/p and T_t/T. Equations (5.10), (5.11), and (5.14) are plotted in Fig. 4.1 for $\gamma = 1.4$ and numerical values for $\gamma = 1.4$, 1.3, and 5/3 are listed in Appendix C. Examination of Fig. 4.1 and Appendix C reveals that for each value of A/A^*, there are two corresponding values of M, one subsonic and the other supersonic for a converging-diverging channel flow. This result is consistent with that shown in Fig. 5.3a, where subsonic acceleration (dV and $dM > 0$ for $dA < 0$) occurs in the converging section and supersonic acceleration (dV and $dM > 0$ for $dA > 0$) in the diverging section. Conversely, as shown in Fig. 5.3b, subsonic deceleration occurs in the diverging section and supersonic deceleration in the converging section. In other words, M is not a single-value function of A/A^*. However, both p and T are single-valued functions of M, and they both decrease with increasing M.

EXAMPLE 5.1 A converging-diverging nozzle is connected to a reservoir containing air at 300 kPa (p_r) and 300 K (T_r).

(a) If air exits the nozzle at a pressure equal to that in the ambient (p_b, also called the "back" pressure) 101.3 kPa, what is the ratio of cross-sectional areas of exit to the throat? Under this condition, what is the exit velocity?

(b) What is the ambient pressure to cause a normal shock wave at the exit plane of such a nozzle? What is the exit velocity?

(c) What is the value of p_b for subsonic exit condition with $M_{th} = 1$ and without shock, given the nozzle determined in (a)?

Solutions –

(a) Assume the flow within the nozzle is isentropic.

$\frac{p_b}{p_t} = \frac{p_e}{p_t} = \frac{p_e}{p_r} = \frac{101.3}{300} = 0.338.$ Therefore, $M_e \approx 1.34$ and $\frac{A}{A^*} \approx 1.084$ and
$\frac{T}{T_t} = \frac{T}{T_r} = 0.7358$ and $T = 220.7$ K.

Because $M_e > 1$, it follows that $M_{th} = 1$, and $A_{th} = A^*$ and therefore
$\frac{A}{A_{th}} \approx 1.084$.

$$V_e = M_e \sqrt{\gamma R T_e} = 1.34 \times \sqrt{1.4 \times 287 \times 220.7} = 399.1 \text{ m/s}$$

(b) Normal shock wave at the exit plane

Let subscripts 1 and 2 designate, respectively, conditions upstream and downstream of the normal shock wave. Then $M_1 = 1.34$ and therefore $M_1 \approx 0.77$, $\frac{p_{t2}}{p_{t1}} \approx 0.972$. Downstream of the normal shock wave $\frac{p_2}{p_{t2}} \approx 0.675$ and $\frac{T_2}{T_{t2}} \approx 0.894$.

Thus, $p_b = p_2 = \frac{p_2}{p_{t2}} \times \frac{p_{t2}}{p_{t1}} \times p_{t1} = 0.675 \times 0.972 \times 300 \text{ kPa} = 196.8 \text{ kPa.}$

$$V_e = V_2 = M_2 \sqrt{\gamma R T_2} = 0.77 \times \sqrt{1.4 \times 287 \times 0.894 \times 300} = 252.8 \text{ m/s}$$

(c) For $\frac{A}{A_{th}} \approx 1.084$, Subsonic value of M_e is $M_e = M_2 = 0.715$, for which $\frac{T_e}{T_t} = 0.907$ and $\frac{p_e}{p_t} = 0.710$.

$$p_e = p_e = 0.710 \times 300 \text{ kPa} = 213.0 \text{ kPa}$$

$$V_e = V_2 = M_2 \sqrt{\gamma R T_2} = 0.715 \times \sqrt{1.4 \times 287 \times 0.907 \times 300} = 236.4 \text{ m/s}$$

Comments –

Because $\frac{p_e}{p_t}$ monotonically decreases with increasing M (see Eqn. (4.9) Appendix C), unlike $\frac{A}{A^*}$, $M_e \approx 1.34$ is the only solution, for which $\frac{A}{A^*}$ is determined. For $\frac{A}{A^*} \approx 1.084$ a subsonic value $M_e \approx 0.715$ is a possible solution. But in this case $\frac{p_e}{p_t} = 0.708$, i.e., $p_e = 212.4$ kPa. For this to occur, however, p_b is 213.0 kPa because a subsonic exit flow can adjust to p_b. Deviation of p_b from 213.0 kPa would result in different exit Mach number; such a scenario will be discussed further in section 5.3.   □

Example 5.1 provides a scenario where a flow resumes subsonic conditions (and deceleration) in the diverging section after having accelerated from subsonic conditions in the converging section to reach sonic conditions at the throat. Whether the existing flow is supersonic or subsonic, because $M_e = 1$, the throat condition serves as a reference and specifically $A_{th} = A^*$. On the other hand, A^* as a reference parameter might not physically exist for a given isentropic channel flow; the following example illustrates this situation.

EXAMPLE 5.2 A converging-diverging nozzle with the exit to throat area ratio $\frac{A_e}{A_{th}} = 1.084$ is connected to a reservoir containing air at 300 kPa and 300 K. If the air exits the nozzle to another reservoir at a pressure of 270 kPa, what is the Mach number entering the second reservoir?

Solution –

Assume an isentropic flow throughout the flow. Therefore $p_t = p_r$ and $\frac{p_e}{p_t} = \frac{p_e}{p_r}$ $= \frac{270}{300} = 0.9$. Therefore $M_e \approx 0.46$ and $\frac{A_e}{A^*} \approx 1.425$. Comparing with $\frac{A_e}{A_{th}} = 1.084$, $A^* < A_{th}$ and $M_{th} \neq 1$. Because $\frac{A_{th}}{A^*} = \frac{1.425}{1.084} \approx 1.315$, $M_{th} \approx 0.51$.

Comments –

(1) For $\frac{A_{th}}{A^*} \approx 1.315$ the supersonic Mach number is $M_{th} \approx 1.68$. However, the supersonic cannot exist at the throat, because M_{th} can at most be equal to unity. Thus, in this problem A^* does not physically exist and $M_{th} < 1$. (2) For exactly the same nozzle as that in Example 5.1, the deviation from $p_b = 213.0$ kPa (now 270 kPa) results in vastly different (sonic vs. subsonic) flow conditions at the throat.   □

EXAMPLE 5.3 In dealing with supersonic jet noise, a fully expanded Mach number, M_{FE}, is an important reference parameter. M_{FE} is defined as the Mach number achieved by expanding the supersonic jet to match the surrounding pressure; that is, the value of M_{FE} might not exist within the nozzle flow field. For an air reservoir pressure of 20 atmospheres, find the value of M_{FE} for expansion into a surrounding having a pressure of 0.8 atmospheres.

Solution –

$$\frac{p_t}{p_\infty} = \frac{20}{0.8} = 25 = \left(1 + \frac{\gamma - 1}{2} M_{FE}^2\right)^{\gamma/(\gamma-1)}$$

By solving the above expression, $M_{FE} = 2.746$.

Comments –

For flight vehicle, the nozzle exit plane pressure is usually higher than that of the surrounding (under-expansion instead of over-expansion to avoid shock waves within the nozzle or at the nozzle exit, both phenomena to be discussed in the next section). Thus, the reference value of M_{FE} is not achieved within the nozzle.   □

5.3 Standing Normal Shock Wave in Converging-Diverging Nozzles

For applications such as rocket propulsion and gas delivery devices, the conditions at the exit of the channel are important. Two questions arise as to what exit conditions are if the ambient pressure (also called the back pressure p_b) falls between those for $M_e > 1$ and $M_e < 1$ (as illustrated in Example 5.1) but with $M_{th} = 1$: (1) Is the back pressure equal to or lower than the pressure of the existing flow, and (2) Is the back pressure higher than the exit plane pressure?

The steady-state momentum equation $dp = -\rho V dV$ indicates that as the flow accelerates ($dV > 0$) or decelerates ($dV < 0$), the static pressure decreases ($dp < 0$) or increases ($dp > 0$) correspondingly. This is true whether the flow is supersonic or subsonic, as no such distinction is needed for applying the principle of conservation

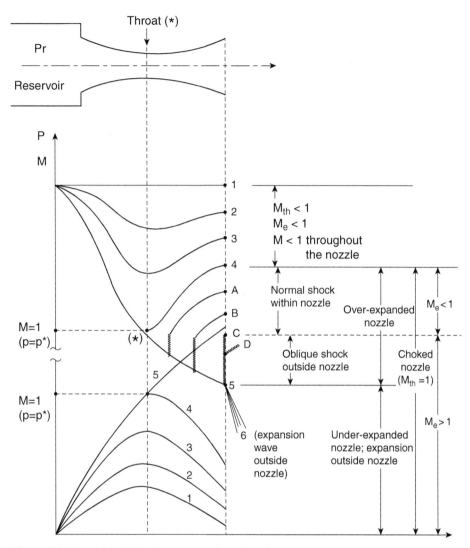

Figure 5.4 Scenarios of flow in a converging-diverging nozzle attached to a reservoir having a pressure p_r. Note that the back pressure (p_b) or the ratio p_b/p_r plays an important role. For $p_b < p_4$, the nozzle is choked with $M_{th} = M^* = 1$ at the throat. Perfect expansion occurs when $p_b = p_5$. For $p_C < p_b < p_4$, normal shocks exist in the diverging section, while for $p_5 < p_b < p_C$, oblique shocks extends downstream of the nozzle exit and are attached to the nozzle lip.

of momentum. The momentum conservation principle can be used to gain knowledge of the pressure variation throughout a converging-diverging nozzle, including the exit plane, to answer the aforementioned two questions.

Consider in Fig. 5.4 a converging-diverging nozzle attached to a reservoir having a pressure p_r, with the profiles of pressure ratio p/p_r and Mach number plotted along the flow direction. For $p_b/p_r = 1$, $dp = -\rho V dV = 0$ everywhere within the nozzle; there is thus no flow and no pressure variation in the nozzle. This condition is described by the horizontal line (curve 1). As p_b/p_r is decreased below unity, a subsonic flow is induced at the entrance to the nozzle. For values of p_b corresponding to curves 2 and 3, subsonic flows exist throughout the nozzle, including the throat. Under conditions represented by curves 2 and 3, the flow accelerates (Mach number increases and the pressure decreases) in the converging section and decelerates (Mach number decreases and the pressure increases) in the diverging section, with minimum pressure and maximum Mach number at the throat. Because p_b is smaller for curve 3, the flow speed and Mach number are higher and the pressure is lower at same locations (i.e., a given x) than curve 2. For curve 4, the value of p_b/p_r is such that $M_{th} = 1$ and $M_e < 1$ (as Example 5.1 illustrates), so that $p_{e4} = p_b$, where p_{e4} is the exit plane pressure for curve 4 (similar subscripts are used for other curves in Fig. 5.4). A further decrease in p_b to a particular p_{e5} (curve 5) would lead to $M_e > 1$. The particular conditions chosen for curves 4 and 5 are such that the exit plane pressure is equal to the ambient pressure—for curve 4 $p_{e4} = p_b$ is able to adjust and recover to p_b because of the subsonic nature in the diverging section. For the case of $M_{e5} > 1$, the flow expands (accelerates) so that $p_{e5} = p_b$; in the this case, the nozzle is said to be *perfectly expanded*. For $p_b = p_6 < p_{e5}$, the flow, being supersonic, cannot adjust to p_b and expansion continues outside the nozzle (see Example 5.2) and the nozzle is *under-expanded* (i.e., the gas is not sufficiently expanded at the exit plane to reach $p_e = p_b$). The perfectly expanded condition is achieved by increasing the area ratio A_e/A^* (for example, by adding further length to the diverging section), or, for a given A_e/A^*, by increasing p_b/p_r (say, by reducing p_r, or increasing p_b, or both as long as the nozzle is still choked). Under-expansion does not occur if the exit flow is subsonic, under which condition the flow can adjust to the ambient condition.

Therefore for a given A_e/A_{th} it is possible to have $M_{th} = 1$ (for which $A_{th} = A^*$ and either $M_e < 1$ or $M_e > 1$ with isentropic conditions throughout the nozzle) depending on the value of p_b. These two possibilities of either subsonic or supersonic exit condition have been numerically illustrated in Example 5.1. These findings help to explain why curves 4 and 5 share a common portion in the converging section and differ in the diverging section. This is because, once the sonic condition is established at the throat, further decreases in pressure in the ambient cannot be felt upstream of the throat (recall the signaling mechanism discussed in Chapter 3). When the condition of $M_{th} = 1$ is established, the nozzle is said to be *choked*. The ratio p_b/p_r corresponding to curve 4 is called the critical pressure ratio below which $M_{th} = 1$.

The nozzle flow discussed so far is isentropic and answers the first and second questions at the beginning of this section. The pressure variation in an isentropic nozzle flow can be further explained by using Eqn. (5.8) as follows:

(1) For $M < 1$ throughout the nozzle length, an increase/decrease in the nozzle cross-sectional area, A, causes p to decrease/increase. This result is seen for curves 1–3 in Fig. 5.4. Curve 4 is a special case where $M_{th} = 1$ and p_b/p_r reaches the minimum at throat and recovers in the diverging section due to deceleration to subsonic speeds. For curves 1 through 4, $p_e = p_b$ due to the subsonic exit flow being able to adjust to the ambient pressure.

(2) For $M > 1$ in the diverging section, $M_{th} = 1$ is required and an increase in A causes p to decrease. Curve 5 in Fig. 5.4 is a special case because $p_b = p_{e5}$ and the nozzle is perfectly expanded. For $p_b = p_6 < p_{e5}$ the pressure profile remains the same as curve 5 due to the supersonic exit flow's inability to adjust to the ambient pressure; the flow continues to expand outside the nozzle exit (curve 6).

How will the nozzle flow behave with values of p_b/p_r between curves 4 and 5? In this range of p_b/p_r values, the flow immediately downstream of the throat is supersonic, but if the flow continues to accelerate/expand toward the nozzle exit, such as that indicated by curve 5, the exit pressure falls below that of the ambient. The adjustment of the exit pressure has to be accomplished through a shock wave due to the supersonic nature of the flow. Let point C in Fig. 5.4 denote the scenario in which supersonic acceleration continues until the flow reaches the exit where the adjustment to p_b requires a normal shock at the exit plane so that $p_b/p_r = p_{eC}/p_r$. For $p_{e5}/p_r < p_b/p_r < p_{eC}/p_r$ (an example is denoted by curve D in Fig. 5.4), the pressure difference between those of the exiting flow and the ambient is not sufficient for a normal shock at the exit; in this case an oblique shock forms attached to the nozzle lip. The deflection by the oblique shock is to direct the flow inwardly toward the nozzle centerline, as the oblique shock theory of Chapter 4 indicates and as, intuitively, a consequence of the higher ambient pressure squeezing the stream tube downstream of the exit. For $p_b/p_r > p_{eC}/p_r$, the normal shock wave moves upstream of the exit as p_b is increased, with curves A and B in Fig. 5.4 as examples (with p_{eA}/p_r and p_{eB}/p_r, respectively). Under these back pressures and the resultant normal shock conditions, the flow becomes subsonic in the diverging section downstream of the normal shock and the flow decelerates toward the exit. Because of the subsonic condition, the flow can adjust to the ambient condition so that $p_e/p_r = p_b/p_r$. Further increase in p_b above p_{eA} eventually moves the normal shock to the throat, where the trivial solution for the shock wave is $M_1 = M_2 = 1$ (where subscripts 1 and 2 denote, respectively, states immediately upstream and downstream of the normal shock) and the flow becomes subsonic throughout the entire nozzle (as shown by curve 4 in Fig. 5.4). In summary, the nozzle exit flow can only adjust to the ambient pressure through shock waves for $p_5/p_r < p_b/p_r < p_4/p_r$. Under these conditions, the flow and the nozzle are said to be over-expanded.

A normal shock within the nozzle causes a loss in stagnation pressure (i.e., $p_{t2} < p_{t1}$). Assuming that the flow is everywhere isentropic except across the shock, the downstream stagnation pressure now serves as a reference value for the subsonic decelerating isentropic flow for the remaining diverging portion of the nozzle. For adiabatic walls, the stagnation temperature remains unchanged ($T_{t2} = T_{t1} = $ constant) throughout the entire nozzle. Does the reference "throat" area, which is also a reference parameter, remain the same as the one upstream of the shock? To answer this question, consider a generic channel where the reference area does not exist in the physical problem, as mentioned in Example 5.2 and now illustrated in Fig. 5.5, where the imaginary extensions and throats are shown. For steady flows, Eqn. (5.13) requires that $\dot{m} = $ constant and $p_{t1}A_1^* = p_{t2}A_2^*$. Thus

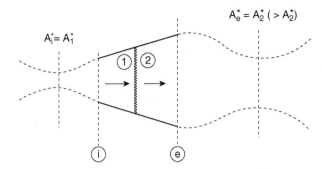

Figure 5.5 Schematic illustrating the imaginary reference cross-sections upstream and downstream of a normal shock. The dashed lines represent the imaginary nozzles.

$$\frac{A_2^*}{A_1^*} = \frac{p_{t1}}{p_{t2}} > 1 \tag{5.15}$$

This result indicates that for the steady state flow, the cross-sectional area of the throat (or the "reference" area) downstream of the normal shock has to be greater than that of the upstream throat. If the channel is without any shock, then $A_1^* = A_2^*$. Implications of this result for supersonic inlet and diffuser designs and the start-up operation of a closed-circuit supersonic wind tunnel will be discussed in the next section.

EXAMPLE 5.4 A converging-diverging nozzle is attached to a reservoir containing air at 300 kPa and 300 K. The areas of the throat and the exit are, respectively, 5 cm^2 and 10 cm^2. (a) What is the range of back pressure for the nozzle to be choked? (b) What is the maximum mass flow rate the nozzle can deliver? (c) What is the range of back pressure for a normal shock wave to exist in the nozzle? (d) What should the back pressure be for a normal shock wave to stand where the cross-sectional area is 7.5 cm^2? (e) What is the back pressure for perfect expansion?

Solution –

(a) Referring to Fig. 5.4, the choked flow exists for $M_{th} = 1$ and $p_b \leq p_{e4}$ (i.e., either $M_e < 1$ or $M_e > 1$). In this case, $A_e/A^* = (10 \text{ cm}^2/5 \text{ cm}^2) = 2$ and therefore $M_e \approx 0.30$ and $p_{e4}/p_t = p_{e4}/p_r = 0.9395$. For a choked nozzle $p_b \leq p_{e4} = 281.9 \ kPa$; this is the back pressure below which $M_{th} = 1$.

(b) When the nozzle is choked, any further decrease in p_b cannot be sensed by the flow upstream of the throat (e.g., curves 4 and 5 share a common portion in the converging section) and the sonic condition at the throat $M_{th} = 1$ gives the maximum mass flow rate. Thus,

$$\dot{m} = \frac{P_t A^*}{\sqrt{RT_t}} \sqrt{\gamma} \left(\frac{\gamma+1}{2}\right)^{(\gamma+1)/2(1-\gamma)}$$

$$= \frac{(300 \times 10^3 \, Pa)\,(5 \times 10^{-4} m^2)}{\sqrt{(287 \ J/kg \cdot K)(300 \ K)}} \sqrt{1.4}\left(\frac{2.4}{2}\right)^{-3} = 0.35\frac{\text{kg}}{\text{s}}$$

(c) The upper limit for the back pressure to cause a normal shock in the nozzle is that the shock stands right at the throat and therefore $p_b = p_{e4} = 281.9$ kPa. The lower limit is p_{e5} in Fig. 5.4. In this case $A_e/A^* = 2$ gives $M_5 \approx 2.20$, $M_{e5} \approx 0.5471$ and the ratio of the stagnation pressure across the shock $p_{t5}/p_r = 0.6281$.   For $M_{e5} \approx 0.5471$, $p_{e5}/p_{t5} \approx 0.815$.   Therefore, $p_b = p_{e5} = (p_{e5}/p_{t5})(p_{t5}/p_r)(p_r) = 0.815 \times 0.6281 \times 300$ kPa $= 153.57$ kPa. Thus for a normal shock to exist in the nozzle is 153.57 kPa $< p_b < 281.9$ kPa.

(d) Let 1 and 2 denote location immediately upstream and downstream of the shock wave. For the shock wave to exist, the flow must be supersonic and must be sonic at the throat. Therefore $A_1/A_1^* = A_1/A_{th} = 1.5$, $M_1 \approx 1.86$, $M_2 \approx 0.60$ and $p_{t2}/p_{t1} = 0.786 = A_1^*/A_2^*$. Therefore, $A_e/A_e^* = A_e/A_2^* = (A_e/A_1^*)(A_1^*/A_2^*) = (2)(0.786) = 1.576$ and $M_e \approx 0.40$ and $p_e/p_{te} = p_e/p_{t2} = 0.896$. Hence $p_b = p_e = (p_e/p_{te})(p_{te}/p_{t1})(p_{t1}) = (0.896)(0.786)(300$ kPa$) = 211.3$ kPa.

(e) For   perfection, $M_2 \approx 2.20$ and $p_b/p_r = p_{e5}/p_r = 0.0935$.   Thus $p_b = 28.1$ kPa. This can also be found using the result in (c) – because $M_5 \approx 2.20$, $p_{e5}/p_5 = 5.480$ and therefore without the normal shock at the exit and for perfect expansion $p_b = p_5 = 153.57$ kPa$/5.48 \approx 28.0$ kPa.   □

5.4 Supersonic Inlet and Diffuser

A supersonic diffuser is a converging-diverging nozzle operating in reverse, for the intake of supersonic air streams. The ideal diffuser operation condition is illustrated in Fig. 5.6a, where the free-stream Mach number M_∞ is the design Mach number, M_D ($M_D = M_\infty$). The design goal is to minimize loss in stagnation pressure (i.e., to maximize the ability of the gases to expand in the downstream nozzle), by achieving isentropic flow throughout the diffuser. Therefore, $A_i/A_D^* = A_i/A_{th}$ is a function of M_D, with A_i being the cross-sectional area of the diffuser inlet and A_D^*, the designed reference area. Operation under the design condition would have the supersonic flow decelerating with no shock wave in the converging section, reaching sonic condition at the throat and then further decelerating in the diverging section to achieve $M_e < 1$ and $p_e = p_b$, where p_b is now the desired pressure (for example, at the inlet of the engine compressor). Therefore $M_{th} = 1$ (the choking condition) and the mass flow rate is limited by the throat cross-sectional area A_{th}. The following examples help to illustrate the point. From another viewpoint, $M_{th} = 1$ is required because $M_{th} > 1$ will lead to flow acceleration in the diverging section and $M_{th} < 1$ means a normal shock is formed in the converging section.

EXAMPLE 5.5 A converging-diverging channel to be used as supersonic diffuser is designed for $M_\infty = M_D = 1.5, p_\infty = 10$ kPa and $p_\infty = 216$ K. The goal is for the flow to decelerate to subsonic condition and then be fed to an engine with a subsonic speed. What are the inlet and exit cross-sectional areas and the exit

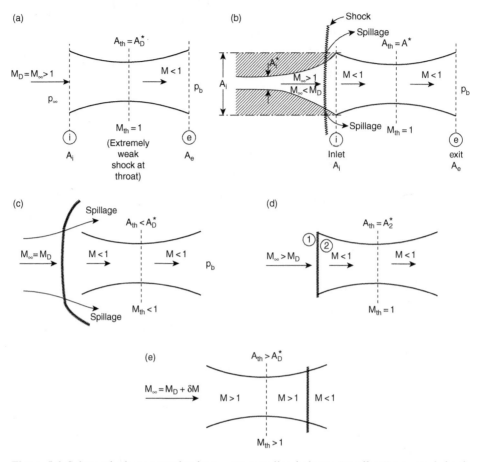

Figure 5.6 Schematic demonstrating how over-speeding helps to "swallow" a normal shock within a converging-diverging nozzle designed for a supersonic inlet Mach number (M_D). The over-speeding needed is δM – (a) "ideal" design condition; (b) the "start-up" problem because $M_\infty < M_D$ when ramping up from rest to M_D; (c) when approaching M_D from $M_\infty < M_D$ a normal shock occurs upstream of the inlet; (d) when M_∞ is slightly larger than M_D the normal is pressed on the nozzle inlet; (e) with $M_\infty = M_D + \delta M$, the shock is swallowed.

flow velocity if the flow is to be isentropic throughout the diffuser with $p_e = 33$ kPa?

Solution –

For $M_i = M_\infty = 1.5$, $A_i/A_i^* \approx 1.176$ and $\frac{p_\infty}{p_{t\infty}} \approx 0.280$ and $\frac{T_\infty}{T_{t\infty}} \approx 0.695$. For subsonic exit conditions, the throat condition must be sonic. Thus $A_{th} = A_i^* \approx 0.85 \, A_i$ or $A_i/A_{th} = 1.176$. Because of the isentropic flow throughout the channel, $p_t = p_{t\infty} = \text{constant}$ and $T_t = T_{t\infty} = \text{constant}$. Thus $\frac{p_e}{p_t} = \left(\frac{p_e}{p_\infty}\right)\left(\frac{p_\infty}{p_t}\right) = \frac{33}{10} \times 0.28 = 0.924$ for which $M_e \approx 0.40$ and $A_e/A_{th} \approx 1.59$ and $T_e/T_t = 0.969$

$$V_e = M_e \sqrt{\gamma R T_e} = 0.40 \times \sqrt{1.4 \times 287 \times \frac{210}{0.695} \times 0.969} \approx 137.2 \; m/s.$$

□

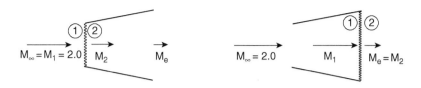

EXAMPLE 5.6 Consider a supersonic inlet/diffuser that is a diverging channel having the exit to inlet area ratio $A_e/A_i = 2.5$ and is designed to operate for a free stream Mach number $M_\infty = 2.0$ with static pressure and temperature equal to 10 kPa and 216 K, respectively. Consider the scenarios where a normal shock wave stands (1) at the inlet and (2) at the exit. What is the loss in stagnation pressure and what is the back pressure in each scenario?

Solution –
Scenario 1 (Normal shock at inlet) –

$M_1 = M_\infty = 2.0$, therefore $A_i/A_1^* = 1.6875$ $p_1/p_{t1} = p_\infty/p_{t\infty} = 0.1278$ and $p_{t2}/p_{t1} = A_1^*/A_2^* = 0.7209$ (a 28% loss in stagnation pressure).

Thus $M_2 = 0.5774$, $p_2/p_{t2} \approx 0.79$.

To find p_b, the M_e is needed, which in turn requires

$$A_e/A_e^* = A_e/A_2^* = (A_e/A_i)(A_i/A_1^*)(A_1^*/A_2^*) = 2.5 \times 1.6875 \times 0.7209 \approx 3.04.$$

Therefore,
$M_e \approx 0.20$ and $p_e/p_{t2} \approx 0.9725$ and

$$p_b = p_e = (p_e/p_{t2})(p_{t2}/p_{t1})(p_{t1}/p_{t\infty})(p_{t\infty}/p_\infty)p_\infty = 0.9725 \times 0.7209 \times 1$$
$$\times(1/0.1278) \times 10 \text{ kPa} = 54.85 \text{ kPa}.$$

$$p_{t1} - p_{t2} = (1 - 0.7209)(p_{t1}/p_1)p_1 = 0.279 \times (1/0.1278) \times 10 \text{ kPa} = 21.8 \text{ kPa}.$$

Scenario 2 (Normal Shock at Exit) $-p_\infty/p_{t\infty} = 0.1278$

$$A_1/A_1^* = (A_e/A_i)(A_i/A_i^*)(A_i^*/A_1^*) = 2.5 \times 1.6875 \times 1 = 4.218$$

$M_1 \approx 3.0$, $p_1/p_{t1} = 0.0272$, $M_2 \approx 0.475$, $p_{t2}/p_{t1} \approx 0.328$ and $p_2/p_{t2} \approx 0.856$

$$p_b = p_2 = (p_2/p_{t2})(p_{t2}/p_{t1})(p_{t1}/p_1)p_1 = 0.858 \times 0.328 \times (1/0.1278) \times 10 \text{ kPa}$$
$$\approx 22.0 \text{ kPa}$$

$$p_{t1} - p_{t2} = (1 - 0.288)(p_{t1}/p_{t\infty})(p_{t\infty}/p_\infty)p_\infty = 0.712 \times 1 \times (1/0.1278) \times 10 \text{ kPa}$$
$$= 55.7 \text{ kPa}$$

Comments –

This is a problem where the imaginary extension of the channel would resemble that shown in Fig. 5.6, where scenario 1 corresponds to a normal shock standing at location i and scenario 2, a normal standing at location e. In fact one can also imagine that the imaginary extension in the upstream direction is connected to a reservoir where $p_r = p_{t\infty} = (p_{t\infty}/p_\infty)p_\infty = 10$ kPa/0.1278 = 72.24 kPa. Then this would become a nozzle problem, where $p_b = 22.0$ kPa causes a normal shock to stand at the nozzle exit, while a higher p_b, equal to 55.4 kPa, would cause the normal shock wave to move upstream, to a location where the cross-sectional area is 1/2.5 = 40% of A_e.   □

For supersonic inlets and diffusers, an important reference cross-sectional area for the supersonic inlet is called the capture area, A_i^* (Hill and Petersen, 1992), and it is equal to the physical throat cross-sectional area, A_{th} (i.e., $A_{th} = A_i^*$).Conceptually, this A_i^* is same as that in Fig. 5.5 and is now shown in Fig. 5.6b. It is the air passing through this fictitious throat that would completely enter and proceed isentropically through the diffuser before entering the desired device at p_b. However, because the supersonic flow cannot sense the diffuser, air passing outside this capture area (i.e., through A_i with $A_i > A_i^*$) will not be able to pass through the physical design throat area A_{th}. The adjustment of the air flow to the mismatch takes place through a normal shock standing at some standoff distance in front of the diffuser inlet, as depicted in Fig. 5.6b. The result is "spillage" of part of the air flow within A_i over the diffuser. The standoff distance is necessary as the subsonic flow downstream of the shock needs a passage for spilling. In Example 5.4 for $M_\infty = 1.50$, part of the air attempting to enter the diffuser will spill – that is the air flow in a cross-sectional area of $(A_i - A_i^*)/A_i \approx 1 - 1/1.176 = 15\%$ spills, represented by the shaded area in the Fig. 5.6b (exaggerated in size for clarity).

During the operation of a supersonic diffuser, there is the so-called *start-up* problem as the flow regime transitions from subsonic to supersonic. At a moment when $M_\infty < M_D$, how would the diffuser function differently from when operated at M_D? For illustration, consider the diffuser geometry in Example 5.5 with and $M_\infty = 1.40$ and $M_D = 1.50$. For $M_\infty = 1.40$, $A_i/A_i^* \approx 1.115$. Therefore, $A_i^*/A_{th} = (A_i^*/A_i)(A_i/A_{th}) = 1.176/1.115 \approx 1.055$, that is, the capture area is larger than the diffuser's throat area but is still smaller than A_i. This means that only the air flow in the stream tube passing the imaginary cross-sectional area of $A_i^* \approx 1.055A_{th}$ can enter the diffuser and be passed through the throat. The air flow in the area of $(A_i - A_i^* = A_i - A_i^* = (1.176 - 1.115)A_{th} = 0.061A_{th})$ will spill, and the spillage occurs with a normal shock at a standoff distance in front of the diffuser inlet. Therefore, operating with $1 < M_\infty < M_D$, the spillage is smaller than operating with $M_\infty = M_D$. However, it is noted that for $M_\infty < M_D$, the mass flow rate through the diffuser is slightly smaller than that for $M_\infty = M_D$ because even for a larger allowable capture area now $A_i^* \approx 1.055A_{th}$ (a 5.5% increase in the allowable capture area), the flow speed ($V_\infty = M_\infty\sqrt{\gamma RT_\infty}$) is less than that for M_D by a factor of 6.7%, corresponding to decrease in M_∞ from 1.5 to 1.4.

Spillage occurs during the transient operation when M_∞ is increased toward M_D. The shock wave then moves closer to the diffuser inlet as $M_\infty = M_D$ is achieved (Fig. 5.6c). Because of the non-zero spillage, $A_i^* < A_i$. To eliminate spillage, A_i^* has to be equal to A_i, then there will be no spillage and the normal shock wave will be standing exactly at the diffuser inlet. Under this condition (shown in Fig. 5.6d), the Mach number downstream of the shock wave (M_2) is desired to be such that $M_{th} = 1$ for the maximum choking mass flow rate allowed by the diffuser. Using numbers provided in Example 5.4, the requirement of $M_{th} = 1$ leads to $M_2 \approx 0.61$ because $A_2/A_2^* = A_2/A_{th} \approx 1.176$. Therefore the Mach number upstream of the normal shock is $M_1 = M_\infty \approx 1.81$. Thus, avoiding spillage would require $M_\infty > M_D$; that is, *over-speeding* is needed. The shock wave, however, is not desirable. Since M_2 is less than unity under this over-speeding condition, the resultant acceleration in the converging section causes pressure to decrease in the flow direction due to subsonic acceleration from location 2 to the throat. This means that the normal shock wave is not stable and cannot stand still in the converging section. By moving further into the converging section, the shock wave is able to adapt a smaller shock strength, because the Mach number upstream of the new shock location decreases. This new conditions begets yet another new downstream location. Eventually, the shock wave moves past the throat and reestablishes itself in the diverging section, as shown in Fig. 5.6e. As a consequence, the shock wave can be *swallowed* with sufficient over-speeding. This is because under the $M_\infty > M_D$ condition, the capture area is smaller than A_i^* in Fig. 5.6b (i.e., $A_{th} > A_D^*$), so A_{th} is greater than needed to decelerate to $M_{th} = 1$ so that $M_{th} > 1$ and the flow in the diverging section is supersonic. The shock wave in the diverging section may be eliminated, depending on the back pressure. However, the captured mass flow rate exceeds that which the diffuser is designed for (again because $A_{th} > A_D^*$). Therefore, once the shock is swallowed, the operation can be returned to the design condition so that $A_{th} = A_D^*$ again, for which an extremely weak shock (if shock cannot be avoided at all, depending on the back pressure) stands at the throat location. Because the shock is nearly a trivial solution for normal shock problem ($M_1 \approx M_2$), the loss in stagnation pressure is nearly zero, which is the desirable result, as illustrated in Fig. 5.6a.

The unstable behavior of a normal shock in the converging section is unlike the normal shock wave in the diverging section of a nozzle. In the post-shock (subsonic) region in the diverging section the flow decelerates and the pressure increases in the flow direction. For the shock wave to move downstream, p_b has to be decreased, as discussed in Chapter 4. The downstream movement of the normal shock wave increases the Mach number upstream of it and increases the shock strength and the pressure in the subsonic region downstream of it. This creates a contradiction in that in subsonic flows, pressure should equilibrate to that in the ambient. Therefore, it is not possible for a normal shock to move downstream in the diverging section.

Over-speeding can be relatively easy to achieve for swallowing the shock wave for moderate values of M_D, but becomes increasingly difficult as M_D is increased. For example, $M_D = 1.8$ and 2.0, the amount of over-speeding ($M_\infty = 3.1$ and $M_\infty > 25$

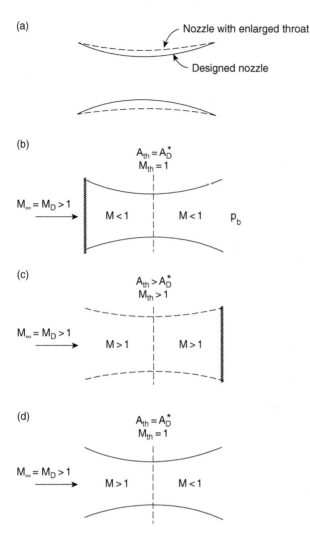

(a)

Nozzle with enlarged throat

Designed nozzle

(b)

$A_{th} = A_D^*$
$M_{th} = 1$

$M_\infty = M_D > 1$

$M < 1$ $M < 1$ p_b

(c)

$A_{th} > A_D^*$
$M_{th} > 1$

$M_\infty = M_D > 1$

$M > 1$ $M > 1$

(d)

$A_{th} = A_D^*$
$M_{th} = 1$

$M_\infty = M_D > 1$

$M > 1$ $M < 1$

Figure 5.7 Swallowing normal shock by enlarging the throat area – (a) solid and dashed lines represent the design and the enlarged operation conditions, respectively; (b) when starting from rest up to $M_\infty = M_D$ normal shocks exist upstream and at the inlet (a start-up problem similar to that described in Fig. 5.7); (c) by enlarging the throat area, the normal shock is swallowed; (d) the throat is restored to the design condition.

are needed, respectively) is left as an end-of-chapter problem (Problem 5.8). The other way to swallow the normal shock is to increase the throat area so that $A_{th}/A_D^* > 1$ and the shock wave moves past the throat, similar to what over-speeding achieves in swallowing the shock wave, as depicted in Fig. 5.7. Once swallowing is achieved, $A_{th}/A_D^* = 1$ can be restored operation at M_D. Such a scheme for swallowing the shock in the starting phase of supersonic diffuser operation is also used for starting supersonic wind tunnels, as will be discussed in the following section.

EXAMPLE 5.7 If a supersonic diffuser is designed for $M_D = 2.2$, during the operation with $M_\infty = 2.2$ how much should the throat open up to swallow the shock wave while maintaining $M_{th} = 1$?

Solution –

For $M_\infty = 2.2$, $A_i/A_i^* = A_i/A_{th} \approx 2.005$. Assume that the shock wave in front of the diffuser lip is a normal shock and let subscripts 1 and 2 denote the upstream and the immediately downstream regions of the shock. Then $M_1 = 2.2$ gives $M_2 = 0.547$. Then $M_{th} = 1$ requires $A_2/A_2^* = A_i/A_{th}' \approx 1.26$. Therefore, $A_{th}'/A_{th} = (A_i/A_{th})/(A_i/A_{th}') = 2.005/1.26 \approx 1.59$. The throat area needs to be increased by 59% to swallow the shock wave. $\square$

5.5 Continuous Closed-Circuit Supersonic Wind Tunnel

For ideal and full pressure recovery in continuous closed-circuit supersonic wind tunnels, double-throated design is preferred (shown in Fig. 5.8). To generate supersonic flow in the test section, a converging-diverging nozzle is necessary, although the diffuser is not. Without the diffuser to properly (or isentropically) slow down the flow exiting the test section, shock waves may occur in the flow passage leading back to the compressor, resulting in loss in stagnation pressure that would require more compressor work to maintain the operation. The deceleration of the supersonic flow using a diffuser is beneficial. An open-circuit (also called blowdown) type supersonic wind tunnel does not "recycle" the gas and therefore has no need for a diffuser.

Similar to the supersonic diffuser for aircraft engines, there is also a start-up problem for the double-throated supersonic wind tunnel. During the start-up, the flows in the nozzle and the diffuser are not in reverse to each other. The following discussion of the start-up operation assumes isentropic conditions everywhere in the wind tunnel except when crossing shock waves and receiving work from the compressor to maintain $p_{ti} = p_{te}$. For a designed Mach number in the test section, M_D, there is a corresponding value of A_{TS}/A_{th1} and $A_{th1} = A_{th2}$ During the steady state operation, the pressure distribution for an isentropic flow from the inlet to the exit of the double-throated section is shown in Fig. 5.8. This means that the flow conditions in the diffuser are exactly the reverse of those in the nozzle, with $M_{th1} = M_{th2} = 1$, $p_i = p_e$ and $p_{ti} = p_{te}$.

At the beginning of the start-up, $p_i > p_b$ has to be established. As p_b/p_i becomes sufficiently low to choke the nozzle, the throat conditions in the nozzle become sonic and $M_{th1} = 1$. A further decrease in p_b/p_i (or p_b/p_{ti}) would result in supersonic conditions, possibly with a normal shock wave, in the diverging section of the nozzle (as described in Chapter 4). The existence of a normal shock would cause loss in stagnation pressure, with the maximum loss occurring when it stands at the exit of the nozzle (i.e., at the inlet to the test section). Under these start-up conditions, the second throat area must be larger than the first for the flow to pass through because

$$\frac{p_{t2}}{p_{t1}} = \frac{A_1^*}{A_2^*} = \frac{A_{th1}}{A_{th2}} < 1$$

At the moment of the maximum loss, p_{t2}/p_{t1} and A_{th2}/A_{th1} is a function only of M_D as the shock has an upstream Mach number equal to M_D. Also at this moment, there

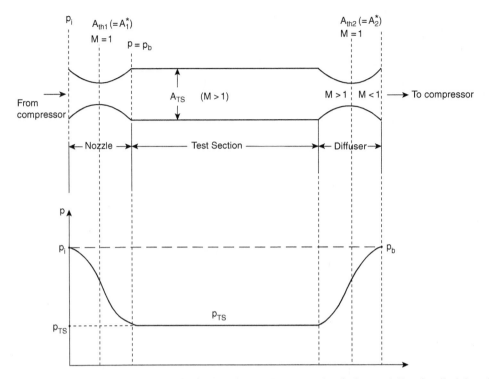

Figure 5.8 Schematic of a double-throated supersonic wind tunnel. For detailed description of the start-up operation, see text. Under the steady-state supersonic operation, isentropic conditions suggest that $A_{th,1} = A_1^* = A_2^* = A_{th,2}$; however, non-isentropic factors such as friction would require $A_{th,2} > A_{th,1}$.

is no preferred location for the shock to stand within the test section due to the isentropic assumption everywhere else and the associated result that any location produces the same shock loss. To ensure supersonic conditions throughout the test section, the shock has to be moved in the downstream direction and to eventually be swallowed into the diverging section of the diffuser. To accomplish swallowing requires further increase in A_{th2}, similar to supersonic inlets described in Section 5.3. Once the swallowing is completed, A_{th2} can be returned to that for the design condition again, *i.e.*, $A_{th2} = A_{th1}$. An example of how to open the diffuser's throat during starting phase has been given in Example 5.7.

5.6 One-Dimensional Flow in a Converging Channel

Flow in a converging channel can be considered a special case of that in a converging-diverging channel. Figure 5.9 shows a converging nozzle attached to a reservoir having a pressure p_r, with pressure distribution through the nozzle length. These results are similar to those in the converging-diverging nozzle, in that the choking condition is reached p_b or p_b/p_r is sufficiently low. Because the nozzle exit has the narrowest cross-sectional area, it is also the throat area under choked conditions,

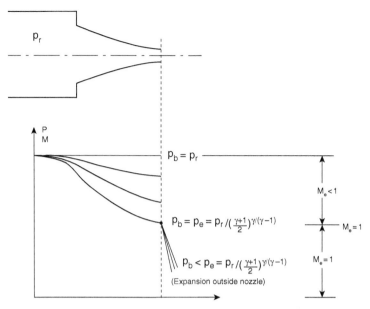

Figure 5.9 Pressure distribution through the length of a converging nozzle attached to a reservoir having a pressure p_r.

that is, $M_{th} = M_e = 1$ and $A_e = A_{th} = A^*$. Under the choked condition, $M_e = 1$ and $p_e/p_r = [(\gamma + 1)/2]^{\gamma/(\gamma+1)}$ (= 0.5283 for $\gamma = 1.4$). This means that for any value of p_b such that $p_b/p_r < [(\gamma + 1)/2]^{\gamma/(\gamma+1)}$, the mass flow rate does not increase further. Such a result differs from those for the converging-diverging nozzle, where the pressure ratio p_b/p_r for choking is dependent on A_e/A_{th}, as Part (a) of Example 5.4 demonstrates. For converging nozzles with $p_b/p_r < [(\gamma + 1)/2]^{\gamma/(\gamma+1)}$, the flow continues to expand outside the nozzle (see Fig. 5.9). The mass flow rate for a choked converging nozzle is thus

$$\dot{m} = \frac{P_t A^*}{\sqrt{RT_t}} \sqrt{\gamma} \left(\frac{\gamma + 1}{2}\right)^{(\gamma+1)/2(1-\gamma)} = \frac{P_r A_e}{\sqrt{RT_r}} \sqrt{\gamma} \left(\frac{\gamma + 1}{2}\right)^{(\gamma+1)/2(1-\gamma)} \quad (5.16)$$

which for a given nozzle and A_e is proportional to p_r. It is natural to expect no supersonic flow or shock wave within the nozzle. This is because a converging nozzle connected to a reservoir cannot generate supersonic flows.

Problems

Problem 5.1 Derive the differential form of momentum conservation, Eqn. (5.3), using Reynold's transport theorem, using the approach shown in Chapter 2.

Problem 5.2 Repeat the calculations for Example 5.4 with a new exit cross-sectional area $A_e = 7.5\text{m}^2$. Observe and explain the difference between these new results and those of Example 5.3.

Problem 5.3 A converging-diverging nozzle is connected to a high pressure air reservoir (at P_r and T_r) to expand the air to an ambient pressure p_b. The exit to throat area ratio A_e/A_{th} is 3. If a normal shock wave stands at a location where the cross-sectional area is equal to $2A_{th}$, what is the value of p_b?

Problem 5.4 A converging-diverging channel as supersonic diffuser is design for $M_\infty = M_D = 2.2, p_\infty = 10$ kPa and $p_\infty = 216$ K. The goal is for the flow to decelerate to a subsonic condition and then be fed to an engine with a subsonic speed. What is the throat area if the flow is to be isentropic throughout the diffuser? What are the exit cross-sectional area and the flow speed if $p_e = 90$ kPa is desired?

Problem 5.5 A converging-diverging nozzle is connected to a reservoir containing air at 500 kPa and 200 °C. The air is to be isentropically expanded to a pressure of 100 kPa at a rate of 15 kg/s. Find the cross-sectional areas where the pressure is 373.2 kPa, 152.8 kPa, and 100.0 kPa. What are the Mach numbers at these locations?

Problem 5.6 A converging-diverging nozzle is connected to a reservoir containing air at 1 MPa and 500 K. The nozzle exit has a cross-sectional area of 700 mm^2. The flow is isentropic everywhere except across a normal shock standing at a location having a local Mach number of 2.5 and a cross-sectional area of 500 mm^2. Find the flowing values: (a) the throat cross-sectional area, (b) the mass flow rate, (c) the exit plane Mach number, velocity, and static pressure, and (d) the entropy change across the shock.

Problem 5.7 A converging-diverging nozzle with an exit to throat area ratio (A_e/A_{th}) of 2.5 is attached to a high-pressure air reservoir having a pressure p_r. Under a certain operation condition, there is a standing normal shock wave at a location having an area ratio $A/A_{th} = 2$. What is the back pressure p_b? What should be the back pressure for the nozzle to be free of normal shock waves?

Problem 5.8 Find the over-speeding necessary to swallow normal shock waves for diffusers designed for flight Mach numbers of 1.8 and 2.0, respectively. If a variable throat is to swallow the shock, how much opening of the throat is needed for these two design Mach numbers?

Problem 5.9 Assume that it is possible to infinitely overspeed. What is the theoretical limit of the design Mach number?

Problem 5.10 A blow-down type supersonic wind tunnel uses the ambient air as its reservoir. The supersonic test condition is established by using a vacuum chamber to draw air through the converging-diverging nozzle, followed by the test section, as here shown. The run time is limited by the time when the chamber achieves a pressure that causes a normal shock wave to appear in the test section. Find the expression for the run time in terms of the test section Mach number, M_{TS}, the throat area, A_{th}, chamber volume, V, and the reservoir condition. If $M_{TS} = 2.5$ is desired, with the chamber volume equal to 20 m^3, the air at 101 kPa and 298 K, the throat area $A_{th} = 1 \times 10^{-2}$ m^2, what is the available time for testing?

Problem 5.11 During the start-up operation of a supersonic wind tunnel, with the test section to throat area ratio $A_{TS}/A_{th1} = 4$, a normal shock wave appears in the test section as shown. How much does the throat of the diffuser (A_{th2}) have to be opened up, compared to its steady state operation value, to swallow the shock wave? If, at this new value of A_{th2}, a normal shock exists at throat 2, what is the pressure downstream of the shock if the test section pressure is 20 kPa?

Problem 5.12 A supersonic diffuser is designed for a test section Mach number $M_{TS} = 2.5$ using air. To swallow the starting shock, how much does the throat cross-sectional area have to be increased? Repeat for $M_{TS} = 3.5$ and make observations by comparing the results of the two design Mach numbers.

Problem 5.13 Derive the following expression for the mass flow rate $\dot{m}$ for an unchoked converging nozzle. That is, for an unchoked converging nozzle

$$\frac{p_r}{p_b} < \left[\frac{\gamma+1}{2}\right]^{\frac{\gamma}{\gamma-1}}$$

$$\dot{m} = \sqrt{\left(\frac{2\gamma}{\gamma-1}\right)\left[\left(\frac{p_r}{p_b}\right)^{\frac{2(\gamma-1)}{\gamma}} - \left(\frac{p_r}{p_b}\right)^{\frac{\gamma-1}{\gamma}}\right]\frac{1}{RT_r}} \cdot (p_b A_e)$$

For comparison the choked nozzle mass flow rate is given by Eqn. (5.16) and is a linear function of p_r and independent of both p_r/p_b and p_b.

Problem 5.14 If a converging-diverging nozzle is designed to expand the flow from a reservoir to vacuum, the velocity is thus the maximum achievable (i.e., V_{max}). Assume an isentropic flow, then show that

$$V_{max} \to \sqrt{\frac{2}{\gamma-1}} a_t$$

where a_t is the speed of sound based on the reservoir temperature. Provide a physical reason why V_{max} does not approach infinity.

Problem 5.15 A converging nozzle is connected to an air reservoir at 500° C and 300 kPa. At the nozzle exit, the air velocity is 300 m/s. Find the Mach number, temperature, pressure, and density at the exit plane.

Problem 5.16 Define a Mach number M' so that $M' \equiv V/\sqrt{\gamma R T_t}$. Show that along a streamline in an isentropic flow:

$$M' = M\left(1 + \frac{\gamma-1}{2}M^2\right)^{-1/2}$$

Then use this expression to find the exit Mach number of Problem 5.15. (This result should help to demonstrate that Mach number based on the stagnation properties can serve as a reference parameter.)

Problem 5.17 To simulate a supersonic flight condition with an air speed of 738.0 m/s at an altitude of 15 km, a wind tunnel is designed to operate at the same Mach number and pressure, but with a temperature of 8° C to avoid condensation of water vapor. What should be the ratio of cross-sectional to throat areas? What should be the reservoir pressure and temperature? Assume an isentropic flow from the reservoir exit to the test section.

Problem 5.18 Use the thrust force equation for a rocket nozzle, following Reynolds transport theorem to give

$$F = \dot{m} V_e + A_e(p_e - p_a)$$

where F, $\dot{m}$, p_a, V_e, A_e, and p_e are the thrust force, mass flow rate, ambient pressure, exit velocity, cross-sectional, and exit pressure, respectively, to show that

(1) $F = p_t A^* \left\{ \left[\left(\frac{2}{\gamma-1} \right) \left(\frac{2}{\gamma-1} \right)^{\frac{\gamma+1}{\gamma-1}} \right] \left[1 - \left(\frac{p_e}{p_t} \right)^{\frac{\gamma+1}{\gamma}} \right] \right\}^{\frac{1}{2}} + A_e(p_e - p_a),$

(2) the maximum thrust for a given rocket chamber pressure (i.e., the total pressure p_t) is achieved with $p_e = p_a$ and

(3) for maximum thrust

$$\frac{A_e}{A^*} = \frac{\left[\frac{2}{\gamma+1} \right]^{\frac{\gamma+1}{2(\gamma-1)}} \left[\frac{p_a}{p_t} \right]^{\frac{\gamma+1}{2\gamma}}}{\left[\frac{2}{\gamma-1} \right]^{\frac{1}{2}} \left[\left(\frac{p_t}{p_a} \right)^{\frac{\gamma-1}{\gamma}} - 1 \right]^{\frac{1}{2}}}.$$

6 One-Dimensional Flows with Friction

Compressible channel flows may very well experience friction and heat transfer in addition to area changes. Wall friction naturally exists, while heat transfer is likely due to large temperature variations as Mach number varies. A typical example is the nozzle flow in the converging-diverging nozzle of a rocket to where there exist simultaneous effects of area change, friction, and heat transfer. In this chapter the effect of friction is analyzed for steady compressible flows. To concentrate on the effect of friction, one-dimensional flow in constant-area channel with adiabatic walls and without external work is considered. As in previous chapters, an ideal gas having constant specific heats (i.e., a perfect gas) is assumed.

6.1 Fanno Line Flow

Consider a steady one-dimensional flow in a channel of constant cross-section area with adiabatic walls and no external work, as shown in Fig. 6.1. Such a flow is called a *Fanno line flow*.

The conservation of mass requires that

$$\rho V = \text{constant} \tag{6.1}$$

As no heat transfer occurs through the adiabatic wall and no external work is done, the energy conservation is

$$h_t = c_p T_t = h + \frac{1}{2} V^2 = \text{constant} \tag{6.2}$$

Because of the friction effect, entropy change occurs. The change is given by Eqn. (2.37):

$$ds = c_v \frac{dT}{T} + R \frac{dv}{v}$$

When expressed in terms of density change, this expression becomes,

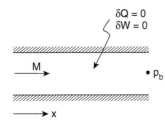

Figure 6.1 Steady one-dimensional flow in a channel of constant cross-section area with adiabatic walls and no external work (i.e., a Fanno-line flow).

$$ds = c_v \frac{dT}{T} - R \frac{d\rho}{\rho} \tag{6.3}$$

Assuming that the state at location 1 is a reference state and that specific heats remain constant throughout the channel, integration of Eqn. (6.3) yields

$$s - s_1 = c_v \ln \frac{T}{T_1} - R \ln \frac{\rho}{\rho_1}$$

By using continuity, Eqn. (6.1), this expression becomes

$$s - s_1 = c_v \ln \frac{T}{T_1} + R \ln \frac{V}{V_1} \tag{6.4}$$

A *T-s diagram* is useful in explaining the flow, and it can be obtained by replacing velocity by temperature using the energy equation, Eqn. (6.2):

$$V = \sqrt{2(h_t - h)} = \sqrt{2c_v(T_t - T)} \tag{6.5}$$

Substituting Eqn. (6.5) into Eqn. (6.4) yields

$$\frac{\Delta s}{c_v} = \frac{s - s_1}{c_v} = \ln\left(\frac{T}{T_1}\right) + \frac{R}{2c_v}\ln\left(\frac{T_t - T}{T_t - T_1}\right) = \ln\left(\frac{T}{T_1}\right) + \frac{\gamma - 1}{2}\ln\left(\frac{T_t - T}{T_t - T_1}\right) + \text{constant} \tag{6.6}$$

where relations $R = c_p - c_v$ and $c_p = \gamma c_v$ are used. By further using these relations, Eqn. (6.6) leads to

$$\frac{\Delta s}{c_p} = \frac{s - s_1}{c_p} = \frac{1}{\gamma}\ln\left(\frac{T}{T_1}\right) + \frac{\gamma - 1}{2\gamma}\ln\left(\frac{T_t - T}{T_t - T_1}\right) = \frac{1}{\gamma}\ln T + \frac{\gamma - 1}{2\gamma}\ln(T_t - T) + \text{constant} \tag{6.7}$$

Equation (6.7) provides a relationship for plotting the *T-s* diagram, which is qualitatively sketched in Fig. 6.2, where the *T-s* curve is the Fanno line for a given mass flow rate. The horizontal axis could be $(s - s_1)/c_p = \Delta s/c_p$ or simply Δs, while the vertical is T. Because the friction effect is irreversible, entropy can only increase when no heat can be taken away from the flow (i.e., $\Delta s = (h_t - h_{t1})/T < 0$ is not possible for the adiabatic channel). Therefore, Δs is also qualitatively related to Δx, the distance traveled by the flow, in Fig. 6.2; the longer the distance travels, the larger

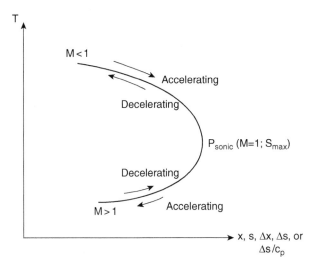

Figure 6.2 The *T-s* curve is the Fanno line for a given mass flow rate.

the entropy increase. The quantitative relation between Δs and Δx will be derived in the next section.

It can be seen in Fig. 6.2 that there exists a maximum entropy along the *T-s* curve denoted as P_{sonic}, where $d\Delta s/dT = 0$ and $M = 1$ (and, thus subscript "sonic"). Point P_{sonic} divides the *T-s* curve into a subsonic branch and a supersonic branch. The sonic condition is established in the following, followed by the discussion of general features of the two branches.

Referring to point P_{sonic} in Fig. 6.2, $d\Delta s/dT = 0$ is obtained by differentiating both sides of Eqn. (6.7):

$$\frac{1}{c_p}\frac{d\Delta s}{dT} = \frac{1}{\gamma}\left(\frac{1}{T} - \frac{\gamma - 1}{2}\frac{1}{T_t - T}\right) = 0$$

The energy conservation requires that $(T_t - T) = V^2/2c_p$, which along with $c_p = \gamma R/(\gamma - 1)$ can be substituted into the above expression to yield

$$\frac{1}{T} - \frac{\gamma - 1}{2}\frac{1}{T_t - T} = \frac{1}{T} - \frac{\gamma - 1}{2}\frac{2}{c_p V^2} = \frac{1}{T}\left(1 - \frac{\gamma - 1}{2}\frac{2}{\gamma - 1}\frac{\gamma R T}{V^2}\right) = \frac{1}{T}\left(1 - M^2\right) = 0$$

Thus,

$$M = 1 \text{ and } s = s_{max} \text{ at point } P_{sonic}. \tag{6.8}$$

Because the wall is adiabatic, T_t is constant along a given *T-s* curve. Since at any location in the channel and the corresponding point on the *T-s* curve (Fig. 6.2),

$$\frac{T}{T_t} = \frac{1}{1 + \frac{\gamma - 1}{2}M^2}$$

It is apparent that the upper branch (with a smaller value of M and, thus, higher temperature T) is the subsonic branch, while the lower branch is supersonic. Because both *s* and *x* increase in the flow direction on the *T-s* diagram, it is concluded (1) that for flows in an adiabatic constant-area channel, a subsonic flow accelerates with increasing *M* toward the sonic condition, i.e., toward point P_{sonic} on the *T-s* curve (with increasing Δx and Δs and decreasing *T*), and (2) that a supersonic flow decelerates with decreasing *M* toward the sonic condition, *i.e.*, toward point P_{sonic} on the *T-s* curve (*also* with increasing Δx and Δs and increasing *T*).

How does this similar qualitative increase in entropy cause very different flow behaviors in subsonic and supersonic flows, with the former accelerating and the latter decelerating? The results from nozzle flow analysis coupled with the development of a boundary layer due to viscosity should help to explain such qualitative differences. Viscosity causes the *no-slip boundary condition* at the wall, resulting in a boundary layer near the wall in which the effect of the wall is felt by the fluid and the fluid velocity within the boundary layer is smaller than that of fluids at a sufficient distance from the wall. That is, the flow outside the boundary layer does not feel the effect of the wall. Because of slower velocities, the mass flow rate within the cross-section of the boundary layer is smaller than what it would be without the effect of viscosity. A *displacement thickness* near the wall can be calculated, as frequently done in boundary layer flows (Fox et al., 2004); it represents a near-wall region within which there is no flow (i.e., a dead zone). As the effect of viscosity accumulates in the flow (*x*) direction, the displacement thickness increases, while steady-state mass flow rate remains the same throughout the channel. For the constant-area channel, this development amounts to a reduction in the effective cross-sectional area, effectively resulting in a converging channel ($dA/dx < 0$). Thus it becomes clear that a subsonic flow accelerates, while a supersonic flow decelerates (as described in Chapter 4), toward the sonic condition at point P_{sonic}.

Because the sonic condition is associated with the maximum entropy on the *T-s* curve, it is not possible for a subsonic flow to accelerate to become supersonic by adding channel length, as the turning around point P_{sonic} would require a decrease in entropy. Neither is it possible to decelerate a supersonic flow to subsonic conditions for the same reason.

For given inlet conditions, the sonic condition can be reached given a sufficiently long channel (or given a sufficient increase in entropy) and a sufficiently low back pressure. This critical length for reaching $M = 1$ (denoted by L^*, following the convention of using the superscript * for sonic conditions) depends on the initial Mach number, as can be seen in Fig. 6.2. For the channel length not exceeding L^*, conditions at any location within the channel can be identified on the *T-s* curve. But what happens if channel length is longer than L^*, as shown by $L_2 > L^*$ in Fig. 6.3a? For subsonic flows, the fluid can sense the extra channel length and would adjust accordingly by reducing the mass flow rate ($\dot{m}_2 < \dot{m}_1$, as shown in Fig. 6.3b), while maintaining $M = 1$ at the end of the channel. The reduction in $\dot{m}$ can be understood by considering that the displacement thickness has increased due to additional channel length, further reducing the effective cross-sectional area at the sonic location. Furthermore, the stagnation pressure becomes lower with the extra length (this

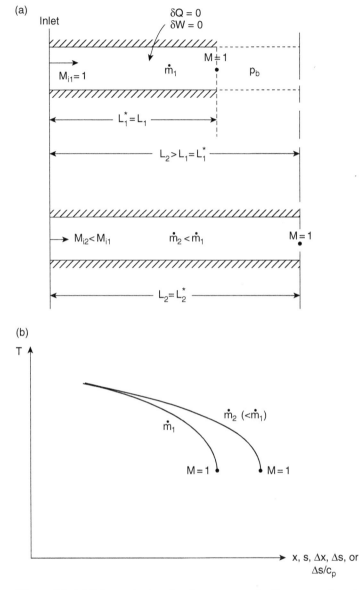

Figure 6.3 (a) Schematic showing the events and adjustment of a subsonic inlet flow when the constant-area duct length exceeds that (L^*) for $M = 1$, the adjustment being a reduced mass flow rate; (b) the adjustment leads to a different Fanno line (the *T-s* curve) with a reduced mass flow rate, $\dot{m}_2 < \dot{m}_1$.

will be shown in the next section). Imagine that the two channels with lengths L_2 and $L_1 = L_1^*$ and exit cross-sectional areas A_2^* and A_1^*, respectively, behave like converging nozzles connected to two reservoirs with stagnation pressures P_{t2} and P_{t1}. Because $L_2 > L^*$, $A_2^* < A_1^*$ and $P_{t2} < P_{t1}$. The mass flow rates equation, Eqn. (5.16), can be used for comparing $\dot{m}_1$ and $\dot{m}_2$:

$$\dot{m}_2 = \frac{P_{t2}A_2^*}{\sqrt{RT_t}}\sqrt{\gamma}\left(\frac{\gamma+1}{2}\right)^{(\gamma+1)/2(1-\gamma)} < \frac{P_{t1}A_1^*}{\sqrt{RT_t}}\sqrt{\gamma}\left(\frac{\gamma+1}{2}\right)^{(\gamma+1)/2(1-\gamma)} = \dot{m}_1 \qquad (6.9)$$

In this case, the channel is choked, meaning that further lowering p_b does not lead to an increase in $\dot{m}$, although increasing the channel length results in further decrease, in $\dot{m}$. Also in this case L_2 becomes the new critical length ($L_2 = L_2^*$) for the new Fanno line of $\dot{m}_2$.

A supersonic flow cannot sense the fact that $L_2 > L^*$. It adjusts to the extra length by means of a normal shock while maintaining the same $\dot{m}$. The flow downstream of the normal shock can sense the remaining length and can adjust to the back pressure because it is subsonic. As a consequence, the location of the shock wave has to be dependent on the back pressure (p_b), because the location needs to be such that the adjustment $p_e = p_b$ is achieved.

The above analysis is based on continuity and conservation of energy, without invoking the momentum equation. It is apparent that friction force affects the flow speed and leads to the formation of the boundary layer and the displacement thickness that reduces the exit cross-sectional area. Without explicitly considering viscous effects, the result of Eqn. (6.9) is only qualitative, as stated, and so is the discussion of the characteristic of the *T-s* diagram. Questions remain regarding (1) how to determine the critical length, L^*, (2) the flow conditions throughout the channel, including those at point P_{sonic} that serve as the reference conditions for a given Fanno line, (3) the location of the normal shock if it exists. The following section describes derivations leading to answers to these questions.

6.2 Equations of Fanno Line Flow and Critical Length L*

Inferring from the results of Fig. 6.2 and the previous discussion, the value of L^* should be dependent on the Mach number at the channel inlet, whether the inlet is subsonic or supersonic. To determine L^* as a function of M, first consider the momentum conservation for the differential control volume in Fig. 6.4. While the continuity is given by Eqn. (6.1), the momentum equation can be derived by applying the Reynolds transport theorem, Eqn. (2.5):

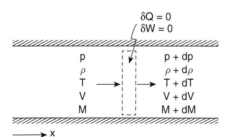

Figure 6.4 Control volume for Fanno line flow analysis.

$$\sum_i \vec{F}_i \Big)_{system} = \frac{\partial}{\partial t} \int_{CV} \rho \vec{V} \, dV + \int_{CS} \vec{V} \left(\rho \vec{V} \cdot d\vec{A} \right)$$

Under the steady-state assumption, this expression is reduced to

$$pA - (p + dp)A - \tau_w A_w = [(\rho + d\rho)(V + dV)(A)](V + dV) - (\rho V A)V = \rho V A \, dV$$

where the continuity $(\rho + d\rho)(V + dV) = \rho V$ is used, and τ_w and A_w are the wall shear stress and the wall surface area, respectively. This expression is further reduced to

$$-A dp - \tau_w A_w = \rho V A \, dV \tag{6.10}$$

A hydraulic diameter, D_h, can be defined to express A_w in terms of A so that Eqn. (6.10) can be further simplified:

$$D_h \equiv \frac{4 \times (\text{Cross-sectional area})}{\text{wetted perimeter}} = \frac{4\,A}{P} \tag{6.11}$$

It can be easily shown that $D_h = D$ for a circular duct with a diameter D, $D_h = W$ for a square with a width W on each side, and $D_h = 2\,WH/(W + H)$ for a rectangular channel with W and H as its width and height, respectively. Therefore,

$$A_w = P dx = \frac{4\,A dx}{D_h}$$

and Eqn. (6.11) becomes

$$-A dp - \tau_w \frac{4\,A dx}{D_h} = \rho V A \, dV \tag{6.12}$$

Define a friction coefficient (also called the Darcy friction coefficient), f, as

$$f \equiv \frac{4\tau_w}{\frac{1}{2}\rho V^2}$$

Equation (6.12) becomes

$$-dp - \frac{1}{2}\rho V^2 f \frac{dx}{D_h} = \rho V dV \tag{6.13}$$

Dividing terms in the above expression through, by either p or ρRT $(= \rho \gamma RT/\gamma = \rho a^2/\gamma)$ yields

$$\frac{dp}{p} + \frac{1}{2}\gamma M^2 f \frac{dx}{D_h} + \gamma M^2 \frac{dV}{V} = 0 \tag{6.14}$$

To express L^* as a function of a Mach number will require that dp/p and dV/V in Eqn. (6.14) be also expressed in terms of a Mach number before integration can be carried out. The ideal gas law can be written as

$$\ln p = RT\ln\rho + \rho T\ln R + \rho R\ln T$$

Differentiating this expression and then dividing the resultant terms on the left-hand side by p and those on the right-hand side by ρRT yields

$$\frac{dp}{p} = \frac{d\rho}{\rho} + \frac{dT}{T} \tag{6.15}$$

The continuity equation, $\rho V = $ constant, can be similarly manipulated to yield

$$\frac{d\rho}{\rho} = -\frac{dV}{V} \tag{6.16}$$

Substituting Eqn. (6.16) into Eqn. (6.15) and then the result for dp/p into Eqn. (6.14) leads to

$$\frac{dT}{T} + \frac{1}{2}\gamma M^2 f\frac{dx}{D_h} - \left(1 - \gamma M^2\right)\frac{dV}{V} = 0 \tag{6.17}$$

Equation (4.8) indicates

$$T\left(1 + \frac{\gamma-1}{2}M^2\right) = T_t \tag{4.8}$$

and $T_t = $ constant for the Fanno line flow. Differentiation of Eqn. (4.6c) produces

$$\frac{dT}{T} = \frac{d\left(1 + \frac{\gamma-1}{2}M^2\right)}{1 + \frac{\gamma-1}{2}M^2} \tag{6.18}$$

Since $V = M\sqrt{\gamma RT}$, similar manipulation results in

$$\frac{dV}{V} = \frac{dM}{M} + \frac{1}{2}\frac{dT}{T} \tag{6.19}$$

Substituting the expressions for dT/T and dV/V into Eqn. (6.17) yields

$$f\frac{dx}{D_h} = \left(\frac{\gamma+1}{2\gamma}\right)\left[\frac{d\left(1 + \frac{\gamma-1}{2}M^2\right)}{1 + \frac{\gamma-1}{2}M^2} - \frac{2\,MdM}{M^2}\right] + \frac{2\,dM}{\gamma\,M^3} \tag{6.20}$$

Integrating Eqn. (6.20) over the length of L^* for the range of M from M to 1 yields

$$\frac{fL^*}{D_h} = \left(\frac{\gamma+1}{2\gamma}\right)\left(\ln\frac{\frac{\gamma+1}{2}M^2}{1 + \frac{\gamma-1}{2}M^2}\right) - \frac{M^2 - 1}{\gamma M^2} \tag{6.21a}$$

Values of fL^*/D_h can be easily calculated for given values of M, which can be subsonic or supersonic, but not vice versa. For convenience, values of fL^*/D_h are tabulated against M in Table B.3. For $M = 1$, $fL^*/D_h = 0$. For $M \to \infty$,

$$\left[\frac{fL^*}{D_h}\right]_{M\to\infty} = \left(\frac{\gamma+1}{2\gamma}\right)\ln\left(\frac{\gamma+1}{\gamma-1}\right) - \frac{1}{\gamma}$$

For $\gamma = 1.4$, this limiting value is 0.822. Thus, for the supersonic branch of the Fanno line, fL^*/D_h falls in the range from 0 to 0.822. For the subsonic branch, fL^*/D_h can be large as $M \to 0$

$$\left[\frac{fL^*}{D_h}\right]_{M\to 0} = \left\{\left(\frac{\gamma+1}{2\gamma}\right)\left[\ln\left(\frac{\gamma+1}{2}M^2\right)\right] + \frac{1}{\gamma M^2}\right\}_{M\to 0}$$

By applying the l'Hospital rule to the ratio of the two terms on the right-hand side of this expression results in $fL^*/D_h \to \infty$ as $M \to 0$.

It is clear that L^* is a function of M, and therefore serves as the reference length over which the flow with M would have to travel to reach the sonic state, point P_{sonic} on the T-s (or T-x) curve shown in Fig. 6.2. A distance short of L^* would indicate that the flow is still subsonic or supersonic. The following two examples demonstrate how to determine L^* and then use it to determine M if the distance traveled is $L < L^*$.

EXAMPLE 6.1 Air flow enters an insulated circular duct with a Mach number of 0.3, and the duct has a diameter of 5 cm and a friction coefficient of 0.02. What is the length of the duct needed for the flow to reach the sonic condition at the exit of the duct? What is the Mach number if the duct length is 1 m?

Solution –
For $M_1 = 0.3$ and $\gamma = 1.4$, $fL_1^*/D_h = 5.2993$. Therefore,
$L_1^* = 5.2993 \times \frac{D_h}{f} = 5.2993 \times \frac{5\ cm}{0.02} = 1,324.8\ cm = 1.325\ m$; for the sonic exit, $L_1^* = L_1$.
For $\frac{fL_2^*}{D_h} = \frac{fL_1^*}{D_h} - \frac{fL_1}{D_h} = 5.2993 - \frac{0.02\times 1000}{5} = 1.2993$. Thus, $M_e \approx 0.475$.
Thus by traveling a distance of 1 m, the $M_1 = 0.3$ flow does not reach the sonic condition. The $M \approx 0.475$ would need a length of 0.325 m, shorter than 1.325 m for the $M = 0.3$ flow, to reach $M_e = 1$. □

EXAMPLE 6.2 Air flow enters an insulated circular duct with a Mach number of 2.5, and the duct has a diameter of 3 cm and a friction coefficient of 0.02. What is the length of the duct needed for the flow to reach the sonic condition at the exit of the duct? What is the exit Mach number if the duct length is 50 cm?

Solution –
For $M_1 = 2.5$ and $\gamma = 1.4$, $fL_1^*/D_h = 0.432$. Therefore,
$L_1^* = 0.432 \times \frac{D_h}{f} = 0.432 \times \frac{3\ cm}{0.02} = 64.8\ cm$; for the sonic exit, $L_1^* = L_1$.
For $\frac{fL_2^*}{D_h} = \frac{fL_1^*}{D_h} - \frac{fL_1}{D_h} = 0.432 - \frac{0.02\times 50}{3} = 0.099$. Thus, $M_e \approx 1.40$.
The flow does not travel a sufficiently long distance to reach the sonic state. □

6.3 Reference State in Fanno Line Flow

The flow that has traveled the length L^* thus reaches sonic conditions. Due to the non-isentropic process, there is loss in stagnation pressure. Therefore p_t does not remain constant in Fanno line flows and cannot be a reference parameter. Because of the adiabatic wall assumption, T_t remains constant and can serve as a reference parameter. It is desirable to find out the values of p_t, p, and T as a functions of a Mach number. Since $T_t = $ constant and $T^* = T_t/[(\gamma + 1)/2]$ is also a constant, temperature at the sonic condition can be a choice for reference. Following similar reasoning, values of p_t and p at $M = 1$ might also serve as references.

Replacing dV/V in Eqn. (6.14) with Eqn. (6.19) leads to

$$\frac{dV}{V} = \frac{dM}{M} + \frac{1}{2}\frac{dT}{T}$$

Substituting this expression along with Eqn. (6.18) for dT/T into Eqn. (6.14) yields

$$\frac{dp}{p} + \frac{1}{2}\gamma M^2 f \frac{dx}{D_h} + \gamma M^2 \frac{dM}{M} - \frac{1}{2}\frac{\gamma M^2(\gamma - 1)M^2}{1 + \frac{\gamma-1}{2}M^2}\frac{dM}{M} = 0 \tag{6.22}$$

The second term of Eqn. (6.22) can be expressed in terms of Mach number by rewriting Eqn. (6.20) as

$$f\frac{dx}{D_h} = \left[\frac{1 - M^2}{1 + \frac{\gamma-1}{2}M^2}\right]\left[\frac{2}{\gamma M^2}\right]\frac{dM}{M} \tag{6.23}$$

Substituting Eqn. (6.23) into Eqn. (6.22) and collecting terms yields

$$\frac{dp}{p} = -\left[\frac{1 + (\gamma - 1)M^2}{1 + \frac{\gamma-1}{2}M^2}\right]\frac{dM}{M} = -\left[\frac{1 + \frac{\gamma-1}{2}M^2}{1 + \frac{\gamma-1}{2}M^2} + \frac{\frac{\gamma-1}{2}M^2}{1 + \frac{\gamma-1}{2}M^2}\right]\frac{dM}{M}$$

$$= -\frac{dM}{M} - \frac{1}{2}\frac{d\left(1 + \frac{\gamma-1}{2}M^2\right)}{1 + \frac{\gamma-1}{2}M^2} \tag{6.24}$$

Integration of this expression from $M = M$ to $M = 1$ leads to

$$\ln\left(\frac{p}{p^*}\right) = \ln\left(\frac{1}{M}\right) + \frac{1}{2}\int_M^1 \frac{d\left(1 + \frac{\gamma-1}{2}M^2\right)}{1 + \frac{\gamma-1}{2}M^2} = \ln\left[\left(\frac{1}{M}\right)\left(\frac{(\gamma + 1)/2}{1 + \frac{\gamma-1}{2}M^2}\right)^{1/2}\right] \tag{6.25}$$

Therefore,

$$\frac{p}{p^*} = \left(\frac{1}{M}\right)\left[\frac{(\gamma + 1)/2}{1 + \frac{\gamma-1}{2}M^2}\right]^{1/2} \tag{6.26}$$

where $p/p^* = 1$ for $M = 1$, as expected. Because at a given location,

$$\frac{p}{p_t} = \frac{1}{\left(1 + \frac{\gamma-1}{2}M^2\right)^{\gamma/(\gamma-1)}}$$

one obtains

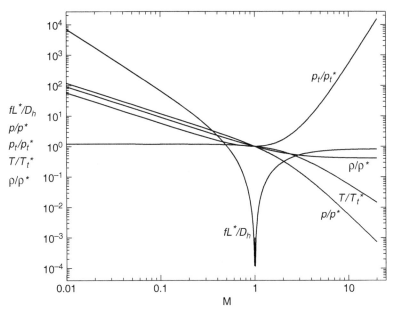

Figure 6.5 Properties of Fanno line flow as a function of M for $\gamma = 1.4$.

$$\frac{p_t}{p_t^*} = \frac{p}{p^*} \frac{\left(1 + \frac{\gamma-1}{2}M^2\right)^{\gamma/(\gamma-1)}}{\left(\frac{\gamma+1}{2}\right)^{\gamma/(\gamma-1)}} = \left(\frac{1}{M}\right)\left[\frac{2}{\gamma+1}\left(1 + \frac{\gamma-1}{2}M^2\right)\right]^{\frac{\gamma+1}{2(\gamma-1)}} \qquad (6.27)$$

By using $T_t = T\left(1 + \frac{\gamma-1}{2}M^2\right) = $ constant,

$$\frac{T}{T^*} = \frac{(\gamma+1)/2}{1 + \frac{\gamma-1}{2}M^2} \qquad (6.28)$$

The continuity equation combined with $V = \sqrt{\gamma RT}$ and $V^* = \sqrt{\gamma RT^*}$ leads to

$$\frac{\rho}{\rho^*} = \frac{V^*}{V} = \left(\frac{1}{M}\right)\left[\frac{2}{\gamma+1}\left(1 + \frac{(\gamma-1)}{2}M^2\right)\right]^{1/2} \qquad (6.29)$$

As expected in Eqns. (6.27) through (6–29), $p_t/p_t^* = 1$, $T/T^* = 1$ and $\rho/\rho^* = 1$ at $M = 1$. The numerical results of these equations are tabulated along with fL^*/D_h in Appendix F. For $\gamma = 1.4$, they are shown graphically in Fig. 6.5.

As with flow across a normal shock, the Fanno line flow is adiabatic and one-dimensional. The equation for entropy increase across the normal shock is applicable to the Fanno line flows. Thus,

$$\frac{s_2 - s_1}{R} = -\ln\frac{p_{t2}}{p_{t1}} \qquad (4.19c)$$

EXAMPLE 6.3 Following Example 6.1, if $p_1 = 150 \, kPa$, $T_1 = 300 \, K$, $L_1 = 10 \, \text{m}$, find the pressure, temperature, speed, and pressure loss at the end of the duct (location 2 or exit).

Solution –

$$M_1 = 0.3, \frac{p_1}{p^*} = 3.619, \frac{T_1}{T^*} = 1.179, \frac{p_1}{p_{t1}} = 0.9395, p_{t1} = 156.7 \text{ kPa}$$

$$\frac{fL_2^*}{D_h} = \frac{fL_1^*}{D_h} - \frac{fL_1}{D_h} = 5.2993 - \frac{0.02 \times 1000}{5} = 1.2993. \text{ Thus, } M_2 = M_e \approx 0.475$$

$$\frac{p_2}{p^*} \approx 2.257, \; p_2 = \frac{p_2}{p^*} \times \frac{p^*}{p_1} \times p_1 = 2.257 \times \frac{1}{3.619} \times 150 \, kPa = 93.5 \, kPa$$

$$\frac{T_2}{T^*} = 1.148, \; T_2 = \frac{T_2}{T^*} \times \frac{T^*}{T_1} \times T_1 = 1.148 \times \frac{1}{1.179} \times 300 \, K = 292.1 \, K$$

$$V_2 = M_2 \sqrt{\gamma R T_2} = 0.475 \times \sqrt{1.4 \times 287 \times 292.1} = 162.7 \frac{m}{s}$$

$V_1 = M_1 \sqrt{\gamma R T_2} = 0.3 \times \sqrt{1.4 \times 287 \times 300} = 104.2 \frac{m}{s}$ ($\rightarrow$ acceleration from location 1 to location 2)

$$\frac{p_{t1}}{p_t^*} = 2.035, \; \frac{p_1}{p_{t1}} = 0.9395 \text{ (locally isentropic)}, \; \frac{p_{t2}}{p_t^*} = 1.391$$

$$p_{t2} = \frac{p_{t2}}{p_t^*} \frac{p_t^*}{p_{t1}} \frac{p_{t1}}{p_1} p_1 = 1.391 \times \frac{1}{2.035} \times \frac{1}{0.9395} \times 150 \, kPa = 109.1 \, kPa$$

$$(p_{t2} - p_{t1}) = p_{t2} - \frac{p_{t1}}{p_1} \times p_1 = 109.1 - \frac{1}{0.9395} \times 150 = -50.6 \text{ kPa}$$

$\square$

6.4 Fanno Line Flow in a Nozzle Followed by a Constant-Area Duct

In practice, the flow in a constant-area duct may be fed by a converging- or a converging-diverging nozzle attached to a reservoir. First consider the case with a converging nozzle followed by a constant-area duct having a length L, as depicted in Fig. 6.6. For convenience, the flow in the converging nozzle is assumed to be isentropic. It is clear that nowhere through the nozzle-duct system can the flow become supersonic for two reasons: First, the flow can at most be sonic at the entrance/inlet to the duct, so $M_i < 1$. Second, the sonic entrance flow to the duct cannot be accelerated to be supersonic without heat transfer away from the fluid to decrease in entropy for the transition, to turn around the point P_{sonic} in the Fanno line, from the upper, subsonic branch to the lower, supersonic branch shown in Fig. 6.2. Therefore, the flow in the constant-area duct shown in Fig. 6.6 can at most be

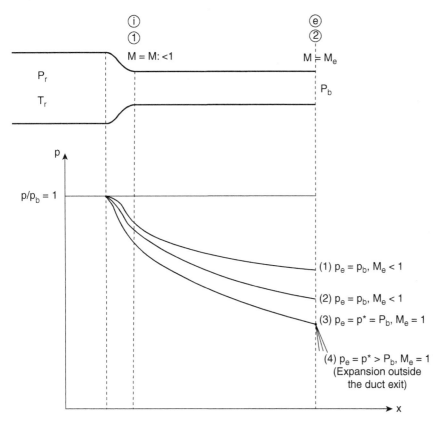

Figure 6.6 Fanno-line flow in a converging nozzle followed by a constant-area duct. For $p_b \leq p_{(3)}$ the sonic exit condition is established at the duct exit and the flow is frictionally choked, that is, the mass flow rate, $\dot{m}$, does not increase for $p_b < p_{(3)}$ and the flow continues to expand outside the duct. For $p_b \geq p_{(4)}$, $\dot{m}$ increases as p_b decreases.

sonic and the only possible sonic location is the exit of the duct (i.e., $M_e = 1$). Associated with this scenario of sonic exit is $p_e = p^*$.

If $M_e < 1$ (an example is represented by Curve 1 in Fig. 6.6), the entire flow throughout the nozzle-duct assembly is subsonic and $p_b = p_e$, but $p_b = p_e > p^*$ and $M_e < 1$. If p_b is reduced correspondingly to Curve 2 in Fig. 6.6, an increase in both M_e and $\dot{m}$ result, still with $p_b = p_e > p^*$ and $M_e < 1$. For conditions represented by Curves 1 and 2, $M_e < 1$. For these two cases, $L < L^*$.

Even lower p_b conditions of Curve 3 in Fig. 6.6 are achieved such that $M_e = 1$, $p_b = p_e = p^*$, and $p_{te} = p_t^*$. In this case $L = L^*$. Further lowering p_b would not cause the flow to become supersonic at exit, as shown by Curve 4 in Fig. 6.6. For Curve 4, $M_e = 1$, $p_b < p_e = p^*$, $p_{te} = p_t^*$, and $L = L^*$ and the exit flow cannot sense further decreases in the back pressure; thus $p_b < p_e = p^*$ and the flow continues to expand outside the duct. Because $L = L^*$, M_i remains the same as that represented by Curve 3, the value of $\dot{m}$ for Curve 4 is thus the same as that for Curve 3. Therefore the duct is choked. In summary, $p_b \leq p_e = p^*$, $p_{te} = p_t^*$, and $L = L^*$ for $M_e = 1$. Once the

$M_e = 1$ condition is reached, the duct is choked by friction. That is, further decreases in p_b do not lead to larger mass flow rates.

For a back pressure of Curve 4, one can afford to add duct length and still achieve $M_e = 1$. This would, however, cause an increase in L^*, as the new physical duct length L is L^*. As a consequence, M_i is reduced and so is $\dot{m}$.

If $p_b = 0$, then $M_e = 1$ and $p_e = p^*$ are guaranteed regardless the length of the duct. Then whatever the duct length is, $L = L^*$. In this case, a larger L leads to a smaller M at the duct inlet and thus a smaller $\dot{m}$. For $L = L^* \to \infty$, $M_i \to 0$ (and $\dot{m} = 0$), as discussed in Section 6.2; this result is expected because no flow can overcome friction over an infinite distance.

EXAMPLE 6.4 A constant-area duct is connected to an air reservoir (at 150 kPa and 300 K) through a converging nozzle. At the junction of the nozzle and the duct, the diameter is 5 cm. The friction coefficient of the duct wall is 0.02, and the duct length is 1 m. What is the maximum back pressure for $M_e = 1$, what is the maximum mass flow rate, and what is the pressure due to the frictional duct?

Solution –

For $M_e = 1$, $p_b \le p_e = p^*$.

Let the location of the nozzle-duct junction be designated as "1."

Using the table in Appendix F, $\frac{fL_1^*}{D_h} = \frac{0.02 \times 1000}{5} = 4$, $M_1 \approx 0.33$, $\frac{p_1}{p^*} \approx 3.287$, $\frac{T_1}{T^*} \approx 1.174$, and $\frac{p_{t1}}{p_t^*} \approx 1.970$.

To find p^*, the duct inlet pressure p_1 is needed. For $M_1 \approx 0.33$, the isentropic relationship provides $\frac{p_1}{p_{t1}} = \frac{p_1}{p_r} \approx 0.922$. Therefore, $p^* = \frac{p^*}{p_1} \times \frac{p_1}{p_r} \times p_r = \frac{1}{3.287} \times 0.922 \times 150 \ kPa = 42.08 \ kPa$.

Therefore, $p_b \le 42.08 \ kPa$ for $M_e = 1$.

$\dot{m} = \rho^* A V^* = \frac{p^*}{RT^*} A \sqrt{\gamma R T^*}$ So, T^* is needed. Following similar process of finding p^*,

$$T^* = \frac{T^*}{T_1} \times \frac{T_1}{T_r} \times T_r = \frac{1}{1.174} \times 0.979 \times 300 \ K = 250.2 \ K$$

Therefore,

$$\dot{m} = \frac{p^*}{RT^*} A \sqrt{\gamma R T^*} = \frac{42.08 \times 10^3}{287 \times 250.2} \times \left[\frac{\pi}{4}(5 \times 10^{-2})^2\right] \sqrt{1.4 \times 287 \times 250.2}$$
$$= 0.365 \ kg/s$$

or

$$\dot{m} = \frac{p_1}{RT_1} A M_1 \sqrt{\gamma R T_1}$$

$$= \frac{150 \times 10^3 \times 0.922}{287 \times 300 \times 0.979} \times \left[\frac{\pi}{4}(5 \times 10^{-2})^2\right] \times 0.33 \times \sqrt{1.4 \times 287 \times 300 \times 0.979}$$

$$= 0.365 \ kg/s$$

Because $p_t^* = p_{te}$ and $p_{t1} = p_r$

$$(p_{te} - p_{t1}) = \frac{p_{te}}{p_t^*} \times \frac{p_t^*}{p_{t1}} \times p_{t1} - p_{t1} = \left(1 \times \frac{1}{1.97} - 1\right)p_r = \left(\frac{1}{1.97} - 1\right) \times 150 \ kPa$$

$$= -73.86 \ kPa$$

□

EXAMPLE 6.5 Repeat the calculation for the mass flow rate and pressure loss in Example 6.4 except now $p_b = 0$ and the duct length is 3.25 m. Discuss the differences in results.

Solution –
Because $p_b = 0 < p^*$ for any realistic value of p^*, so the duct is choked and $M_e = 1$.

According to the table in Appendix F, $\frac{fL_1^*}{D_h} = \frac{0.02 \times 3250}{5} = 13$, $M_1 \approx 0.21$, $\frac{p_1}{p^*} \approx 5.205$, $\frac{T_1}{T^*} \approx 1.189$, and $\frac{p_{t1}}{p_t^*} \approx 2.836$.

$$p^* = \frac{p^*}{p_1} \times \frac{p_1}{p_r} \times p_r = \frac{1}{5.205} \times 0.970 \times 150 \ kPa = 27.95 \ kPa$$

$$T^* = \frac{T^*}{T_1} \times \frac{T_1}{T_r} \times T_r = \frac{1}{1.189} \times 0.991 \times 300 \ K = 250.0 \ K$$

$$\dot{m} = \frac{p^*}{RT^*} A \sqrt{\gamma R T^*} = \frac{27.95 \times 10^3}{287 \times 250.0} \times \left[\frac{\pi}{4}\left(5 \times 10^{-2}\right)^2\right]\sqrt{1.4 \times 287 \times 250.0}$$

$$= 0.242 \ kg/s$$

$$p_t^* = p_{te}, \ (p_{te} - p_{t1}) = \left(\frac{1}{2.836} - 1\right)p_{t1} = \left(\frac{1}{2.836} - 1\right)p_r$$

$$= \left(\frac{1}{2.836} - 1\right) \times 150 \ kPa = -97.11 \ kPa$$

Discussion –
The mass flow rate is reduced from that in Example 6.4 due to the added duct length. This trend is consistent with the result of the *T-s* diagram shown in Fig. 6.2; that is, when the duct is already choked (as in Example 6.4 also for $p_b = 0$) further addition of the duct length should lead to a reduced mass flow rate and a new *T-s* curve. This reduction is accompanied by an increase in the loss in stagnation pressure, which according to $\frac{s_2 - s_1}{R} = -\ln\frac{p_{t2}}{p_{t1}}$ also leads to

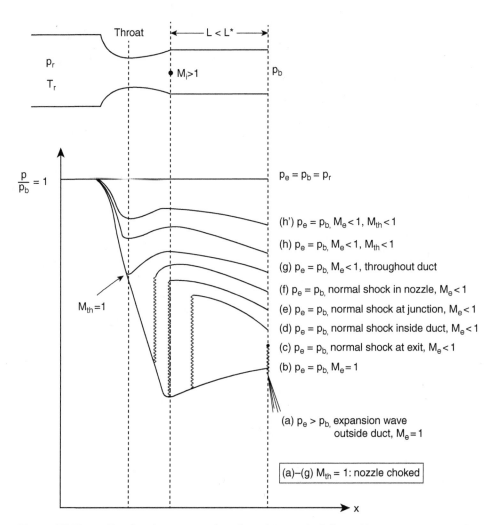

Figure 6.7 Fanno line flow in a converging-diverging nozzle followed by a constant-area duct with $L < L^*$. Unlike in Fig. 6.5, supersonic flow can exist in the duct for $p_b \leq p_e$ as no normal shock exists in the diverging section of the nozzle (p_e is the ambient pressure corresponding to Curve e). For $p_b \leq p_g$, the nozzle-duct assembly is choked and $\dot{m}$ does not change. For $p_b \leq p_{(b)}$, the exit Mach number $M_e = 1$.

a higher increase in entropy at the duct exit. As a result the new *T-s* curve is to the right of the one with shorter duct length, as clearly shown in Fig. 6.2.   □

The case of a constant-area duct connected to a reservoir through a converging-diverging nozzle is shown in Figs. 6.7 (for $L < L^*$) and 6.8 (for $L > L^*$). The flow in the converging-diverging section is again assumed to be isentropic, while the effect of wall friction is present in the constant-area duct.

For $p_b = p_r$ there is no flow and the pressure is the same and equal to p_b throughout the entire channel. As the back pressure is lowered ($p_b < p_r$), a flow is induced. For the cases where subsonic flow enters the duct ($M_i < 1$), the flow would

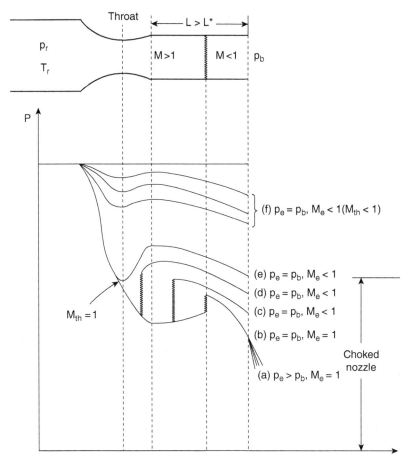

Figure 6.8 Fanno line flow in a converging-diverging nozzle followed by a constant-area duct with $L > L^*$. Supersonic flow can exist in the duct; however, unlike in Fig. 6.9, normal shock always exists even if the ambient is vacuum. Only for $p_b \leq p_{(b)}$ is the exit Mach number $M_e = 1$.

remain subsonic throughout the flow, and at most become sonic at the duct exit ($M_e = 1$). Such a scenario is similar to that of a converging nozzle-duct assembly already discussed. Thus, the case of supersonic flow at the inlet of the duct (i.e., at the nozzle exit) is more relevant for gaining new insight.

For the flow to be supersonic at the nozzle exit, the throat Mach number must be unity, that is, $M_{th} = 1$. As a consequence, the flow does not sense any further lowering of p_b and the mass flow rate reaches the maximum allowed:

$$\dot{m} = \frac{P_r A^*}{\sqrt{RT_r}} \sqrt{\gamma} \left(\frac{\gamma + 1}{2}\right)^{(\gamma+1)/2(1-\gamma)} \tag{5.13}$$

where $A^* = A_{th}$. In this case the nozzle, rather than the duct, is choked and $M_i > 1$. Three scenarios of exit condition are possible: (1) $L < L^*$, $M_e > 1$, and $p_e \gtrless p_b$; (2) $L < L^*$ and if the flow at the duct exit has to adjust to p_b, the adjustment has to be done through a shock wave and the shock wave location is determined by p_b so that $p_e = p_b$; and (3) that if $L > L^*$, there would always be a shock wave in the duct, as discussed earlier for the characteristics of the *T-s* diagram of Fig. 6.2, and the shock wave location is determined by p_b so that $p_e = p_b$.

For the first scenario ($L < L^*$, $M_e > 1$, and $p_e \geq p_b$) for the given M_i, the pressure variation in the nozzle-duct assembly is shown by Curves (a) and (b) in Fig. 6.7. Curve (a) represents the case of $p_e > p_b$, while Curve (b) represents the special condition of $p_b = p_e$, with $M_e > 1$ for both cases. For back pressure of Curve (a), the flow continues to expand outside the duct. The flow does not have to (and cannot because $M_e > 1$) adjust to p_b.

The second scenario requires an increase in p_b above conditions represented by Curve (b). Such an increase in p_b would lead to oblique shocks at the duct exit (i.e., conditions between Curves [b] and [c]) until the back pressure is sufficiently high to cause a normal shock at the exit (Curve [c]). Downstream of the shock, the exit pressure adjusts to the ambient value, that is, $p_e = p_b$.

Further increases in p_b above that of Curve (c) result in moving the normal shock upstream of the duct exit, with Curve (d) as an example. For sufficiently high values of p_b, the normal shock moves into the diverging section of the nozzle and the entire flow in the duct is subsonic. The back pressure for Curve (e) causes the normal shock to stand right at the junction of the nozzle and the duct, while p_b for Curve (f) moves the shock inside the diverging section of the nozzle. For Curves (f), (g), and (h), $M_i < 1$, enabling the adjustment for $p_e = p_b$.

In summary, for $M_i > 1$ and $L < L^*$, the location of normal shock, if it occurs, in the nozzle-duct assembly depends on the back pressure. If the back pressure is sufficiently low, no normal shock exists at all. This conclusion is similar to the existence and the location of a normal shock wave in a choked converging-diverging frictionless nozzle, as discussed in chapter in Chapter 5.

EXAMPLE 6.6 A constant-area duct is connected to a converging-diverging nozzle, which is in turn connected to a reservoir of air at 400 kPa and 400 K. The area ratio of the nozzle is 2.0 while the length to diameter ratio of the duct is 5 and the friction coefficient is 0.02. What is the range of back pressure for a normal shock to exist in the duct?

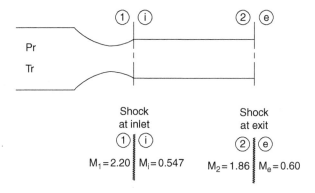

Solution –

For a normal shock to exist in the duct implies that the Mach number at the nozzle exit must be greater than one. As shown in Fig. E6.6, $\frac{A_1}{A^*} = \frac{A_1}{A_{th}} = 2$. Therefore, $M_1 = M_i = 2.20$, for which $\frac{fL_1^*}{D_h} = 0.3609$. For the given duct $\frac{fL_1}{D_h} = 0.02 \times 5 = 0.10 < \frac{fL_1^*}{D_h} = 0.3609$. Therefore, $\frac{fL_2^*}{D_h} = \frac{fL_e^*}{D_h} = \frac{fL_1^*}{D_h} - \frac{fL_1}{D_h} = 0.2609$ and $M_2 \approx 1.86$. For this Mach number at the exit the pressure ratio across a normal shock is $\frac{p_b}{p_2} \approx 3.870$. However, what is p_2?

$$p_2 = \frac{p_2}{p^*} \times \frac{p^*}{p_1} \times \frac{p_1}{p_{t1}} \times \frac{p_{t1}}{p_r} \times p_r = 0.4528 \times \frac{1}{0.3549} \times 0.0935 \times 1 \times 400 \ kPa$$

$$= 47.72 \text{ kPa}$$

$$p_b = 187.1 \ kPa$$

For the normal shock to exist at the inlet to the duct, $M_1 = 2.20$ and due to the normal shock $M_i \approx 0.547$ for which $\frac{fL_1^*}{D_h} = 0.730 > \frac{fL_1}{D_h} = 0.10$ and $\frac{p_i}{p^*} \approx 1.95$. Therefore, $\frac{fL_2^*}{D_h} = \frac{fL_e^*}{D_h} = \frac{fL_1^*}{D_h} - \frac{fL_1}{D_h} = 0.630$. Thus $M_e = M_2 \approx 0.565$.

$$p_b = p_e = \frac{p_e}{p^*} \times \frac{p^*}{p_i} \times \frac{p_i}{p_1} \times \frac{p_1}{p_{t1}} \times \frac{p_{t1}}{p_r} \times p_r$$

$$\approx 1.88 \times \frac{1}{1.95} \times 5.341 \times 0.0935 \times 1 \times 400 \ kPa = 192.58 \ kPa$$

Therefore, for the range of back pressure 47.72 kPa $< p_b <$ 192.58 kPa, a normal shock exists in the duct, with the higher and lower limits for it to stand at the duct inlet and at the duct exit, respectively. ☐

In the third scenario, with $M_i > 1$ and $L > L^*$, a normal shock exists in the nozzle-duct assembly. Figure 6.8 shows the pressure variation in the nozzle-duct assembly as the back pressure is varied. For extremely low back pressures (such as a vacuum), the flow accelerates downstream of the normal shock with decreasing pressure. Because of the low back pressure, the flow will expand and achieve sonic condition at the exit. Note that even for a vacuum back pressure, the subsonic flow downstream of the shock cannot become supersonic, as discussed earlier. The sonic exit condition is still possible with p_b for Curves (a) and (b). Curve (b) is a special case with $M_e = 1$ and $p_e = p_b$ while Curve (a) has $M_e = 1$ and $p_e > p_b$. For back pressures below curve (b), the location of normal shock does not change to a further downstream location, because the sonic flow at the duct exit prevents the flow from sensing any lower back pressure and adjusts accordingly. For back pressure higher than that for Curve (b), the normal shock wave moves upstream toward the nozzle, as depicted by Curve (c). For sufficiently higher p_b, the normal shock wave moves into the diverging section of the nozzle, as shown by Curve (d), and the entire flow in the duct is subsonic. For p_b corresponding to that of Curve (e), the shock wave disappears with $M_{th} = 1$, i.e., a choked nozzle. For p_b below that for Curve (e), no

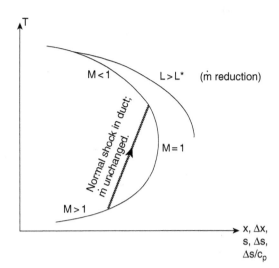

Figure 6.9 The transition from supersonic to subsonic conditions due to shock wave on a given Fanno line (i.e., maintaining the same mass flow rate as the nozzle is already choked). Note that the shock within the duct can be caused by either $L > L^*$ or by sufficiently high back pressure with $L < L^*$. Also shown is the mass reduction due to duct lengthening for the subsonic Fanno line.

more increase in the mass flow rate is possible, as the flow in the nozzle cannot sense such a condition. For p_b above Curve (b), the exit Mach number is less than unity, $M_e < 1$ and $p_e = p_b$.

Figure 6.9 shows the transition from supersonic to subsonic conditions due to the shock wave on a given Fanno line, that is, maintaining the same mass flow rate as the nozzle is already choked. It is noted that the transition is in the direction of increased entropy and the subsonic flow accelerates toward point P_{sonic}. Depending on the value of p_b, the exit flow may reach the sonic condition as discussed above.

EXAMPLE 6.7 For conditions given in Example 6.6, except that the length to diameter ratio is now 25, find the pressure range for a normal shock to exist in the duct.

Solution –
Because $\frac{fL_1}{D_h} = 0.02 \times 25 = 0.50 > \frac{fL_1^*}{D_h} = 0.3609$, there will be a normal shock in the nozzle-duct assembly even if the back pressure is a vacuum (in fact this is the case for back pressure below that corresponds to Curve (b) in Fig. 6.8).

The question first of all becomes: at what back pressure does the shock stands at the duct inlet?
$\frac{A_1}{A^*} = \frac{A_1}{A_{th}} = 2$. Therefore, the Mach number at the nozzle exit is $M_1 = 2.20$, for which $\frac{p_1}{p_{t1}} = \frac{p_1}{p_r} = 0.0935$ at the nozzle exit and $\frac{p_i}{p_1} = 5.80$ across the shock. After the shock the Mach number at the duct inlet is $M_i = 0.5471$ for which $\frac{fL_i^*}{D_h} = 0.747$ and $\frac{p_i}{p^*} \approx 1.95$. Therefore, $\frac{fL_e^*}{D_h} = \frac{fL_i^*}{D_h} - \frac{fL_1^*}{D_h} = 0.747 - 0.50 = 0.247$, and $M_e \approx 0.68$, for which $\frac{p_e}{p^*} \approx 1.541$.

$$p_b = p_e = \frac{p_e}{p^*} \times \frac{p^*}{p_i} \times \frac{p_i}{p_1} \times \frac{p_1}{p_r} \times p_r = 1.541 \times \frac{1}{1.95} \times 5.480 \times 0.0953 \times 400 \ kPa$$

$$= 165.1 \ kPa.$$

Therefore, for a normal shock wave to exist in the duct, $0 < p_b < 165.1 \ kPa$.

Comments –

This upper limit of back pressure is higher than that in Example 6.6, simply because for the same shock location at the nozzle-duct junction, there is a longer duct now for the subsonic duct for to accelerate, resulting in a lower exit-plane pressure and, thus, a lower back pressure for the normal shock to stand at the duct inlet. Also differing from Example 6.6 is that the current example has a long duct to guarantee a normal shock within the duct. □

Problems

Problem 6.1 Show the steps leading to Eqns. (6.21) and (6.23).

Problem 6.2 A perfectly insulated circular pipe (2.5 m in length and 5 cm in diameter; friction coefficient is 0.02) is connected to an air reservoir (at 700 kPa and 450 K) through a converging nozzle. What is the maximum flow rate that can be delivered by this nozzle-pipe assembly? What is the exit velocity under this condition? Under what back pressure can this maximum mass flow rate be delivered?

Problem 6.3 Referring to Problem 6.2, what is the flow rate if the back pressure is 450 kPa? What are the velocities at the duct inlet and exit?

Problem 6.4 A constant-area duct is connected to a converging-diverging nozzle (resembling those depicted in Figs. 6.6 and 6.7), which is in turn connected to a reservoir of air at 400 kPa and 400 K. The area ratio of the nozzle is 2.0 while the length to diameter ratio of the duct is 150 and the friction coefficient is 0.02. Find the maximum back pressure for the nozzle to be choked. Under this condition, what is the loss in stagnation pressure?

Problem 6.5 For the same conditions given in Problem 6.4, except that the duct length to diameter ratio is 25, find the back pressure below which a normal shock exists and that the exit Mach number is 1 (downstream of the duct inlet). Also find the location of the shock.

Problem 6.6 For the conditions given in Problem 6.4 (air reservoir pressure at 400 kPa, temperature at 400 K; nozzle area ratio of 2, duct length to diameter ratio of 150), find the back pressure for a normal shock to exist in the diverging section of the nozzle where the area ratio is1.30. What is the exit velocity?

Problem 6.7 Following Example 6.2, if the stagnation pressure and stagnation temperature entering the duct are 500 kPa and 800 K, respectively, what are the exit pressure and velocity and what is the pressure loss?

Problem 6.8 Following the same conditions given in Example 6.7, except with a vacuum back pressure. Determine the location of the normal shock.

Problem 6.9 Consider an air flow in an adiabatic constant-area rectangular duct (dimensions: 2 cm x 4 cm) connected to a reservoir (pressure = 300 kPa) through a converging-diverging nozzle (with an exit-to-throat area ratio of 1.688). The duct length is 50 cm and the friction factor f is 0.01.

(a) Show that supersonic exit is possible.

(b) What is the back pressure if no normal shock is to stand at the duct exit?

(c) What is the highest back pressure for the nozzle-duct assembly to be choked?

(d) What happens if the back pressure falls within the range between those determined in (b) and (c)?

7 One-Dimensional Flows with Heat Transfer

Heat transfer in a compressible flow occurs because of the temperature difference between the flow and the surroundings across the non-adiabatic channel walls. Heat addition to the fluid can also arise from chemical, heat-releasing reactions that convert chemical enthalpy to thermal enthalpy. The flow in the converging-diverging nozzle of a rocket is a good example, as the high-temperature flow may lose energy to the wall of the nozzle, while some chemical reactions may continue into the nozzle as they are not completed in the combustion chamber. Steady compressible one-dimensional flow in a frictionless constant-area channel is considered to focus on the significance of heat transfer. Similar to the Fanno line flow, the constant-area channel flow with gas supplied by a nozzle (converging or converging-diverging) connected to a reservoir will be discussed. As in the Fanno line flow, a perfect gas is assumed.

7.1 Rayleigh Line Flow

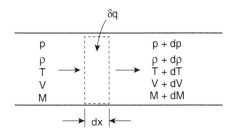

Figure 7.1 Steady one-dimensional flow in the channel of a constant cross-section area with frictionless walls and heat addition (i.e., a Rayleigh-line flow).

The steady, inviscid one-dimensional compressible flow in the channel of a constant cross-section area, with heat transfer to the fluid and no external work, constitutes the *Rayleigh line flow*. To derive governing equations of the Rayleigh line flow consider the differential control volume shown in Fig. 7.1. The conservation of mass requires that

$$\rho V = (\rho + d\rho)(V + dV) = \text{constant} \qquad (7.1)$$

By differentiating both sides of $\rho V = \text{constant}$ and the dividing through by ρV, or by neglecting the second-order term in the equality $\rho V = (\rho + d\rho)(V + dV)$, it is easy to show that

$$\frac{d\rho}{\rho} + \frac{dV}{V} = 0 \qquad (7.2)$$

Because of the frictionless wall, the momentum conservation equation takes the form of Eqn. (2.24)

$$p + \rho V^2 = \text{constant}. \qquad (2.24)$$

Noting that $\rho V = \text{constant}$, the momentum conservation equation simplifies to

$$dp + \rho V dV = 0 \qquad (7.3)$$

The energy conservation law with $c_p = \text{constant}$ is

$$dh_t = d(c_p T_t) = c_p dT_t = dh + d\left(\frac{1}{2}V^2\right) = c_p dT + V dV = \delta q \qquad (7.4)$$

where δq is the energy added to the fluid per unit mass. Equations (7.2), (7.3), and (7.4) are the governing equations for the Rayleigh line flow. Just as for the Fanno line flow, it is desirable to express these equations in terms of a Mach number.

With $p = \rho RT$ and $a^2 = \gamma RT$, Eqn. (2.24) becomes

$$p(1 + \gamma M^2) = \text{constant} \qquad (7.5)$$

Because $V = M\sqrt{\gamma RT}$, and $p = \rho RT$

$$T = \frac{p}{\rho R} = \frac{p}{(\dot{m}/AV)R} = \frac{pM\sqrt{\gamma RT}}{(\dot{m}/A)R}$$

Therefore, with $\dot{m}/A = \rho V = \text{constant}$,

$$\sqrt{T} = (\text{constant})pM$$

By using Eqn. (7.5) for p, this expression for T becomes

$$T = (\text{constant})\frac{M^2}{(1 + \gamma M^2)^2} \qquad (7.6)$$

Equations (7.4), (7.5), and (7.6) provide the relations for p, T, and T_t between any two points in the Raleigh line flow:

$$\frac{p_2}{p_1} = \frac{1 + \gamma M_1^2}{1 + \gamma M_2^2} \qquad (7.7)$$

$$\frac{T_2}{T_1} = \frac{M_2^2(1 + \gamma M_1^2)^2}{M_1^2(1 + \gamma M_2^2)^2} \tag{7.8}$$

$$q = c_p(T_{t2} - T_{t1}) \tag{7.9}$$

As in the Fanno line flow, it is convenient to adopt the properties at $M = 1$ as the reference values. Thus, with location 1 designated as the reference state $(M_1 = 1)$ and location 2 as an arbitrary location, the above equations can be rewritten as

$$\frac{p}{p^*} = \frac{1 + \gamma}{1 + \gamma M^2} \tag{7.10}$$

$$\frac{T}{T^*} = \frac{(1 + \gamma)^2 M^2}{(1 + \gamma M^2)^2} \tag{7.11}$$

The continuity requirement yields

$$\frac{V}{V^*} = \frac{\rho^*}{\rho} = \frac{p^*/RT^*}{p/RT} = \frac{(1 + \gamma)M^2}{1 + \gamma M^2} \tag{7.12}$$

Locally isentropic relations

$$p_t = p\left(1 + \frac{\gamma - 1}{2}M^2\right)^{\gamma/(\gamma-1)}$$

and

$$T_t = T\left(1 + \frac{\gamma - 1}{2}M^2\right)$$

lead to

$$\frac{p_t}{p_t^*} = \left(\frac{1 + \gamma}{1 + \gamma M^2}\right)\left[\left(\frac{2}{\gamma + 1}\right)\left(1 + \frac{\gamma - 1}{2}M^2\right)\right]^{\gamma/(\gamma-1)} \tag{7.13}$$

and

$$\frac{T_t}{T_t^*} = \frac{2(1 + \gamma)M^2}{(1 + \gamma M^2)^2}\left(1 + \frac{\gamma - 1}{2}M^2\right) \tag{7.14}$$

Numerical values of these ratios expressed in Eqns. (7.10) to (7.14) as functions of a Mach number are tabulated in Appendix G. The results are plotted in Fig. 7.2 for $\gamma = 1.4$. For a better understanding of these relations, it is desirable to construct the *T-s* diagram for a given mass flow rate, much like the one for the Fanno line flow in Chapter 6.

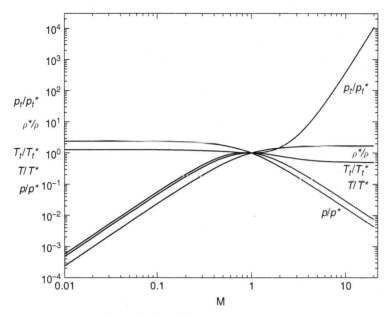

Figure 7.2 Properties of Rayleigh line flow as a function of M for $\gamma = 1.4$.

For an ideal gas,

$$ds = c_p \frac{dT}{T} - R \frac{dp}{p}$$

Since $R/c_p = (\gamma - 1)/\gamma$

$$\frac{s - s_1}{c_p} = \ln \frac{T}{T_1} - \frac{(\gamma - 1)}{\gamma} \ln \frac{p}{p_1} \tag{7.15a}$$

or, alternatively,

$$\frac{s - s^*}{c_p} = \ln \frac{T}{T^*} - \frac{(\gamma - 1)}{\gamma} \ln \frac{p}{p^*} \tag{7.15b}$$

To obtain the *T-s* relation, the pressure term can be expressed in terms of temperature. Combining Eqns. (7.7) and (7.8) yields

$$\frac{p}{p_1} = \frac{M_1}{M} \sqrt{\frac{T}{T_1}} \tag{7.15c}$$

Utilizing Eqns. (7.8) and (7.9) again produces

$$M^2 = \frac{1}{\gamma} \left[(1 + \gamma M_1^2) \frac{p_1}{p} - 1 \right] \tag{7.15d}$$

Substituting M from Eqn. (7.15d) into Eqn. (7.15c) yields

$$\frac{p}{p_1} \sqrt{\frac{1}{\gamma} \left[(1 + \gamma M_1^2) \frac{p_1}{p} - 1 \right]} = M_1 \sqrt{\frac{T}{T_1}} \tag{7.15e}$$

Squaring this expression and rearranging the result leads to

$$\left(\frac{p}{p_1}\right)^2 - \left(1 + \gamma M_1^2\right)\left(\frac{p}{p_1}\right) + \gamma M_1^2 \frac{T}{T_1} = 0 \tag{7.15f}$$

Solving this quadratic equation results in

$$\frac{p}{p_1} = \frac{1}{2}\left[\left(1 + \gamma M_1^2\right) \pm \sqrt{\left(1 + \gamma M_1^2\right)^2 - 4\gamma M_1^2\left(\frac{T}{T_1}\right)}\right] \tag{7.15g}$$

Thus Eqn. (7.15b) for entropy change becomes

$$\frac{s - s_1}{c_p} = \ln\frac{T}{T^*} - \frac{(\gamma - 1)}{\gamma}\ln\left\{\frac{1}{2}\left[\left(1 + \gamma M_1^2\right) \pm \sqrt{\left(1 + \gamma M_1^2\right)^2 - 4\gamma M_1^2\left(\frac{T}{T^*}\right)}\right]\right\} \tag{7.16}$$

It is noted that the following inequality has to be satisfied for Eqn. (7.16):

$$\frac{T}{T_1} \leq \frac{\left(1 + \gamma M_1^2\right)^2}{4\gamma M_1^2}$$

If the reference condition 1 is replaced by the sonic condition, then

$$\frac{s - s^*}{c_p} = \ln\frac{T}{T^*} - \frac{(\gamma - 1)}{\gamma}\ln\left\{\frac{1}{2}\left[\left(1 + \gamma\right) \pm \sqrt{\left(1 + \gamma\right)^2 - 4\gamma\left(\frac{T}{T^*}\right)}\right]\right\} \tag{7.17}$$

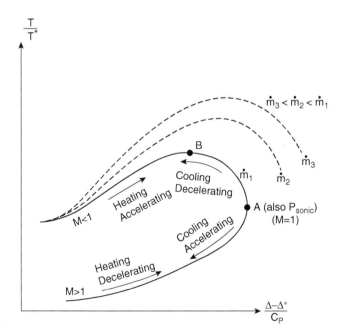

Figure 7.3 Each *T-s* curve is the *Rayleigh line* for a given mass flow rate. The solid-line Rayleigh line curve (with mass flow rate $\dot{m}_1$) demonstrates the effects of subsonic and supersonic heating and cooling on temperature and Mach number. The dashed-line curves show that for a given inlet condition, the effect of subsonic heating rates exceeding that required for $M_e = 1$(point A on the solid curve), resulting in reduced mass flow rates: $\dot{m}_3 < \dot{m}_2 < \dot{m}_1$.

The *T-s* relationship described by Eqn. (7.17) is qualitatively sketched in Fig. 7.3, where the + and the − signs in Eqn. (7.16) are used respectively for the upper and the lower branches, separated by point B, where *T* is the maximum ($T = T_{max}$) along the *T-s* curve. Similar to the Fanno line flow, there is a point for maximum entropy that also corresponds to $M = 1$ conditions, as will be shown in the following, and is designated as P_{sonic}.

At P_{sonic}, $ds/dT = 0$. The continuity and momentum conservation throughout the flow require

$$\frac{d\rho}{\rho} + \frac{dV}{V} = 0 \tag{7.2}$$

$$dp + \rho V dV = 0 \tag{2.24}$$

Combining these two equations yields

$$V^2 = \frac{dp}{d\rho}$$

At point P_{sonic}, there is no variation in entropy. Thus

$$V^2 = \left(\frac{dp}{d\rho}\right)_s = a^2$$

That is, $M_{psonic} = 1$ at the point of maximum entropy, with the point already denoted as P_{sonic} for easy identification. The condition for P_{Tmax} is $dT = 0$. Thus, by using Eqn. (7.11),

$$dT = d\left[T^* \frac{(1+\gamma)^2 M^2}{(1+\gamma M^2)^2}\right] = \frac{1}{(1+\gamma M^2)^2}\left[2(1+\gamma)^2 M(1+\gamma M^2)^2\right.$$

$$\left. -(1+\gamma)^2 M^2 4\gamma M(1+\gamma M^2)\right] dM = 0$$

Therefore,

$$M_{Tmax} = \sqrt{\frac{1}{\gamma}} \tag{7.18}$$

and

$$\left(\frac{T}{T^*}\right)_{max} = \frac{(1+\gamma)^2}{4\gamma} \tag{7.19}$$

It is apparent that $M_{Tmax} < 1$, and for air ($\gamma = 1.4$), $M_{Tmax} = 0.845$ and $(T/T^*)_{max} = 0.735$.

To understand more clearly the general characteristics of the T-s curve, it is instructive to express the entropic change as a function of a Mach number. Substituting Eqns. (7.8) and (7.9) into Eqn. (7.15a) yields

$$\frac{s - s_1}{c_p} = \ln \frac{T}{T_1} - \frac{(\gamma - 1)}{\gamma} \ln \frac{p}{p_1} = \ln \left[\frac{M^2 (1 + \gamma M_1^2)^2}{M_1^2 (1 + \gamma M^2)^2} \right] - \frac{(\gamma - 1)}{\gamma} \ln \left[\frac{1 + \gamma M_1^2}{1 + \gamma M^2} \right]$$

$$= \ln \left[\left(\frac{M}{M_1} \right)^2 \left(\frac{1 + \gamma M_1^2}{1 + \gamma M^2} \right)^{(\gamma+1)/\gamma} \right] \qquad (7.20)$$

Because the maximum temperature occurs under the subsonic condition, the upper branch of T-s curve shown in Fig. 7.3 is the subsonic branch, while the lower branch is the supersonic branch. On the subsonic branch, heat addition increases the entropy and accelerates the flow toward the sonic condition (point P_{sonic}); on the supersonic branch heat addition increases entropy and leads to deceleration toward P_{sonic}. There are alternative, qualitative ways of ascertaining the flow acceleration/deceleration resulting from heat transfer; Problem 7.11 provides the instruction and exercise for obtaining the results.

7.2 Reference State in Rayleigh Line Flow

If $M_1 = 1$ is the reference state, then Eqn. (7.20) becomes

$$\frac{s - s^*}{c_p} = \ln \left[M^2 \left(\frac{1 + \gamma}{1 + \gamma M^2} \right)^{(\gamma+1)/\gamma} \right] \qquad (7.21)$$

It can be seen that because $(\gamma + 1)/\gamma > 1$, $\left[M^2 \left(\frac{1+\gamma}{1+\gamma M^2} \right)^{(\gamma+1)/\gamma} \right]$ and therefore $\frac{s-s^*}{c_p}$ are increasing (decreasing) functions of M for $M < 1$ ($M > 1$). It can be concluded that, for a given flow rate (i.e., a given T-s curve) with heat addition, an initially subsonic flow accelerates, while an initially supersonic flow decelerates toward the sonic condition (point P_{sonic} on the T-s curve). With heat removal from the flow, the opposite occurs. Because $\frac{T}{T^*} = \frac{(1+\gamma)^2 M^2}{(1+\gamma M^2)^2} \sim \frac{1}{M^2}$, it is also concluded that the subsonic branch of the T-s curve has larger values of $\frac{T}{T^*}$ than the supersonic branch; so the subsonic regime corresponds to the upper branch. This conclusion is consistent with the result shown in Eqn. (7.18) in that P_{Tmax} is on the upper branch and $M_{Tmax} = \sqrt{(1/\gamma)} < 1$. As can be seen in Fig. 7.3, for $M < \sqrt{(1/\gamma)}$ on the subsonic branch of the T-s curve, T/T^* increases with increasing M by heat addition (or simply, heating); for $1 > M > \sqrt{(1/\gamma)}$, T/T^* decreases with increasing M still by heating. For $M > 1$ (the lower, supersonic branch), T/T^* always increases as M is decreased by heating. (A quantitative description of the trends of the combined effects of Mach number and heat transfer can be derived and is left as an end-of-chapter exercise, Problem 7.11.)

As with the Fanno line flow, similar questions are asked for the Rayleigh line flow as to what occurs when the amount of heat addition exceeds that required to accelerate/decelerate a subsonic/supersonic flow to the sonic condition. For subsonic Rayleigh line flow (which can be generated by connecting the constant-area duct to a reservoir by a converging nozzle), the inlet flow to the channel is subsonic because of the converging nozzle. By heating, the flow in the constant-area channel cannot become supersonic, according to the result shown in Fig. 7.3. If heat addition exceeding that required for $M = 1$, the mass flow rate will be reduced as the flow is affected by the downstream heating and a new Rayleigh line is established, as shown in Fig. 7.3. In Fig. 7.3, $\dot{m}_2 < \dot{m}_1$, where $\dot{m}_1$ is for subsonic flow with the heat addition and $\dot{m}_2$ is the mass flow rate due to excessive heat addition. Even more heat addition results in yet another Rayleigh line with $\dot{m}_3 < \dot{m}_2$. Therefore, a duct flow can be choked by heat addition, and the maximum amount of heat that can be added to the flow for a given Rayleigh line (i.e., a given $\dot{m}$) is determined by the attainment of the sonic condition.

Similar to the Fanno line flow, whether the choking condition occurs also depends on the back pressure. For example, if the back pressure is sufficiently low, then the exit Mach number $M_e = 1$, and the stagnation and static temperatures attained at the exit are T_t^* and T^*, respectively. This is illustrated by curves 4–6 in

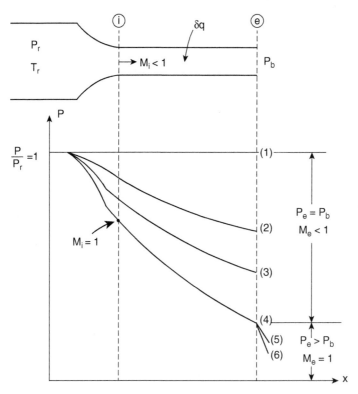

Figure 7.4 Rayleigh-line flow in a converging nozzle followed by a constant-area duct. For $p_b \leq p_{(4)}$ the sonic exit condition is established at the duct exit and the flow is thermally choked; that is, the mass flow rate, $\dot{m}$, does not increase for $p_b \leq p_{(4)}$; for $p_b \geq p_{(4)}$, $\dot{m}$ increases as p_b decreases. For $p_b < p_{(4)}$, the flow continues to expand outside the duct.

Fig. 7.4. Therefore, the flow is subsonic throughout the nozzle-channel system, with the possibility of achieving sonic exit conditions if p_b is sufficient. Curve 4 corresponds to the maximum back pressure that would choke the flow (with $M_e = 1$ and $p_e = p_b = p^*$) for the given mass flow rate and heat addition. Above this back pressure, the exit condition is subsonic, represented by curves 1–3. For back pressure lower than that of curve 4 (i.e., curves 5 and 6), $M_e = 1$, $p_e > p_b$, and the flow continues to expand outside the duct exit.

For a supersonic Rayleigh flow, the adjustment to excessive heating results in a normal shock wave in the duct. The location of the normal shock wave depends on the back pressure and the distribution of the head addition along the length of the duct. Because the amount of heat transfer varies with the flow speed and the gas temperature, unlike the friction coefficient that is constant throughout the Fanno line flow, the determination of the shock location is more complicated than that in the Fanno line flow. This scenario is schematically shown in Fig. 7.5.

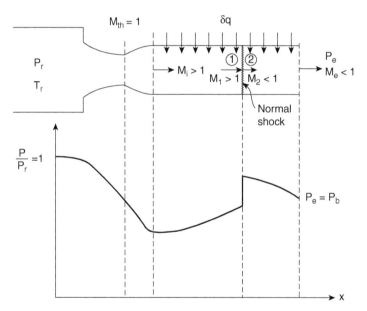

Figure 7.5 Rayleigh-line flow in a converging-diverging nozzle followed by a constant-area duct with duct inlet Mach number $M_i > 1$ and $\delta q > 0$ such that a normal shock stands within the duct. Note that $M_e < 1$ and $p_e = p_b$.

There exists a significant difference between the Fanno and Rayleigh line flows regarding entropy changes. In a Fanno line flow, entropy can only increase in the direction of the flow. However, entropy can either decrease or increase in a Rayleigh line flow, depending on whether heat addition or heat loss takes place in the fluid. One distinct example is that if heat loss occurs in the converging nozzle-duct assembly, a favorable back pressure would result in a sonic condition at the entrance flow to the duct, which would accelerate to become a supersonic flow throughout the duct. For such an assembly of a converging nozzle and a constant-area duct, the Fanno line flow cannot reach supersonic conditions, as already discussed in Chapter 6.

The net result of heat transfer is best seen in the change in stagnation temperature, as indicated by Eqn. (7.9): $q = c_p(T_{t2} - T_{t1})$. The reference stagnation temperature T_t^* at $M = 1$ can be calculated using Eqn. (7.14), based on conditions at a given location where $M \neq 1$ (*i.e.*, T_{t1}). The new Mach number can then be found by the ratio of (T_{t2}/T_t^*), followed by new flow parameters (at the new location 2) calculated using Eqns. (7.10) through (7.13).

EXAMPLE 7.1 Air supplied from a reservoir through a converging nozzle enters a frictionless circular duct with a Mach number M_1 of 0.5 and a temperature and a pressure of 300 K and 300 kPa, respectively. Over the length, heat was added to the flow at 100 kJ/kg. Find the mass flow rate and the exit velocity at the exit. What is the back pressure? What is the change in entropy? Assume the specific heat of air is 1.004 kJ/kg·K.

Solution –

For $M_1 = 0.5$, $\frac{T_1}{T_{t1}} = 0.9542 = \frac{300\,k}{T_{t1}}$; $\frac{p_1}{p_{t1}} = 0.843 = \frac{300\,kPa}{p_{t1}}$

Therefore, $T_{t1} = 314.4$ K and $p_{t1} = 355.9$ kPa. Also using the tabulated values of Appendix G, $\frac{T_{t1}}{T_t^*} = 0.6914$, $\frac{p_{t1}}{p_t^*} = 1.1141$.

Therefore $T_t^* = 454.7$ k and $p_t^* = 319.5$ kPa.

$q = 100$ kJ/kg $= c_p(T_{t2} - T_{t1}) = 1.004$ kJ/kg·K $\times (T_{t2} - 314.4$ K$)$.

Therefore, $T_{t2} = 414.0$ K.

$\frac{T_{t1}}{T_t^*} = 0.9105$. So, $M_2 \approx 0.70$, for which $\frac{T_2}{T_{t2}} = 0.9107$, $\frac{p_2}{p_{t2}} = 0.7209$, and $\frac{p_{t2}}{p_t^*} = 1.0431$

$$p_{t2} = \frac{p_{t2}}{p_t^*} \times \frac{p_t^*}{p_{t1}} \times p_{t1} = 1.0431 \times \frac{1}{1.1141} \times 355.9 \text{ kPa} = 333.2 \text{ kPa}$$

$T_2 = 373.3$ K and $V_2 = M_2\sqrt{\gamma R T_2} = 0.70 \times \sqrt{1.4 \times 287 \times 373.3} = 271.1$ m/s.

$p_2 = \frac{p_2}{p_{t2}} \times p_{t2} = 0.7209 \times 333.2$ kPa $= 240.2$ kPa. Therefore,

$p_2 = p_b = 240.2$ kPa

The mass flow rate is

$$\frac{\dot{m}}{A} = \rho_2 V_2 = \frac{p_2}{RT_2}\sqrt{\gamma R T_2} = \frac{240.2 \times 10^3}{287 \times 373.3} \times \sqrt{1.4 \times 287 \times 373.3} = 605.0 \frac{\text{kg}}{\text{m}^2 \cdot \text{s}}$$

Alternatively, $M_1 = 0.5$ leads to $\frac{p_1}{p^*} = 1.7778$. $M_2 \approx 0.70$ gives $\frac{p_2}{p^*} = 1.4235$

Then, $p_2 = p_b = \frac{p_2}{p^*}\frac{p^*}{p_1}p_1 = 1.4235 \times \frac{1}{1.7778} \times 300 = 240.2$ *kPa*

For steady-state flow, mass flow rate can be calculated at any location. For now at the duct entrance,

$$\frac{\dot{m}}{A} = \rho_1 V_1 = \frac{p_1}{RT_1}M_1\sqrt{\gamma R T_1} = \frac{300 \times 10^3}{287 \times 300} \times 0.5 \times \sqrt{1.4 \times 287 \times 300}$$

$$= 604.9 \text{ kg/s} \cdot \text{m}^2$$

The above conditions correspond to the conditions represented by either curve 1 or curve 2 in Fig. 7.3.

$$\frac{s_2 - s_1}{c_p} = \ln\left[\left(\frac{M_2}{M_1}\right)^2 \left(\frac{1 + \gamma M_1^2}{1 + \gamma M_2^2}\right)^{(\gamma+1)/\gamma}\right] = \ln\left[\left(\frac{0.7}{0.5}\right)^2 \left(\frac{1 + 1.4 \times 0.25}{1 + 1.4 \times 0.49}\right)^{2.4/1.4}\right]$$

$$= 0.292$$

□

EXAMPLE 7.2 Use the nozzle-duct assembly from Example 7.1 with reservoir pressure and temperature equal to 355.9 *kPa* and 314.4 K, respectively, and $q = 100$ kJ/kg. What is the mass flow rate and what is the exit velocity if $P_b = 270$ *kPa*?

Solution –

Unlike Example 7.1, where M_1 is known to allow easy determination of the reference conditions, the current problem is solved by iterations.

First guess: $M_1 = 0.42$.

Therefore tabulated values of Appendix G indicate $\frac{T_{t1}}{T_t^*} = 0.5638$ and $\frac{p_{t1}}{p_t^*} = 1.148$.

Therefore $T_t^* = 557.6$ *k* and $p_t^* = 310.0$ kPa.

$q = 100$ kJ/kg $= c_p(T_{t2} - T_{t1}) = 1.004$ kJ/kg·K $\times (T_{t2} - 314.4$ K$)$.

Therefore, $T_{t2} = 414.0$ K.

$\frac{T_{t2}}{T_t^*} = 0.7425$ So, $M_2 \approx 0.54$, for which

$$\frac{T_2}{T_{t2}} = 0.9449, \quad \frac{p_2}{p_{t2}} = 0.8201, \quad \frac{p_{t2}}{p_t^*} = 1.0979, \quad \text{and} \quad \frac{p_2}{p^*} = 1.7043$$

$$P_b = p_2 = \frac{p_2}{p_{t2}} \times \frac{p_{t2}}{p_t^*} \times \frac{p_t^*}{p_{t1}} \times p_{t1} = 0.8201 \times 1.0979 \times \frac{1}{1.148} \times 355.9\text{kPa}$$

$$= 279.1 \text{ kPa}$$

Second guess: $M_1 = 0.44$.

Therefore, $\frac{T_{t1}}{T_t^*} = 0.5979$ and $\frac{p_{t1}}{p_t^*} = 1.1394$.

Therefore, $T_t^* = 525.8$ *k* and $p_t^* = 312.4$ kPa.

$q = 100$ kJ/kg $= c_p(T_{t2} - T_{t1}) = 1.004$ kJ/kg·K $\times (T_{t2} - 314.4$ K$)$.

Therefore, $T_{t2} = 414.0$ K.

$\frac{T_{t2}}{T_t^*} = 0.7874$. So, $M_2 \approx 0.57$, for which $\frac{T_2}{T_{t2}} = 0.940$, $\frac{p_2}{p_{t2}} = 0.80$, $\frac{p_{t2}}{p_t^*} = 1.086$, and $\frac{p_2}{p^*} = 1.65$

$$p_2 = \frac{p_2}{p_{t2}} \times \frac{p_{t2}}{p_t^*} \times \frac{p_t^*}{p_{t1}} \times p_{t1} = 0.80 \times 1.086 \times \frac{1}{1.139} \times 355.9 \text{ kPa} = 271.5 \text{ kPa}$$

For answers to the questions, $M_1 = 0.44$ and $M_2 \approx 0.57$.

$\frac{p_1}{p_{t1}} \approx 0.8755$ and $\frac{T_1}{T_{t1}} = 0.9627$; thus, $p_1 = 311.6$ *kPa* and $T_1 = 302.7$ *K*

$$\frac{\dot{m}}{A} = \rho_1 V_1 = \frac{p_1}{RT_1} M_1 \sqrt{\gamma R T_1} = \frac{311.6 \times 10^3}{287 \times 302.7} \times 0.44 \times \sqrt{1.4 \times 287 \times 302.7}$$

$$= 550.4 \text{ kg/s·m}^2$$

$$V_2 = M_2\sqrt{\gamma RT_2} = 0.57 \times \sqrt{1.4 \times 287 \times 0.94 \times 414} = 225.4 \ m/s$$

Comments –
When the duct is not choked, an increase in the back pressure leads to decreases in both the mass flow rate and exit velocity. Therefore, the results in Examples 7.1 and 7.2 can be qualitatively represented by curves 2 and 1, respectively, in Fig. 7.4. $\square$

EXAMPLE 7.3 Repeat Example 7.1 except that, instead of heat addition, now heat is taken away from the flow at 100 kJ/kg.

Solution –
For $M_1 = 0.5$, $\frac{T_1}{T_{t1}} = 0.9542 = \frac{300 \ K}{T_{t1}}$, $\frac{p_1}{p_{t1}} = 0.843 = \frac{300 \ kPa}{p_{t1}}$
Therefore, $T_{t1} = 314.4$ K and $p_{t1} = 355.9$ kPa Also, $\frac{T_{t1}}{T_t^*} = 0.6914$, $\frac{p_{t1}}{p_t^*} = 1.1141$.
Therefore, $T_t^* = 454.7 \ K$ and $p_t^* = 319.5$ kPa.
$q = -100 \ kJ/kg = c_p(T_{t2} - T_{t1}) = 1.004 \ kJ/kg \cdot K \times (T_{t2} - 314.4 \ K)$.
Therefore, $T_{t2} = 214.8$ K.
$\frac{T_{t2}}{T_t^*} = 0.4724$ So, $M_2 \approx 0.37$, for which $\frac{T_2}{T_{t2}} = 0.973$, $\frac{p_2}{p_{t2}} = 0.910$, and $\frac{p_{t2}}{p_t^*} = 1.169$

$$\frac{p_2}{p_{t2}} = 0.91$$

$$p_{t2} = \frac{p_{t2}}{p_t^*} \times \frac{p_t^*}{p_{t1}} \times p_{t1} = 1.169 \times \frac{1}{1.1141} \times 355.9 \ kPa = 373.4 \ kPa$$

$$p_2 = 0.91 \times 373.4 \ kPa = 339.8 \ kPa$$

$T_2 = 209.0$ K and $V_2 = M_2\sqrt{\gamma RT_2} = 0.37 \times \sqrt{1.4 \times 287 \times 209.0} = 107.2$ m/s.

$$\frac{\dot{m}}{A} = \rho_2 V_2 = \frac{p_2}{RT_2}\sqrt{\gamma RT_2} = \frac{339.8 \times 10^3}{287 \times 209.0} \times \sqrt{1.4 \times 287 \times 209.0} = 605.8 \ \frac{kg}{m^2 \cdot s}$$

$$\frac{s_2 - s_1}{c_p} = \ln\left[\left(\frac{M_2}{M_1}\right)^2 \left(\frac{1 + \gamma M_1^2}{1 + \gamma M_2^2}\right)^{(\gamma+1)/\gamma}\right] = \ln\left[\left(\frac{0.37}{0.5}\right)^2 \left(\frac{1 + 1.4 \times 0.25}{1 + 1.4 \times 0.1369}\right)^{2.4/1.4}\right]$$

$$= -0.388$$

Comments –
Cooling results in an increase in the stagnation pressure and a decrease in entropy at the exit, contrary to heating as shown in Example 7.1. However, the mass flow rate remains the same as it can be determined by the conditions at the duct inlet, which are the same, but can now exit to a higher back pressure. These results are consistent with the *T-s* curve in Fig. 7.2. $\square$

EXAMPLE 7.4 To choke the nozzle-duct assembly in Example 7.1 with the same *reservoir* (not inlet) conditions, what is the amount of heat addition to cause the flow to be sonic? What is the maximum back pressure if the heat addition is (1) 300 kJ/kg and (2) 500 kJ/kg?

Solution –

Again,

The reservoir conditions are $T_r = T_{t1} = 314.4$ K and $p_r = p_{t1} = 355.9$ kPa

To reach sonic conditions at the duct exit, the total amount of heat addition is

$$q = c_p\left(T_t^* - T_{t1}\right) = 1.004\,\frac{\text{kJ}}{\text{kg}\cdot\text{K}} \times (454.7\text{ K} - 314.4\text{ K}) = 140.9\,\frac{\text{kJ}}{\text{kg}}$$

By adding a larger amount of heat that this value results in the duct being choked at the exit and $M_e = M_2 = 1$.

(1) For $q = 300$ kJ/kg > 140.9 kJ/kg, so $M_e = 1$, but M_1 will no longer be 0.5 and there will be different values of T_t^* and p_t^* from that in Example 7.1.

$$T_t^* = T_{t1} + \frac{q}{c_p} = 314.4\text{ K} + \frac{300\text{ kJ/kg}}{1.004\text{ kJ/kg}\cdot\text{K}} = 613.2\text{ K}$$

Therefore, $\frac{T_{t1}}{T_t^*} = \frac{314.4}{613.2} = 0.5127$ and the new Mach number is $M_1 \approx 0.39$, for which $\frac{p_1}{p_{t1}} \approx 0.90$, $\frac{p_{t1}}{p_t^*} = 1.16$, and $p_t^* = 306.8$ kPa.

$\frac{T_1}{T_{t1}} \approx 0.97$. Therefore, $p_1 = 0.90 \times 355.9\ kPa = 319.5$ kPa and $T_1 = 305.0\ K$

$$\frac{\dot{m}}{A} = \rho_1 V_1 = \frac{p_1}{RT_1} M_1\sqrt{\gamma RT_1} = \frac{319.5 \times 10^3}{287 \times 305.0} \times 0.39 \times \sqrt{1.4 \times 287 \times 305.0}$$

$$= 498.3\text{ kg/s}\cdot\text{m}^2.$$

Now because $M_e = 1$, $\frac{p_e}{p_{te}} = \frac{p^*}{p_t^*} = 0.5283$. For $M_1 \approx 0.39$, $\frac{p_{t1}}{p_t^*} \approx 1.16$.

Therefore, $p_e = \frac{p^*}{p_t^*} \times \frac{p_t^*}{p_{t1}} \times p_{t1} = 0.5283 \times \frac{1}{1.16} \times 355.9$ kPa $= 162.1$ kPa

For $\frac{\dot{m}}{A} = 498.3$ kg/s$\cdot$m^2 and $q = 300$ kJ/kg , $p_b < 161.9$ kPa will ensure a choked duct (corresponding to curve 3 in Fig. 7.3).

Alternatively, $\frac{p_1}{p^*} = 1.979$.

$p_e = p^* = \frac{p^*}{p_1} \times p_1 = \frac{1}{1.979} \times 319.5\ kPa = 161.4\ kPa \approx 161.9$ kPa (as expected – the two approaches for the static should yield the same result).

(2) For $q = 500$ kJ/kg, $T_t^* = T_{t1} + \frac{q}{c_p} = 314.4$ K $+ \frac{500\text{ kJ/kg}}{1.004\text{ kJ/kg}\cdot\text{K}} = 812.4$ K

Therefore, $\frac{T_{t1}}{T_t^*} = \frac{314.4}{812.4} = 0.387$ and the new Mach number is $M_1 \approx 0.32$, for which $\frac{p_1}{p_{t1}} \approx 0.9315$ and $\frac{T_1}{T_{t1}} \approx 0.980$. Therefore, $p_1 = 0.9315 \times 355.9\ kPa = 331.5\ kPa$ and $T_1 = 308.1\ K$

$$\frac{\dot{m}}{A} = \rho_1 V_1 = \frac{p_1}{RT_1} M_1\sqrt{\gamma RT_1} = \frac{331.5 \times 10^3}{287 \times 308.1} \times 0.32 \times \sqrt{1.4 \times 287 \times 308.1}$$

$$= 422.1\text{ kg/s}\cdot\text{m}^2.$$

Now because $M_e = 1$, $\frac{p_e}{p_{te}} = \frac{p^*}{p_t^*} = 0.5283$. For $M_1 \approx 0.32$, $\frac{p_{t1}}{p_t^*} \approx 1.190$.

Therefore, $p_e = \frac{p^*}{p_t^*} \times \frac{p_t^*}{p_{t1}} \times p_{t1} = 0.5283 \times \frac{1}{1.190} \times 355.9$ kPa $= 158.0$ kPa

For $\frac{\dot{m}}{A} = 422.1$ kg/s$\cdot$m^2 and $q = 500$ kJ/kg, $p_b < 158.0$ kPa will ensure a choked duct (corresponding to curve 3 in Fig. 7.3).

Alternatively, $\frac{p_1}{p^*} = 2.0991$ and

$p_e = p^* = \frac{p^*}{p_1} \times p_1 = \frac{1}{2.0991} \times 331.5\ kPa \approx 158.0$ kPa

Comments –

1. This example illustrates that increasing the amount of heat addition, the choked condition results in a reduction in the mass flow rate and the point P_{sonic} moves to the right, as depicted in Fig. 7.3.

2. In Examples 7.1 and 7.2, specifying M_1 for the given reservoir pressure and temperature (albeit calculated based on the condition at the entrance of the duct) indirectly specifies p_e, which is equal to p_b, and vice versa. This is because the amount of heat addition is not sufficient to accelerate the flow to sonic conditions at the duct exit and the exit pressure has to equilibrate to that in the surrounding. In the current example, there exists a corresponding back pressure for a given amount of heat addition. Below this critical back pressure, both M_1 and $\dot{m}/A$ do not change further, because the duct is choked. $\square$

EXAMPLE 7.5 Air supplied from a reservoir (the reservoir temperature and pressure are 400 K and 400 kPa, respectively) through a converging nozzle enters a frictionless circular duct, as depicted in Fig. 7.3. Over the length, heat was added to the flow at 500 kJ/kg. Find the mass flow rate and the exit velocity if the back pressure is 275 kPa. Assume the specific heat of air is 1.004 kJ/kg·K.

Solution –

First of all, it is important to decide whether the duct is thermally choked by heat transfer. If it is, then $M_e = 1$ and $p_e = p^* = p_b = 275$ kPa is the maximum back pressure for choking (curve 4 in Fig. 7.3). Therefore,

$\frac{p_e}{p_{te}} = 0.5238 = \frac{p_e}{p_t^*}$. Thus $p_t^* = 515$ kPa and $p_e = p^*$. Since during subsonic acceleration p_t can only decrease, it cannot exceed the reservoir. Therefore, the duct must not be choked and $M_e < 1$ (the condition resembles that represented by curve 1 or 2 in Fig. 7.3).

One way to solve the problem is to guess the duct inlet Mach number and then, based on the change in T_t, to find M_e and p_e. If $p_e = p_b$, the solution is found. Another path to a solution is to assume that the exit condition is sonic. There are two approaches. The first is described as follows:

First guess – the inlet Mach number $M_i = 0.3$, then $\frac{p_{ti}}{p_t^*} = 1.1985$ and $\frac{T_{ti}}{T_t^*} = 0.3469$ and therefore $T_t^* = 400\ K/0.3469 = 1,153.1\ K$. $T_{te} = T_{ti} + q/c_p = 400 + 500/1.004 = 898.0\ K$. $\frac{T_{te}}{T_t^*} = \frac{898}{1,153} = 0.7788$.

Therefore, $M_e \approx 0.56$, $\frac{p_e}{p_{te}} = 0.8082$ and $\frac{p_{te}}{p_t^*} = 1.0901$.

Therefore, $p_e = \frac{p_e}{p_{te}} \times \frac{p_{te}}{p_t^*} \times \frac{p_t^*}{p_{ti}} \times p_{ti} = 0.8082 \times 1.0901 \times \frac{1}{1.1985} \times 400\ kPa = 294\ kPa > p_b$

Second guess – the inlet Mach number $M_i = 0.34$, then $\frac{p_{ti}}{p_t^*} = 1.1822$ and $\frac{T_{ti}}{T_t^*} = 0.4206$ and therefore $T_t^* = 400\ K/0.4206 = 951.0\ K$. $T_{te} = T_{ti} + q/c_p = 400 + 500/1.004 = 898.0\ K$. $\frac{T_{te}}{T_t^*} = \frac{898}{951.0} = 0.9443$.

Therefore, $M_e \approx 0.63$, $\frac{p_e}{p_{te}} = 0.765$ and $\frac{p_{te}}{p_t^*} = 1.065$.

Therefore, $p_e = \frac{p_e}{p_{te}} \times \frac{p_{te}}{p_t^*} \times \frac{p_t^*}{p_{ti}} \times p_{ti} = 0.765 \times 1.065 \times \frac{1}{1.1822} \times 400\ kPa = 275.6\ kPa \approx p_b$.

Thus $M_i = 0.34$, $M_e \approx 0.63$, $\frac{p_i}{p_{ti}} = 0.9231$, $\frac{p_e}{p_{te}} = 0.765$, $\frac{T_i}{T_{ti}} = 0.9774$, $\frac{T_e}{T_{te}} = 0.923$, $\frac{p_i}{p^*} = 2.0657$, and $\frac{p_e}{p^*} = 1.543$.

$$\frac{\dot{m}}{A} = \rho_i V_i = \frac{p_i}{RT_i} M_i \sqrt{\gamma R T_i} = \frac{0.9321 \times 400 \times 10^3}{287 \times 0.9774 \times 400} \times 0.32$$

$$\times \sqrt{1.4 \times 287 \times 0.9774 \times 400} = 447.8 \ kg/s \cdot m^2$$

$$V_e = M_e \sqrt{\gamma R T_e} = M_e \sqrt{\gamma R \frac{T_e}{T_{te}} T_e} = 0.34 \times \sqrt{1.4 \times 287 \times 0.923 \times 898.0}$$

$$= 196.2 \ m/s.$$

The second approach to a solution is to assume that the exit condition is sonic with $p_e = p^* = p_b = 275$ kPa, and $p_t^* = p_{te}$. Under this assumption $p_t^* = p_{te} = p_e/0.5283 = 525.0$kPa $> p_r$, which is not possible. Therefore, $M_e < 1$ and the previous solution must be followed. □

EXAMPLE 7.6 A model supersonic combustor has the air flow at the inlet at Mach number 2.5 and temperature and pressure at 226 K and 21 kPa, respectively. Assume a Rayleigh line flow in the combustor and the average specific heat is $c_p = 1.004 \cdot kJ/kg \cdot K$. What amount of heat addition is needed to have the Mach number to be unity at the combustor exit? Under this condition, what is the loss in stagnation pressure across the combustor? If the JP8 fuel adds 1,900 kJ/kg to the gases, what may happen in the combustor?

Solution –
For $M_i = 2.5$, $\frac{T_i}{T_t^*} = 0.4444$, $\frac{p_i}{p_{ti}} = 0.0585$, $\frac{p_{ti}}{p_t^*} = 2.2218$, and $\frac{T_{ti}}{T_t^*} = 0.7101$
Therefore, $T_{ti} = T_i/0.444 = 226/0.444 = 508.6 \ K$, $T_t^* = T_{ti}/0.7101 = 716.2 \ K$.

$$c_p(T_{te} - T_{ti}) = 1.004 \times (716.2 \ K - 508.6 \ K) = 208.4 \ kJ/kg.$$

Under the choked condition,

$$p_{te} - p_{ti} = p_t^* - p_{ti} = \left(\frac{p_t^*}{p_{ti}} - 1\right) \frac{p_{ti}}{p_i} \times p_i = \left(\frac{1}{2.2218} - 1\right) \times \frac{1}{0.0585} \times 21 \ kPa$$

$$= -197.4 \ kPa$$

For $q = 1,900$ kJ/kg, $T_{te} = 2401.0 \ K$, then $\frac{T_{te}}{T_t^*} = \frac{2,401.0}{716.2} > 1$. So the heat addition is more than needed to decelerate the flow to sonic conditions. Therefore, there must be a normal shock in the combustor. □

EXAMPLE 7.7 A constant area duct is connected to an air reservoir (at 700 kPa and 400 K) through a converging-diverging nozzle. The area ratio of the duct inlet to the throat is 2.0 and the heat addition throughout the duct is 50 kJ/kg. Determine the range of back pressure for a normal shock wave to stand within the duct.

Solution –

Assume that the nozzle is isentropic. Therefore, a 2.0 area ratio leads to $M_i = 2.20$, $P_{ti} = P_r = 700\ kPa$, $T_{ti} = T_r = 400\ k$, and $\frac{P_{ti}}{p_t^*} = 1.7434$, and $\frac{T_{ti}}{T_t^*} = 0.7561$.

(Let subscripts 1 and 2 denote locations immediately upstream and downstream of the shock, respectively.)

First scenario – The normal shock stands at the exit plane

Therefore, $\frac{T_{t1}}{T_t^*} = \frac{T_{ti} + q/c_p}{(T_t^*/T_{ti})T_{ti}} = \frac{400 + 50/1.004}{400/0.7561} = 0.8502$ and so $M_1 = 1.74$, $\frac{P_{t1}}{p_t^*} = 1.2692$.

Across the shock wave, $M_2 = 0.6305$, $\frac{P_{t2}}{P_{t1}} = 0.8389$.

Also $\frac{T_2}{T_{t2}} \approx 0.9265$ and $\frac{P_2}{p_{t2}} \approx 0.7653$

$$p_2 = p_b = \frac{p_2}{p_{t2}} \times \frac{P_{t2}}{P_{t1}} \times \frac{P_{t1}}{p_t^*} \times \frac{p_t^*}{P_{ti}} \times P_{ti} = 0.7653 \times 0.8389 \times 1.2692 \times \frac{1}{1.7434}$$

$$\times 700\ kPa = 327.2\ kPa$$

Second scenario – The normal shock stands at the inlet of the duct

$M_i = M_1 = 2.20$, so $M_2 = 0.5471$ and $\frac{T_{t2}}{T_t^*} = \frac{T_{t1}}{T_t^*} = \frac{T_r}{T_t^*} = 0.756$, $\frac{P_{t2}}{p_t^*} \approx 1.095$

$\frac{T_{te}}{T_t^*} = \frac{T_{t2} + q/c_p}{(T_t^*/T_{t2})T_{t2}} = \frac{400 + 50/1.004}{400/0.756} = 0.8502$, and therefore, $M_e \approx 0.627$, $\frac{P_{te}}{p_t^*} \approx 1.068$ and

$\frac{p_e}{p_{te}} = 0.927$

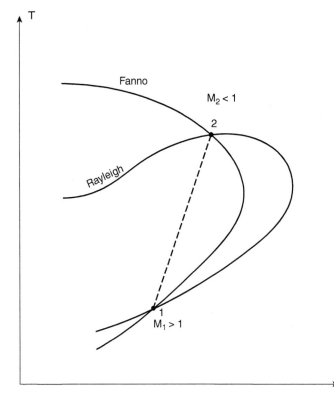

Figure 7.6 The intersection of the Rayleigh and Fanno lines determines the normal shock states. The dashed line signifies that the process from supersonic (point 1) to subsonic (point 2) is non-isentropic.

Across the shock $\frac{P_{t2}}{P_{t1}} = 0.6281$.

Therefore,

$$p_e = p_b = \frac{p_e}{p_{te}} \times \frac{P_{te}}{p_t^*} \times \frac{p_t^*}{P_{t2}} \times \frac{P_{t2}}{P_{t1}} \times P_{t1}$$

$$= 0.927 \times 1.068 \times \frac{1}{1.095} \times 0.6281 \times 700 \; kPa = 397.5 \; kPa$$

The range of back pressure for the normal shock to stand within the duct is

$$327.2 \; kPa \leq p_b \leq 397.5 \; kPa$$

(As the back pressure increases, the location of the normal shock moves upstream toward the nozzle.)   □

7.3 Normal Shock on Rayleigh and Fanno Line *T-s* Diagrams

Recall that to develop the *T-s* diagram for Fanno line flow, the continuity and energy equations were used:

$$\rho V = \text{constant} \tag{6.1}$$

$$h_t = c_p T_t = h + \frac{1}{2} V^2 = \text{constant} \tag{6.2}$$

For the Rayleigh line flow, the continuity, Eqn. (6.1) and momentum equations were used:

$$p + \rho V^2 = \text{constant} \tag{7.24}$$

These three conservation equations were used to derive normal shock relationships. For a constant-area channel flow with heat transfer and wall friction, if a normal shock exists in the duct, then the states immediately upstream and downstream of the shock must appear at the intersection of the of the *T-s* curves of the Fanno and Rayleigh lines. This result for a given mass flow rate is schematically shown in Fig. 7.6. Since entropy increases across the shock, state 2 (downstream; $M_2 < 1$) lies to the upper right of state 1 (upstream; $M_2 > 1$) with $T_2 > T_1$ and $s_2 > s_1$, due to the non-isentropic nature of shocks. The line connecting the two states is dashed because the shock process is not a thermodynamic equilibrium process and its path is not exactly known.

Although the two *T-s* curves in Fig. 7.6 are for a given mass flow rate, it does not specify the exact combination of the heat addition and friction that leads to the normal shock conditions. For the conditions immediately upstream and downstream of the normal shock, the three governing equations have to be considered simultaneously. The following section is devoted to such a purpose.

7.4 Flow with Friction and Heat Addition

The differential forms of conservation of mass and momentum for the constant-area duct are:

$$\frac{d\rho}{\rho} + \frac{dV}{V} = 0 \tag{6.16}$$

$$\frac{dp}{p} + \frac{1}{2}\gamma M^2 f \frac{dx}{D_h} + \gamma M^2 \frac{dV}{V} = 0 \tag{6.14}$$

For energy conservation,

$$\delta q = c_p dT_t$$

Since $T_t = T\left(1 + \frac{\gamma-1}{2}M^2\right)$, the following expression is found:

$$\frac{\delta q}{c_p T} = \left(1 + \frac{\gamma-1}{2}M^2\right)\frac{dT}{T} + (\gamma-1)MdM \tag{7.22}$$

For an ideal gas,

$$\frac{dp}{p} = \frac{d\rho}{\rho} + \frac{dT}{T} \tag{6.15}$$

After substituting the continuity into this expression,

$$\frac{dp}{p} = -\frac{dV}{V} + \frac{dT}{T} \tag{7.23a}$$

By using $M = \frac{V}{\sqrt{\gamma RT}}$, it is readily shown that

$$\frac{dV}{V} = \frac{dM}{M} + \frac{1}{2}\frac{dT}{T} \tag{7.23b}$$

Thus

$$\frac{dp}{p} = -\frac{dM}{M} + \frac{1}{2}\frac{dT}{T} \tag{7.23c}$$

Substitution of these two expressions into Eqn. (6.14) yields

$$\frac{1}{2}\left(1 + \gamma M^2\right)\frac{dT}{T} + \left(\gamma M^2 - 1\right)\frac{dM}{M} + \frac{1}{2}\gamma M^2 f \frac{dx}{D_h} = 0 \tag{7.23d}$$

After using dT/T from Eqn. (7.22) and some algebraic manipulations, one obtains

$$\frac{1}{2}\left(1 + \gamma M^2\right)\frac{\delta q}{c_p T_t} + \frac{1}{2}\gamma M^2 f \frac{dx}{D_h} = \left[\frac{(1-M^2)}{1 + \frac{(\gamma-1)}{2}M^2}\right]\frac{dM}{M} \tag{7.24}$$

Therefore, with $\delta q = c_p dT_t$

$$\frac{dM}{dx} = M \left[\frac{1 + \frac{(\gamma-1)}{2} M^2}{(1 - M^2)} \right] \left[\frac{1}{2} \left(1 + \gamma M^2 \right) \frac{1}{T_t} \frac{dT_t}{dx} + \frac{1}{2D_h} \gamma M^2 f \right] \tag{7.25}$$

Once T_t (i.e., δq) as a function of x is known, Mach number at any location along the duct can be determined by integrating Eqn. (7.25).

7.5 Flow with Friction, Heat Transfer, and Area Change

The flow in a rocket nozzle simultaneously experiences friction, heat transfer, and area change (Hill and Peterson, 1991; Sutton and Biblarz, 2001). The effects of friction and area change are apparent. Heat transfer may result from the remaining chemical reactions due to the shift in thermodynamic equilibrium as the nozzle flow creates changes in both temperature and pressure, to which chemical reaction and equilibrium conditions are sensitive. Heat transfer may take place simply due to the temperature difference between the gas and the nozzle wall, with the latter being exposed to the surroundings that may have vastly different temperatures.

Assume that the flow is one-dimensional, and the continuity, momentum, and energy equations are, respectively,

$$\frac{d\rho}{\rho} + \frac{dV}{V} + \frac{dA}{A} = 0 \tag{5.2}$$

$$\frac{dp}{p} + \frac{1}{2} \gamma M^2 f \frac{dx}{D_h} + \gamma M^2 \frac{dV}{V} = 0 \tag{6.14}$$

$$\text{or } dp + \rho V dV = 0 \tag{7.3}$$

$$\frac{\delta q}{c_p T} = \left(1 + \frac{\gamma - 1}{2} M^2 \right) \frac{dT}{T} + (\gamma - 1) M dM \tag{7.22}$$

Because of the area change, Eqn. (6.15) is modified to be

$$\frac{dp}{p} = - \left(\frac{dV}{V} + \frac{dA}{A} \right) + \frac{dT}{T} \tag{7.26}$$

Substituting Eqn. (7.26) into Eqn. (6.14), or simply adding dA/A to Eqn. (7.23d), yields

$$\frac{1}{2} \left(1 + \gamma M^2 \right) \frac{dT}{T} + \left(\gamma M^2 - 1 \right) \frac{dM}{M} + \frac{1}{2} \gamma M^2 f \frac{dx}{D_h} - \frac{dA}{A} = 0 \tag{7.27}$$

Similarly,

$$\frac{1}{2}\left(1 + \gamma M^2\right)\frac{\delta q}{c_p T_t} + \frac{1}{2}\gamma M^2 f\frac{dx}{D_h} - \frac{dA}{A} = \left[\frac{\left(1 - M^2\right)}{1 + \frac{(\gamma-1)}{2}M^2}\right]\frac{dM}{M} \qquad (7.28a)$$

or

$$\frac{dM^2}{M^2} = \left[\frac{1 + \frac{(\gamma-1)}{2}M^2}{\left(1 - M^2\right)}\right]\left[-2\frac{dA}{A} + \left(1 + \gamma M^2\right)\frac{dT_t}{T_t} + \gamma M^2 f\frac{dx}{D_h}\right] \qquad (7.28b)$$

Eqn. (7.28a) can be rewritten for Mach number variation along the flow direction, as

$$\frac{dM}{dx} = M\left[\frac{1 + \frac{(\gamma-1)}{2}M^2}{\left(1 - M^2\right)}\right]\left[-\frac{1}{A}\frac{dA}{dx} + \frac{1}{2}\left(1 + \gamma M^2\right)\frac{1}{T_t}\frac{dT_t}{dx} + \frac{1}{2}\frac{\gamma M^2 f}{D_h}\right] \qquad (7.29)$$

It is clear that from Eqn. (7.29) that the effects of area change, heat transfer, and friction all contribute to determining the location where sonic conditions exist. This is because for the right-hand side of Eqn. (7.29) to be meaningful for $M = 1$, the following condition must be satisfied:

$$\frac{dA}{A} = \frac{1}{2}\left[\left(1 + \gamma M^2\right)\frac{dT_t}{T_t} + \gamma M^2 f\frac{dx}{D_h}\right]_{M=1} \qquad (7.30)$$

That is, the location for $M = 1$ is no longer at the throat (i.e., $dA/dx = 0$) when non-isentropic effects such as friction and heat transfer are included. The exact location for $M = 1$ depends on the relative magnitudes of the heat transfer and friction. Since the friction term is always positive, its contribution is to shift the sonic location to where $dA/dx > 0$, that is, the diverging section downstream of the throat. Depending on whether heat is lost/gained by the fluid, the effect of heat transfer is to shift the sonic location upstream/downstream of the throat. If the sum of contributions from friction and heat transfer is positive/negative, the sonic location is situated downstream/upstream of the throat.

For a constant-area duct flow with heat transfer and friction, Eqn. (7.29) is reduced to Eqn. (7.25). The general solution of Eqn. (7.29) for $M = M(x)$ may be obtained using numerical integration methods for given dA/dx, dT_t/dx, and f as functions of x.

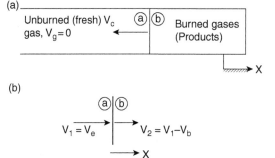

Figure 7.7 (a) A long tube with one closed end containing a reactive mixture initially at rest; the figure shows a combustion/pressure wave propagating into the quiescent mixture sometime after ignition is initiated at the closed end; (b) the flow field as observed by riding on the wave.

7.6 Detonation and Deflagration Waves

A special case of heat addition involves heat-releasing chemical reactions (Glassman and Yetter, 2008). Consider a long tube with one end closed that is filled with reactive mixture, where the fresh gas mixture is initially at rest. After the combustion reaction is initiated at the closed end, the energy release raises the temperature of the burned gas, igniting the adjacent layer of fresh mixture. A self-sustaining combustion wave is generated, as shown in Fig. 7.7a. At the same time, the thermal expansion of the burned gas (due to combustion) results in a pressure wave that pushes into the fresh mixture while carrying the hot product gas to the right, similar to a piston motion that may lead to a shock wave. For reactions with high activation energies, the combustion wave is thin, the temperature gradient is large across the combustion wave, and the pressure and combustion waves coincide. After an initial transient period, the pressure wave propagates with a steady-state speed, denoted by V_c.

It is useful to know about the propagation speed of the combustion wave V_c. Assuming a steady-state one-dimensional wave propagation, Eqns. (4.66) through (4.69) for conservation of mass, momentum, and energy can be rewritten in the coordinate system following the motion of the wave (Fig. 7.7b) as

$$\rho_1 V_1 = \rho_2 V_2 \tag{7.31}$$

$$\rho_1 V_1^2 + p_1 = \rho_2 V_2^2 + p_2 \tag{7.32}$$

$$c_p T_1 + q + \frac{V_1^2}{2} = c_p T_2 + \frac{V_2^2}{2} \tag{7.33}$$

where q is the energy added (or chemical enthalpy) per unit mass and $V_1 = V_c$. Because the equation of state $p_1 = \rho_1 R T_1$ relates known variables, the only state equation is, assuming a constant gas constant,

$$p_2 = \rho_2 R T_2 \tag{7.34}$$

There are four equations, Eqns. (7.31) through (7.34), for five unknown variables: V_1, V_2, p_2, T_2, and ρ_2. Therefore, there exist an infinite number of solutions to these four equations. Physical reasoning dictates that there must exist an intrinsic variable (an eigenvalue) in the mixture system, which is expected to be the wave propagation speed that is a function of the variables of the unburned gas. It is desirable to develop a relationship to demonstrate the existence of this unique variable before attempting to obtain solutions of these four equations. Similar processes in Chapter 4 leading to the Rankine-Hugoniot relation can be followed. In the following, R, c_p, and γ are assumed to be constant throughout the entire flow field (i.e., in both the burned and unburned regions). Combining Eqns. (7.31) and (7.32), one finds the following two expressions:

$$V_1^2 = \frac{1}{\rho_1^2}\left(\frac{p_2 - p_1}{\frac{1}{\rho_1} - \frac{1}{\rho_2}}\right) \text{ or } V_1^2 = v_1^2\left(\frac{p_2 - p_1}{v_1 - v_2}\right) \tag{7.35}$$

and

$$V_2^2 = \frac{1}{\rho_2^2}\left(\frac{p_2 - p_1}{\frac{1}{\rho_1} - \frac{1}{\rho_2}}\right) \text{ or } V_2^2 = v_2^2\left(\frac{p_2 - p_1}{v_1 - v_2}\right) \tag{7.36}$$

Dividing through these two expression, respectively, by their speed of sound $(a^2 = \gamma RT = \gamma P/\rho)$ leads to

$$\gamma M_1^2 = \frac{p_2/p_1 - 1}{1 - \rho_1/\rho_2} \tag{7.37}$$

and

$$\gamma M_2^2 = \frac{1 - \rho_1/\rho_2}{p_2/p_1 - 1} \tag{7.38}$$

Both Equations (7.37) and (7.38) suggest that solutions to Eqns. (7.31) through (7.34) only exist for either (i) $p_2/p_1 > 1$ and $\rho_2/\rho_1 > 1$ (i.e., compression; these waves are called *detonation* waves) or (ii) $p_2/p_1 < 1$ and $\rho_2/\rho_1 < 1$ (i.e., expansion; these waves are called *deflagration* waves).

Recalling that $c_p = \gamma R/(\gamma - 1)$, the energy equation, Eqn. (7.33) can be written as

$$\frac{\gamma}{\gamma - 1}R(T_2 - T_1) - \frac{1}{2}(V_1^2 - V_2^2) = q$$

Substituting Eqns. (7.35) and (7.36) into this expression leads to

$$\frac{\gamma}{\gamma - 1}\left(\frac{p_2}{T_2} - \frac{p_1}{T_1}\right) - \frac{1}{2}(p_2 - p_1)\left(\frac{1}{\rho_1} + \frac{1}{\rho_2}\right) = q \tag{7.39}$$

Multiplying both sides of Eqn. (7.39) by ρ_2 and rearranging terms leads to

$$\frac{p_2}{p_1} = \frac{\left(\frac{\gamma+1}{\gamma-1} + \frac{2\,q}{RT_1}\right)\frac{\rho_2}{\rho_1} - 1}{\left(\frac{\gamma+1}{\gamma-1}\right) - \frac{\rho_2}{\rho_1}} \tag{7.40a}$$

or alternatively,

$$\frac{p_2}{p_1} = \frac{\left(\frac{\gamma+1}{\gamma-1} + \frac{2\,q}{RT_1}\right) - \frac{v_2}{v_1}}{\left(\frac{\gamma+1}{\gamma-1}\right)\frac{v_2}{v_1} - 1} \tag{7.40b}$$

It is readily seen that for $q = 0$ these two equations reduce to the Rankine-Hugoniot (or simply, Hugoniot) relationship for shock waves derived in Chapter 4, Eqn. (4.72). The detailed derivation of Eqn. (7.40) is left as an exercise (i.e., Problem 7.10). In addition to the thermodynamic quantities contained in the shock adiabat described in Chapter 4, Eqns. (7.40a) and (7.40b) contains the heat release per unit mass. The function $p_2 = p_2(\rho_2, q)$ or $p_2 = p_2(v_2, q)$ is called the detonation adiabat and will be discussed in more detail in the following.

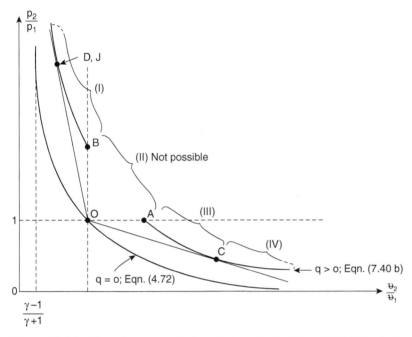

Figure 7.8 The Hugoniot curve with $\delta q > 0$ plotted using Eqn. (7.40b) Physical reasoning leads to two steady-state solutions: Chapman-Jouget detonation wave (point J) and deflagration wave (regimes III).

The Hugoniot relationship of p_2/p_1 vs. v_2/v_1 is plotted using Eqn. (7.40b) (see Fig. 7.8). It can be seen that the pressure increase due to chemical reaction is higher than that without heat addition; Eqn. (4.72) is also seen in Fig. 7.8 for comparison. A different Hugoniot curve has to be drawn for a different value of q.

The propagating wave causes the burned gas to move with a speed $V_b = V_s - V_2 = V_1 - V_2$. Using Eqns. (7.35) and (7.36) yields

$$V_b = \sqrt{(v_1 - v_2)(p_2 - p_1)} \qquad (7.41)$$

The discussion above pertaining to Eqns. (7.38) and (7.39) excludes portion of the Rankine-Hugoniot curve, i.e., region II between points A and B in Fig. 7.7, as possible solutions. There are two points on the curve that require particular attention – tangency points C and J, called Chapmen-Jouget points. Points C and J are the intersections between the curve and the tangent lines drawn from the initial point (point I). For these two points, an angle, α_J, is defined as

$$\tan \alpha_J \equiv \frac{(p_2 - p_1)}{(v_1 - v_2)} \quad \text{or} \quad \tan \alpha_J \equiv \left(\frac{1}{\rho_1} - \frac{1}{\rho_2}\right)(p_2 - p_1) \qquad (7.42)$$

Therefore, Eqn. (7.35) can be rewritten as

$$V_1 = \textbf{v}_1 \sqrt{\tan \alpha_J} \qquad (7.43)$$

Along the Hugoniot curve,

$$T ds_2 = de_2 + p_2 d(1/\rho_2) \qquad (7.44)$$

It is desirable to express e_2 in terms of p and $\textbf{v}$(or ρ), which has been used so far. The energy equation, Eqn. (7.33), can be rewritten as

$$h_{t1} + \frac{V_1^2}{2} = h_{t2} + \frac{V_2^2}{2}$$

where h_{t1} and h_{t2} equal the total energy, that is, the sum of thermal energy and chemical (rather than kinetic) energy of the fresh and burned gases, respectively, such that

$$h_t = h + h^o = c_p T + h^o$$

where h^o is the baseline reference enthalpy in the standard state of gas species, and

$$q = h_1^o - h_2^o$$

Equation (7.33) then becomes

$$h_1 + h_1^o + \frac{V_1^2}{2} = h_2 + h_2^o + q + \frac{V_2^2}{2} \qquad (7.45)$$

Therefore

$$h_2 - h_1 = \frac{V_1^2}{2} - \frac{V_2^2}{2}$$

Substituting part of the result from Eqn. (7.39):

$$\frac{V_1^2}{2} - \frac{V_2^2}{2} = \frac{1}{2}(p_2 - p_1)\left(\frac{1}{\rho_1} + \frac{1}{\rho_2}\right) \qquad (7.46)$$

and $h = e + (p/\rho)$ into Eqn. (7.44) yields

$$e_2 - e_1 = \frac{1}{2}(p_2 + p_1)\left(\frac{1}{\rho_1} - \frac{1}{\rho_2}\right) \qquad (7.47)$$

Differentiating the above expression, while noting that state 1 is given, leads to

$$de_2 = -\frac{1}{2}(p_2 + p_1)d\left(\frac{1}{\rho_2}\right) + \frac{1}{2}\left(\frac{1}{\rho_1} - \frac{1}{\rho_2}\right)dp_2$$

With the above expression, Eqn. (7.44) becomes

$$T ds_2 = \frac{1}{2}(p_2 - p_1)d\left(\frac{1}{\rho_2}\right) + \frac{1}{2}\left(\frac{1}{\rho_1} - \frac{1}{\rho_2}\right)dp_2$$

Therefore, along the Hugoniot curve,

$$T\frac{ds_2}{d\left(\frac{1}{\rho_2}\right)} = \frac{1}{2}\left(\frac{1}{\rho_1} - \frac{1}{\rho_2}\right)\left[\frac{(p_2 - p_1)}{\left(\frac{1}{\rho_1} - \frac{1}{\rho_2}\right)} + \frac{dp_2}{d\left(\frac{1}{\rho_2}\right)}\right] \tag{7.48}$$

Because region II has been eliminated from possible solutions, then in regions I, III, and IV

$$\frac{(p_2 - p_1)}{\left(\frac{1}{\rho_1} - \frac{1}{\rho_2}\right)} > 0$$

At tangency points C and J, the above inequality holds. Since at points C and J,

$$\frac{dp_2}{d\left(\frac{1}{\rho_2}\right)} = -\frac{(p_2 - p_1)}{\left(\frac{1}{\rho_1} - \frac{1}{\rho_2}\right)} \tag{7.49}$$

one can conclude that

$$\left[\frac{ds_2}{d\left(\frac{1}{\rho_2}\right)}\right]_{K,J} = 0 \tag{7.50}$$

Differentiating both sides of Eqn. (7.48) with respective to $1/\rho_2$ for constant temperature leads to

$$\frac{d^2s_2}{d\left(\frac{1}{\rho_2}\right)^2} = \frac{1}{2\,T}\left(\frac{1}{\rho_1} - \frac{1}{\rho_2}\right)\frac{d^2p_2}{d\left(\frac{1}{\rho_2}\right)^2} \tag{7.51}$$

The derivative of p_2 in the above equation can be obtained from Eqn. (7.40a). After some differential and algebraic manipulations (this is left as an end-of-chapter exercise, Problem 7.10),

$$\frac{d^2s_2}{d\left(\frac{1}{\rho_2}\right)^2} = -\frac{p_1}{T}\left(\frac{p_2}{p_1} - 1\right)\left(\frac{\rho_2^2}{\rho_1}\right)\left\{\frac{\left(\frac{\gamma+1}{\gamma-1}\right)\left[1 - \left(\frac{\gamma+1}{\gamma-1}\right)\left(\frac{\gamma+1}{\gamma-1} + \frac{2\,q}{RT_1}\right)\right]}{\left(\frac{\gamma+1}{\gamma-1} - \frac{p_2}{p_1}\right)^3}\right\} \tag{7.52}$$

Some useful observations of Eqn. (7.52) can be made as follows –

(1) The numerator in the bracelet of Eqn. (7.52) is always < 0 for $q \geq 0$.
(2) In region of III and IV of Fig. 7.8 (expansion wave), which includes point C, $p_2 < p_1$ and $\rho_2/\rho_1 < 1$ making $d^2s_2/d(1/\rho_2)^2 < 0$. Therefore,

$$\left[\frac{d^2s_2}{d\left(\frac{1}{\rho_2}\right)^2}\right]_{III\,\&IV} < 0 \tag{7.53a}$$

(3) In region I, including point J, $p_2 > p_1$ and $\rho_2/\rho_1 > 1$, the displacement of the detonation adiabat above that of the shock adiabat gives the result $\Psi_2/\Psi_1 > (\gamma - 1)/(\gamma + 1)$ or $\rho_2/\rho_1 < (\gamma + 1)/(\gamma - 1)$. Therefore,

$$\left[\frac{d^2 s_2}{d\left(\frac{1}{\rho_2}\right)^2} \right]_I > 0 \tag{7.53b}$$

As suggested by Eqn. (7.50), the processes near points C and J are essentially isentropic. Therefore the local speed of sound can be determined as

$$a_2^2 = \left(\frac{\partial p_2}{\partial \rho_2} \right)_s = -\frac{1}{\rho_2^2} \left[\frac{\partial p_2}{\partial \left(\frac{1}{\rho_2}\right)} \right]_s \tag{7.54}$$

By substituting Eqn. (7.49) for point C and J and equating the result with Eqn. (7.36),

$$(a_2^2)_{KJ} = -\frac{1}{\rho_2^2} \left[\frac{\partial p_2}{\partial \left(\frac{1}{\rho_2}\right)} \right]_s = \frac{1}{\rho_2^2} \frac{(p_2 - p_1)}{\left(\frac{1}{\rho_1} - \frac{1}{\rho_2}\right)} = V_2^2 \tag{7.55}$$

That is, at points C and J, the burned gas has a sonic velocity relative to the wave:

$$M_{2J} = M_{2,C} = 1 \tag{7.66}$$

Because $V_1 > V_{2,C}$ and $T_2 > T_{1,C}$ (and $a_2 > a_C$), $M_{1,J} > 1$. In fact the compression wave in region I has a Mach number $(M_1)_I > 1$. Therefore, the wave in this region is a shock wave supported by the chemical heat-releasing reaction. So far the analysis of Eqns. (7.31) through (7.34) has produced the possible solutions represented by region I, III, and IV, including point J. In the following, physical reasoning will help to further eliminate some of these solutions.

Region I above point J is the strong detonation region because $(p_2/p_1) > (p_2/p_1)_J$. Because of the higher compression ratio, $T_2 > T_{2J}$ and $V_2 < V_{2J}$ and $M_2 < M_{2J} = 1$. (The fact $M_2 < M_{2J}$ is also consistent with entropy being minimum at J.) In this region, the subsonic downstream condition enables rarefaction and dissipative effects to overtake and weaken the shock wave, and move down the Hugoniot curve toward point J. Because $M_{2J} = 1$, downstream effects cannot reach the shock wave. For the portion of the region between B and J, $(p_2/p_1) < (p_2/p_1)_J$, $T_2 > T_{2J}$, and $M_2 > M_{2J} = 1$; it is the weak detonation region. However, the flow immediately downstream of the shock is subsonic and heat addition in a one-dimensional flow on a Rayleigh line cannot be accelerated past the sonic condition. Therefore, for region I, point J is the only possible solution. Thus, conditions represented by J correspond to a self-sustaining steady-state detonation wave, where a supersonic wave is supported by a heat-releasing reaction and leaves the burned gas traveling at a sonic speed relative to the wave.

In region IV (i.e., to the right of point C), p_2 is only slightly less than p_1, and ρ_2 is much less than ρ_1 (and $V_2 > V_{1,K}$), representing more expansion and acceleration downstream of the expansion than those at K. The wave is an expansion wave and is thus subsonic. Due to this excessive expansion it is the strong deflagration region, with $T_2 < T_{1,K}$ and $a_2 < a_K$, leading to $M_2 > M_{1,K} = 1$. It is not possible to accelerate a subsonic flow to supersonic speed by heat addition in a constant area duct. Region IV is thus not physically possible.

Region III represents the solution where a subsonic expansion wave is supported by heat release that leaves the burned gas accelerating to a higher speed than the unburned gas, relative to the wave. This region is called the weak deflagration region where $(p_2/p_1) < (p_2/p_1)_C$ and $\rho_2 < \rho_C$.

As a consequence, the steady-state solution on the Hugoniot curve of Fig. 7.8 consists of point J and region III. The wave propagation corresponding to point J is called the *Chapman-Jouquet* (or simply *C-J*) *detonation wave*. Region I above point J, although not a steady-state solution, is possible during the transient period before it quickly resettles to J due to external weakening effects. Region III comprises solutions for subsonic propagation of pressure and combustion waves, called *deflagration waves*.

It is of interest to determine the unique detonation speed of the steadily propagating C-J detonation wave. Assuming that γ and R remain across the wave, the continuity, momentum, and energy equations, Eqn. (7.31) through (7.33), can be rewritten as

$$\frac{p_2}{p_1} = \frac{M_1}{M_2}\sqrt{\frac{T_2}{T_1}} \tag{7.67}$$

$$\frac{p_2}{p_1} = \frac{1 + \gamma M_1^2}{1 + \gamma M_2^2} \tag{7.77}$$

$$T_{t2} = T_{t1} + \frac{q}{c_p} \tag{7.78}$$

where $T_t = h + V^2/2$ is the stagnation temperature. At J, the isentropic relationship $T = T_t/[1 + (\gamma - 1)M^2/2]$ applies, as permitted by Eqn. (7.50). Combining Eqns. (7.67) and (7.77) leads to

$$\frac{T_2}{T_1} = \frac{M_2^2(1 + \gamma M_1^2)^2}{M_1^2(1 + \gamma M_2^2)^2} \tag{7.79}$$

$$1 + \frac{q}{c_p T_1\left[1 + \frac{(\gamma-1)}{2}M_1^2\right]} = \left[\frac{M_2^2(1 + \gamma M_1^2)^2}{M_1^2(1 + \gamma M_2^2)^2}\right]\left[\frac{1 + \frac{(\gamma-1)}{2}M_2^2}{1 + \frac{(\gamma-1)}{2}M_1^2}\right] \tag{7.80}$$

For C-J detonation ($M_2 = 1$), with the chemical heat release term on the left-hand side of Eqn. (7.81) is significantly larger than unity,

$$\frac{q}{c_p T_1 \left[1 + \frac{(\gamma-1)}{2} M_1^2\right]} = \left[\frac{(1 + \gamma M_1^2)^2}{M_1^2 (1+\gamma)^2}\right] \left[\frac{\frac{(\gamma+1)}{2}}{1 + \frac{(\gamma-1)}{2} M_1^2}\right] \tag{7.81}$$

The steady C-J detonation wave travels at a Mach number ranging from approximately 5 and 10 for most stoichiometric fuel-oxygen mixtures. Therefore, with the assumption of $M_1^2 = 1$, Eqn. (7.81) becomes

$$\frac{q}{c_p T_{t1}} \approx \frac{\gamma^2}{(\gamma+1)(\gamma-1)} \tag{7.82}$$

and

$$M_1 \approx \sqrt{2(\gamma+1)\frac{q}{c_p T_1}} \tag{7.83}$$

However, examination of Eqn. (7.81) suggests that the largest error associated with the $M_1^2 = 1$ assumption might very well arise from the term $[1 + (\gamma - 1)M_1^2/2]$; as for $M_1 = 5$ and $\gamma = 1.4$, $(\gamma - 1)M_1^2/2 = 5$, which is not much larger than 1. A rearrangement by eliminating the $[1 + (\gamma - 1)M_1^2/2]$ term from both sides of Eqn. (7.81) leads to

$$M_1 \approx \frac{1}{\gamma}\sqrt{2(\gamma+1)\frac{q}{c_p T_1}} \tag{7.84}$$

This value of M_1 is smaller than that of Eqn. (7.83) by a factor of $1/\gamma$.

EXAMPLE 7.8 Calculate the C-J detonation wave Mach number of the stoichiometric hydrogen-oxygen mixture, initially at 1 atmosphere and 300 K. Also (i) find the speed of the burned gas relative to the laboratory and the pressure downstream of the wave. The heat capacity of the fresh mixture and the water vapor is $\bar{c}_p = 2.4$ kJ/kg·K and the heat from the reaction is 119,960 kJ/kgH$_2$. Assume the product is water vapor ($\gamma = 1.327$ and $R = 0.461$ kJ/kg·K and $c_p = 2.2$ kJ/kg·K); (ii) the density ratio using the $M^2 = 1$ assumption and the Hugoniot relationship, Eqn. (7.40a).

Solutions –
The value R of the fresh mixture is ($H_2 + \frac{1}{2}O_2$), which has an average molecular weight of 12 kg/kmol and thus 13,329 kJ per kilogram of the mixture. Therefore, $\bar{R} = (8.314 \text{ kJ/kmol·K})/(12 \text{ kg/kmol}) = 0.6928$ kJ/kg·K.

$$M_1 = \frac{1}{\gamma}\sqrt{2(\gamma+1)\frac{q}{c_p T_1}} = \frac{1}{1.4} \times \sqrt{2 \times 2.4 \times \frac{13,329}{2.4 \times 300}} = 6.73$$

$$\frac{p_2}{p_1} = \frac{1 + \gamma M_1^2}{1 + \gamma M_2^2} = \frac{1 + 1.4 \times 6.604^2}{1 + 1.4} = 25.86$$

$$\frac{T_2}{T_1} = \frac{M_2^2 \left(1 + \gamma M_1^2\right)^2}{M_1^2 \left(1 + \gamma M_2^2\right)^2} = \frac{\left(1 + 1.4 \times 6.73^2\right)^2}{6.73^2 \times (1 + 1.4)^2} = 15.90$$

Assuming constant γ and R,

$$\frac{\rho_2}{\rho_1} = \frac{p_2}{p_1} \times \frac{T_1}{T_2} = 1.63$$

$$V_2 = M_2 \sqrt{\gamma_2 R_2 T_2} = 1 \times \sqrt{1.327 \times 461 \times 15.90 \times 300} = 1,708.34 \text{ m/s}$$

$$V_1 = M_1 \sqrt{\gamma_1 R_1 T_1} = 6.73 \times \sqrt{1.4 \times 693 \times 300} = 3,630.8 \text{ m/s}$$

For a moving detonation wave, the speed of burned gas relative to the laboratory is

$$V_b = V_1 - V_2 = 1,922.5 \text{ m/s}$$

$$M_b = \frac{V_b}{\sqrt{\gamma_2 R_2 T_2}} = \frac{1,922.5}{\sqrt{1.327 \times 461 \times 15.90 \times 300}} \approx 1.125$$

(The burned gas moves at supersonic speed to a laboratory observer, although at a sonic speed relative to the wave.)

Comments –
As comparisons, a shock wave with $M_1 = 6.73$ and $\gamma = 1.4$ without chemical heat release causes $p_2/p_1 \approx 52$ and $T_2/T_1 \approx 8.5$. The reason the detonation wave produces a higher value is easy to understand: the chemical reaction adds energy to the gas. The lower p_2/p_1 by detonation might be attributed to the extra increase in entropy and thus smaller p_{t2}; coupled with the sonic condition, the equation yields a lower value. $\qquad\square$

Another way for determining V_1 is to take advantage of the fact that $M_2 = 1$ and that continuity requires that $V_1 = (\rho_2/\rho_1)V_2$.

$$V_1 = \left(\frac{\rho_2}{\rho_1}\right)\sqrt{\gamma_2 R_2 T_2} = \left(\frac{\rho_2}{\rho_1}\right)\sqrt{\gamma_2 R_2 \frac{T_{t2}}{1 + \frac{\gamma_2 - 1}{2M_2^2}}} = \left(\frac{\rho_2}{\rho_1}\right)\sqrt{\frac{2\gamma_2 R_2}{\gamma_2 + 1} T_{t2}}$$

$$= \left(\frac{\rho_2}{\rho_1}\right)\sqrt{\frac{2\gamma_2 R_2}{\gamma_2 + 1}\left(T_{t1} + \frac{q}{c_{p,2}}\right)} \qquad (7.85)$$

Equation (7.85) requires first determining ρ_2/ρ_1. This density ratio is close to 1.8 for many fuels under stoichiometric conditions (Glassman and Yetter, 2008).

EXAMPLE 7.9 Use Eqn. (7.85) to determine the detonation wave speed of the mixture of Example 7.8.

$$V_1 = \left(\frac{\rho_2}{\rho_1}\right)\sqrt{\frac{2\gamma_2 R_2}{\gamma_2 + 1}\left(T_{t1} + \frac{q}{c_{p,2}}\right)}$$

Since the determination of T_{t1} requires knowledge of M_1, which is itself to be determined, it is appropriate to use

$$V_1 = \left(\frac{\rho_2}{\rho_1}\right)\sqrt{\frac{2\gamma_2 R_2}{\gamma_2 + 1} T_{t2}}$$

$$T_{t2} = \frac{1}{c_{p,2}}\left(c_{p,1} T_1 + q\right) = \frac{1}{2.2}\left(2.4 \times 300 + 13,329\right) \approx 6,386\text{K}$$

$$V_1 = 1.8 \times \sqrt{\frac{2 \times 1.327 \times 461 \times 6,386}{2.327}} \approx 3,300 \text{ m/s}$$

$$M_1 = \frac{V_1}{\sqrt{\gamma_1 R_1 T_1}} = \frac{3,300}{\sqrt{1.4 \times 692.8 \times 300}} \approx 6.12$$

$$V_2 = M_2\sqrt{\gamma_2 R_2 T_2} = 1 \times \sqrt{1.327 \times 461 \times \frac{6,368}{1 + \frac{1.327-1}{2} \times 1}} \approx 1,830 \text{ m/s}$$

Comments –
The results of Examples 7.8 and 7.9 agree within 10% for V_1 and M_1, and approximately 7% for V_2.   □

Problems

Problem 7.1 Derive Eqn. (7.15d).

Problem 7.2 Show the steps leading to Eqn. (7.24).

Problem 7.3 An air flow enters a frictionless constant-area duct with $M_1 = 0.3$, $T_1 = 25°\text{C}$, and $\rho_1 = 2.20 \text{ kg/m}^3$ and exits the duct with $M_2 = 0.7$. Find the heat transfer and changes in entropy and stagnation pressure between the duct entrance and exit.

Problem 7.4 An air flow enters a frictionless constant-area duct with $M_1 = 2.0$ and exits the duct with $M_2 = 2.5$, $T_2 = 300$ K, and $p_2 = 101.3$ kPa. Find the heat transfer and changes in entropy and stagnation pressure between the duct entrance and exit.

Problem 7.5 A circular pipe (5 cm in diameter) is connected to an air reservoir (at 700 kPa and 450 K) through a converging nozzle. What is the maximum flow rate that can be delivered by this nozzle-pipe assembly if the cooling of the air (i.e., heat loss) is 50 kJ/kg and the back pressure is 0 kPa (i.e., vacuum)? Determine the pressure and the Mach number at the exit plane under these conditions?

Problem 7.6 A circular pipe (5 cm in diameter) is connected to an air reservoir (at 700 kPa and 450 K) through a converging nozzle. The cooling rate of the air is 50 kJ/kg. Determine the back pressure when a normal shock stands at the exit.

Problem 7.7 A long constant-area frictionless duct is attached to an air reservoir (at 300 kPa and 300 K) through a converging-diverging nozzle with smooth geometrical transition in shapes. Assume the area ratio of the nozzle exit to the throat is 3. Assuming there is no shock wave in the duct-nozzle assembly, find the amount of heat transfer for the duct exit condition to be sonic. Also find the exit plane pressure, the entropy change, and pressure loss. Assume that $c_p = 1.004$ kJ/kg·K throughout the entire flow.

Problem 7.8 A long frictionless circular duct, having a diameter of 2 cm, is connected to an air reservoir (at 500 kPa and 400 K) by a converging nozzle. It is to deliver air into another reservoir that is a vacuum. Determine the mass flow rate and the change in stagnation pressure if (1) the heat is added to the air through the duct wall at 50.2 kJ/kg, and (2) if heat is taken away from the air (i.e., cooling) at 50.2 kJ/kg. Assume that $c_p = 1.004$ kJ/kg·K throughout the entire flow.

Problem 7.9 A frictionless circular duct is connected to an air reservoir (at 300 kPa and 300 K) through a converging-diverging nozzle. The duct inlet to nozzle throat area ratio is 2.0 and the heat addition to the flow through the duct wall is 50 kJ/kg. Assume that $c_p = 1.004$ kJ/kg·K throughout the entire flow. (1) What is the maximum back pressure to avoid a normal shock through the duct-nozzle device? Under this condition, what is the duct exit velocity? (2) What would be the combination of the amount of heat addition and back pressure for the duct exit condition to be sonic?

Problem 7.10 Derive in details Eqn. (7.40).

Problem 7.11

(a) Use the differential forms of mass, momentum, and the ideal gas law to show that

$$\frac{dT_t}{T} = \frac{dV}{V}(1 - M^2)$$

That is, the Rayleigh line flows heat addition/removal in the subsonic regime, leading to flow acceleration/deceleration, and to deceleration/acceleration in the supersonic regime.

(b) Use the differential forms of mass, momentum, and energy equations, combined with results in Part (a) to show that for a Rayleigh line

$$\frac{dT}{T} = \left(\frac{1 - \gamma M^2}{1 - M^2}\right)\frac{dT_t}{T}$$

and that the maximum temperature occurs at $M = \sqrt{1/\gamma}$, that is, on the subsonic branch.

Problem 7.12 Show all steps leading to Eqn. (7.52).

Problem 7.13 The fresh reactant mixture of Example 7.8 is now diluted with nitrogen or helium gases, so that it is $(H_2 + \frac{1}{2}O_2 + 5N_2)$ and $(H_2 + \frac{1}{2}O_2 + 5He)$. Find the detonation wave speeds and pressure and temperature ratios of these two mixtures. Compare with the results of Example 7.8, especially those of V_1 and M_1, and try to draw conclusions.

Problem 7.14 The following expression for pressure ratio applies to both detonation and shock waves:

$$\frac{p_2}{p_1} = \frac{1 + \gamma M_1^2}{1 + \gamma M_2^2} \qquad\qquad \text{Eqn. (4.18a)}$$

(1) Assuming that $\gamma_2 = \gamma_1 = \gamma$ across the C-J detonation wave, compare the magnitudes of pressure ratios of the detonation and the shock waves with the same M_1.

(2) Find the difference between the pressure ratios of these two waves.

(3) For the same very large M_1, show that the pressure ratio across the shock is approximately twice that across the detonation wave.

Problem 7.15 Assume that a jet combustor is a constant-area duct where viscous effects are negligible. If the combustion leads to a stagnation temperature equal to 4.5 and the desired Mach number of 0.9, determine (a) the duct inlet Mach number, and (b) the amount of heat added. Foe convenience, assume and throughout the combustion and flow processes and an inlet temperature of 250 K.

8 Equations of Multidimensional Frictionless Flow Subject to Small Perturbation

The one-dimensional theory discussed in previous chapters demonstrates essential features of gas dynamics signaling mechanism, compression, expansion, shock wave and its formation, flow turning, effects of area changes, friction, and heat transfer, with applications to lift and drag on airfoils and channel flows. In general, there are effects of multidimensional and unsteady nature, such as curved shocks, shock, and expansion wave propagation. While simplifications to one-dimensionality may serve as good approximations, many practical flows are inherently multidimensional and unsteady. To focus on the multidimensional and unsteady effects, this chapter will explore inviscid flows without heat transfer.

8.1 Differential Control Volume Approach for Mass Conservation

The conservation of mass (continuity) equation for an arbitrary control volume is given in Chapter by Eqn. (2.4):

$$0 = \frac{\partial}{\partial t}\int_{CV}\rho dV + \int_{CS}(\rho\vec{V}\cdot d\vec{A}) \text{ or } \frac{\partial}{\partial t}\int_{CV}\rho dV = -\int_{CS}(\rho\vec{V}\cdot d\vec{A}) \qquad (2.4)$$

Consider the differential control volume having an infinitesimal size shown in Fig. 8.1, at whose center all the fluid properties are defined. The time rate of change within the control volume (or CV) can be rewritten as

$$\frac{\partial}{\partial t}\int_{CV}\rho dV = \frac{\partial\rho}{\partial t}dxdydz$$

Let the velocity field be represented by the vector, $\vec{V} = \vec{i}u + \vec{j}v + \vec{k}w$. The mass flux term across the boundary of the boundary (i.e., the control surface, or CS) in Eqn. (2.4) becomes

$$\int_{CS}(\rho\vec{V}\cdot d\vec{A}) = \left[\left(\rho u + \frac{\partial(\rho u)}{\partial x}\frac{dx}{2}\right) - \left(\rho u - \frac{\partial(\rho u)}{\partial x}\frac{dx}{2}\right)\right]dydz + \left[\left(\rho v + \frac{\partial(\rho v)}{\partial y}\frac{dy}{2}\right)\right.$$

$$\left. - \left(\rho v - \frac{\partial(\rho v)}{\partial y}\frac{dy}{2}\right)\right]dxdz + \left[\left(\rho w + \frac{\partial(\rho w)}{\partial z}\frac{dz}{2}\right) - \left(\rho w - \frac{\partial(\rho w)}{\partial z}\frac{dz}{2}\right)\right]dxdy$$

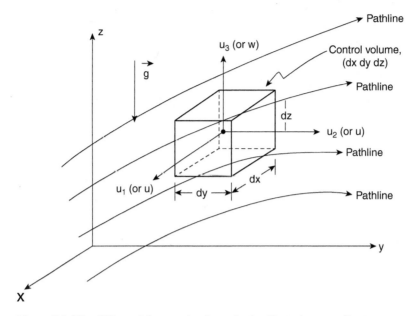

Figure 8.1 The differential control volume in the Cartesian coordinate system.

Therefore,

$$\int_{CS} (\rho \vec{V} \cdot d\vec{A}) = \left[\frac{\partial(\rho u)}{\partial x} + \frac{\partial(\rho v)}{\partial y} + \frac{\partial(\rho w)}{\partial z} \right] dxdydz$$

Thus the mass conservation for the differential control volume becomes

$$\frac{\partial \rho}{\partial t} + \frac{\partial(\rho u)}{\partial x} + \frac{\partial(\rho v)}{\partial y} + \frac{\partial(\rho w)}{\partial z} = 0 \qquad (8.1a)$$

or, by using the divergence operator $\nabla \equiv \vec{i} \frac{\partial}{\partial x} + \vec{j} \frac{\partial}{\partial y} + \vec{k} \frac{\partial}{\partial z}$,

$$\frac{\partial \rho}{\partial t} + \nabla \cdot (\rho \vec{V}) = 0 \qquad (8.1b)$$

The continuity equation can also be rewritten by using the substantial, or Eulerian, differential operator

$$\frac{D}{Dt} \equiv \frac{\partial}{\partial t} + \vec{V} \cdot \nabla = \frac{\partial}{\partial t} + \left(u \frac{\partial}{\partial x} + v \frac{\partial}{\partial y} + w \frac{\partial}{\partial z} \right) \qquad (8.1c)$$

where the first term on the right-hand side accounts for unsteadiness and $\vec{V} \cdot \nabla$ represents the convective effects or the changes a fluid particle experiences due to convection. Thus, the Eulerian derivative represents changes in fluid particles due to the combined unsteady and convective effects. Equation (8.1b) then becomes

$$\frac{\partial \rho}{\partial t} + \vec{V} \cdot \nabla \rho + \rho \nabla \cdot \vec{V} = 0$$

and therefore

$$\frac{D\rho}{Dt} + \rho \nabla \cdot \vec{V} = 0 \tag{8.1d}$$

It is noted that for steady flows, $\partial \rho / \partial t = 0$, $\rho \nabla \cdot \vec{V} = -\vec{V} \cdot \nabla \rho$ so that $D\rho/Dt = \vec{V} \cdot \nabla \rho \neq 0$ in general except for incompressible flows where $\rho = $ constant.

For steady flows, Eqn. (8.1b) reduces to

$$\nabla \cdot (\rho \vec{V}) = 0 \tag{8.1e}$$

If the flow is incompressible, ρ remains unchanged for time and location, and both Eqns. (8.1b) and (8.1d) become

$$\nabla \cdot \vec{V} = 0 \tag{8.1f}$$

In cylindrical coordinates, the del operator, ∇, is expressed as

$$\nabla = \vec{e}_r \frac{\partial}{\partial r} + \vec{e}_\theta \frac{1}{r} \frac{\partial V_r}{\partial \theta} + \vec{k} \frac{\partial}{\partial z} \tag{8.1e} \ (8.2)??$$

where $\vec{e}_r$, $\vec{e}_\theta$, and $\vec{k}$ are unit vectors in the radial, circumferential, and axial directions, respectively.

8.2 Conservation of Momentum

The control volume form of momentum conservation has been given in Chapter 2 as

$$\sum_i \vec{F}_i \bigg)_{system} = \frac{\partial}{\partial t} \int_{CV} \rho \vec{V} d\mathcal{V} + \int_{CS} \vec{V} (\rho \vec{V} \cdot d\vec{A}) \tag{2.5}$$

Due to the vector nature of momentum, it is convenient and instructive to first obtain the result in the x-direction for the differential CV shown in Fig. 8.1, as follows.

$$\frac{\partial}{\partial t} \int_{CV} \rho u d\mathcal{V} = \frac{\partial(\rho u)}{\partial t} dx dy dz$$

$$\int_{CS} u(\rho \vec{V} \cdot d\vec{A}) = \left[\left(u + \frac{\partial u}{\partial x} \frac{dx}{2} \right) \left(\rho u + \frac{\partial(\rho u)}{\partial x} \frac{dx}{2} \right) - \left(u - \frac{\partial u}{\partial x} \frac{dx}{2} \right) \right.$$

$$\left. \left(\rho u - \frac{\partial(\rho u)}{\partial x} \frac{dx}{2} \right) \right] dy dz + \left[\left(u + \frac{\partial u}{\partial y} \frac{dy}{2} \right) \left(\rho v + \frac{\partial(\rho v)}{\partial y} \frac{dy}{2} \right) \right.$$

$$\left. - \left(u - \frac{\partial u}{\partial y} \frac{dy}{2} \right) \left(\rho v - \frac{\partial(\rho v)}{\partial y} \frac{dy}{2} \right) \right] dx dz + \left[\left(u + \frac{\partial u}{\partial z} \frac{dz}{2} \right) \right.$$

$$\left. \left(\rho w + \frac{\partial(\rho w)}{\partial z} \frac{dz}{2} \right) - \left(u - \frac{\partial u}{\partial z} \frac{dz}{2} \right) \left(\rho w - \frac{\partial(\rho w)}{\partial z} \frac{dz}{2} \right) \right] dx dy$$

For frictionless flow, the forces acting on the fluid particles can be decomposed into two groups: one due to pressure (p) and the other due to all body forces $(\vec{f}$, including gravitation and electromagnetic forces, etc.). Therefore, in the x-direction this becomes

$$\sum_i \vec{F}_i\Big)_{system} = \left(p - \frac{\partial p}{\partial x}\frac{dx}{2}\right)dydz - \left(p + \frac{\partial p}{\partial x}\frac{dx}{2}\right)dydz + \rho f_x dxdydz$$

$$= -\frac{\partial p}{\partial x}dxdydz + \rho f_x dxdydz$$

Combining the above three results while neglecting the second- and higher-order terms, the differential equation for conservation in the x-direction is

$$-\frac{\partial p}{\partial x} + \rho f_x = \frac{\partial(\rho u)}{\partial t} + u\frac{\partial(\rho u)}{\partial x} + u\frac{\partial(\rho v)}{\partial y} + u\frac{\partial(\rho w)}{\partial z} + \rho u\frac{\partial u}{\partial x} + \rho v\frac{\partial u}{\partial y} + \rho w\frac{\partial u}{\partial z}$$

Substituting the continuity equation, Eqn. (8.1a), and dividing through the x-momentum equation by ρ yields

$$-\frac{1}{\rho}\frac{\partial p}{\partial x} + f_x = \frac{\partial u}{\partial t} + + \left(u\frac{\partial u}{\partial x} + v\frac{\partial u}{\partial y} + w\frac{\partial u}{\partial z}\right) \tag{8.2a}$$

Similarly, the y- and z-momentum equations can be found, respectively, as

$$-\frac{1}{\rho}\frac{\partial p}{\partial y} + f_y = \frac{\partial v}{\partial t} + \left(u\frac{\partial v}{\partial x} + v\frac{\partial v}{\partial y} + w\frac{\partial v}{\partial z}\right) \tag{8.2b}$$

$$-\frac{1}{\rho}\frac{\partial p}{\partial z} + f_z = \frac{\partial w}{\partial t} + \left(u\frac{\partial w}{\partial x} + v\frac{\partial w}{\partial y} + w\frac{\partial w}{\partial z}\right) \tag{8.2c}$$

In vector notation, these three components of the momentum equation can be combined as

$$\frac{D\vec{V}}{Dt} = \frac{\partial \vec{V}}{\partial t} + (\vec{V}\cdot\nabla)\vec{V} = -\frac{1}{\rho}\nabla p + \vec{f} \tag{8.2d}$$

If the frictional and gravitational forces can be neglected, Eqn. (8.2d) is reduced to

$$\frac{D\vec{V}}{Dt} = -\frac{1}{\rho}\nabla p \tag{8.2e}$$

Equation (8.2e) is the well-known *Euler's equation* for inviscid flow without body forces.

8.3 Conservation of Energy

The principle of conservation of energy, Eqn. (2.19) for a frictionless flow with no work done by external mechanisms can be rewritten as

$$\dot{Q} - \dot{W} = \frac{\partial}{\partial t} \int_{CV} \left(e + \frac{1}{2} V^2 \right) \rho d\mathcal{V} + \int_{CS} \left(h + \frac{1}{2} V^2 \right) \left(\rho \vec{V} \cdot d\vec{A} \right) \qquad (8.3)$$

where e denotes internal energy and $h_t = (h + V^2/2) = \left(h + \vec{V} \cdot \vec{V}/2 \right)$ (with $V^2/2 = (u^2 + v^2 + w^2)/2 = \vec{V} \cdot \vec{V}/2$) is the stagnation enthalpy. For the control volume shown in Fig. 8.1,

$$\dot{Q} - \dot{W} = \rho \dot{q} dx dy dz + \rho \left(\vec{f} \cdot \vec{V} \right) dx dy dz = \left[\rho \dot{q} + \rho \left(\vec{f} \cdot \vec{V} \right) \right] dx dy dz \qquad (8.4a)$$

where $\dot{q}$ is the heat transfer rate per unit mass of the gas. Following the similar procedure for mass conservation, the right-hand side terms of Eqn. (8.3) can be rewritten as follows. For brevity and convenience in derivation, let $\dot{e} = (e + V^2/2) = \left(e + \vec{V} \cdot \vec{V}/2 \right)$. Then

$$\frac{\partial}{\partial t} \int_{CV} \left(e + \frac{1}{2} V^2 \right) \rho d\mathcal{V} = \frac{\partial(\rho \dot{e})}{\partial t} dx dy dz \qquad (8.4b)$$

$$\begin{aligned}
&\int_{CS} \left(h + \frac{1}{2} V^2 \right) \left(\rho \vec{V} \cdot d\vec{A} \right) \\
&= \left[\left(h_t + \frac{\partial h_t}{\partial x} \frac{dx}{2} \right) \left(\rho u + \frac{\partial(\rho u)}{\partial x} \frac{dx}{2} \right) - \left(h_t - \frac{\partial h_t}{\partial x} \frac{dx}{2} \right) \left(\rho u - \frac{\partial(\rho u)}{\partial x} \frac{dx}{2} \right) \right] dy dz \\
&+ \left[\left(h_t + \frac{\partial h_t}{\partial y} \frac{dy}{2} \right) \left(\rho v + \frac{\partial(\rho v)}{\partial y} \frac{dy}{2} \right) - \left(h_t - \frac{\partial h_t}{\partial y} \frac{dy}{2} \right) \left(\rho v - \frac{\partial(\rho v)}{\partial y} \frac{dy}{2} \right) \right] dx dz \\
&+ \left[\left(h_t + \frac{\partial h_t}{\partial z} \frac{dz}{2} \right) \left(\rho w + \frac{\partial(\rho w)}{\partial z} \frac{dz}{2} \right) - \left(h_t - \frac{\partial h_t}{\partial z} \frac{dz}{2} \right) \left(\rho w - \frac{\partial(\rho w)}{\partial z} \frac{dz}{2} \right) \right]
\end{aligned}$$
$$(8.4c)$$

Expanding and neglecting the second- and higher-order terms of Eqn. (8.4c), and then combining it with Eqns. (8.3a) and (8.3b) leads to

$$\rho \dot{q} + \rho \vec{f} \cdot \vec{V} = \frac{\partial(\rho \dot{e})}{\partial t} + \frac{\partial(\rho u h_t)}{\partial x} + \frac{\partial(\rho v h_t)}{\partial y} + \frac{\partial(\rho w h_t)}{\partial z} = \frac{\partial(\rho \dot{e})}{\partial t} + \nabla \cdot \left(\rho h_t \vec{V} \right) \qquad (8.5a)$$

or

$$\rho \dot{q} + \rho \vec{f} \cdot \vec{V} = \frac{\partial}{\partial t} \left(\rho e + \frac{1}{2} \rho V^2 \right) + \nabla \cdot \left[\rho \left(h + \frac{1}{2} V^2 \right) \vec{V} \right] \qquad (8.5b)$$

Because

$$h = e + p/\rho,$$

$$\rho \dot{q} + \rho \vec{f} \cdot \vec{V} = \frac{\partial}{\partial t} \left[\rho \left(e + \frac{1}{2} V^2 \right) \right] + \nabla \cdot \left[\rho \left(e + \frac{1}{2} V^2 \right) \vec{V} \right] + \nabla \cdot \left(p \vec{V} \right) \qquad (8.5c)$$

The first two bracketed terms on the right-hand side of Eqn. (8.5c) represent, respectively, the rate of storage of the combined internal and kinetic energy within the CV and rate of convection of the sum outside of the CV. The third term, $\nabla \cdot \left(p\vec{V} \right)$, is the net "flow work" per unit volume that is needed to push the fluid particle against the pressure so as to leave the CV.

This equation can also be rewritten as

$$\rho\dot{q} + \rho\vec{f}\cdot\vec{V} = \left[\rho\frac{\partial e}{\partial t} + \rho\vec{V}\cdot\nabla e\right] + e\left[\frac{\partial\rho}{\partial t} + \nabla\cdot\left(\rho\vec{V}\right)\right] + \left(\frac{1}{2}V^2\right)\left[\frac{\partial\rho}{\partial t} + \nabla\cdot\left(\rho\vec{V}\right)\right]$$

$$+ \left[\rho\frac{\partial}{\partial t}\left(\frac{1}{2}V^2\right) + \rho\vec{V}\cdot\nabla\left(\frac{1}{2}V^2\right)\right] + \nabla\cdot\left(p\vec{V}\right) \qquad (8.5\text{d})$$

The first bracket in the above expression becomes De/Dt. With the help of continuity, Eqn. (8.1b), the second and third brackets on the right-hand side are identically zero and the fourth bracket becomes

$$\rho\vec{V}\cdot\frac{\partial\vec{V}}{\partial t} + \rho\vec{V}\cdot\nabla\left(\frac{1}{2}V^2\right) = \rho\vec{V}\cdot\left[\frac{\partial\vec{V}}{\partial t} + \nabla\left(\frac{1}{2}V^2\right)\right] = \rho\vec{V}\cdot\left[\frac{\partial\vec{V}}{\partial t} + \vec{V}\cdot\nabla\vec{V}\right]$$

$$= \rho\vec{V}\cdot\frac{D\vec{V}}{Dt}$$

Equation (8.5c) then becomes

$$\rho\dot{q} + \rho\vec{f}\cdot\vec{V} = \rho\frac{De}{Dt} + \rho\vec{V}\cdot\frac{D\vec{V}}{Dt} + \nabla\cdot\left(p\vec{V}\right) \qquad (8.5\text{e})$$

Neglecting $\vec{f}$ and assuming the change in elevation is small, the momentum equation, Eqn. (8.2e), dictates that $D\vec{V}/Dt = -\nabla p/\rho$, which can be substituted into Eqn. (8.5e) to yield

$$\rho\dot{q} = \rho\frac{De}{Dt} + \rho\vec{V}\cdot\left(-\frac{\nabla p}{\rho}\right) + p\nabla\cdot\vec{V} + \vec{V}\cdot\nabla p \qquad (8.5\text{f})$$

Therefore

$$\rho\dot{q} = \rho\frac{De}{Dt} + p\nabla\cdot\vec{V} \qquad (8.5\text{g})$$

Therefore the conservation of energy indicates that the rate of heat transfer to the CV is equal to the sum of the increase in the internal energy of the gas passing through the CV and the work done by the gas expansion (if $\nabla\cdot\vec{V} > 0$), or on the gas (i.e., compression and $\nabla\cdot\vec{V} < 0$).

To express the energy equation in terms of stagnation enthalpy, h_t, Eqn. (8.5c) can be rewritten by incorporating a pressure term, as

$$\rho\dot{q} = \frac{\partial}{\partial t}\left[\rho\left(e + \frac{p}{\rho} + \frac{1}{2}V^2\right)\right] + \nabla\cdot\left[\rho\left(e + \frac{p}{\rho} + \frac{1}{2}V^2\right)\vec{V}\right] + \nabla\cdot\left(p\vec{V}\right) - \frac{\partial p}{\partial t} - \nabla\cdot\left(p\vec{V}\right)$$

$$(8.5\text{h})$$

With the help of continuity, Eqn. (8.1b), this expression simplifies to

$$\rho \dot{q} = \rho \frac{Dh_t}{Dt} - \frac{\partial p}{\partial t} \tag{8.5i}$$

It is also possible to express the energy conservation equation in terms of $h \, (= e + p/\rho)$:

$$\frac{De}{Dt} = \frac{Dh}{Dt} - \frac{D(p/\rho)}{Dt} = \frac{Dh}{Dt} - \frac{1}{\rho}\frac{Dp}{Dt} + \frac{p}{\rho^2}\frac{D\rho}{Dt}$$

The continuity, Eqn. (8.1d), provides $D\rho/Dt = -\rho\nabla \cdot \vec{V}$ so that the above expression becomes

$$\frac{De}{Dt} = \frac{Dh}{Dt} - \frac{1}{\rho}\frac{Dp}{Dt} - \frac{p}{\rho}\nabla \cdot \vec{V}$$

Substituting this expression for De/Dt into Eqn. (8.5g) yields

$$\rho \dot{q} = \rho \frac{Dh}{Dt} - \frac{Dp}{Dt} \tag{8.5j}$$

For steady, adiabatic flows neglecting the effects of friction and body/gravitational forces, $Dh_t/Dt = 0$. That is, the stagnation enthalpy of a gas particle remains constant along its path. In steady flows, the path line coincides with the streamline. Therefore under these conditions,

$$h_t = h + \frac{V^2}{2} = e + \frac{p}{\rho} + \frac{V^2}{2} = \text{constant} \tag{8.5k}$$

along a streamline. The constant in Eqn. (8.5k) in general varies from one to the other streamline, as expected in multidimensional flows. This result is the same as that for isentropic flows discussed in Chapter 4.

It is noteworthy that Eqn. (8.5j) can be derived alternatively by starting with Eqn. (2.10b), which is the first law of thermodynamics for a system:

$$\delta Q - \delta W = dE$$

For an ideal gas the only work is due to volume change and thus $\delta W = pd\mathcal{V}$. In the intensive form

$$\delta q = de + pd\mathcal{v} = d(h - p\mathcal{v}) + pd\mathcal{v} = dh - \mathcal{v}dp$$

By following the change of a gas particle, one essentially is tracking the change of a system. Thus, the time rate change regarding energy can be written using substantial derivatives for the variation and the exact differentials in the above expression and the following is arrived at, with $\mathcal{v} = 1/\rho$,

$$\dot{q} = \frac{Dh}{Dt} - \frac{1}{\rho}\frac{Dp}{Dt} \tag{8.5l}$$

which is identical to Eqn. (8.5j).

8.4 The Entropy Equation

The entropy equation for a system undergoing reversible changes is

$$Tds = de + pdѵ \tag{2.33}$$

By following similar procedures leading to Eqn. (8.5l), one obtains

$$T\frac{Ds}{Dt} = \frac{De}{Dt} + p\frac{Dѵ}{Dt} \tag{8.6}$$

For an isentropic flow, the fluid particle experiences no irreversibility or heat transfer (i.e., the adiabatic condition) and receives no work from, and does no work on, the surroundings. Thus according to Eqn. (8.5l),

$$\frac{Ds}{Dt} = \frac{D}{Dt}\left(e + \frac{p}{\rho}\right) - \frac{1}{\rho}\frac{Dp}{Dt} = \frac{Dh}{Dt} - \frac{1}{\rho}\frac{Dp}{Dt} = \dot{q} \tag{8.7}$$

Eqn. (8.7) can easily be shown by substituting Eqn. (8.5l) with $\dot{q} = 0$ into Eqn. (8.6). Since the substantial derivative tracks changes of a given fluid particle, Eqn. (8.7) indicates that the entropy remains constant along the fluid path. For a steady flow, this translates into constant entropy along a streamline.

8.5 The Substantial (Eulerian) Derivative

The physical meaning of the Eulerian operator deserves further explanation. The first term $\partial/\partial t$ on the right-hand side of Eqn. (8.1c) denotes the time rate of change of the property (mass, momentum, energy, etc.) due to unsteadiness within the differential control volume shown in Fig. 8.1. The $\vec{V}\cdot\nabla$ term represents the change due to the gradient of the property in the direction of the flow while passing through the control volume and is also called the convective derivative, as the inner (or "dot") product projects the gradient onto the direction of the flow, with the projected magnitude being the time rate of change (note $\vec{V}\cdot\nabla$ has the unit of 1/time). One would visualize that for a given gradient in the direction of flow, the larger the velocity, the larger the rate of change. Then, increasing the flow speed by a factor of two implies twice as much power needs to be added to the fluid to maintain the same temperature rise (i.e., same temperature gradient).

Similarly, $\rho D\vec{V}/Dt = \rho\partial\vec{V}/\partial t + \rho\vec{V}\cdot\nabla\vec{V}$ represents the force experienced by the fluid particle due to the unsteadiness within the control volume and the velocity gradient across the control volume (the latter being the convective acceleration). Therefore, even if flow is a steady one, the fluid element accelerates/decelerates (with an increase/decrease in its momentum) while moving from one point to the other (for example, in a converging-diverging nozzle). For flows with negligible friction and body forces, Eqn. (8.2e) indicates that the acceleration of the fluid particle is due to the pressure gradient.

8.6 Fluid Rotation, Vorticity, and Circulation

A special class of fluid motion allows great simplification of governing equations. It is the irrotational flow. To discuss irrotational fluid motion, it is necessary to introduce vorticity. In general, vorticity in a flow arises from rotational motion of the fluid particle. For example, a fluid particle with a rigid-body rotation possesses vorticity. A velocity or momentum boundary layer near a surface comprises a vorticity field resulting from the velocity gradient due to the effect of fluid viscosity. A flow of "idealized" fluid with no viscosity (i.e., an inviscid fluid) does not generate a boundary layer near surfaces, although it may possess vorticity due to rotation. Free vortical motion of real fluids, of which hurricanes are among the examples, results in no vorticity. Therefore, there is no definite relationship between fluid viscosity and vorticity, as the former is a property of the fluid and the latter a property of the flow. For further discussion on vorticity, it serves to first demonstrate how vorticity can be calculated from the velocity field (a property of the flow, not the fluid).

To understand and characterize rotational fluid motion, consider a differential fluid element OACB shown in Fig. 8.2, where the velocity at point O is $(\vec{i}u + \vec{j}v)$ and the fluid element at the instant rotates about the z-axis. By following the fluid element, one can describe its rotational and deformational motion as if the fluid element is "pinned" at point O. Also shown in Fig. 8.2 are the line segments (now dashed lines) that have undergone a general motion for over an infinitesimal time interval (dt), with the fluid element now represented by OA'C'B'. The line segments OA and OB have turned through angles $d\theta_A$ counterclockwise and $d\theta_B$ clockwise, respectively, to the locations represented by OA' and OB'. These angles are related to the velocity of the fluid element as follows:

$$d\theta_A = \frac{1}{dx}\left[\left(v + \frac{\partial v}{\partial x}dx\right)dt - vdt\right] = \frac{\partial v}{\partial x}dt$$

$$d\theta_B = \frac{1}{dy}\left[\left(u + \frac{\partial u}{\partial y}dy\right)dt - udt\right] = \frac{\partial u}{\partial y}dt$$

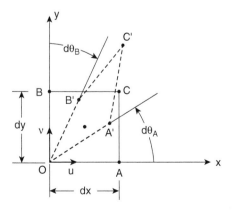

Figure 8.2 Schematic showing simultaneous rotation and deformation of fluid element *OACD* in a two-dimensional *xy*-plane.

For the differential fluid element, the fluid rotational speed about the z-axis (ω_z) is the average of the angular motion of the segments OA and OB. By observing the right-hand rule,

$$\omega_z = \frac{1}{2}\frac{d(\theta_A - \theta_B)}{dt} = \frac{1}{2}\left(\frac{\partial v}{\partial x} - \frac{\partial u}{\partial y}\right)$$

Similarly, the angular speeds about the y- and x-axes are, respectively,

$$\omega_y = \frac{1}{2}\left(\frac{\partial u}{\partial z} - \frac{\partial v}{\partial x}\right)$$

and

$$\omega_x = \frac{1}{2}\left(\frac{\partial w}{\partial y} - \frac{\partial v}{\partial z}\right)$$

Therefore in Cartesian coordinates, fluid rotational velocity is

$$\vec{\omega} = \frac{1}{2}\left[\vec{i}\left(\frac{\partial w}{\partial y} - \frac{\partial v}{\partial z}\right) + \vec{j}\left(\frac{\partial u}{\partial z} - \frac{\partial v}{\partial x}\right) + \vec{k}\left(\frac{\partial v}{\partial x} - \frac{\partial u}{\partial y}\right)\right] \tag{8.8}$$

In cylindrical coordinates (r, θ, z), $\vec{\omega}$ is given by

$$\vec{\omega} = \vec{e}_r\left(\frac{1}{r}\frac{\partial V_z}{\partial \theta} - \frac{\partial V_\theta}{\partial z}\right) + \vec{e}_\theta\left(\frac{\partial V_r}{\partial z} - \frac{\partial V_z}{\partial r}\right) + \vec{k}\left(\frac{1}{r}\frac{\partial r V_\theta}{\partial r} - \frac{1}{r}\frac{\partial V_r}{\partial \theta}\right) \tag{8.9}$$

By using the curl operator ($\nabla \times$),

$$\vec{\omega} = \frac{1}{2}\nabla \times \vec{V} \tag{8.10}$$

It is customary to define vorticity as twice the fluid rotation velocity,

$$\vec{\zeta} = 2\vec{\omega} = \nabla \times \vec{V} \tag{8.11}$$

A flow field with $\vec{\omega} = 0$ (and thus $\vec{\zeta} = 0$) is said to be irrotational. The reader can readily show that one-dimensional flows are inherently irrotational.

The fluid particle shown in Fig. 8.2 has also experienced an angular deformation rate on the xy-plane (ϵ_{xy}) equal to

$$\epsilon_{xy} \equiv \frac{1}{2}\frac{d(\theta_A + \theta_B)}{dt} = \frac{1}{2}\left(\frac{\partial v}{\partial x} + \frac{\partial u}{\partial y}\right) \tag{8.12a}$$

Similarly, the angular deformation rates on the yz- and zx-planes are, respectively,

$$\epsilon_{yz} = \frac{1}{2}\left(\frac{\partial w}{\partial y} + \frac{\partial v}{\partial z}\right) \tag{8.12b}$$

and

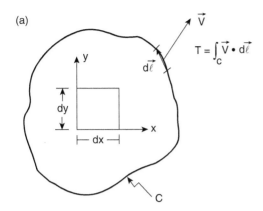

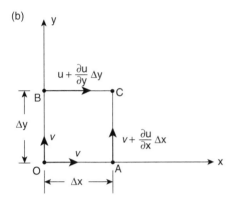

Figure 8.3 (a) Schematic showing flow circulation, Γ, over an arbitrary closed contour C; (b) a special case of a closed contour – a parallelogram $OACB$.

$$\epsilon_{zx} = \frac{1}{2}\left(\frac{\partial w}{\partial x} + \frac{\partial u}{\partial z}\right) \tag{8.12c}$$

If OA and OB rotate through the same angle and in different senses from their original positions (i.e., respectively from the x- and y-axes), then $d\theta_A = d\theta_B$ and the fluid particle would experience only angular deformation (besides convection, which is the translational motion of point O in Fig. 8.2). In this case $\partial v/\partial x = \partial u/\partial y$, $\partial u/\partial z = \partial w/\partial x$, and $\partial w/\partial y = \partial v/\partial z$; similarly $\epsilon_{xy} = \partial u/\partial y$, $\epsilon_{yz} = \partial v/\partial z$, and $\epsilon_{zx} = \partial w/\partial x$. In general, $d\theta_A \neq d\theta_B$ and the fluid motion results in a combination of rotation and deformation, that is, $\vec{\omega} \neq 0$, $\epsilon_{xy} \neq 0$, $\epsilon_{yz} \neq 0$, and $\epsilon_{zx} \neq 0$.

A flow property closely related to rotation is circulation, Γ, defined as

$$\Gamma \equiv \oint_C \vec{V} \cdot d\vec{l} \tag{8.13}$$

where the integration is carried out over a closed curve C and $d\vec{l}$ is the differential line segment along C pointing in the counterclockwise direction, as shown in Fig. 8.3a. Consider a special closed curve that is a rectangle lying within C with its detailed velocity components and geometric characteristics shown in Fig. 8.3b. Let the velocity be $\vec{V} = \vec{i}u + \vec{j}v$ at point O. Then the circulation over this small enclosed "rectangular curve" is $\Delta\Gamma$ given by

$$\Delta\Gamma = udx + \left(v + \frac{\partial v}{\partial x}dx\right)\Delta y - \left(u + \frac{\partial u}{\partial y}dy\right)dx - vdy = \left(\frac{\partial v}{\partial x} - \frac{\partial u}{\partial y}\right)dxdy = 2\omega_z dxdy$$

(8.14)

Summing up the contribution from all elemental areas like that of the closed curve OACB, one obtains

$$\Gamma \equiv \oint_C \vec{V} \cdot d\vec{l} = \int\int 2\omega_z dxdy = \int\int \zeta_z dxdy = \int\int_A \zeta_z dA$$

(8.15a)

Equation (8.12) is also the *Stokes Theorem* that relates the line integral to the surface integral and vice versa. For a two-dimensional flow in the xy-plane, Eqn. (8.14) can be rewritten as

$$\zeta_z = \lim_{dA \to 0} \frac{d\Gamma}{dA}$$

(8.15b)

EXAMPLE 8.1 Find the vorticity distribution of an incompressible Couette flow between two infinite flat plates with the velocity distribution given as

$$\vec{V} = \vec{i}u = \vec{i}\frac{h^2}{2\mu}\left(\frac{dp}{dx}\right)\left[\left(\frac{y}{h}\right)^2 - \frac{1}{4}\right]$$

where h is the gap between the two plates with $y = 0$ at the mid-plane, μ is the fluid viscosity, and dp/dx is the pressure gradient that drives the flow.

Solution –
For this velocity distribution there is only velocity in the x-direction; then Eqns. (8.8) and (8.11) lead to

$$\vec{\zeta} = \vec{j}\left(\frac{\partial u}{\partial z} - \frac{\partial v}{\partial x}\right) + \vec{k}\left(\frac{\partial v}{\partial x} - \frac{\partial u}{\partial y}\right) = \vec{j}\frac{\partial u}{\partial z} - \vec{k}\frac{\partial u}{\partial y} = -\vec{k}\frac{y}{\mu}\left(\frac{dp}{dx}\right)$$

It can be seen that $\vec{\zeta} = 0$ at the mid-plane, as expected due to the symmetry of the velocity distribution while it reaches the maximum at $y = \pm h/2$.

Comments –
The fully developed velocity distribution between two parallel, infinite flat plates arises from the merge of the viscous (velocity/momentum) boundary layers over the plates. Once the two boundary layers merge after some entrance length, the velocity distribution is no longer a function of x because of the requirement for the plates to be infinite. Because the magnitude of the velocity everywhere except at the surface increases with the pressure gradient, one expects vorticity to likewise increase due to an increase in $\partial u/\partial y$ (which in this case increases linearly with dp/dx). This example illustrates that viscosity causes vorticity and that a parallel flow is not necessarily irrotational.   □

EXAMPLE 8.2 Find (1) the vorticity field, and (2) circulation over an arbitrary closed circular curve C with a radius R and centered at $r = 0$ for the flow field: $\vec{V} = \vec{e}_\theta(C_1 r^n)$.

Solution –
Eqn. (8.9) yields

$$\vec{\omega} = \vec{k}\left(\frac{1}{r}\frac{\partial r V_\theta}{\partial r}\right) = \vec{k}\left(\frac{C_1}{r}\frac{\partial r^{n+1}}{\partial r}\right) = \vec{k}\left[(n+1)C_1 r^{n-1}\right]$$

$$\Gamma \equiv \oint_C \vec{V} \cdot d\vec{l} = \int_0^{2\pi} V_\theta(r d\theta) = \int_0^{2\pi} (C_1 r^n)(r d\theta) = 2\pi r^{n+1} C_1$$

A few special values of n might be of interest.

For $n = -1$, $V_\theta = C_1/r$ and $\vec{\omega} = 0$ (except at $r = 0$, which is a singular point) and the flow field is irrotational and $\Gamma = 2\pi C_1 = $ constant everywhere, independent of the radius of the circular curve C.

For the rigid-body fluid rotation, $n = 1$, $V_\theta = C_1 r$, $\vec{\omega} = \vec{k}(2C_1)$, and $\Gamma = 2\pi r^2 C_1 = 2C_1 A$. The circulation increases linearly with the area of the enclosing circle.

For $n \neq -1$, $\vec{V} = \vec{e}_\theta(C_1 r^n)$ and the flow is rotational and the circulation varies with the diameter of the enclosing circle.

Comments –
For the steady flow of this example,

$$\nabla \cdot \vec{V} = \left(\vec{e}_r \frac{\partial}{\partial r} + \vec{e}_\theta \frac{1}{r}\frac{\partial V_r}{\partial \theta} + \vec{k}\frac{\partial}{\partial z}\right) \cdot \left(\vec{e}_\theta(C_1 r^n)\right) = 0$$

Thus the flow is incompressible regardless of the value of n.

The rigid-body rotation ($n = 1$) is a "forced" vortex, while the rotational motion with $n = -1$ is sometimes called a "free vortex." The former is typical near the eye of a tornado and the latter is found further away from the eye. A free vortex is in fact *free of vorticity*. A circular fluid motion does not necessarily make the flow rotational.

Since the only velocity gradient is $\frac{1}{r}\frac{\partial r V_\theta}{\partial r}$, for $n = -1$ this is zero and thus there is no shear stress throughout the flow field. In this case no assumption needs to be made as to whether or not the fluid is inviscid; the effect of fluid viscosity can simply be neglected. This example also demonstrates that rotational motion and viscosity do not necessarily generate vorticity. □

8.7 Crocco's Theorem and Shock-Induced Vorticity

Examples 8.1 and 8.2 demonstrate how vorticity can be generated and calculated in incompressible flows. For compressible flows, as in incompressible flows, vorticity can be calculated for the given, or known, velocity field. As these examples suggest, vorticity could arise from non-uniform velocity fields, Because of the varying shock strength of a bow shock across which non-uniform velocity changes occur, leading to velocity gradients and thus vorticity. Therefore, vorticity generation occurs due to a curved shock wave even if the upstream supersonic flow is uniform in velocity and thermodynamic properties. Can non-uniformity in other flow properties, such as the entropy field across a detached shock wave that is not uniform along the wave, be related to vorticity? A relationship between the entropy change and vorticity generation is useful in allowing calculating vorticity from a state variable, entropy, rather than a vector field of velocity. To focus on the effects of curvature and a non-uniform entropy field on vorticity generation, neglect the force term $(\vec{f})$ in Eqn. (8.2d) so that

$$\frac{\partial \vec{V}}{\partial t} + (\vec{V} \cdot \nabla)\vec{V} = -\frac{1}{\rho}\nabla p \tag{8.16}$$

Replacing the differentials in the combined first and second laws of thermodynamics, Eqn. (2.33), with the gradient operator yields

$$T\nabla s = \nabla h - v\nabla p = \nabla h - \frac{1}{\rho}\nabla p$$

which resembles Eqn. (8.6). Combining this expression with Eqn. (8.16) results in

$$T\nabla s = \nabla h + \frac{\partial \vec{V}}{\partial t} + (\vec{V} \cdot \nabla)\vec{V} \tag{8.17}$$

Incorporating the relationship

$$h_t = h + V^2/2$$

and the vector identity

$$\vec{V} \times \left(\nabla \times \vec{V}\right) = \nabla\left(\frac{V^2}{2}\right) - \left(\vec{V} \cdot \nabla\right)\vec{V}$$

into Eqn. (8.17) yields

$$T\nabla s = \nabla h_t - \vec{V} \times \left(\nabla \times \vec{V}\right) + \frac{\partial \vec{V}}{\partial t} \tag{8.18}$$

Equation (8.18) was due to L. Crocco and is called *Crocco's theorem* (Shapiro, 1953; Liepmann and Roshko, 1957). For steady flows,

$$T\nabla s = \nabla h_t - \vec{V} \times \left(\nabla \times \vec{V}\right) \tag{8.19}$$

Examination of Eqn. (8.19) leads to several physical interpretations.

(1) In general $T\nabla s - \nabla h_t \neq 0$, and because $\vec{V} \neq 0$, $\nabla \times \vec{V} = \vec{\zeta} \neq 0$ and the flow is rotational.

(2) In a uniform flow field (such as the one upstream of the shock system) because $\nabla s = \nabla h_t = 0$ and $\vec{V} \neq 0$, $(\nabla \times \vec{V}) = \vec{\zeta} = 0$. A uniform flow is thus irrotational.

(3) Thirdly and very importantly, $\nabla s \neq 0$ downstream of the shock wave due to its curvature (i.e., non-uniform shock strength) and $\nabla h_t = 0$ due to adiabatic conditions even with shock wave curvature (see Chapter 4). From Crocco's theorem, $\nabla \times \vec{V} = \vec{\zeta} \neq 0$; that is, the flow field behind a curved shock is rotational, even if the upstream flow field is uniform and irrotational. The rotational motion and vorticity is therefore induced by the curved shock. The entropy layer is also the layer within which velocity gradients exist; this conclusion is reached even if the fluid is assumed to be inviscid.

8.8 Velocity Potential and Potential Flow

In an irrotational flow $\omega_x = \omega_y = \omega_z = 0$ and

$$\frac{\partial v}{\partial x} - \frac{\partial u}{\partial y} = \frac{\partial u}{\partial z} - \frac{\partial w}{\partial x} = \frac{\partial w}{\partial y} - \frac{\partial v}{\partial z} = 0 \tag{8.20}$$

The flow can be described by defining a scalar function $\phi = \phi(x, y, z)$ such that

$$u = \frac{\partial \phi}{\partial x} \quad v = \frac{\partial \phi}{\partial y} \quad \text{and} \quad w = \frac{\partial \phi}{\partial z} \tag{8.21a}$$

or alternatively,

$$\vec{V} = \nabla \phi \tag{8.21b}$$

Substituting Eqn. (8.21a) into Eqn. (8.20) yields

$$\frac{\partial^2 \phi}{\partial x \partial y} - \frac{\partial^2 \phi}{\partial y \partial x} \equiv 0 \quad \frac{\partial^2 \phi}{\partial z \partial x} - \frac{\partial^2 \phi}{\partial x \partial z} \equiv 0 \quad \text{and} \quad \frac{\partial^2 \phi}{\partial y \partial z} - \frac{\partial^2 \phi}{\partial z \partial y} \equiv 0$$

which satisfies the irrotationality condition. The scalar function ϕ is called *the velocity potential* and, for this reason, irrotational flow is also called *potential flow*.

8.9 Governing Equations in Terms of Velocity Potential

The governing equations for irrotational flows can be summarized as

$$\frac{D\rho}{Dt} + \rho \nabla \cdot \vec{V} = \frac{\partial \rho}{\partial t} + \nabla \cdot \left(\rho \vec{V} \right) = 0 \tag{8.1d}$$

$$\frac{D\vec{V}}{Dt} = \rho \frac{\partial \vec{V}}{\partial t} + \left(\rho \vec{V} \cdot \nabla \right) \vec{V} = -\nabla p \tag{8.2e}$$

$$\nabla h_t = 0 \tag{8.22}$$

For a steady irrotational flow, the momentum equation can be derived in terms of velocity potential and becomes

$$\left(\rho \vec{V} \cdot \nabla\right) \vec{V} = -\nabla p \tag{8.23}$$

Recalling the speed of sound defined as

$$a^2 \equiv \left(\frac{\partial p}{\partial \rho}\right)_s \tag{3.6}$$

Therefore,

$$dp = a^2 d\rho \quad \text{or} \quad \nabla p = a^2 \nabla \rho \tag{8.23}$$

and

$$\left(\vec{V} \cdot \nabla\right) \vec{V} = -\frac{a^2}{\rho} \nabla \rho \tag{8.24}$$

The steady-state continuity, Eqn. (8.1d), becomes

$$\rho \nabla \cdot \vec{V} + \vec{V} \cdot \nabla \rho = 0$$

Substituting this relationship for $\nabla \rho / \rho$ in Eqn. (8.24) yields

$$a^2 \nabla \cdot \vec{V} - \vec{V} \cdot \left(\vec{V} \cdot \nabla\right) \vec{V} = 0 \tag{8.25}$$

In the above derivations, the vector identity

$$\vec{V} \times \left(\nabla \times \vec{V}\right) = \nabla \left(\frac{V^2}{2}\right) - \left(\vec{V} \cdot \nabla\right) \vec{V}$$

And because the flow is irrotational, $\left(\nabla \times \vec{V}\right) = 0$ and

$$\left(\vec{V} \cdot \nabla\right) \vec{V} = \nabla \left(\frac{V^2}{2}\right)$$

Substituting this expression and Eqn. (8.21a) into Eqn. (8.25) produces

$$a^2 \nabla \cdot \left(\vec{i}\frac{\partial \phi}{\partial x} + \vec{j}\frac{\partial \phi}{\partial y} + \vec{k}\frac{\partial \phi}{\partial z}\right) - \left(\vec{i}\frac{\partial \phi}{\partial x} + \vec{j}\frac{\partial \phi}{\partial y} + \vec{k}\frac{\partial \phi}{\partial z}\right) \cdot \left(\frac{\partial \phi}{\partial x}\frac{\partial}{\partial x} + \frac{\partial \phi}{\partial y}\frac{\partial}{\partial y} + \frac{\partial \phi}{\partial z}\frac{\partial}{\partial z}\right)$$
$$\left(\vec{i}\frac{\partial \phi}{\partial x} + \vec{j}\frac{\partial \phi}{\partial y} + \vec{k}\frac{\partial \phi}{\partial z}\right) = 0 \tag{8.26}$$

The first term on the left-hand side of Eqn. (8.26) can also be written as $a^2 \nabla^2 \phi$, where ∇^2 is the Laplacian operator and $\nabla^2 = \left(\partial^2/\partial x^2 + \partial^2/\partial y^2 + \partial^2/\partial z^2\right)$ in Cartesian coordinates.

8.10 Flow with Small Velocity Perturbations; Linearized Theory

Flows over a thin airfoil (illustrated in Chapter 4) or a slender body constitute examples of a class of flows called *linearized flow*, because they are physically only slightly perturbed from the uniform flow, and the nonlinear terms involving products of derivatives of ϕ in Eqn. (8.26) are negligibly small compared with linear terms.

Consider a slight perturbation to a uniform flow having only velocity component U_1 parallel to the x-axis, assuming the perturbation is so small that it can be considered as an isentropic process. Thus, due to the isentropic and irrotational nature of the perturbation, the velocity field can be written using the new velocity potential ϕ by substituting

$$\phi \rightarrow \phi + U_1 x \qquad (8.27)$$

so that

$$\vec{V} = \vec{i}\left(U_1 + \frac{\partial \phi}{\partial x}\right) + \vec{j}\frac{\partial \phi}{\partial y} + \vec{k}\frac{\partial \phi}{\partial z} \qquad (8.28)$$

In Eqn. (8.28), ϕ is now the *perturbation velocity potential*. The velocities due to perturbation can be written as $\vec{i}u + \vec{j}v + \vec{k}w$ such that

$$u' = \frac{\partial \phi}{\partial x}, \quad v' = \frac{\partial \phi}{\partial y}, \quad w' = \frac{\partial \phi}{\partial z} \qquad (8.29)$$

For small perturbations,

$$u' = \frac{\partial \phi}{\partial x} \ll U_1', = \frac{\partial \phi}{\partial y} \ll U_1, \text{and } w' = \frac{\partial \phi}{\partial z} \ll U_1 \quad \text{or} \quad \left(\frac{u'}{U_1}\right)^2, \left(\frac{v'}{U_1}\right)^2,$$

$$\text{and } \left(\frac{w'}{U_1}\right)^2 \ll \ll 1 \qquad (8.30)$$

Writing Eqn. (8.26) in terms of velocity,

$$a^2 \nabla \cdot \left(\vec{i}(U_1 + u') + \vec{j}v' + \vec{k}w'\right) - \left(\vec{i}(U_1 + u') + \vec{j}v' + \vec{k}w'\right) \cdot$$

$$\left((U_1 + u')\frac{\partial}{\partial x} + v'\frac{\partial}{\partial y} + w'\frac{\partial}{\partial z}\right)\left(\vec{i}(U_1 + u') + \vec{j}v' + \vec{k}w'\right) = 0 \qquad (8.31)$$

Expanding Eqn. (8.29) leads to

$$a^2\left(\frac{\partial u'}{\partial x} + \frac{\partial v'}{\partial y} + \frac{\partial w'}{\partial x}\right) = \left[(U_1 + u')^2\frac{\partial u'}{\partial x} + (U_1 + u')v'\frac{\partial u'}{\partial y} + (U_1 + u')w'\frac{\partial u'}{\partial z}\right]$$

$$+ \left[v'(U_1 + u')\frac{\partial v'}{\partial x} + v'^2\frac{\partial v'}{\partial y} + v'w'\frac{\partial v'}{\partial z}\right] + \left[w'(U_1 + u')\frac{\partial w'}{\partial x} + w'v'\frac{\partial w'}{\partial y} + w'^2\frac{\partial w'}{\partial z}\right]$$

$$= (U_1 + u')^2 \frac{\partial u'}{\partial x} + v'^2 \frac{\partial v'}{\partial y} + w'^2 \frac{\partial w'}{\partial z} + (U_1 + u')v' \left(\frac{\partial u'}{\partial y} + \frac{\partial v'}{\partial x} \right)$$
$$+ w'v' \left(\frac{\partial w'}{\partial y} + \frac{\partial v'}{\partial z} \right) + w'(U_1 + u') \left(\frac{\partial w'}{\partial x} + \frac{\partial u'}{\partial z} \right) \tag{8.32}$$

By neglecting terms containing products of perturbation velocities, one obtains

$$a^2 \left(\frac{\partial u'}{\partial x} + \frac{\partial v'}{\partial y} + \frac{\partial w'}{\partial z} \right) = (U_1^2 + 2U_1 u') \frac{\partial u'}{\partial x} + U_1 v' \left(\frac{\partial u'}{\partial y} + \frac{\partial v'}{\partial x} \right) + U_1 w' \left(\frac{\partial w'}{\partial x} + \frac{\partial u'}{\partial z} \right) \tag{8.33}$$

It is desirable to use the free-stream value of speed of sound, a_1, to pair with the free-stream velocity U_1, so that Eqn. (8.33) would only consist of the free-stream and perturbation velocities. Recall Eqn. (8.20), that is, the stagnation enthalpy of the irrotational flow is constant throughout the entire flow field. Thus

$$h_t = h + \frac{1}{2} \left[(U_1 + u')^2 + v'^2 + w'^2 \right] = \text{constant} \tag{8.34}$$

However,

$$h = c_v T + p\mathrm{v} = c_p T + RT = \gamma c_v T = \frac{c_p}{R} a^2 = \frac{a^2}{R/c_p} = \frac{a^2}{\gamma - 1} \tag{8.35}$$

Similarly,

$$h_1 = \frac{a_1^2}{\gamma - 1} \tag{8.36}$$

Combining Eqns. (8.34) – (8.36) yields

$$\frac{1}{2} \left[(U_1 + u')^2 + v'^2 + w'^2 \right] + \frac{a^2}{\gamma - 1} = \frac{U_1^2}{2} + \frac{a_1^2}{\gamma - 1} \tag{8.37a}$$

By again neglecting the second-order terms and rearranging, Eqn. (8.37a) becomes

$$a^2 = a_1^2 - (\gamma - 1) U_1 u' \tag{8.37b}$$

With this, Eqn. (8.33) can be written as

$$(a_1^2 - U_1^2) \frac{\partial u'}{\partial x} + a_1^2 \frac{\partial v'}{\partial y} + a_1^2 \frac{\partial w'}{\partial z} = (\gamma + 1) U_1 u' \frac{\partial u'}{\partial x} + (\gamma - 1) U_1 u' \left(\frac{\partial v'}{\partial y} + \frac{\partial w'}{\partial z} \right)$$
$$+ U_1 v' \left(\frac{\partial u'}{\partial y} + \frac{\partial v'}{\partial z} \right) + U_1 w' \left(\frac{\partial u'}{\partial z} + \frac{\partial w'}{\partial x} \right) \tag{8.38a}$$

Dividing through by a_1^2 and noting $M_1 \equiv U_1/a_1$, this equation becomes

$$\left(1 - M_1^2\right)\frac{\partial u'}{\partial x} + \frac{\partial v'}{\partial y} + \frac{\partial w'}{\partial z} = M_1^2(\gamma + 1)\frac{u'}{U_1}\frac{\partial u'}{\partial x} + M_1^2(\gamma - 1)\frac{u'}{U_1}\left(\frac{\partial v'}{\partial y} + \frac{\partial w'}{\partial z}\right)$$

$$+ M_1^2\frac{v'}{U_1}\left(\frac{\partial u'}{\partial y} + \frac{\partial v'}{\partial x}\right) + M_1^2\frac{w'}{U_1}\left(\frac{\partial u'}{\partial z} + \frac{\partial w'}{\partial x}\right) \tag{8.38b}$$

Further simplification of Eqn. (8.38b), considering that (u'/U_1), (v'/U_1), and $(w'/U_1) \ll 1$, leads to

$$\left(1 - M_1^2\right)\frac{\partial u'}{\partial x} + \frac{\partial v'}{\partial y} + \frac{\partial w'}{\partial z} = 0 \tag{8.39}$$

Equation (8.39) is linear, providing the linearized theory for flows is subject to small perturbations. Judicial decisions need to be made when comparing magnitudes of various terms in Eqn. (8.38b). For example, in transonic flow $(M_1 \to 1)$, $M_1^2(\gamma + 1)(u'/U_1)$ is of the order of u'/U_1; however, it may not be necessarily smaller than $(1 - M_1^2)$ that approaches zero. Therefore, the $\partial u'/\partial x$ on the right-had side of Eqn. (8.38b) cannot be neglected and for small perturbation, the following equation is applicable for subsonic, transonic, and supersonic flows:

$$\left(1 - M_1^2\right)\frac{\partial u'}{\partial x} + \frac{\partial v'}{\partial y} + \frac{\partial w'}{\partial z} = M_1^2(\gamma + 1)\frac{u'}{U_1}\frac{\partial u'}{\partial x} \tag{8.40}$$

Substituting Eqn. (8.29) into Eqn. (8.40) yields

$$\left(1 - M_1^2\right)\frac{\partial^2 \phi}{\partial x^2} + \frac{\partial^2 \phi}{\partial y^2} + \frac{\partial^2 \phi}{\partial z^2} = \frac{M_1^2(\gamma + 1)}{U_1}\frac{\partial \phi}{\partial x}\frac{\partial^2 \phi}{\partial x^2} \tag{8.41}$$

Thus for transonic flow, Eqn. (8.41) remains nonlinear. For supersonic and subsonic flows,

$$\left(1 - M_1^2\right)\frac{\partial^2 \phi}{\partial x^2} + \frac{\partial^2 \phi}{\partial y^2} + \frac{\partial^2 \phi}{\partial z^2} = 0 \tag{8.42}$$

which is linear.

8.11 Boundary Conditions for the Linearized Theory

For flow over a body, two types of boundary conditions are necessary. As in boundary layer flows, both surface conditions and conditions far away from the surface are necessary. Consider that the small perturbation is due to a solid surface defined by the following function in Cartesian coordinates:

$$g(x, y, z) = 0 \tag{8.43}$$

on which the normal unit vector is

$$\vec{n} = \frac{\nabla g}{|\nabla g|} \tag{8.44}$$

This is because the flow is tangent to the surface, that is, $\vec{V} \cdot \vec{n} = 0$. Thus

$$\vec{V} \cdot \nabla g = 0 \tag{8.45}$$

Over the surface $\vec{V} = \vec{\imath}(U_1 + u') + \vec{\jmath}v' + \vec{k}w'$; therefore

$$(U_1 + u')\frac{\partial g}{\partial x} + v'\frac{\partial g}{\partial y} + w'\frac{\partial g}{\partial z} = 0 \tag{8.46}$$

For two dimensional flows, the tangency requires that

$$\frac{v'}{U_1 + u'} \approx \frac{v'}{U_1} = \frac{dy}{dx} = -\frac{\partial g/\partial x}{\partial g/\partial y} \tag{8.47}$$

Because of the small perturbation (i.e., $y \approx 0$ on the surface),

$$v'(x,y) \approx v'(x,0) = U_1\left(\frac{dy}{dx}\right)_s \tag{8.48}$$

where the subscript s denotes that the derivative is evaluated at the surface. The other boundary conditions, far away from the surface, are

$$u' = v' = w' = 0 \quad \text{or} < \infty \; (\text{i.e, finite}) \quad \text{for } y \to \infty \tag{8.49}$$

Problems

Problem 8.1 Use Eqn. (8.51) to show that for isentropic flow

$$\frac{Ds}{Dt} = 0$$

Problem 8.2 Carry out the detailed derivation of Eqn. (8.26).

Problem 8.3 The velocity of a steady incompressible laminar boundary layer over a flat plate can be represented by an approximation to the Blasius solution as:

$$u(y) = U_\infty \sin\left(\frac{\pi y}{2\delta}\right)$$

where U_∞ is the freestream velocity in the x-direction and δ is the local thickness of the boundary layer. Find the vorticity field within the boundary layer and show that the magnitude of vorticity is proportional to the average velocity gradient across the boundary and that it is zero at the edge of the boundary layer.

Problem 8.4 Consider a stream tube enclosed by a material curve C which moves and evolves with the fluid motion as shown here. With the help of Eqns. (8.2e) and (8.13), that is,

$$\Gamma \equiv \oint_C \vec{V} \cdot d\vec{l} \text{ and } \frac{D\vec{V}}{Dt} = -\frac{1}{\rho} \nabla p$$

respectively, show that for an incompressible flow

$$\frac{D\Gamma}{Dt} = 0$$

that is, the circulation (and the irrotationality/rotationality) is conserved and this expression is called *Kelvin's theorem*.

9 Applications of Small Perturbation Theory

The small perturbation, or linearized, theory derived in Chapter 8 finds a number of useful applications. This chapter discusses applications such as flows over very thin airfoils, where the velocity perturbation is caused by geometry of the surface, and acoustic wave propagation. Relationships among perturbation quantities are derived. Perturbations of pressure, temperature, and density are small compared with free-stream values, while velocity perturbation is much smaller than the local acoustic speed. Flows of a perfect gas are assumed with no effects of viscous and body forces and heat transfer.

9.1 Pressure Coefficient

In Chapter 4, the pressure coefficient was defined while discussing the shock-expansion theory:

$$C_p \equiv \frac{p - p_1}{\frac{1}{2}\rho_1 V_1^2} = \frac{p - p_1}{\frac{1}{2}\frac{\gamma p_1}{\gamma R T_1} V_1^2} = \frac{2}{\gamma M_1^2}\frac{p - p_1}{p_1} = \frac{2}{\gamma M_1^2}\left(\frac{p}{p_1} - 1\right) \qquad (4.57)$$

where subscript 1 denotes free stream conditions. For a flow subjected to a small perturbation, such as by a thin airfoil as shown in Fig. 9.1, it is desirable to express the value in terms of the perturbation. Specifically, the term p/p_1 needs to be replaced with perturbation quantities. Recall that h_t = constant throughout the flow field. Thus

$$h_t = h_1 + \frac{U_1^2}{2} = h + \frac{\left(U_1 + u'\right)^2 + v'^2 + w'^2}{2} = \text{constant} \qquad (9.1)$$

For a small perturbation, the heat capacity c_p can be assumed to remain constant. Dividing Eqn. (9.1) through by c_p leads to

$$T - T_1 = \frac{U_1^2}{2c_p} - \frac{\left(U_1 + u'\right)^2 + v'^2 + w'^2}{2c_p}$$

Substituting $c_p = \gamma R/(\gamma - 1)$ into this equation and dividing through by T_1 yields

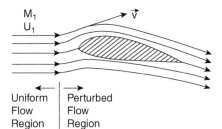

Figure 9.1 Schematic of streamline pattern and flow regions of a flow past an airfoil.

$$\frac{T}{T_1} - 1 = \frac{\gamma - 1}{2a_1^2}[U_1^2 - \left(U_1 + u'\right)^2 - v'^2 - w'^2] \tag{9.2}$$

By using the small perturbation conditions given by Eqn. (8.30) and the isentropic relationship between pressure and temperature, Eqn. (4.4),

$$\left(\frac{u'}{U_1}\right)^2, \left(\frac{v'}{U_1}\right)^2, \left(\frac{w'}{U_1}\right)^2 \ll 1 \tag{8.30}$$

$$\frac{p}{p_1} = \left(\frac{T}{T_1}\right)^{\gamma/(\gamma-1)} \tag{4.4}$$

Substituting Eqn. (9.2) into Eqn. (9.4) while preserving the first-order term of u'/U_1, it can be shown that (this is left as Problem 9.1)

$$C_p = -\frac{2u'}{U_1} \tag{9.3}$$

Thus the pressure coefficient depends linearly and only on the x component of the perturbation velocity.

9.2 Linearized Two-Dimensional Flow Past a Wavy Wall

It is of interest to find the pressure coefficient on a two-dimensional arbitrarily shaped surface (e.g., of an airfoil) in the xy-plane (as shown in Fig. 9.2) where the surface protrusion is very small compared with the characteristic dimension in the x direction. Because the protrusion (or elevation) of the surface y can be represented by a Fourier series of sinusoidal functions of x, it is instructive to consider a representative function given by

$$y = f(x) = \epsilon \sin \kappa x \tag{9.4}$$

where $\kappa = 2\pi/L$ is the wave number with ϵ and L (with $\epsilon \ll L$) denoting, respectively, the amplitude and the wavelength of the representative wavy wall (as shown in Fig. 9.3) and κ is thus the wave number.

For two-dimensional subsonic or supersonic flow and M_1 not close to unity,

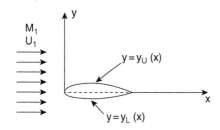

Figure 9.2 A two-dimensional, arbitrarily shaped surface in the xy-plane (an airfoil is shown).

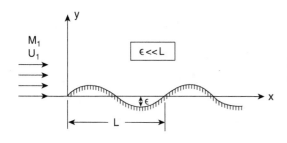

Figure 9.3 A two-dimensional periodic wavy surface in the xy-plane.

$$\left(1 - M_1^2\right)\frac{\partial^2 \phi}{\partial x^2} + \frac{\partial^2 \phi}{\partial y^2} = 0 \tag{9.5}$$

where ϕ is the perturbation velocity potential. Equation (9.5) is linear, and therefore solution for various Fourier components (with correspondingly different values of κ in Eqn. (9.4)) can be superimposed to form the solution for an arbitrary function. The boundary conditions are

$$u' = \frac{\partial \phi}{\partial x}, \ v' = \frac{\partial \phi}{\partial y} \ < \infty \ \text{for } y \to \infty \tag{9.6a}$$

and

$$v'(x, \ 0) = \left(\frac{\partial \phi}{\partial y}\right)_{y=0} = U_1\left(\frac{\partial y}{\partial x}\right)_s = U_1 \kappa \epsilon \cos\kappa x \tag{9.6b}$$

9.3 Subsonic Flow Past a Wavy Wall

The method of separation of variables is applicable for a linear equation such as Eqn. (9.5) by setting

$$\phi(x, \ y) = F(x)G(y)$$

Eqn. (9.5) becomes

$$\left(1 - M_1^2\right)F''G + FG'' = 0 \tag{9.7a}$$

or

$$\frac{F''}{F} + \frac{G''}{(1 - M_1^2)G} = 0 \tag{9.7b}$$

The two terms in Eqn. (9.7b) are only functions of only x and y, respectively. For it to hold, the following must be satisfied:

$$\frac{F''}{F} = -k^2 \tag{9.8a}$$

and thus

$$\frac{G''}{G} = (1 - M_1^2)k^2 \tag{9.8b}$$

where k is a real number. The signs in front of the constant k^2 are so chosen with the expectation that the variation at the wall surface should be sinusoidal in x. The solutions to Eqn. (9.8a) and (9.8b) are, respectively,

$$F = C_1 \cos kx + C_2 \sin kx \tag{9.9a}$$

and

$$G = C_3 exp\left(ky\sqrt{1 - M_1^2}\right) + C_4 exp\left(-ky\sqrt{1 - M_1^2}\right) \tag{9.9b}$$

where C_1 through C_4 are constants. These results lead to

$$\phi(x, y) = F(x)G(y) = [C_1 \cos kx + C_2 \sin kx]\left[C_3 exp\left(ky\sqrt{1 - M_1^2}\right)\right.$$

$$\left. + C_4 exp\left(-ky\sqrt{1 - M_1^2}\right)\right] \tag{9.10}$$

Thus

$$v'(x, y) = \frac{\partial \phi}{\partial y}$$

$$= [C_1 \cos kx + C_2 \sin kx]\left[C_3 k\sqrt{1 - M_1^2}exp\left(ky\sqrt{1 - M_1^2}\right)\right.$$

$$\left. - C_4 k\sqrt{1 - M_1^2}exp\left(-ky\sqrt{1 - M_1^2}\right)\right] \tag{9.11}$$

Because $v'(x, y) = 0$ or is finite as $y \to \infty$, C_3 has to be zero. $\phi(x, y)$ is thus reduced to

$$\phi(x, y) = [C_1 \cos kx + C_2 \sin kx]\left[C_4 exp\left(-ky\sqrt{1 - M_1^2}\right)\right] \tag{9.12}$$

Applying the boundary condition, Eqn. (9.6), one obtains

$$U_1 a\epsilon \cos \alpha x = [C_1 \cos kx + C_2 \sin kx]\left[-C_4 k\sqrt{1 - M_1^2}\right]$$

Therefore $C_2 C_4$ must also be zero. For $G(y)$ to be a nontrivial solution, only $C_2 = 0$ can be required. The perturbation velocity potential thus takes the form of

$$\phi(x, y) = F(x)G(y) = C'[\cos kx]\left[exp\left(-ky\sqrt{1 - M_1^2}\right)\right]$$

where C' is a new constant. Applying Eqn. (9.6b) again, one finds

$$U_1 \kappa \epsilon \cos \kappa x = -C'k\sqrt{1 - M_1^2} \cos kx$$

$$k = \kappa \quad \text{and} \quad C = -\frac{U_1 \epsilon}{\sqrt{1 - M_1^2}}$$

The final form of the perturbation velocity is

$$\phi(x, y) = -\frac{U_1 \epsilon}{\sqrt{1 - M_1^2}}[\cos \kappa x]\left[exp\left(-\kappa y\sqrt{1 - M_1^2}\right)\right] \qquad (9.13)$$

It is now clear that if the sign for k^2 in Eqn. (9.8a) were chosen positive $F(x)$ would be exponential in x. Then as $x \to \infty$, $F(x)$ would either become ∞ (a violation of finite value) or zero (a trivial solution) and both cases violates the periodic nature of the surface.

With $\phi(x, y)$ given by Eqn. (9.13), the velocity field is described by

$$u'(x, y) = \frac{\partial \phi}{\partial x} = \frac{U_1 \epsilon \kappa}{\sqrt{1 - M_1^2}}[\sin \kappa x]\left[exp\left(-\kappa y\sqrt{1 - M_1^2}\right)\right] \qquad (9.14a)$$

and

$$v'(x, y) = \frac{\partial \phi}{\partial y} = U_1 \epsilon \kappa[\cos \kappa x]\left[exp\left(-\kappa y\sqrt{1 - M_1^2}\right)\right] \qquad (9.14b)$$

It can be seen that the magnitudes of u' and v' differ by a factor of $\sqrt{1 - M_1^2}$ and they are $\pi/2\kappa$ ($= L/4$) out of phase in the direction of x. The pressure coefficient is thus

$$C_p = -\frac{2u'}{U_1} = -\frac{2\epsilon \kappa}{\sqrt{1 - M_1^2}}[\sin \kappa x]\left[exp\left(-\kappa y\sqrt{1 - M_1^2}\right)\right] \qquad (9.14c)$$

and it is shifted by and phase angle π from the wall surface and u'. The results of u', v', and C_p are shown in Fig. 9.4, which show that each of them decreases with increasing y (i.e., away from the surface), resulting in attenuations in the far field. The attenuation satisfies the boundary condition set by Eqn. (9.6a), i.e., $u' < \infty$ and

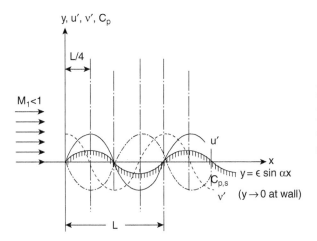

Figure 9.4 The results of u', v', and $C_{p,s}$ of a subsonic flow past the two-dimensional periodic wavy surface in the xy-plane shown in Fig. 9.3.

$v' < \infty$ for $y \to \infty$. It is clear that the attenuations of u', v', and C_p become weaker as M_1 is increased. It will be shown in Section 9.4 that for supersonic flow over similar airfoils, the linear theory predicts no attenuations.

Both components of the perturbation velocity component and the pressure coefficient decay exponentially in the direction away from the surface; however, the decay is weaker as M_1 in the exponent approaches unity.

The pressure coefficient at the surface, $C_{p,s}$ (i.e., evaluated at $y \to 0$ for the wavy wall), is of interest as the drag and lift forces relate directly to it. For $y = 0$, Eqn. (9.14c) becomes

$$C_{p,s} = -\frac{2\epsilon\kappa}{\sqrt{1 - M_1^2}} \sin \kappa x \qquad (9.15)$$

As shown in Fig. 9.3, the wall pressure coefficient is out of phase with the wall surface by an angle π and the pressure is symmetric about the crest and trough of the wavy surface, resulting in no drag forces. The pressure coefficient increases with an increasing free-stream Mach number; the increase is proportional to $1/\sqrt{1 - M_1^2}$, which is the *Prandtl-Glauert factor* and the *Prandtl-Glauert rule* can be written as

$$C_p = \frac{C_{p,o}}{\sqrt{1 - M_1^2}} \qquad (9.16)$$

where $C_{p,o}$ is the pressure coefficient of the incompressible flow (as $M_1 \to 0$ and $C_{p,o} = -2\epsilon\kappa \sin \kappa x$, given by Eqn. (9.15)). The Prandtl-Glauert rule predicts that for a given perturbation, the pressure coefficient (and drag coefficient) increases dramatically as it approaches unity, which has been experimentally observed.

Discussion of the Prandtle-Glauert factor and rule can be used to rewrite Eqn. (9.14c) in the following form:

$$\frac{C_p\sqrt{1-M_1^2}}{2\epsilon\kappa} = f\left(\kappa x, \kappa y\sqrt{1-M_1^2}\right) \qquad (9.17)$$

This indicates that when the pressure coefficient and x and y coordinates are properly scaled, the number of variables are reduced from six (x, y, C_p, M_1, ϵ, and κ) in Eqn. (9.14c) to the three scaled ones (or functional groups, also called *similarity variables*) that appear in Eqn. (9.17): $C_p\sqrt{1-M_1^2}/2\epsilon\kappa$, κx, and $\kappa y\sqrt{1-M_1^2}$. Different combinations of the six original, dimensional variables that give the same value for each of the three similarity variables yield one single curve for the same family of shapes and a wide range and Mach numbers.

It is appropriate to examine the conditions under which the linearized results, Eqns. (9.14) through (9.17), are applicable in terms of similarity variables. The assumption that both u'/U_1 and $v'/U_1 \ll 1$ requires that in Eqns. (9.14a) and (9.14b)

$$\frac{\epsilon\kappa}{\sqrt{1-M_1^2}} \ll 1 \qquad (9.18)$$

As noted in Section 8.10, that transonic flow equation is nonlinear. The linearized equation that is used to derive Eqn. (9.14) is based on the following assumption for Eqn. (8.40):

$$\left(1-M_1^2\right) \gg M_1^2(\gamma+1)\frac{u'}{U_1}$$

which also means that M_1 is not close to being unity. Considering Eqn. (9.14a) and the above inequality leads to

$$\frac{u'}{U_1} = O\left(\frac{\epsilon\kappa}{\sqrt{1-M_1^2}}\right) \ll \frac{1-M_1^2}{M_1^2(\gamma+1)}$$

Therefore,

$$\left(1-M_1^2\right) \gg \frac{M_1^2(\gamma+1)\epsilon\kappa}{\sqrt{1-M_1^2}} \quad \text{or} \quad \frac{M_1^2(\gamma+1)\epsilon\kappa}{\left(1-M_1^2\right)^{3/2}} \ll 1 \qquad (9.19)$$

Equation (9.19) suggests that M_1 has to be sufficiently smaller than unity for the linearized theory to be applicable. Such a value of M_1 depends on γ and $\epsilon\kappa$ ($= 2\,\pi\epsilon/L$). It can be shown that for local sonic conditions to occur, the free stream Mach number is such that

$$\frac{M_1^2(\gamma+1)\epsilon\kappa}{\left(1-M_1^2\right)^{3/2}} = 1 \qquad (9.20)$$

For this condition, the Mach is called the critical Mach number, and $M_1 = M_{cr}$.

EXAMPLE 9.1 For the wavy surface given by Eqn. (9.4), estimate the limiting value of M_1 for the linearized theory to be applicable for $L/\epsilon = 100$ and $\gamma = 1.4$. For the estimated value of M_1, find the maximum local velocity and Mach number. Also find the value of M_1 for the local velocity to be sonic.

Solution –
For the applicability of the linearized theory,

$$\frac{2\,\pi M_1^2(\gamma + 1)\epsilon/L}{\left(1 - M_1^2\right)^{3/2}} \ll 1$$

Some typical functional values, $f(M_1,\ \gamma, \epsilon/L)$ of the left-hand side of this inequality as a function of M_1 are given below (assuming $\gamma = 1.4$):

M_1	$f(M_1,\ \gamma, \epsilon/\lambda) = \dfrac{2\,\pi M_1^2(\gamma + 1)\epsilon/L}{\left(1 - M_1^2\right)^{3/2}}$
0.4	0.0313
0.5	0.0580
0.6	0.1060
0.7	0.2029
0.8	0.4468
0.85	0.7453
0.87	0.9523
0.873	0.9929
0.874	1.0039
0.88	1.0898
0.9	1.4748

Thus the estimated value of $M_1 = 0.6$ can be chosen as $f(M_1,\ \gamma, \epsilon/L) \approx 0.1 \ll 1$. For this Mach umber, the local maximum velocity is on the surface $y = 0$ and at $x = \pi/2\alpha$, with $\alpha = 2\,\pi/L$

$$U^2 = \left(U_1 + u'\right)^2 + v'2 \approx U_1^2 + 2U_1 u' = U_1^2\left[1 + \frac{4\,\pi\epsilon/L}{\sqrt{1 - M_1^2}}\right]$$

At this surface location, the local speed of sound is given by Eqn. (8.37b):

$$a^2 = a_1^2 - (\gamma - 1)U_1 u' = U_1^2\left[\frac{1}{M_1^2} - \frac{2\,\pi(\gamma - 1)\epsilon/L}{\sqrt{1 - M_1^2}}\right]$$

Therefore,

$$M^2 = \frac{U^2}{a^2} = \left[1 + \frac{4\,\pi\epsilon/L}{\sqrt{1 - M_1^2}}\right] \Bigg/ \left[\frac{1}{M_1^2} - \frac{2\,\pi(\gamma - 1)\epsilon/L}{\sqrt{1 - M_1^2}}\right] = 0.4213$$

The maximum local Mach number is

$$M = 0.649$$

which is approximately 10% larger than M_1. The maximum local velocity is

$$U_{max} = U_1 + u'_{max} = U_1 \left[1 + \frac{2\,\pi\epsilon/L}{\sqrt{1 - M_1^2}} \right] = 1.0785 U_1$$

From the tabulated values of M_1, $M_{cr} \approx 0.873$.

Comments –

It appears that at the crest of the surface, the flow accelerates (i.e., $u' > 0$) and the pressure and temperature should decreased based on Bernoulli's principle. Therefore, the speed of sound decreases and the 7.85% increase in the local velocity over the free-stream value results in an approximately 10% increase in Mach number. The reader is left to find that the Mach number at the trough is approximately 0.554 and $U_{min} = 0.9215 U_1$.   □

Example 9.1 illustrates how the local Mach number can be calculated for a given free-stream Mach number and the body shape. In general the local Mach number is, as described in Example 9.1:

$$M^2 = \frac{U^2}{a^2} \approx \frac{U_1^2 + 2U_1 u'}{a_1^2 - (\gamma - 1)U_1 u'} = \frac{M_1^2\left(1 + 2\frac{u'}{U_1}\right)}{1 - M_1^2(\gamma - 1)\frac{u'}{U_1}} = \frac{1 + \frac{4\,\pi\epsilon/L}{\sqrt{1-M_1^2}}}{\frac{1}{M_1^2} - \frac{2\,\pi(\gamma-1)\epsilon/L}{\sqrt{1-M_1^2}}} \tag{9.21a}$$

This relationship can be used to further improve the Prandtl-Glauert rule, Eqn. (9.16), which is known to under-predict the pressure coefficient because the free-stream instead of the local Mach number is used. By replacing M_1 with M, to more fully incorporate the effects of compressibility, and setting $C_{p,o} = -2u'/U_1$, Eqn. (9.21a) becomes

$$M^2 = \frac{M_1^2(1 - C_{p,o})}{1 + M_1^2\left(\frac{\gamma-1}{2}\right)C_{p,o}} \tag{9.21b}$$

Substituting this into Eqn. (9.16) leads to

$$C_p = \frac{C_{p,o}}{\sqrt{1 - M_1^2} + \left[\frac{M_1^2\left(1 + \frac{\gamma-1}{2}M_1^2\right)}{\sqrt{1-M_1^2}}\right]\frac{C_{p,o}}{2}} \tag{9.22}$$

This result has shown improved agreement with experimental data (Anderson, 2002). It is easy to see that as $C_{p,o}$ becomes small, Eqn. (9.22) recovers the Prandtl-Glauert rule.

In arriving at the Prandtl-Glauert rule, Eqn. (9.16), coordinate transformation is needed so that the pressure coefficient from incompressible flow, $C_{p,o}$, can be used to predict C_p. (A similar transform is also useful for treating supersonic flows in Section 9.4.) Let $\beta \equiv \sqrt{1 - M_1^2}$. Then the linearized equation in two dimensions becomes

$$\beta^2 \frac{\partial^2 \phi}{\partial x^2} + \frac{\partial^2 \phi}{\partial y^2} = 0 \tag{9.23}$$

By introducing a transformed coordinate system (ξ, η) such that

$$\xi = x; \ \eta = \beta y \tag{9.24}$$

a surface described by $y = f(x)$ should be rewritten as $\eta = f^*(\xi)$. By further introducing a transformed perturbation velocity potential $\breve{\phi}$ such that

$$\breve{\phi}(\xi, \eta) = \beta \phi(x, y) \tag{9.25}$$

and using the chain rule of differentiation, Eqn. (9.25) becomes

$$\frac{\partial^2 \breve{\phi}}{\partial \xi^2} + \frac{\partial^2 \breve{\phi}}{\partial \eta^2} = 0 \tag{9.26}$$

Equation (9.26) is the Laplace equation for incompressible flows. The derivation provides the theoretical foundation for relating ϕ to $\breve{\phi}$ and thus C_p to $C_{p,o}$ in the following manner:

$$C_p = -\frac{2u'}{U_1} = -\frac{2}{U_1}\frac{\partial \phi}{\partial x} = -\frac{1}{\beta}\frac{2}{U_1}\frac{\partial \breve{\phi}}{\partial \xi} = -\frac{1}{\beta}\frac{2\breve{u}'}{U_1} = -\frac{C_{p,o}}{\sqrt{1 - M_1^2}}$$

where $\breve{u}'$ is the perturbation velocity in the corresponding incompressible flow over the *same shape* of the surface. To ascertain the same shape, compare v' and $\breve{v}'$ as follows. Combining Eqns. (9.6a) and (9.25) leads to

$$v' = \frac{\partial \phi}{\partial y} = \frac{1}{\beta}\frac{\partial \breve{\phi}}{\partial y} = \frac{\partial \breve{\phi}}{\partial \eta}\left(= \breve{v}'\right) = U_1 \frac{\partial f^*}{\partial x} = U_1 \frac{\partial f^*}{\partial \xi}$$

Thus

$$\frac{\partial f}{\partial x} = \frac{\partial f^*}{\partial \xi} \tag{9.27}$$

which indicates that the surfaces described by $y = f(x)$ and $= f^*(\xi)$ in the original (compressible flow) and the transformed (incompressible flow) coordinate systems have the same slope everywhere for $\xi = x$, and they are therefore of the same shape. The above derivation also indicates that

$$v = \breve{v}' \tag{9.28}$$

and

$$u' = \frac{\breve{u}'}{\sqrt{1 - M_1^2}} \tag{9.29}$$

Therefore the effect of compressibility lies in the fact that as M_1 increases, u' and C_p increase, strengthening the perturbation. Equations (9.14a) through (9.14c) indicate that for a given M_1, velocity perturbations (u' and v') and C_p decrease in the direction away from the wall and as M_1 increases toward unity, the attenuation also decreases.

EXAMPLE 9.2 Consider a hypothetical, thin airfoil, with its upper surface described by Eqn. (9.4), $y = \epsilon \sin \kappa x$, and a flat lower surface (where $c/\epsilon = 50$; c is the chord length). The free-stream flow is subsonic with Mach number M_1 and is parallel to the lower surface. Find the lift and drag coefficients.

Solutions –

The upper surface is given by

$$y = f(x) = \epsilon \sin \kappa x \text{ for } 0 \le x \le c.$$

For this airfoil, $\kappa = \pi/c$ (i.e., $L = 2\,c$). Since $\epsilon/c \ll 1$, the linearized theory is applicable. The pressure coefficient on the upper surface can be given by Eqn. (9.14c):

$$C_{p,U} = \frac{u'}{U} = -\frac{\pi\epsilon/c}{\sqrt{1 - M_1^2}} \sin \frac{\pi x}{c}$$

The pressure coefficient on the lower surface is zero. Because only the pressure component in the y-direction contributes to the lift, the projected area is proportional to dx on the surface. Therefore,

$$C_l = \frac{1}{c} \int_0^c (C_{p,L} - C_{p,U}) dx = \frac{1}{c} \frac{\pi\epsilon/c}{\sqrt{1 - M_1^2}} \int_0^c \sin \frac{\pi x}{c} dx = \frac{2\epsilon/c}{\sqrt{1 - M_1^2}}$$

For the drag coefficient, only the surface area projected onto the y-axis contributes to the drag. This area is proportional to dy. Because

$$dy = \frac{\pi\epsilon}{c} \cos \frac{\pi x}{c} dx$$

Noting that the lower surface does not contribute any drag forces,

$$C_d = \frac{1}{c} \int_0^\epsilon C_{p,U} dy = \frac{1}{c} \frac{\pi\epsilon}{c} \frac{\pi\epsilon/c}{\sqrt{1 - M_1^2}} \int_0^c \sin \frac{\pi x}{c} \cos \frac{\pi x}{c} dx$$

$$= \frac{1}{c} \frac{\pi\epsilon}{c} \frac{\pi\epsilon/c}{\sqrt{1 - M_1^2}} \int_0^c \sin \frac{2\,\pi x}{c} dx = 0$$

Comments –

This result regarding the drag coefficient is similar to the result discussed following Eqn. (9.15), in that the pressure on the upper surface is symmetric about the crest of the airfoil and should generate no drag. For a closed two-dimensional body in an inviscid incompressible flow, **d'Alembert's** paradox indicates no drag force is produced. This example illustrates that the same paradox is extended to include subsonic compressible flow. □

9.4 Supersonic Flow Past a Wavy Wall

For supersonic flow over the wavy wall (Fig. 9.5), Eqn. (8.42) is written as

$$\left(M_1^2 - 1\right)\frac{\partial^2 \phi}{\partial x^2} - \frac{\partial^2 \phi}{\partial y^2} = 0 \tag{9.30}$$

or

$$\frac{\partial^2 \phi}{\partial x^2} - \frac{1}{\lambda^2}\frac{\partial^2 \phi}{\partial y^2} = 0 \tag{9.31}$$

where $M_1^2 - 1 \equiv \lambda^2 > 0$. Equation (9.31) is a linear hyperbolic wave equation and its solutions can be represented by either one of the two arbitrary functions, $f(x - \lambda y)$ and $g(x + \lambda y)$. The general solution is therefore

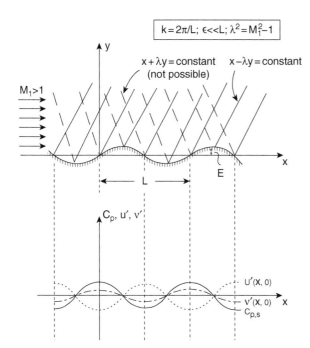

Figure 9.5 The results of u', v', and $C_{p,s}$ of a subsonic flow past the two-dimensional periodic wavy surface in the xy-plane shown in Fig. 9.3.

$$\phi(x, y) = f(x - \lambda y) + g(x + \lambda y) \tag{9.32}$$

With the wavy wall described by $y = \epsilon \sin \kappa x$, the boundary conditions are the same as those given by Eqns. (9.6a) and (9.6b) for the subsonic case. For similar reasons for the waves described in Section 4.8, the disturbance produced by the wavy wall lying to the right of the flow generates a family of left-running (the "−" family) waves for which $(x - \lambda y) = $ constant. Thus one can focus on the particular solution $f(x - \lambda y)$ and set $g = 0$.

The boundary condition, Eqn. (9.6), requires that

$$v'(x, 0) = \left(\frac{\partial \phi}{\partial y}\right)_{y=0} = -\lambda \left[\frac{df(x - \lambda y)}{dy}\right]_{y=0} = U_1 \kappa \epsilon \cos \kappa x$$

Thus

$$f(x) = -\frac{U_1 \epsilon}{\lambda} \sin \kappa x \text{ and } f(x - \lambda y) = -\frac{U_1 \epsilon}{\lambda} \sin \kappa (x - \lambda y)$$

The perturbation velocity potential becomes

$$\phi(x, y) = -\frac{U_1 \epsilon}{\sqrt{M_1^2 - 1}} \sin \kappa \left(x - y\sqrt{M_1^2 - 1}\right) \tag{9.33a}$$

and throughout the flow field on the xy-plane:

$$u'(x, y) = \frac{\partial \phi}{\partial x} = -\frac{U_1 \kappa \epsilon}{\sqrt{M_1^2 - 1}} \cos \kappa \left(x - y\sqrt{M_1^2 - 1}\right) \tag{9.33b}$$

$$v'(x, y) = \frac{\partial \phi}{\partial y} = U_1 \kappa \epsilon \cos \kappa \left(x - y\sqrt{M_1^2 - 1}\right) \tag{9.33c}$$

$$C_p(x, y) = -\frac{2 u'}{U_1} = \frac{2\kappa \epsilon}{\sqrt{M_1^2 - 1}} \cos \kappa \left(x - y\sqrt{M_1^2 - 1}\right) \tag{9.34}$$

Compared to Eqn. (9.17) for subsonic similarity rule, the six variables $(x, y, C_p, M_1, \epsilon,$ and $\kappa)$ for supersonic flow have been grouped into two, instead of three, scaled similarity variables and they are related by the following functional form

$$\frac{C_p \sqrt{M_1^2 - 1}}{2\epsilon\kappa} = f\left(\kappa \left[x - y\sqrt{M_1^2 - 1}\right]\right) \tag{9.35}$$

The solution $\phi(x, y)$ to the wave equation, Eqn. (9.31), $f(x - \lambda y)$, $u'(x, y)$, $v'(x, y)$, and $C_p(x, y)$ are constant along the lines with

$$x - y\sqrt{M_1^2 - 1} = \text{constant} \tag{9.37}$$

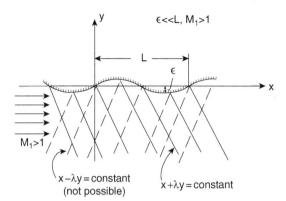

Figure 9.6 Schematic showing the possible and non-possible characteristics resulting from a supersonic flow past a wavy wall.

Therefore their magnitudes do not decrease with y, unlike in subsonic flows. Along these lines $dy/dx = 1/\sqrt{M_1^2 - 1} = \tan\mu$, where $\mu = \sin^{-1}(1/M_1)$ is the Mach angle discussed in Chapter 4. Therefore they are the Mach lines, inclined at the Mach angle, measured counterclockwise from the free-stream flow. They are also called characteristics. Because they extend in the downstream direction and to the left of the free-stream flow (i.e., wall lying to the right of the flow), they are called the left-running characteristics. Similarly $g(x + \lambda y)$ forms the right-running characteristics that should extend from the wall lying to the left of the flow, which does not exist for the wavy wall shown in Fig. 9.3. This is the rationale for the selection of $f(x - \lambda y)$ as the solution to Eqn. (9.31) for the wavy surface shown in Fig. 9.5, with representative characteristics $(x - \lambda y) = $ constant. It is readily clear that for the wavy wall shown in Fig. 9.6 only characteristics $x + \lambda y = $ constant exists. It is important to bear in mind that along each characteristic (or Mach line), the disturbances or perturbations such as those given by Eqns. (9.33) and (9.34) are constant and that along these characteristics, disturbances propagate without attenuation away from the wall.

The pressure coefficient on the surface $(y = 0)$ is

$$C_{p,s} = \frac{2\kappa\epsilon}{\sqrt{M_1^2 - 1}} \cos\kappa x \qquad (9.37)$$

The results of Eqns. (9.33) and (9.34) are plotted in Fig. 9.5b. It can be seen that $C_{p,s}$ is not symmetric about the crest and trough of the wavy wall in supersonic flows. The drag coefficient over a wavelength is therefore not zero as in the case of subsonic flows. This non-zero wave drag has been previously discussed in Sections 4.14 through 4.16 on shock and expansion waves.

EXAMPLE 9.3 For the airfoil of Example 9.2, find the lift and drag coefficients.

Solutions –

$$C_l = \frac{1}{c}\int_0^c (C_{p,L} - C_{p,U})\,dx = -\frac{1}{c}\frac{\pi\epsilon/c}{\sqrt{M_1^2 - 1}}\int_0^c \cos\frac{\pi x}{c}\,dx = 0$$

Because the lower surface does not contribute any drag forces,

$$C_d = \frac{1}{c}\int_0^\epsilon C_{p,U}\,dy = \frac{1}{c}\frac{\pi\epsilon}{c}\frac{2\,\pi\epsilon/c}{\sqrt{M_1^2-1}}\int_0^c\left(\cos\frac{\pi x}{c}\right)^2 dx$$

$$= \frac{1}{c}\frac{\pi\epsilon}{c}\frac{2\,\pi\epsilon/c}{\sqrt{M_1^2-1}}\int_0^c\frac{1}{2}\left(1+\cos\frac{2\,\pi x}{c}\right)dx = \frac{(\pi\epsilon/c)^2}{\sqrt{M_1^2-1}}$$

Comments –

Compared with the subsonic results over the same airfoil shape in Example 9.2, there is now drag force on an airfoil with a symmetric shape in supersonic flows. This is similar to the results in Chapter 4, where the shock-expansion theory predicts drag forces for symmetric airfoils. The drag force is a wave in nature and is called the wave drag. There is no **d'Alembert's** paradox for drag forces in supersonic flows.

It is interesting to note that no lift is generated with the supersonic flow. This is because the pressure coefficients are of the same magnitude but opposite signs at equal distance from the crest of the upper surface of the airfoil.   □

Equation (9.37) can easily be shown to be equal to

$$C_{p,s} = \frac{2}{\sqrt{M_1^2-1}}\left(\frac{dy}{dx}\right)_s \tag{9.38}$$

Following Example 9.3, it is easily shown that

$$C_d = \frac{2}{\sqrt{M_1^2-1}}\overline{\left(\frac{dy}{dx}\right)_s^2} = \frac{2}{\sqrt{M_1^2-1}}\frac{1}{c}\int_0^c\left(\frac{dy}{dx}\right)_s^2 dx \tag{9.39}$$

Equations (9.38) and (9.39) indicate that according to the small-perturbation theory, the local pressure coefficient is proportional to the slope of surface and the drag coefficient to the mean square of the slope over the length of the airfoil (or the wavy wall), for which case the lower surface does not contribute to the drag force like the airfoil in Example 9.3.

9.5 Thin Airfoil in Supersonic Flow

Consider a thin airfoil of an arbitrary shape in supersonic flow at an angle of attack, as shown in Fig. 9.7. The general solution of the wave equation, Eqn. (9.31), is again given by Eqn. (9.32):

$$\phi(x,\ y) = f(x-\lambda y) + g(x+\lambda y)$$

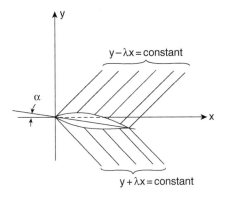

Figure 9.7 Characteristics resulting from a supersonic flow past an airfoil; α is the angle of attack.

Because the disturbance propagates along the characteristics in the downstream direction, the upper/lower surface generates only left/right running characteristics. Consequently

$$\phi(x,\ y) = f(x - \lambda y) \quad \text{upper surface}$$

$$\phi(x,\ y) = g(x + \lambda y) \quad \text{lower surface}$$

The pressure coefficient on the upper surface is

$$C_{p,s,U} = -\left(\frac{2u'}{U_1}\right)_U = -\frac{2}{U_1}\left(\frac{\partial \phi}{\partial x}\right)_{y=0+} = -\frac{2}{U_1}f'(x) \quad \text{upper surface}$$

Similarly,

$$C_{p,s,L} = -\left(\frac{2u'}{U_1}\right)_U = -\frac{2}{U_1}\left(\frac{\partial \phi}{\partial x}\right)_{y=0-} = -\frac{2}{U_1}g'(x) \quad \text{lower surface}$$

It is desirable to express these coefficients in terms of the airfoil shape, as in Eqn. (9.37), thus calling for the use of boundary conditions. At the upper surface,

$$v'_U(x,\ 0) = U_1\left(\frac{dy}{dx}\right)_U = \left(\frac{d\phi}{dy}\right)_{y=0^+} = -\lambda f'(x)$$

yielding

$$f'(x) = -\frac{U_1}{\lambda}\left(\frac{dy}{dx}\right)_{y=0^+}$$

Following similar processes,

$$g'(x) = \frac{U_1}{\lambda}\left(\frac{dy}{dx}\right)_L$$

Therefore, Eqns. (4.61a) and (4.61b), for pressure coefficients on a thin airfoil at small angles of attack using the weak shock-expansion theory, are recovered:

$$C_{p,s,U} = \frac{2}{\sqrt{M_1^2 - 1}} \left(\frac{dy}{dx} \right)_U \tag{8.37a}$$

$$C_{p,s,L} = \frac{2}{\sqrt{M_1^2 - 1}} \left(-\frac{dy}{dx} \right)_L \tag{8.37b}$$

EXAMPLE 9.4 Consider a flat plate airfoil at a small angle of attack $\alpha \ll 1$. Find the lift and drag coefficients using the small-perturbation theory.

Solution –

By directly using Eqn. (9.37),

$$C_{p,s,U} = \frac{2}{\sqrt{M_1^2 - 1}} \left(\frac{dy}{dx} \right)_U = \frac{-2\alpha}{\sqrt{M_1^2 - 1}}$$

$$C_{p,s,L} = \frac{2}{\sqrt{M_1^2 - 1}} \left(-\frac{dy}{dx} \right)_U = \frac{2\alpha}{\sqrt{M_1^2 - 1}}$$

$$C_l = \frac{1}{c} \int_0^c (C_{p,s,L} - C_{p,s,U}) \cos\alpha\, dx = \frac{1}{c} \int_0^c \frac{4\alpha}{\sqrt{M_1^2 - 1}} dx = \frac{4\alpha \cos\alpha}{\sqrt{M_1^2 - 1}}$$

$$\approx \frac{4\alpha}{\sqrt{M_1^2 - 1}} \quad \text{for } \alpha \ll 1$$

$$C_d = \frac{1}{c} \int_0^c (C_{p,s,L} - C_{p,s,U}) \sin\alpha\, dx = \frac{1}{c} \int_0^c \frac{4\alpha}{\sqrt{M_1^2 - 1}} dx = \frac{4\alpha \sin\alpha}{\sqrt{M_1^2 - 1}}$$

$$\approx \frac{4\alpha^2}{\sqrt{M_1^2 - 1}} \quad \text{for } \alpha \ll 1$$

Comments –

One finds interesting an comparison of the results of Eqn. (4.63), which was derived using the weak shock-expansion theory, to an airfoil of arbitrary shape. For a flat plate, the following contributions due to camber and thickness disappear from Eqn. (4.63): $\overline{\alpha_c^2} = \left(\frac{dh}{dx} \right)^2 = 0$, leading to identical results of this example. Effects of camber and thickness of the thin airfoil has been discussed in Chapter 4, to which the reader is referred. □

9.6 Physical Acoustics in One Dimension

Because the propagation of sound is assumed to be isentropic, the small disturbances associated with sound propagation can be described by the Euler equation. Similar to the small perturbation theory developed for wavy walls, it is desirable to find governing equations for acoustic motion in terms of a perturbation velocity potential. Many details of theoretical development of physical acoustics can be found in Whitham (1974). This section describes the essential features of acoustic wave propagation.

Recall and rewrite the continuity and Euler equations, Eqn. (8.1d) and (8.2e), respectively:

$$\frac{\partial \rho}{\partial t} + \nabla \cdot \left(\rho \vec{V} \right) = \frac{\partial \rho}{\partial t} + \rho \nabla \cdot \vec{V} + \left(\vec{V} \cdot \nabla \right) \rho = 0 \tag{8.1d}$$

$$\rho \frac{\partial \vec{V}}{\partial t} + \left(\rho \vec{V} \cdot \nabla \right) \vec{V} = -\nabla p \tag{8.2e}$$

For small perturbations in a quiescent media, it is convenient to rewrite the properties as the following:

$$\rho = \rho_1 + \rho' \tag{9.40a}$$

$$p = p_1 + p' \tag{9.40b}$$

$$\vec{V} = \vec{V}_1 + \vec{v'} \tag{9.40c}$$

where the subscript 1 and the prime denote, respectively, the undisturbed state and the deviations from the undisturbed state. Consider sound propagation in a quiescent media ($\vec{V}_1 = 0$). Substituting Eqn. (9.40) into Eqns. (8.1d) and (8.2e) and neglecting the second- and higher-order terms in all products lead to

$$\frac{\partial \rho'}{\partial t} + \rho_1 \nabla \cdot \vec{v'} = 0 \tag{9.41}$$

and

$$\rho_1 \frac{\partial \vec{v'}}{\partial t} = -\nabla p' = a_1^2 \nabla \rho' \tag{9.42}$$

where the definition of sound speed, $\nabla p' = \nabla p \equiv a_1^2 \nabla \rho = a_1^2 \nabla \rho'$, is used. It is useful to define a non-dimensional density perturbation H as:

$$H = \frac{\rho'}{\rho_1} \tag{9.43a}$$

Therefore the pressure perturbation becomes

$$p' = a_1^2 \rho' = \rho_1 a_1^2 H \tag{9.43b}$$

To satisfy the requirement of small disturbances,

$$|H| \ll 1; \quad |p'| \ll \rho_1 a_1^2 = \gamma p_1; \quad |\vec{v}'| \ll a_1 \tag{9.44}$$

where $a_1^2 = \gamma R T_1 = \gamma p_1/\rho_1$ is used. Thus the governing equations for perturbed properties become

$$\frac{\partial H}{\partial t} + \nabla \cdot \vec{v}' = 0 \tag{9.45}$$

$$\frac{\partial \vec{v}'}{\partial t} = a_1^2 \nabla H \tag{9.46}$$

By using

$$\nabla \cdot \frac{\partial \vec{v}'}{\partial t} = \frac{\partial}{\partial t} \nabla \cdot \vec{v}'$$

Eqn. (9.45) and (9.46) can be combined into

$$\frac{\partial^2 H}{\partial t^2} = a_1^2 \nabla^2 H \tag{9.47}$$

Due to the isentropic nature of acoustic disturbances, a perturbation velocity potential, $\phi(x, y, z, t)$, can be defined in a similar manner as in flow over the wavy wall discussed earlier in this chapter. Therefore

$$\vec{v}' = \nabla \phi \tag{9.48}$$

Substituting Eqn. (9.48) into Eqn. (9.46) yields

$$\nabla \left(\frac{\partial \phi}{\partial t} + a_1^2 H \right) = 0$$

This expression indicates that the term in the parenthesis is a function of time, which is written as

$$\frac{\partial \phi}{\partial t} + a_1^2 H = \sigma(t)$$

where $\sigma(t)$ is a function of time that can assume a value of zero without affecting the result of $\vec{v}' = \nabla \phi$. Then

$$H = -\frac{1}{a_1^2} \frac{\partial \phi}{\partial t} \tag{9.49}$$

Substituting Eqns. (9.48) and (9.49) into Eqn. (9.45) leads to

$$\frac{\partial^2 \phi}{\partial t^2} - a_1^2 \nabla^2 \phi = 0 \tag{9.50}$$

which is a wave equation for unsteady propagation. For one-dimensional acoustic motion, the perturbation velocity potential is $\phi(x, t)$, and Eqn. (9.50) becomes

$$\frac{\partial^2 \phi}{\partial t^2} - a_1^2 \frac{\partial^2 \phi}{\partial x^2} = 0$$

Similar to the solutions of the steady supersonic wave equation, the general solution of Eqn. (9.50) is of the following form:

$$\phi = F(x - a_1 t) + G(x + a_1 t) \tag{9.51}$$

The functions $F(x - a_1 t)$ and $G(x + a_1 t)$ are constant for constant values of $(x - a_1 t)$ and $(x + a_1 t)$, respectively. Therefore $(x - a_1 t) =$ constant and $(x + a_1 t) =$ constant are also called *characteristics*. Unlike the characteristics in the steady supersonic waves, the current characteristics move with time (i.e., waves). The F family moves in the $+x$ direction while the G family in the $-x$ direction, and both families move with propagation speed a_1. Combining Eqn. (9.48) and (9.51) gives

$$u' = F' + G' \tag{9.52}$$

where F' denotes the derivative of F with respective to $\eta \equiv (x - a_1 t)$ and G', the derivative of G with respective to $\zeta \equiv (x + a_1 t)$, so that

$$F' = \frac{\partial F}{\partial x} = \frac{\partial F}{\partial \eta} \frac{\partial \eta}{\partial x} \text{ and } G' = \frac{\partial G}{\partial x} = \frac{\partial G}{\partial \eta} \frac{\partial \eta}{\partial x}$$

where $\partial \eta / \partial x = \partial \zeta / \partial x = 1$. Then the non-dimensional density perturbation becomes

$$H = -\frac{1}{a_1^2} \frac{\partial \phi}{\partial t} = -\frac{1}{a_1^2} \left(\frac{\partial F}{\partial \eta} \frac{\partial \eta}{\partial t} + \frac{\partial G}{\partial \zeta} \frac{\partial \zeta}{\partial t} \right) = \frac{1}{a_1} \left(F' - G' \right) \tag{9.53}$$

To relate H and u', one can define

$$f(x - a_1 t) \equiv \frac{1}{a_1} F' \text{ and } g(x + a_1 t) \equiv -\frac{1}{a_1} G'$$

So that

$$u' = a_1 (f - g) \text{ and } H = f + g \tag{9.54}$$

Because $|H| \ll 1$ and $u' \ll a_1$ and amplitudes of both f and g are $\ll 1$, which implies $|f - g| \ll 1$.

Waves that propagate in only one direction are *simple waves* (recall from Chapter 4 that in steady supersonic flows either compression or expansion waves

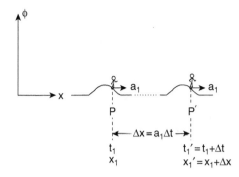

Figure 9.8 An acoustic wave traveling in the $+x$ direction (i.e., the right-running or the F family wave).

are simple waves and the associated flow is a simple flow; a non-simple flow region is one where these two waves, and their characteristics, intersect). For convenience, consider waves traveling in the $+x$ direction (i.e., the right-running or the F family). Figure 9.8 illustrates such wave propagation, where the waveform, that is, the general solution $\phi = F(x - a_1 t)$, is preserved. In Fig. 9.7 are both points P and P' and same value of ϕ, which is because at P' $(x'_1 - a_1 t'_1) = (x_1 + \Delta x) - a_1 (t_1 + \Delta t) = (x_1 - a_1 t_1)$ due to $\Delta x = a_1 \Delta t$.

For the F-family wave,

$$H = f(x - a_1 t) \tag{9.55a}$$

and Eqn. (9.43b) gives

$$p' = \rho_1 a_1^2 f(x - a_1 t) \tag{9.55b}$$

$$u' = a_1 f(x - a_1 t) \tag{9.55c}$$

It is clear that the perturbed variables (H, ρ', p', u', etc.) of Eqn. (9.55) all satisfy the wave equation, and they are constant along the characteristic (i.e., $(x - a_1 t) =$ constant). That is, for an observer riding on the wave (Fig. 9.8), all perturbed variables are constant. On the other hand, a stationary observer feels the perturbation only during the wave's passage. Because these variables are related by $f(x - a_1 t)$, they have the same form and sign, as indicated by Eqns. (9.55a), (9.55b), and (9.55c). When properly normalized, they also have the same magnitude:

$$H = \frac{p'}{\rho_1 a_1^2} = \frac{u'}{a_1} = f(x - a_1 t) \tag{9.55d}$$

Consequently for a compression wave, $H = f(x - a_1 t) > 0$, $p' > 0$, and $u' > 0$; that is, the fluid passed by the wave moves in the same direction as the wave. Following similar reasoning, a rarefaction (or expansion) wave causes the fluid passed by the wave to move in the opposite direction as $f(x - a_1 t) < 0$. These phenomena have been described in Chapter 3, where infinitesimal movements of the piston-tube device generate compression/expansion waves that travel at the acoustic speed and cause the fluid to move in the same/opposite direction.

It is of interest to be able to predict the trajectory of the fluid particle affected by the traveling wave. Let x_1 and ξ be, respectively, the initial unperturbed position and the displacement of the fluid particle. Since the choice of x_1 is arbitrary, the solution for it represents trajectories of any given fluid particle throughout the fluid. Then the particle position is

$$x(t) = x_1 + \xi(x_1, \, t) \tag{9.56}$$

Following Eqn. (9.54c), the perturbed velocity of the particle becomes

$$u' = a_1 f(x_1 + \xi - a_1 t) \tag{9.57}$$

Therefore,

$$\xi(x_1, \, t) = a_1 \int_0^t f(x_1 + \xi - a_1 \tau) d\tau$$

where τ is the running variable for time. Because the displacement is due to the passing of the wave, $d\xi = u' d\tau$, one can expect $\xi \approx u'\tau \ll a_1 \tau$. Then the above expression becomes

$$\xi(x_1, \, t) = a_1 \int_0^t f(x_1 - a_1 \tau) d\tau \tag{9.58}$$

Similar to H, ρ', p', and u', $\xi(x_1, \, t)$ also satisfies the wave equation. By using a new variable $\eta = (x - a_1 \tau)$, Eqn. (9.58) becomes

$$\xi(x_1, \, t) = -\int_{x_1}^{x_1 - a_1 t} f(\eta) d\eta \tag{9.59}$$

Further manipulation leads to

$$\xi(x_1, \, t) = -\int_0^{x_1 - a_1 t} f(\eta) d\eta + \int_0^{x_1} f(\eta) d\eta \tag{9.60}$$

Since the wave (disturbance) will be on $\eta = x_1 - a_1 \tau = 0, f(\eta) = 0$ for $\eta > 0$. For the wave propagating in the $+x$ direction, $x_1 > 0$. Therefore, the first integral results in $\xi(x_1, \, t) = \xi(x_1 - a_1 t)$.

Figure 9.9 demonstrates the particle displacement (or trajectory) during the passing of a perturbation wave. The following example describes the effect of a sinusoidal wave.

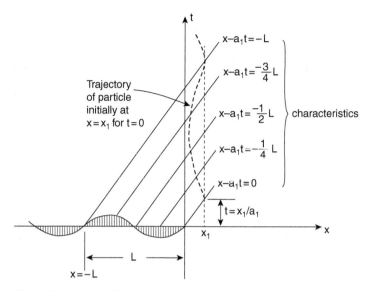

Figure 9.9 Particle displacement (or trajectory) during the passing of a perturbation (or acoustic) wave.

EXAMPLE 9.5

Find the expression of fluid particle displacement in a long piston-tube device (in the $+x$ direction, with the piston position initially at $x = 0$). The piston movement causes a density disturbance described by

$$H = \epsilon \sin \kappa (x - a_1 t)$$

where $\epsilon \ll 1$ and $\kappa = 2\pi/L$ is the wave number, with L being the period, or wavelength, of the wave.

Solutions –

Following Eqn. (9.55),

$$u' = a_1 f(x - a_1 t) = a_1 H = a_1 \epsilon \sin \kappa (x - a_1 t)$$

For a fluid particle initially located at an arbitrary location $x = x_1$ in the undisturbed medium, the disturbed particle location is

$$\xi(x_1,\ t) = a_1 \int_0^t \epsilon \sin \kappa (x_1 - a_1 \tau) f d\tau = a_1 \epsilon \int_{x_1/a_1}^t \sin\ \kappa (x_1 - a_1 \tau) d\tau$$

$$+ a_1 \int_0^{x_1/a_1} \epsilon \sin \kappa (x_1 - a_1 \tau) d\tau$$

where the second integral is equal to 0 because between $t = 0$ and x_1/a_1 no disturbance has arrived at x_1 yet. Thus

$$\xi(x_1, \ t) = \frac{\epsilon}{\kappa}[-\cos\kappa(x_1 - a_1\tau)]_{x_1/a_1}^t = \frac{\epsilon L}{2\pi}\left[1 - \cos\frac{2\pi}{L}(x_1 - a_1t)\right]$$

or

$$\xi(x_1, \ t) = -\int_0^{x_1-a_1t} f(\eta)d\eta = -\int_0^{x_1-a_1t} \epsilon \sin\kappa\eta \, d\eta = -\frac{\epsilon}{\kappa}[\cos\kappa\eta]_0^{x_1-a_1t}$$

$$= \frac{\epsilon L}{2\pi}\left[1 - \cos\frac{2\pi}{L}(x_1 - a_1t)\right]$$

The particle displacement (or trajectory) is plotted in Fig. 9.9. □

Problems

Problem 9.1 Derive Eqn. (9.3); also show as an intermediate step:

$$\frac{p}{p_1} \approx 1 - \gamma M_1^2 \frac{u'}{U_1}$$

Problem 9.2 Following the condition given in Example 9.1, find the pressure fluctuation over the surface.

Problem 9.3 Show that for the two-dimensional flow, with free-stream velocity U_1 parallel to the x-axis, over the sinusoidal surface, $y = \epsilon \sin kx$ would reach sonic for the free-stream Mach number given by

$$\frac{M_1^2(\gamma + 1)\epsilon k}{(1 - M_1^2)^{3/2}} = 1 \ \text{or} \ \frac{2M_1^2(\gamma + 1)\pi k/L}{(1 - M_1^2)^{3/2}} = 1$$

Problem 9.4 Show steps leading to Eqn. (9.22) for small M_1 and show that for very weak perturbations, Eqn. (9.22) simplifies to Eqn. (9.16).

Problem 9.5 Carry out the derivation leading to Eqn. (9.26).

Problem 9.6 An airfoil shown in the figure is placed in a supersonic flow $(M_1 > 1)$ at an angle of attack, $\alpha \ll 1$ (i.e., $2\,t/c \ll 1$), to the free-stream. Find the lift and drag coefficients using the small perturbation theory.

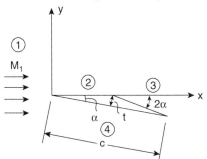

Problem 9.7 A flat-bottomed airfoil depicted in the figure has its upper surface described by the following expression:

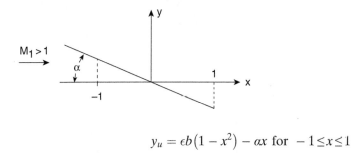

$$y_u = \epsilon b(1 - x^2) - \alpha x \text{ for } -1 \leq x \leq 1$$

where the effect of the angle of attack α is included (α measured using the flat lower surface, as shown in the figure), $\alpha = \epsilon \alpha_o$ with $\epsilon \ll 1$, and b is a constant.

(a) Use the small disturbance theory to find C_p, u, and v *around* the airfoil.

(b) Sketch lines of

$$\frac{u}{[2\epsilon U_\infty b/\lambda]} = \text{constant}$$

showing the values of the constant on lines through the leading edges on both the upper and the lower surfaces, and the line through the midpoint of the airfoil on the upper surface.

(c) Indicate whether the waves of the leading edge are compression or expansion waves.

Problem 9.8 Show that for acoustic waves

$$u' = \pm \frac{p'}{\rho_1 a_1}$$

where the + and the − signs are for the right- (the F family) and left-running (the G family) characteristics, respectively.

Problem 9.9 For an acoustic wave propagating in an ideal gas that experiences a small local density disturbance, as expressed by $H = (\rho - \rho_1)/\rho_1 = \rho'/\rho_1$, with $|H| \ll 1$, where the subscript 1 denotes the undisturbed state:

(a) Find the change in local speed of sound, $(a - a_1)/a_1$, in terms of γ and H.

(b) If the pressure is assumed constant, find an expression of $(a - a_1)/a_1$. Assess the validity of the constant-pressure assumption.

Problem 9.10 Following Problem 9.9, find an expression of $(a - a_1)/a_1$ in terms of γ and K, where $K = (p - p_1)/p_1 = p'/p_1$, with $|K| \ll 1$.

Figure for Problem 9.11.

Problem 9.11 The piston motion in the piston-cylinder device is described by the following equation:

$$u_p = \epsilon a_1 \sin \omega t, \quad 0 \leq \omega t \leq 2\pi$$
$$= 0, \qquad\qquad \omega t > 2\pi$$

where $\omega = a_1 \kappa$, a_1 is the acoustic speed of the undisturbed gas and κ is the spatial frequency or the wave number.

(a) Find the velocity distribution in regions I and II.

(b) Would it be possible to arrange the conditions such that no wave in region III is reflected from the piston end of the tube?

Problem 9.12 Derive Eqn. (9.26).

10 Method of Characteristics for Two Independent Spatial Variables

Chapter 9 described solving the governing equations for small perturbations, where two types of characteristics were discussed: two-dimensional steady supersonic flow and unsteady one-dimensional flow. It was also shown that flow properties are constant along characteristics (or characteristic lines) and there exist two families of characteristics: the left-running and the right-running. Examples given in Chapter 9 consist of one single family of characteristics, where only simple wave/flow regions exist. Practical applications, such as flow in two-dimensional channels, often see both the right- and left-running characteristics intersect to form non-simple wave (or non-simple flow) regions. In unsteady one-dimensional flow, where a wall or surface produces reflected waves propagating in opposite direction to the original disturbances, non-simple flow regions form. In these non-simple flow regions, the method of characteristics (MOC) can be used to solve the flow problem. The MOC takes advantage of the constant properties along the characteristics to solve the nonlinear governing equations for which analytic solutions are rare. This chapter focuses on this method applied to two-dimensional flows. Chapter 11 will be devoted to the MOC in one-dimensional unsteady flows.

10.1 Characteristics in Steady Two-Dimensional Irrotational Flow

For the irrotational flow field, it is desirable to define a velocity potential Φ such that

$$\vec{V} = \vec{i}u + \vec{j}v + \vec{k}w = \vec{i}\frac{\partial \Phi}{\partial x} + \vec{j}\frac{\partial \Phi}{\partial y} + \vec{k}\frac{\partial \Phi}{\partial z} \tag{10.1}$$

Note that the velocity components, u, v, and w are not necessarily small perturbations as those described Chapter 9. The goal is to find a solution for the partial differential equations governing Φ, as in the case for the perturbation velocity potential ϕ. Recall the continuity, irrotationailty, and momentum (i.e., Euler) equations:

$$\nabla \cdot \left(\rho \vec{V} \right) = 0 \tag{8.1e}$$

$$\nabla \times \vec{V} = 0 \tag{8.11}$$

$$\left(\rho\vec{V}\cdot\nabla\right)\vec{V} = -\nabla p \quad\text{or}\quad \nabla\left(\frac{V^2}{2}\right) + \frac{1}{\rho}\nabla p = 0 \tag{8.22}$$

In terms of the velocity potential, the continuity and momentum equations become

$$\rho\left(\frac{\partial^2 \Phi}{\partial x^2} + \frac{\partial^2 \Phi}{\partial y^2} + \frac{\partial^2 \Phi}{\partial z^2}\right) + \left(\frac{\partial \Phi}{\partial x}\frac{\partial \rho}{\partial x} + \frac{\partial \Phi}{\partial y}\frac{\partial \rho}{\partial y} + \frac{\partial \Phi}{\partial z}\frac{\partial \rho}{\partial z}\right) = 0 \tag{10.2}$$

$$dp = -\frac{\rho}{2}d\left[\left(\frac{\partial \Phi}{\partial x}\right)^2 + \left(\frac{\partial \Phi}{\partial y}\right)^2 + \left(\frac{\partial \Phi}{\partial z}\right)^2\right] \tag{10.3}$$

To arrive at an equation for Φ, it is desirable to eliminate ρ from these equations. The irrotational flow is also homentropic ($\nabla \cdot s = 0$), for which $a^2 = (\partial p/\partial \rho)_s$. Then Eqn. (10.3) becomes

$$d\rho = -\frac{\rho}{2a^2}d\left[\left(\frac{\partial \Phi}{\partial x}\right)^2 + \left(\frac{\partial \Phi}{\partial y}\right)^2 + \left(\frac{\partial \Phi}{\partial z}\right)^2\right] \tag{10.4}$$

Therefore,

$$\frac{\partial \rho}{\partial x} = -\frac{\rho}{a^2}\left[\frac{\partial \Phi}{\partial x}\frac{\partial^2 \Phi}{\partial x^2} + \frac{\partial \Phi}{\partial y}\frac{\partial^2 \Phi}{\partial x\partial y} + \frac{\partial \Phi}{\partial z}\frac{\partial^2 \Phi}{\partial x\partial z}\right] \tag{10.5a}$$

$$\frac{\partial \rho}{\partial y} = -\frac{\rho}{a^2}\left[\frac{\partial \Phi}{\partial x}\frac{\partial^2 \Phi}{\partial y\partial x} + \frac{\partial \Phi}{\partial y}\frac{\partial^2 \Phi}{\partial y^2} + \frac{\partial \Phi}{\partial z}\frac{\partial^2 \Phi}{\partial z\partial y}\right] \tag{10.5b}$$

$$\frac{\partial \rho}{\partial z} = -\frac{\rho}{a^2}\left[\frac{\partial \Phi}{\partial x}\frac{\partial^2 \Phi}{\partial z\partial x} + \frac{\partial \Phi}{\partial y}\frac{\partial^2 \Phi}{\partial z\partial y} + \frac{\partial \Phi}{\partial z}\frac{\partial^2 \Phi}{\partial z^2}\right] \tag{10.5c}$$

Substituting these derivatives into Eqn. (10.2) yields

$$\left[1 - \frac{1}{a^2}\left(\frac{\partial \Phi}{\partial x}\right)^2\right]\frac{\partial^2 \Phi}{\partial x^2} + \left[1 - \frac{1}{a^2}\left(\frac{\partial \Phi}{\partial y}\right)^2\right]\frac{\partial^2 \Phi}{\partial y^2} + \left[1 - \frac{1}{a^2}\left(\frac{\partial \Phi}{\partial z}\right)^2\right]\frac{\partial^2 \Phi}{\partial z^2}$$

$$-\frac{2}{a^2}\left[\frac{\partial \Phi}{\partial x}\frac{\partial \Phi}{\partial y}\frac{\partial^2 \Phi}{\partial x\partial y}\right] - \frac{2}{a^2}\left[\frac{\partial \Phi}{\partial x}\frac{\partial \Phi}{\partial z}\frac{\partial^2 \Phi}{\partial x\partial z}\right] - \frac{2}{a^2}\left[\frac{\partial \Phi}{\partial y}\frac{\partial \Phi}{\partial z}\frac{\partial^2 \Phi}{\partial y\partial z}\right] = 0 \tag{10.6}$$

Equation (10.6) is *the velocity potential equation*. It is nonlinear, as opposed to the linear small perturbation equations for subsonic and supersonic flows, Eqn. (8.42).

It is further desirable to express the speed of sound a in Eqn. (10.6) in terms of the velocity potential. Taking advantage of Eqn. (8.37a) while noting that U_1 is now absorbed in $u = U_1 + u'$ and taking the state 1 to be the reference state, which is also the stagnation state ($U_1 = 0$ in Eqn. (8.37a)). Therefore, Eqn. (8.37) adapted for the velocity potential is

$$a^2 = a_1^2 - \frac{\gamma - 1}{2} \left[\left(\frac{\partial \Phi}{\partial x} \right)^2 + \left(\frac{\partial \Phi}{\partial y} \right)^2 + \left(\frac{\partial \Phi}{\partial z} \right)^2 \right] \tag{10.7}$$

where a_1 is the speed of sound of some reference state.

The solution procedure for the flow field involves (i) solving Eqns. (10.6) and (10.7) for Φ subject to boundary conditions, (ii) determining the velocity field by using Eqn. (10.1), $\vec{V} = \nabla\Phi$, (iii) calculating the speed of sound and local Mach number using Eqn. (10.7) and the velocity determined, and finally (iv) calculating all other state variables, p, T, and ρ using the isentropic relationships derived in Chapter 4.

For a quick validation of Eqn. (10.7) consider incompressible flows, for which $a \ll u, v$, and w and $a = a_1$. Then Eqn. (10.7) is reduced to the Laplace's equation:

$$\frac{\partial^2 \Phi}{\partial x^2} + \frac{\partial^2 \Phi}{\partial y^2} + \frac{\partial^2 \Phi}{\partial z^2} = 0 \tag{10.8}$$

For compressible two-dimensional flows, Eqn. (10.6) becomes

$$\left[1 - \frac{1}{a^2} \left(\frac{\partial \Phi}{\partial x} \right)^2 \right] \frac{\partial^2 \Phi}{\partial x^2} + \left[1 - \frac{1}{a^2} \left(\frac{\partial \Phi}{\partial y} \right)^2 \right] \frac{\partial^2 \Phi}{\partial y^2} - \frac{2}{a^2} \left[\frac{\partial \Phi}{\partial x} \frac{\partial \Phi}{\partial y} \frac{\partial^2 \Phi}{\partial x \partial y} \right] = 0 \tag{10.9}$$

or

$$\left[1 - \frac{u^2}{a^2} \right] \frac{\partial u}{\partial x} + \left[1 - \frac{v^2}{a^2} \right] \frac{\partial v}{\partial y} - \frac{2\,uv}{a^2} \frac{\partial u}{\partial y} = 0 \tag{10.10}$$

Equations (10.9a) and (10.10) remain nonlinear, and analytical solutions to nonlinear equations in general do not exist. A numerical approach to solving the nonlinear equation is to use the method of characteristics (MOC). The mathematical theory of MOC indicates (i) that two-dimensional steady flows characteristics, or Mach lines, exist only for $M > 1$, that is, $(u^2 + v^2)/a^2 > 1$ for which the equation is hyperbolic (Courant and Friderichs, 1948), (ii) that the normal derivatives of dependent variables on a characteristic may be discontinuous although they must be continuous, and (iii) that the dependent variables must satisfy a relation called the compatibility relation (Zucrow and Hoffman, 1976; Thompson, 1972).

So far the concept of characteristics has been encountered while dealing with problems of isentropic turning (Chapter 4) and acoustic propagation (Chapter 9). Systematic approaches indeed leads to the following *compatibility relations* (Thompson, 1972):

$$\nu(M) + \delta = Q = \text{constant (along right-running or } \xi \text{ characteristics)} \tag{4.52a}$$

$$\nu(M) - \delta = R = \text{constant} \quad (\text{along left-running or } \eta \text{ characteristics}) \quad (4.52\text{b})$$

where $\nu(M)$ and δ are, respectively, the local Prandtl-Meyer function/angle and flow angle (or wall angle if bounded) relative to a reference coordinate. These compatibility relations are derived in the following by using a systematic approach, where the streamline coordinates are used. When deriving or using Eqns. (4.52a) and (4.52b), the wall surface is considered to be streamlined.

Cautions are needed for the difference between the angle δ used in Chapter 4 and here for the method of characteristics. In Chapter 4, δ is the turning angle of the wall surface relative to the flow. For the method of characteristics, δ is the general flow angle measured from a reference coordinate (e.g., the Cartesian coordinate system) and $\delta > 0$ (or $\delta < 0$) if it is counterclockwise (or clockwise). Therefore, for MOC, if a wall turns away from the flow and in a counterclockwise direction, the flow angle is considered positive.

Recall that the Prandtl-Meyer function is

$$\nu(M) = \int \frac{\sqrt{M^2 - 1}}{1 + \frac{\gamma-1}{2}M^2} \frac{dM}{M} = \sqrt{\frac{\gamma+1}{\gamma-1}} \tan^{-1} \sqrt{\frac{\gamma-1}{\gamma+1}(M^2 - 1)} - \tan^{-1}\sqrt{M^2 - 1}$$

$$(4.47\text{f})$$

Because the flow direction in general changes throughout the flow with respect to the reference coordinate, it is more convenient to adopt the *natural coordinate system*, in

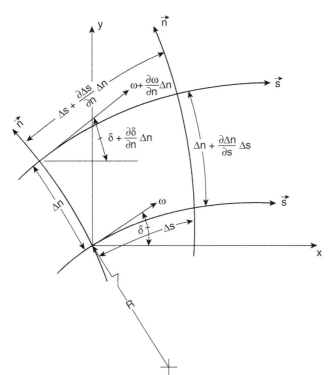

Figure 10.1 Schematic showing the flow and the coordinate systems in the xy-plane; the flow direction is along the s-coordinate and is counterclockwise from the x-coordinate resulting in a flow angle $\delta > 0$. The n-coordinate is normal to the flow everywhere.

which the two coordinates are in the streamline direction and the other one is normal to it, that is, (s, n), as the velocity component in the normal direction is identically zero as the definition of streamlines requires. The flow and the coordinate systems are depicted in Fig. 10.1, where the flow direction is the s-coordinate and, as depicted, is counterclockwise from the x-coordinate resulting a flow angle $\delta > 0$. The irrotationality condition, Eqn. (8.11), in the natural coordinate system is developed as follows. For the differential control volume shown in Fig. 10.1, the circulation $\Delta\Gamma$ can be calculated by following the procedure leading to Eqn. (8.14). Let ω be the velocity component along the streamline. Thus

$$\Delta\Gamma = +\omega\Delta s - \left(\omega + \frac{\partial\omega}{\partial n}\Delta n\right)\left(\Delta s + \frac{\partial s}{\partial n}\Delta n\right) \tag{10.11}$$

Geometric consideration of Fig. 10.1 leads to $\partial(\Delta s)/\partial n = -\Delta s/R$, with $R = \Delta s/\Delta\delta$ being the radius of the streamline curvature. Then $\Delta\Gamma$ becomes

$$\Delta\Gamma = -\frac{\partial\omega}{\partial n}\Delta n\Delta s - \omega\frac{\partial\Delta s}{\partial n}\Delta n - \frac{\partial\omega}{\partial n}\frac{\partial\Delta s}{\partial n}(\Delta n)^2$$

Neglecting the third-order term,

$$\Delta\Gamma = -\frac{\partial\omega}{\partial n}\Delta n\Delta s + \omega\frac{\partial\delta}{\partial s}\Delta s\Delta n = -\left(\frac{\partial\omega}{\partial n} - \omega\frac{\partial\delta}{\partial s}\right)\Delta s\Delta n = -\left(\frac{\partial\omega}{\partial n} - \omega\frac{\partial\delta}{\partial s}\right)\Delta A$$

Irrotationality ($\Delta\Gamma = 0$) requires:

$$\frac{\partial\omega}{\partial n} - \omega\frac{\partial\delta}{\partial s} = 0 \tag{10.12}$$

Adapting Eqn. (4.47a) by replacing V in the Cartesian coordinate system with ω in the natural coordinate system, one arrives at

$$\sqrt{M^2 - 1}\frac{d\omega}{\omega} = d\nu \tag{10.13}$$

With this expression and $\tan\mu = 1/\sqrt{M^2 - 1}$, Eqn. (10.12) becomes

$$\tan\mu\frac{\partial\nu}{\partial n} - \frac{\partial\delta}{\partial s} = 0 \tag{10.14}$$

By using Fig. 10.1 again, one can rewrite the continuity and momentum equations, Eqn. (8.1e) and (8.22), along the streamline as

$$\rho u\Delta n = \text{constant} \tag{10.15}$$

$$u\frac{\partial u}{\partial s} + \frac{1}{\rho}\frac{\partial p}{\partial s} = 0 \tag{10.16}$$

It is left as an exercise (Problem 10.1) to combine Eqn. (10.15) and (10.16) by similarly using the definitions of the Mach wave angle and the Prandtl-Meyer function to show that

$$\frac{\partial \nu}{\partial s} - \tan \mu \frac{\partial \delta}{\partial n} = 0 \tag{10.17}$$

The difference and the sum of Eqn. (10.14) from Eqn. (10.17), respectively, are

$$\frac{\partial}{\partial s}(\nu + \delta) - \tan \mu \frac{\partial}{\partial n}(\nu + \delta) = 0 \tag{10.18a}$$

$$\frac{\partial}{\partial s}(\nu - \delta) + \tan \mu \frac{\partial}{\partial n}(\nu - \delta) = 0 \tag{10.18b}$$

Thus Eqns. (10.18a) and (10.18b) result from combining continuity, momentum conservation along the streamline, and the requirement for irrotationailty in the natural coordinate system. The compatibility relation has yet to be found on the characteristics (or Mach lines). To do so, an attempt is made by examining the possibility of finding the derivatives of $(\nu - \delta)$ and $(\nu + \delta)$ in the direction of the characteristics.

(a)

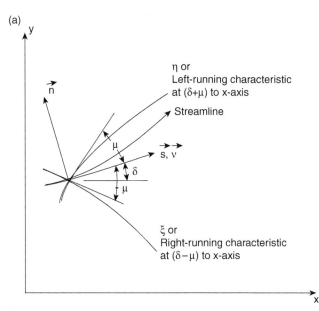

Figure 10.2 (a) The axes ξ and η are at the angles $-\mu$ and $+\mu$ to the streamline, where $\mu = \sin^{-1}(1/M)$ and therefore the axes ξ and η are Mach lines; (b) details of the ξ and η coordinates relative to the flow (or s and n) coordinates.

(b)

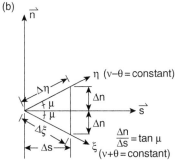

The coordinate system (ξ, η) relative to the streamline is shown in Fig. 10.2, where δ is the local inclination angle of the flow (or streamline) with respective to a reference coordinate (here the x axis). The axes ξ and η are at the angles $-\mu$ and $+\mu$ to the streamline, where $\mu = \sin^{-1}(1/M)$ and therefore the axes ξ and η are Mach lines. The derivatives of an arbitrary scalar function f in the coordinate system (ξ, η) can be found as follows. The change in f from point 1 to point 2 in Fig. 10.2a can be found either simply along the ξ direction or along the natural/streamline coordinates; the two results should be equal. Therefore

$$\Delta f = \frac{\partial f}{\partial \xi} \Delta \xi = \frac{\partial f}{\partial s} \Delta s + \frac{\partial f}{\partial n} \Delta n$$

Dividing this equation through by Δs leads to

$$\frac{\partial f}{\partial \xi} \frac{\Delta \xi}{\Delta s} = \frac{\partial f}{\partial s} + \frac{\partial f}{\partial n} \frac{\Delta n}{\Delta s}$$

From the geometry shown in Fig. 10.2b, this expression becomes

$$\sec \mu \frac{\partial f}{\partial \xi} = \frac{\partial f}{\partial s} - \tan \mu \frac{\partial f}{\partial n} \qquad (10.19a)$$

Similarly, the derivative of f in the η direction can be shown to be

$$\sec \mu \frac{\partial f}{\partial \eta} = \frac{\partial f}{\partial s} + \tan \mu \frac{\partial f}{\partial n} \qquad (10.19b)$$

Comparing Eqns. (10.18) and (10.19), one concludes that (note that $\sec \mu \neq 0$)

$$\frac{\partial}{\partial \xi}(\nu + \delta) = 0 \qquad (10.20a)$$

$$\frac{\partial}{\partial \eta}(\nu - \delta) = 0 \qquad (10.20b)$$

These results indicate that

$$\nu + \delta = Q = \text{constant along a } \xi \text{ characteristic (right-running Mach line)} \quad (10.21a)$$

$$\nu - \delta = R = \text{constant along an } \eta \text{ characteristic (left-running Mach line)} \quad (10.21b)$$

Equations (10.21a) and (10.21b) are the *compatibility relations* along characteristics. They provide solution algorithms for two-dimensional, steady, isentropic flow fields. They constitute two equations for two dependent variables, ν and δ – the Prandtl-Meyer angle (a function only of Mach number) and the flow inclination from the reference coordinate (x-axis in the Cartesian coordinate system. Thus the original three differential equations (8.1e, 8.11, and 8.22) for three variables (u, v, and ρ) have been reduced to two algebraic equations with two variables. In Eqn. (10.21a) and

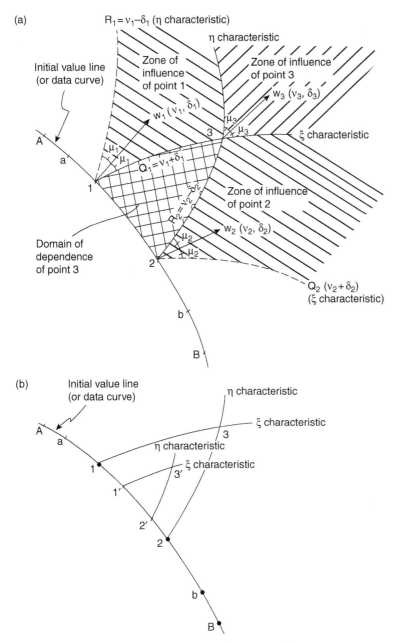

Figure 10.3 (a) Schematic showing zones (or domains) of dependence and influence of various points in the flow field and characteristics with their Riemann invariants; (b) choosing a smaller zone of influence (region 1'-2'-3') than region 1-2-3 in Fig. 10.3a improves computational accuracy.

(10.21b), $\nu + \delta$ and $\nu - \delta$ are called *Riemann invariants*. The results thus obtained, ν and δ, can be used to determine the velocity field, with ν providing Mach number M (and speed) and δ, the flow direction. Thermodynamic properties as functions of M can then be calculated using isentropic relationships or found in Appendix C.

The next section illustrates the numerical procedure of the method of characteristics. It is now informative to observe that Eqns. (10.21a) and (10.21b) are identical to Eqns. (4.52a) and (4.52b), respectively, describing simple wave region due to isentropic turning.

10.2 Method of Characteristics

The procedure of the method of characteristics is illustrated as follows. Suppose that the values of ν and δ are already known along the curve AB shown in Fig. 10.3a, and it is desirable to find the flow properties at an *arbitrary* point 3 downstream of the curve. The curve AB is also called the *data curve* or the *initial value line*. Because of the hyperbolic nature of the problem, it is possible to determine flow properties using only the information originating from some upstream region. The characteristics intersecting at point 3 can be traced back to the data curve such that they are, respectively, the ξ and η characteristics coming from points 1 and 2. Because these characteristics are Mach waves, point 3 lies at the edge of the *zones of influence* (or zones of action, as described in Section 3.2) of points 1 and 2. Therefore point 3 is also under influence of all points between 1 and 2 on the data line, but not under the influence of those outside of the segment connecting 1 and 2, such as points a and b, as shown in Fig. 10.3a. The domain represented by the triangular region of 1-2-3 is called the *domain of dependence* for point 3. Similarly, the region downstream of point 3 and bounded by the ξ and η characteristics emanating from it is its zone of influence. One can easily see that the zone of influence of point 3 is also under the influence of points 1 and 2, as the point 3 is the intersection of the characteristics from those two points. Extension of such reasoning to the whole flow field can then be made.

> **EXAMPLE 10.1** Assuming that points 1 and 2 in Fig. 10.3a have the following Mach numbers and flow angle (the latter being counterclockwise from the reference x-axis):
>
> at point $1 - M_1 = 2.0$ and $\delta_1 = 10°$
> at point $2 - M_2 = 1.96$ and $\delta_2 = 9°$
>
> find the Mach number and flow angle at point 3.

Solutions –
Point 3 is at the intersection of the ξ and η characteristics from points 1 and 2, respectively. Therefore, at point 3:

$$\nu_3 + \delta_3 = \nu_1 + \delta_1 \quad \text{and} \quad \nu_3 - \delta_3 = \nu_2 - \delta_2$$

$$\nu_3 = \frac{1}{2}(\nu_1 + \nu_2) + \frac{1}{2}(\delta_1 - \delta_2)$$

$$\delta_3 = \frac{1}{2}(\nu_1 - \nu_2) + \frac{1}{2}(\delta_1 + \delta_2)$$

From Appendix C, $\nu_1 = 26.380°$ and $\nu_2 = 25.271°$.

Therefore, $\nu_3 = 25.271°$ and $M_2 \approx 1.98$ and $\delta_3 \approx 9.555°$

As shown in Example 10.1, the values of variables at point 3 fall between those at points 1 and 2; however, they are not exactly equal to the arithmetic averages of the upstream values. If the coordinates of points 1 and 2 are known, then the location of point 3 can be easily calculated by assuming straight line segments for the characteristics and using the Mach wave angles μ_1 and μ_2 for slopes of the straight lines. ☐

Since the location of the characteristics depends on the choice of the downstream location (i.e., the knowledge of the location of characteristics are not known a priori), it is possible to choose point 3' as shown in Fig. 10.3b. In this case point 3' is at the edges of the zones of influences of points 1' and 2'; thus conditions at points 1 and 2 do not affect those at 3'. Such a new choice as shown in Fig. 10.3b is meant to improve accuracy, as the mesh size is decreased. Smaller mesh sizes also make better approximations of the characteristics as straight line segments. Further decreases in the mesh size greatly reduce the error in the prediction of the flow field.

A typical characteristic network is shown in Fig. 10.4, bounded by an upper wall and a lower free boundary. The characteristic network thus involves characteristics originating from the wall and from the free boundary. Following Liepmann and Roshko [1957], Fig. 10.5 describes the treatment of all the possibilities.

Figure 10.5a shows the scheme for data tabulation for calculating toward point 3. For the interior point (Fig. 10.5b), the data are known and carried by the ξ and η characteristics emanating from points 1 (the constant Q_1) and 2 (the constant R_2), respectively. In this case, $\nu_3 = (Q_1 + R_2)/2$ and $\delta_3 = (Q_1 - R_2)/2$ are to be, and can be, calculated, as demonstrated by Example 10.1. They are in general

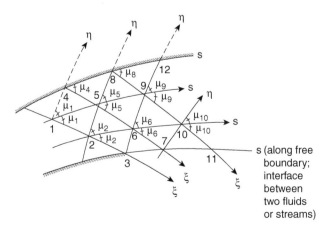

Figure 10.4 A typical characteristic network bounded by an upper wall and a lower free boundary – interior points: 1, 2, 5, 6, 9, 10; free boundary points: 3, 7, 11; wall points: 4, 8, 12.

s (along free boundary; interface between two fluids or streams)

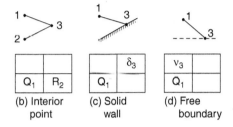

(a) Table

(b) Interior point　(c) Solid wall　(d) Free boundary

Figure 10.5 Known data for calculation toward point 3 – a general quart chart showing possible known data (part a); data known for calculation toward an interior point (part b), a wall point (part c), and a free boundary point (part d).
(Source: Liepmann, H. W., and Roshko, A., Elements of Gasdynamics, John Wiley and Sons, New York, 1957)

$$\nu = \frac{1}{2}(Q + R) \tag{12.22a}$$

$$\delta = \frac{1}{2}(Q - R) \tag{12.22b}$$

For point 3 on a wall (Fig. 10.5c), the flow angle δ_3 is known, as it is determined by the slope of the solid boundary as well as the constant. The constant Q_1 is known (i.e., the data, or the initial value, is supplied by point 1). Then the relationship $\nu_3 + \delta_3 = \nu_1 + \delta_1 = Q_1$ yields ν_3, completing the calculation for the flow at point 3.

An example of a point at the free boundary is that at the interface of the jet fluid and the ambient fluid outside the supersonic nozzle (such as point 7 in Fig. 10.4). At such points the local Mach number (M_3) and the Prandtl-Meyer angle (ν_3) are determined by the isentropic relationship using the ratio of the stagnation to the ambient pressures:

$$\frac{p_t}{p} = \left(1 + \frac{\gamma - 1}{2} M_3^2\right)^{\frac{\gamma}{\gamma - 1}}$$

Then $\nu_3 + \delta_3 = \nu_1 + \delta_1 = Q_1$ yields δ_3, providing the necessary ν_3 and δ_3 for the flow conditions at point 3.

10.3 Classification of Flow Regions

The two-dimensional steady supersonic flow can be categorized into three types. The first type is the general, or non-simple region, where ξ and η characteristics intersect and each of the ξ and η characteristics is curved (see Fig. 10.4 serving as an example). Each of these characteristics corresponds to one value of Q or R (as shown in Fig. 10.6). The values of ν and δ at the intersection of the ξ and η characteristics are found by using Eqns. (10.22a) and (10.22b).

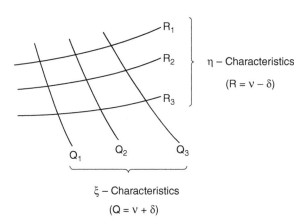

η – Characteristics

(R = v − δ)

Figure 10.6 Characteristic curves and their associated Riemann invariants in a general, or non-simple (wave) region.

ξ – Characteristics

(Q = v + δ)

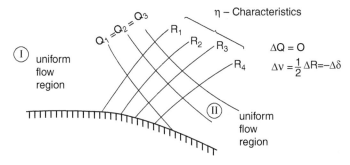

η – Characteristics

$\Delta Q = 0$

$\Delta v = \frac{1}{2}\Delta R = -\Delta \delta$

Figure 10.7 Characteristic curves and their associated Riemann invariants in a simple (wave) region.

Along a ξ-characteristic, Q = constant, and Eqns. (10.22a) and (10.22b) indicate that the changes in v and δ depend on changes in R following the flow crossing the η characteristics. Therefore, the changes in v and δ are related by the following equation:

$$\Delta v = \frac{1}{2}(\Delta Q + \Delta R) = \frac{1}{2}\Delta R = -\Delta \delta \qquad (10.23a)$$

Similarly, along an η-characteristic, R = constant, so that

$$\Delta v = \frac{1}{2}(\Delta Q + \Delta R) = \frac{1}{2}\Delta Q = \Delta \delta \qquad (10.23b)$$

The second type is the simple region, or simple wave in which either Q or R is constant throughout the entire region. Figure 10.7 illustrates this type of flow for which $Q = Q_o$ = constant along all the ξ-characteristics originating from region I and extending into region II. Therefore Q = constant and $\Delta Q = 0$ throughout the entire flow field. The changes in v and δ depend on changes in R following the flow crossing the η-characteristics (i.e., going along the ξ -characteristics). Along an η-characteristic, $\Delta R = 0$ and therefore v and δ are constant. As a consequence, both

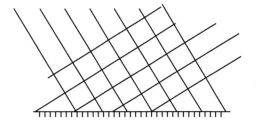

Figure 10.8 Characteristic curves and their associated Riemann invariants in a uniform region. Note $Q_1 = Q_2 = Q_3$, $R_1 = R_2 = R_3$, and all characteristics are straight lines.

the Mach number and the flow inclination are constant along an η-characteristic. Note that the characteristic is the Mach wave at an angle of $\sin^{-1}(1/M)$ to the flow. Thus in a simple region (simple wave), the η-characteristics (or the ξ-characteristics if R = constant) are straight lines with uniform conditions. It can be seen in Fig. 10.7 that flow in regions I and II are uniform.

The third type is the uniform flow region, in which Q = constant and R = constant. Not only along each of the ξ- and η-characteristics but also throughout the entire simple wave region, ν and δ are constant. Therefore, all characteristics are straight lines. Figure 10.8 illustrates the simple flow over a flat surface.

EXAMPLE 10.2 Consider an air flow at $M = 4$ turns around a $10°$ rounded convex corner, as shown in the figure. Find the Mach number and flow direction when the flow has turned through $6°$. What is the Mach number when the flow completes the $10°$ turn. Assume no flow separation due to the turning.

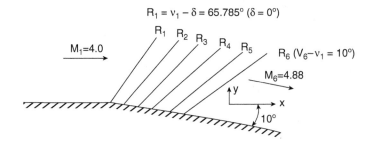

Solutions –
As shown in the figure to the right, Q = constant throughout the corner region. Equation (12.23a) indicates

$$\nu_4 - \nu_1 = \Delta\nu = \frac{1}{2}\Delta R = -\Delta\delta = -(\delta_4 - \delta_1) = -\left(-6° - 0°\right) = 6°$$

$M_1 = 4$, then $\nu_1 = 65.785°$
$\qquad \nu_4 = 65.785° + 6° = 71.785°$ and $M_4 \approx 4.50$
$\qquad$ Similarly, $\nu_6 - \nu_1 = \Delta\nu = \frac{1}{2}\Delta R = -\Delta\delta = -(\delta_6 - \delta_1) = -\left(-10° - 0°\right) = 10°$
$\nu_6 = 65.785° + 10° = 75.785°$ and $M_6 \approx 4.88$

Comments –

These results are the same as those of Example 4.13, where Prandtl-Meyer expansion theory was used.   □

10.4 Characteristics and Their Approximations by Weak Waves

Example 10.2 provides a case where a series of Prandtl-Meyer expansion waves for small turning angles can serve as approximation of the method of characteristics, although characteristics and waves are not the same. That is, a continuous flow over a slowly turning (i.e., curved) wall can be approximated by a series of flat segments, which are successively turned away from the flow through finite, small angles. Over the curved wall (Fig. 10.9a), as many η-characteristics as desired can be drawn for those originating from it. With the flat-segment approximation (Fig. 10.9b), the Prandtl-Meyer expansion fans divide the flow field into regions of uniform flow. Because these regions are simple regions, the changes in the flow (Mach number and direction) are found using Eqn. (10.23), which can be generalized as

$$\Delta\nu = \pm\Delta\delta \qquad (10.24)$$

where the $+$ and $-$ signs are associated for the lower wall (generating η-characteristic) and the upper wall (generating ξ-characteristic), respectively. The turning angle δ is positive/negative if it is in the counterclockwise/clockwise from the flow direction. (So in Fig. 10.9 the lower wall has $\delta < 0$ and δ is increasingly negative as the flow turns; therefore $\Delta\delta < 0$ resulting in $\Delta\nu > 0$ and acceleration/expansion.) Further approximation involves representing each expansion fan in Fig. 10.9b by a single line, as shown in Fig. 10.9c. Thus the continuous flow change through the expansion fan becomes discontinuous. The angle of the single line is the average of the Mach wave angles of the uniform flow regions upstream divided by it. (Because of the uniform flows, the characteristics in each of the simple regions are parallel straight lines, as shown in Fig. 10.8.)

Recall for the shock-expansion theory in Chapter 4, Eqn. (4.47b) can be written as $\pm\delta +$ constant $= \nu(M)$. The positive/negative sign in Eqn. (4.47a) applies when the flow turning is away from/into the flow direction, for both upper and lower walls. Therefore, if the upper wall turns away from the flow direction, $\Delta\nu$ is positive. By following the convention of Eqn. (10.24), an upper wall turning away from the flow direction (generating ξ-characteristic) has a positive value of $\Delta\delta$ as the flow crossing the ξ–characteristics (and along the η–characteristics) would observe $\nu - \delta = R =$ constant so that $\Delta\nu = \Delta\delta > 0$, resulting in expansion and acceleration.

Similarly, if the upper wall turns into the flow, Eqn. (4.47a) indicate $\Delta\nu < 0$, resulting in flow deceleration and compression. If the lower wall turns into the flow $\Delta\delta > 0$ and $\Delta\nu = -\Delta\delta < 0$, also resulting in flow deceleration and compression. By comparison, Eqn. (10.24) indicates that the lower wall generates η–characteristics, and along ξ characteristics $\Delta\nu = -\Delta\delta < 0$ leads to compression and deceleration.

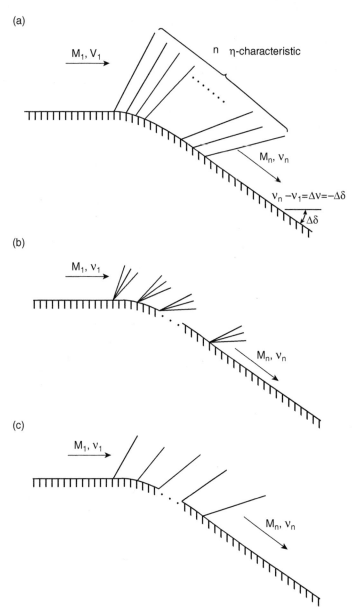

Figure 10.9 A supersonic flow experiences Prandtl-Meyer expansion: (a) over a slowly turning surface; (b) over a series of flat-segment surfaces, as an approximation of the slowly turning surface; (c) over a further refined surface approximation consisting of a larger number of smaller flat-segment surfaces, where the expansion can be represented by a single line (a Mach wave).

10.5 Application of the Method of Characteristics: Two-Dimensional Supersonic Nozzle Design

To achieve supersonic internal flows, such as in the wind tunnel and in the converging-diverging nozzle, some flow conditions are desirable. In the wind tunnel, uniform,

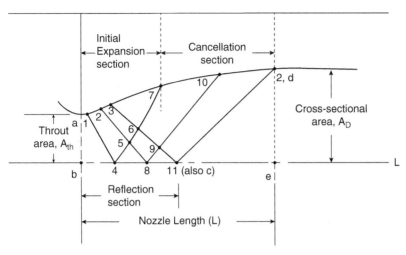

Figure 10.10 Schematic showing the application of the method of characteristics to solving the flow field within a nozzle, symmetric with respective to the centerline. (The grid numbers are those used in Example 10.3.)

parallel supersonic flows are needed in the test section. The exit flow of a converging-diverging nozzle for the propulsion purpose is supersonic and parallel to the axis of the nozzle. In both cases, $M > 1$ and $\Delta \delta = 0$ at the exit plane of the nozzle. In addition, the wall is designed so that no shock waves exists throughout the nozzle. Therefore isentropic flow assumption is assumed so that the method of characteristics can be used to design the flow field and the shape of the diverging section of the nozzle. Consider the nozzle sketched in Fig. 10.10, which is symmetric with respect to the mid-plane, and where the grid system formed by the intersecting characteristics (or waves) and the nodes is also shown. The accuracy of calculations involved in the calculations increases with the increasing number of characteristics and nodes that are chosen.

The following points are noteworthy regarding the constraints given by the design Mach number (M_D) and $\Delta \delta = 0$ at the nozzle exit plane.

(i) An initial expansion section is necessary to accelerate the sonic flow at the throat (located at the dashed line ab in Fig. 10.10) to be supersonic in the diverging section of the nozzle. The initial expansion waves are reflected from the plane of symmetry of the nozzle.

(ii) The maximum expansion angle (δ_{max}) can be arbitrarily chosen; however, along with the number of the waves chosen, the furthest centerline location (point c and thus the segment bc in Fig. 10.10) for wave reflection would then be determined accordingly. The length of the segment bc is the reflection length, L_r.

(iii) Because the region downstream of the single-line characteristic cd (the last single-line expansion wave) is a uniform flow region, the flow has the designed Mach number and is uniform and parallel to the plane of symmetry.

(iv) In the cancellation section, the wall is shaped so that the exit flow is parallel to the plane of symmetry. That is, $\delta_d = 0$. Point d thus determines the length (L) to achieve uniform flow at the nozzle exit (exit plane denoted by the dashed line de) that has M_D and is parallel to the plane of symmetry. The length is expected to be dependent upon the number and the lengths of flat segments to achieve the initial expansion.

(v) To minimize L, the initial expansion region can be shrunk to a sharp corner. If extension of the nozzle length (such as for the test section of the supersonic wind tunnel) is necessary, the additional wall needs to satisfy $\Delta\delta = 0$ at the nozzle exit to avoid formation of shock waves. One can imagine that for the upper wall shown in Fig. 10.10, further expansion ($\Delta\nu = \Delta\delta > 0$) results in the exit plane pressure being lower than that of the surroundings, which in turn causes the shock wave. On the other hand, if the nozzle extension is accompanied with $\Delta\nu < 0$, then either compression or shock wave occurs and the flow deflected without achieving the design condition.

(vi) The line segment cd has an inclination angle equal to the Prandtl-Meyer angle ($\nu_c = \nu_d$) for the design Mach number M_D (with the "design" Prandtl-Meyer angle $\nu_D = \sin^{-1}(1/M_D)$). Therefore, $\nu_c = \nu_d = \nu_D$. For the ξ-characteristics that runs from point 3 to point c,

$$\nu_3 + \delta_3 = Q_3 = \text{constant} = \nu_c + \delta_c$$

However, for the initial expansion (from $M = 1$), δ_3 is also the Prandtl-Meyer angle, so $\delta_3 = \nu_3$. Because of the symmetry, $\delta_c = 0$. Therefore,

$$\delta_3 = \frac{\nu_c}{2} = \frac{\nu_d}{2} = \frac{\nu_D}{2} \tag{10.25}$$

That is, the initial expansion angle has to be half of that of the designed Prandtl-Meyer angle. Furthermore, along the η-characteristics, line cd, the equality $\nu_c - \delta_c = Q_c = \nu_d - \delta_d$ indicates that $\delta_d = \delta_c = 0$ and that the wall at point d has to be parallel to the plane of symmetry.

(vii) The height of the nozzle exit must satisfy the value of A_e/A^* corresponding to that for M_D.

The following example illustrates the procedures that one can follow to design a supersonic nozzle using the method of characteristics with the single-line, finite-wave approximation.

EXAMPLE 10.3 Design the shape of a two-dimensional supersonic nozzle for air flow to achieve the exit Mach number of 2.0. Use the grid system shown in Fig. 10.10.

Solution –
Referring to the numbering system in Fig. 10.10, for $M_D = 2.0$, $\nu_c (= \nu_{11}) = \nu_d (= \nu_{12}) = \nu_D = 26.380°$. Therefore, $\delta_3 = \nu_D/2 = 13.190°$ according to Eqn. (12.25). For the initial expansion, three linear segments are chosen so

that each of them results in an equal amount of flow deflection, $\Delta\delta = \delta_3/3 = 4.397°$. Therefore, $\delta_1 = 4.397°$, $\delta_2 = 8.794°$, and $\delta_3 = 13.190°$. In addition, $\delta_{12} = \delta_d = 0$. Following Eqn. (10.21),

$$\nu + \delta = Q = \text{constant along } \xi - \text{characteristics (line segments 1-4, 2-5-8,}$$

$$\text{and 3-5-9-11)}$$

$$\nu - \delta = R = \text{constant along } \eta - \text{characteristics (line segments 4-5-6-7,}$$

$$\text{8-9-10, and 11-12)}$$

In the cancellation region, the wall is shaped so that $\delta_d = 0$. This condition can be achieved by having wall deflection angle of $4.397°$ each at points 7, 10, and 12. The results using the method of characteristics are tabulated below.

Point	$\nu + \delta = Q$	$\nu - \delta = R$	$\nu = \frac{1}{2}(Q + R)$	$\delta = \frac{1}{2}(Q - R)$	M	μ
1	$8.974°$	$0°$	$4.397°$*	$4.397°$* (wall)	1.23	$54.4°$
2	$17.948°$	$0°$	$8.974°$*	$8.974°$* (wall)	1.40	$45.6°$
3	$26.380°$	$0°$	$13.190°$*	$13.190°$* (wall)	1.54	$40.5°$
4	$8.974°(= Q_1)$*	$8.974°$	$8.974°$	$0°$* (symmetry)	1.40	$45.6°$
5	$17.948°(= Q_2)$*	$8.974°(= R_4)$*	$13.461°$	$4.487°$	1.55	$40.2°$
6	$26.380°(= Q_3)$*	$8.974°(= R_5)$*	$17.677°$	$8.703°$	1.70	$36.0°$
7	$26.922°$	$8.974°(= R_6)$*	$17.948°$	$8.974°(= \delta_2)$*	1.70	$36.0°$
8	$17.948°(= Q_5)$*	$17.948°$	$17.948°$	$0°$*(symmetry)	1.70	$36.0°$
9	$26.380°(= Q_6)$*	$17.948°(= R_8)$*	$22.164°$	$4.216°$	1.85	$32.7°$
10	$22.688°$	$17.948°(= R_9)$*	$22.318°$	$4.370°(= \delta_1)$*	1.86	$32.7°$
11	$26.380°(= Q_9)$*	$26.380°$	$26.380°$	$0°$* (design)	2.00	$30.0°$
12	$26.380°$	$26.380°(= R_{11})$*	$26.380°$	$0°$*(Symmetry)	2.00	$30.0°$

* Values known prior to calculations for the point.

The height of the nozzle is determined by successively determining the points of intersection between the wall extended from the upstream intersection and the incident wave. For example, point 10 is located where the straight wall extension from point 7 meets the incident wave represented by line 9–10. The extension from point 10 then intersects with wave represented by the straight line 10–11 (or cd) at point 12, thus leading to the height of the nozzle. The final check of the result is done by comparing the isentropic value of A_e/A^* ($= 1.688$ for $M_D = 2.0$). The accuracy improves as the number of linear segments of the wall is increased ($\Delta\delta$ decreased) in the initial expansion region. □

The difference between predictions by using the one-dimensional isentropic nozzle flow theory and two-dimensional method of characteristics can be seen from the results of Example 10.3. For instance, the Mach number downstream of the wave cd is uniformly equal to 2.00 ($= M_D$), while its value falls between 1.86 and 2.00. Thus, on a given cross-sectional plane between points 10 and 12, the Mach number is not uniform, in contrast to the one-dimensional prediction.

Problems

Problem 10.1 Derive Eqn. (10.17) using the continuity and momentum equations in the natural coordinate system. (*Hint*: use the definitions of the Mach wave angle, the Prandtl-Meyer angle/function, and the speed of sound.)

Problem 10.2 A supersonic flow of air in a variable channel is shown in the figure. The upper wall of the channel is curved, with the curvature starting at point A and ending at point E, while the lower wall is flat. The lines are the supposed characteristics. The incoming flow ($M = 1.5$ at the inlet plane AA') is uniform and parallel to the lower wall (and to the upper wall upstream of point A). The inclination angles along the upper wall are $\delta_A = 0, \delta_B = -4°, \delta_C = -7°, \delta_D = -10°$, and $\delta_E = 0$, with the negative angles signifying turning away from the flow. Assume the walls are perfectly insulated and frictionless.

(a) In which region does the incoming flow remain uniform and parallel?

(b) In which region does the Prandtl-Meyer solution hold?

(c) Find Mach numbers at points a, b, c, d, and P. What is the Mach number at point E (which will be the Mach number downstream of it)?

11 Unsteady One-Dimensional Flows and Nonlinear Waves

Previous chapters focused on steady flows, with the exception of linear physical acoustic wave propagation in Chapter 9. This chapter concerns gas dynamic phenomena that are unsteady in nature. To focus on the effect of unsteadiness, flows dependent on only one spatial dimension are considered in this chapter. Three types of flow are of particular interest: moving normal shock and expansion waves, formation of shocks, and interactions between waves. The acoustic wave propagation treated in Section 9.6 is due to infinitesimal disturbances (or amplitudes), and thus is a linear wave. A linear wave is characterized by a constant propagation speed, for which the small perturbation theory adequately describes the phenomenon, and the resultant effect due to interactions between multiple waves is the sum of the individual effects. Waves of finite amplitudes, whether shock or expansion, are nonlinear waves. The resultant effects of nonlinear wave interaction cannot be simply added. With the two independent variables – one spatial and one temporal – the method of characteristics can be used for solving the unsteady wave propagation problem, with the characteristics being in the x-t domain, instead of the x-y domain in the steady two-dimensional flows.

11.1 Introduction

Two scenarios of unsteady wave propagation arise: simple and non-simple flow regimes. The first scenario is exemplified by a wave (compression or expansion) moving through a quiescent or uniformly flowing medium contained in a piston-cylinder device with an infinitely long tube, as shown in Fig. 11.1a. In the scenario depicted by Fig. 11.1a, the medium ahead of the shock, propagating with an incident speed (V_{SI}) is quiescent $(V_a = 0)$ while the medium behind the shock travels at a speed V_b. There is only one wave propagating in the medium and no interaction between waves; these types of waves are called *simple waves*. The flow region due to the passing simple wave is called the *simple region* (or simple flow regime). An "intuitive" approach is illustrated in the following section for moving normal shocks by considering the relative velocity between the wave and the medium, followed by using the method of characteristics (MOC). The MOC not only solves

275

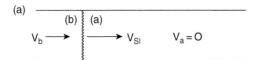

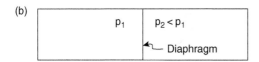

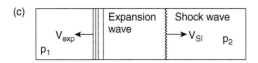

Figure 11.1 Unsteady one-dimensional wave formation and its propagation – (a) a normal shock wave propagating into a quiescent gas; (b) a diaphragm separating high- and low-pressure gases in a container; (c) after the rupture of the diaphragm, a normal shock wave forms and propagates into the low-pressure gas while an expansion forms and propagates into the high-pressure gas; (d) the normal shock and expansion waves have reflected from the end walls of the container and are now propagating into each other.

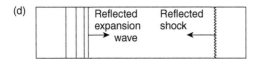

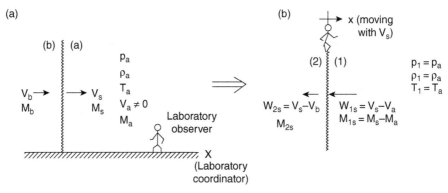

Figure 11.2 (a) A normal shock wave propagates relative to the laboratory coordinate (or to an observer fixed to the earth, or the *X*-coordinate); (b) the flow as seen by an observer riding on the shock wave affixed with the *x*-coordinate.

the unsteady wave propagation problem, but also allows tracking the formation of shock waves.

The second scenario results from interaction between two like waves (either both are compression waves or both are expansion waves) or between two unlike waves of moving waves. Wave interaction results in the non-simple flow regime. A typical example can be visualized in a long tube that is divided into high- and low-pressure sections, as separated by a diaphragm (Fig. 11.1b). Upon the rupture of the diaphragm, a compression wave travels into the low-pressure gas while an expansion wave travels into the high-pressure gas (Fig. 11.1c), where the gas ahead of the respective waves possesses the pressure prior to the rupture of diaphragm.

The pressure (and temperature) in the region between the parting waves is to be determined. For simplicity, regions immediately downstream of these two parting waves resemble those left behind by simple waves, but how and where the two regions would interface resulting from waves of two different nature, compression and expansion, will be discussed later in this chapter. Sometime later, the reflected compression and expansion waves travel in opposite directions (Fig. 11.1d) and later pass each other, resulting in a complicated interaction pattern. As a consequence the pressure and temperature in regions behind the reflected waves cannot be readily determined.

Tracking the states throughout the tube is tedious and time consuming, especially after multiple reflections have occurred. The method of characteristics is used to describe the flow field due to the non-simple nature of the flow regime.

11.2 A Normal Shock Traveling at a Constant Speed through a Medium

In Section 4.3, the phenomenon of a traveling wave was qualitatively described. It was noted that a compression wave "carries along" with it the gas it passes, while the expansion wave "repels" the gas away from it that it passes from the laboratory observer's point of view. For a quantitative description of these phenomena, first consider a traveling normal shock wave moving with a velocity V_s (the subscript s pertains to the shock here and throughout this chapter) into a gas medium that is flowing at a velocity of V_a with respective to the laboratory. The gas behind the shock wave follows the wave with a velocity of V_b with respect to the laboratory, as depicted in Fig. 11.2a, where the coordinate X is fixed to the earth (or the laboratory). For an observer traveling with the same velocity as the shock wave (i.e., "riding" on the wave), the gas is moving into the shock wave with a velocity of $W_{1s} = V_s - V_a$, leaving the gas behind the wave with a velocity of $W_{2s} = V_s - V_b$, with the subscripts 1 and 2 denoting states ahead of and behind the shock, respectively, as shown in Fig. 11.2b, where the transformed coordinate x is fixed to the shock wave and the observer and the shock wave is stationary; the problem becomes a steady one. Equation (4.11) for the stationary shock wave is applicable and can be rewritten as

$$dp = -\rho W_s dW_s \qquad (11.1)$$

which explains the carrying-along and repelling action of shock and expansion waves, respectively in Chapter 4. In the case of a moving shock, the gas behind the shock is being carried along, with a velocity V_b relative to the X-coordinate. The velocity V_b is called the *induced mass motion*. In the transformed x-coordinate system, all equations derived in Chapter 4 are applicable, with the requirements: (1) all velocities are relative to the shock wave, and (2) the thermodynamic properties in the gas ahead of the shock are static properties as the shock wave acts to compress the gas it passes.

It is important to recall that all static properties are defined as measured by a probe moving at the absolute flow velocity and are independent of the velocity of the observer, whether riding on the wave or standing in the laboratory. Therefore, by referring to Figs. 11.2a and 11.2b,

$$\frac{p_b}{p_a} = \frac{p_2}{p_1} \text{ and } \frac{T_b}{T_a} = \frac{T_2}{T_1} \tag{11.2}$$

A special case involves $V_a = 0$ (i.e., quiescent gas ahead of the shock), for which $p_a = p_1, p_b = p_2, T_a = T_1$, and $T_b = T_2$ and which automatically satisfies Eqn. (11.2). On the other hand, the stagnation properties are measured by bringing the gas to rest. In the X-coordinate system, the gas at state a is at rest and thus $T_{ta} = T_a$ and $p_{ta} = p_a$, while in the x-coordinate system the gas at state 1 needs to be brought to rest for measuring its stagnation properties. Consequently, $p_{t1} > p_{ta}$ and $T_{t1} > T_{ta}$ for the simple reason that the gas in state 1 has the relative velocity $W_{1s} = (V_s - V_a)$ or, alternatively, that the shock (i.e., the moving observer) has a compressive or ramming effect on the gas. As a consequence, $W_{2s} = (V_s - V_b) < (V_s - V_a) = W_{1s}$ and for continuity $W_{2s}/W_{1s} = \rho_1/\rho_2 = \rho_a/\rho_b < 1$. One can imagine that if the shock wave has an *isentropic* ramming effect (which is certainly not true due to the non-isentropic shock compression), an imaginary stagnation pressure exists in the gas ahead of the shock wave, with its value associated with the speed of the moving shock. This imaginary stagnation pressure will decrease across the shock, much like what happens in a gas flow passing through a stationary shock – because the compression due to a shock wave is non-isentropic.

The following example illustrates the use of coordinate transformation in solving moving shock problems.

EXAMPLE 11.1 A normal shock wave is traveling into quiescent air (with $T_1 = 216$ K and $p_1 = 19.4$ kPa) and causing a pressure ratio of 4.5. What are the air velocity, Mach number, pressure, and stagnation pressure and temperature downstream of the shock as seen by a laboratory observer? Compare the results with those of the stationary shock with the same Mach number and static properties.

Solution –
Referring to Fig. 11.2, one finds the same results for this moving shock problem for the following quantities. Note that for quiescent air ahead of the shock, $V_a = 0$, $W_{1s} = V_s$, and $M_{1s} = M_a$, where M_{1s} is the air flow Mach number relative to the moving shock.
  For $p_2/p_1 = 4.5$

$$M_a = 2.0$$

For $V_a = 0$,

$$M_{1s} = M_a = 2.0$$

$$W_{1s} = V_s = M_{1s}\sqrt{\gamma R T_a} = 2.0 \times \sqrt{1.4 \times 287 \times 216} = 590.2 \text{ m/s}$$

Therefore,

$$T_{t1}/T_1 = 1.80$$

$T_{t1} = T_{t2} = 388.80$ K (due to no heat transfer to the one-dimensional steady shock wave)

$M_{2s} = 0.5774$ (this relative to the shock wave)

$T_2/T_1 = 1.6875$; $T_2 = 364.5$ K; $\rho_2/\rho_1 = 2.667$; $p_{t2}/p_{t1} = 0.7209$

$$W_{2s} = (V_s - V_b) = M_{2s}\sqrt{\gamma R T_2} = (0.5774)\sqrt{(1.4)(288 J/kg \cdot K)(1.6875 \times 216 K)}$$
$$= 221.4 \text{ m/s}.$$

W_{2s} can also be found by using the continuity equation:

$$W_{2s} = W_{1s}\left(\frac{\rho_1}{\rho_2}\right) = M_{1s}\sqrt{\gamma R T_1}\left(\frac{\rho_1}{\rho_2}\right)\left(\frac{T_2}{T_1}\right)$$

$$= 2 \times \sqrt{(1.4)(288 J/kg \cdot K)(216 K)} \times \frac{1}{4.5} \times 1.6875$$

$$W_{2s} = 221.4 \text{ m/s}$$

$$p_{t1} = p_1\left(1 + \frac{\gamma - 1}{2} M_1^2\right)^{\gamma/(\gamma-1)} = (19.4 \text{ kPa})(7.8247) = 151.80 \text{ kPa}$$

(p_{t1} is the "imaginary" stagnation pressure.)

$$p_{t2} = 0.7209 p_{t1} = 109.43 \text{ kPa}$$

(This is the stagnation pressure downstream of a stationary shock with $M_{1s} = 2$.) Note that $(V_s - V_b) = 221.4$ m/s is the gas velocity relative to the moving shock. It is in the $-x$ direction in the transformed coordinate system. Then

$$V_s = M_{1s}\sqrt{\gamma R T_1} = 2.0 \times \sqrt{(1.4)(288 J/kg \cdot K)(216 K)} = 590.2 \text{ m/s}$$

(Check $-\frac{W_{2s}}{W_{1s}} = \frac{221.4}{590.2} = 0.375 \approx \frac{\rho_2}{\rho_1}$ with $\frac{\rho_1}{\rho_2} = 2.667$ given in the above.)

$$V_b = V_s - (V_s - V_b) = 590.2 \text{ m/s} - 221.4 \text{ m/s} = 368.8 \text{ m/s}$$

Therefore, to the laboratory observer, the gas behind the shock moves in the +X direction, that is, it is being carried along by the wave. The Mach number of the gas flow behind the shock is M_b:

$$M_b = V_b/\sqrt{\gamma R T_2} = 368.8 \text{ m/s}/\sqrt{(1.4)(288 J/kg \cdot K)(1.6875 \times 216 K)} = 0.962$$

To find the stagnation temperature T_{tb} one needs to first find T_b.

$\frac{T_b}{T_a} = \frac{T_2}{T_1}$; $T_a = T_1$, and $T_b = T_2$.

$T_b = T_2 = 364.5$ K

$\frac{T_{tb}}{T_b} = 1 + \frac{\gamma-1}{2} M_b^2 \approx 1.19$

Therefore,

$$T_{tb} = \frac{T_{tb}}{T_b} T_b = 1.19 \times 364.5 \ K = 433.76 \ K$$

The stagnation pressure behind the traveling shock is then from the table or $\frac{p_b}{p_a} = \frac{p_2}{p_1}$; $p_a = p_1$, and $p_b = p_2$ for $V_a = 0$.

$$p_b = 4.5 p_a = 4.5 p_1$$

$$\frac{p_{tb}}{p_b} == \left(1 + \frac{\gamma-1}{2} M_b^2\right)^{\gamma/(\gamma-1)} \approx 1.81. \text{ Therefore,}$$

$$p_{tb} = \frac{p_{tb}}{p_b} \frac{p_b}{p_a} p_a = 1.81 \times 4.5 \times 19.4 \ \text{kPa} = 157.14 \ \text{kPa}$$

Note that for the conditions in this example, $M_b \neq M_{2s}$ and $p_{tb} > p_{t2}$ and $T_{tb} > T_{t2} = T_{t1}$, because the stagnation properties are measured by bringing the gas flow to rest with respective to the observer, and, thus, post-shock stagnation quantities are different, in the X- and x-coordinate systems. One may make a comparison of these results with those of Example 4.3, where a stationary shock faces an incoming flow at $M_1 = 2.0$ with the same T_1 and p_1 producing $p_{t2} = 109.43 \ kPa$ downstream of the shock and $T_{t2} = T_{t1} = T_{ta} = 388.8 \ K.$ □

The results of Example 11.1 suggest that the stagnation temperature and pressure downstream for a moving shock are greater than those across a stationary shock with the same strength, β (or same Mach number), as shown by Eqn. (4.18d):

$$\beta \equiv \frac{p_2 - p_1}{p_1} = \frac{2\gamma}{\gamma + 1}\left(M_1^2 - 1\right) \tag{4.18d}$$

This comparison of stagnation properties behind the shock is opposite to that ahead of the shock. Why are there such fundamental, qualitative differences in stagnation properties in flows with steady and moving shock waves? To answer this question, it is helpful to consider for examples the one-dimensional nozzle flow with a standing normal shock and the moving shock wave due to the moving piston in a piston-cylinder device. In the case of the steady/standing shock, *no work* is done on the gas as it passes through the shock; this is inherently implied in the derivations leading to all the shock relations shown in Chapter 4, by the fact that the stagnation enthalpy remains constant across the shock, $h_{t2} = h_{t1}$. In the case of unsteady/moving shocks, a piston *moving* into the gas is *doing work* on the gas, resulting in a higher stagnation temperature, and $h_{tb} > h_{t1}$ (and $T_{tb} > T_{t1}$), for the same Mach number. Because of the work, the stagnation pressure also increases over that downstream of the steady shock caused by the same Mach number. One can also see that by the coordinate transformation from the X-coordinate to the x-coordinate, energy/work is added/done to the entire system; that is, moving the originally stagnant air, as in the state a in Fig. 11.2a (i.e., moving the entire reference frame) requires energy. In the piston-cylinder system, the energy/work is added/done by moving the piston. The work done by the piston in Example 11.1 results in $p_{tb} > p_{t1}$, even though the non-isentropic shock process should cause a decrease in stagnation pressure.

Before leaving this section, consider a more general situation of Example 11.2 below, where $V_a \neq 0$ instead of the quiescent air of Example 11.1.

EXAMPLE 11.2 Consider all the same conditions as in Example 11.1, except that $V_a = -100$ m/s (in the X-coordinate system, as shown in Fig. 11.2a), that is, the air is moving in the opposite direction of the shock wave.

Solution –

For the same pressure ratio $p_2/p_1 = 4.5$

$$M_{1s} = M_a = 2.0$$

$$M_{2s} = 0.5774$$

In the x-coordinate system, $W_{1s} = M_{1s}\sqrt{\gamma R T_1} = 590.2$ m/s $= (V_s - V_a)$

$$V_s = 490.2 \ m/s$$

$$V_s - V_b = M_{2s}\sqrt{\gamma R T_2} = 0.5774 \times \sqrt{(1.4)(288 \ J/kg \cdot K)(1.6875 \times 216 \ K)}$$
$$= 221.35 \ \text{m/s}$$

$$V_b = V_s - (V_s - V_b) = 268.9 \ \text{m/s}$$

This $V_b = 268.9$ m/s is smaller than that in Example 11.1 by exactly 100 m/s, consistent with the fact that the transformed x-coordinate does not have to advance at the same large velocity as in the case where $V_b = 0$. That is for $V_a = -100$ m/s, all gas velocities is reduced by 100 m/s in the X direction.

The static temperature and pressure remain the same as those in Example 11.1. However, the stagnation properties need to be calculated using the new Mach number M_b for the new V_b.

$$M_b = V_b/\sqrt{\gamma R T_2} = 268.9 \ \text{m/s}/\sqrt{(1.4)(288 \ J/kg \cdot K)(1.6875 \times 216 \ K)} = 0.701$$

$$\frac{p_{tb}}{p_b} == \left(1 + \frac{\gamma-1}{2}M_b^2\right)^{\gamma/(\gamma-1)} \approx 1.388. \ \text{Therefore,}$$

$$p_{tb} = \frac{p_{tb}}{p_b}\frac{p_b}{p_a}p_a = 1.388 \times 4.5 \times 19.4 \ \text{kPa} = 121.2 \ \text{kPa}$$

$$\frac{T_{tb}}{T_b} = 1 + \frac{\gamma-1}{2}M_b^2 \approx 1.098$$

$$T_{tb} = \frac{T_{tb}}{T_b}T_b = 1.098 \times 364.5 \ K = 400.3 \ K$$

The values of p_{tb} and T_{tb} are different from those obtained for $V_a = 0$ in Example 11.1.

Comments –

If $V_a = 590.2$ m/s, then $W_{1s} = 0$, resulting in a stationary shock. Then the results of Example 4.3 are recovered. In this case there is no need to add energy to move the air in the state a in Fig. 11.2a for the shock to be stationary.   □

11.3 General Relations for a Traveling Normal Shock at a Constant Speed

The previous section described an intuitive approach to the traveling normal shock problem by invoking the stationary normal shock relations, with coordinate transformation and proper accounting of relative velocity with respect to both the laboratory and the shock itself. Such an intuitive approach resorts to and reflects understanding of the physical event of a moving shock. Expanding on this understanding, the following describes a formulation of general scenarios of the relative velocity between the shock and the gas ahead of it.

Consider the transformed stationary shock problem depicted in Fig. 11.2b. The conservation laws for mass, momentum, and energy are

$$\rho_1(V_s - V_a) = \rho_2(V_s - V_b) \tag{11.3}$$

$$p_1 + \rho_1(V_s - V_a)^2 = p_2 + \rho_2(V_s - V_b)^2 \tag{11.4}$$

$$h_1 + \frac{(V_s - V_a)^2}{2} = h_2 + \frac{(V_s - V_b)^2}{2} \tag{11.5}$$

Equations (11.3) through (11.5) are identical with those for stationary normal shocks given in Chapter 4. Therefore, with similar algebraic manipulations as detailed in Chapter 4, the Rankine-Hugoniot equation for a traveling normal shock can be obtained as

$$(V_s - V_a)^2 - (V_s - V_b)^2 = W_{1s}^2 - W_{2s}^2 = (p_2 - p_1)(v_1 - v_2) \tag{11.6a}$$

$$h_2 - h_1 = \frac{1}{2}(p_2 - p_1)(v_1 - v_2) \tag{11.6b}$$

which is identical with Eqn. (4.71a) for a stationary normal shock. As expected from the reasoning given in Section 4.18, the Rankine-Hugoniot relationship relates thermodynamic quantities that do not depend on whether or not the shock wave is traveling. For a calorically perfect gas, $h = c_p T$ and $v = RT/p$. With $c_p = \gamma R/(\gamma - 1)$ Eqn. (11.6) becomes

$$\frac{T_2}{T_1} = \frac{p_2}{p_1} \left[\frac{\frac{\gamma+1}{\gamma-1} + \frac{p_2}{p_1}}{1 + \left(\frac{\gamma+1}{\gamma-1}\right)\left(\frac{p_2}{p_1}\right)} \right] \tag{11.7}$$

or

$$\frac{\rho_2}{\rho_1} = \frac{p_2}{p_1}\frac{T_1}{T_2} = \frac{\left(\frac{\gamma+1}{\gamma-1}\right)\left(\frac{p_2}{p_1}\right) + 1}{\frac{\gamma+1}{\gamma-1} + \frac{p_2}{p_1}} \tag{11.8}$$

Equations (11.7) and (11.8) are the same as Eqns. (4.72) and (4.73) for a stationary normal shock. For the transformed stationary coordinate shown in Fig. 11.2b, Eqn. (4.18b) is applicable, with the Mach number now being M_{1s}:

$$\frac{p_2}{p_1} = \frac{2\gamma M_{1s}^2}{\gamma + 1} - \frac{\gamma - 1}{\gamma + 1} \tag{11.9}$$

Therefore,

$$M_{1s} = \sqrt{\frac{\gamma + 1}{2\gamma}\left(\frac{p_2}{p_1} - 1\right) + 1} \tag{11.10}$$

and

$$W_{1s} = V_s - V_a = M_{1s}a_1 = a_1\sqrt{\frac{\gamma + 1}{2\gamma}\left(\frac{p_2}{p_1} - 1\right) + 1} \tag{11.11}$$

By using the continuity equation, Eqn. (11.3), one finds

$$W_{2s} = V_s - V_b = \frac{\rho_1}{\rho_2}(V_s - V_a)\frac{\rho_1}{\rho_2}W_{1s} \tag{11.12}$$

and

$$V_b = \left(1 - \frac{\rho_1}{\rho_2}\right)V_s + \frac{\rho_1}{\rho_2}V_a\left(\text{also} = \frac{\rho_1}{\rho_2}W_{1s}\right) \tag{11.13}$$

For the special case where $V_a = 0$, Eqn. (11.11) is reduced to

$$W_{1s} = V_s = a_1\sqrt{\frac{\gamma + 1}{2\gamma}\left(\frac{p_2}{p_1} - 1\right) + 1} = a_1\sqrt{\frac{\gamma + 1}{2\gamma}\beta + 1} \text{ (for } V_a = 0) \tag{11.14}$$

Similarly by incorporating Eqn. (4.18e) for $\beta = (p_2/p_1 - 1)$, Eqn. (11.13) becomes

$$V_b = \left(1 - \frac{\rho_1}{\rho_2}\right)V_s = \frac{a_1}{\gamma}\left(\frac{p_2}{p_1} - 1\right)\sqrt{\frac{\frac{2\gamma}{\gamma+1}}{\frac{p_2}{p_1} + \frac{\gamma-1}{\gamma+1}}} = \frac{2a_1}{\gamma + 1}\left(M_{1s} - \frac{1}{M_{1s}}\right) \text{ (for } V_a = 0)$$

$$\tag{11.15}$$

The derivation of Eqn. (11.15) is left as an exercise problem (Problem 11.1). Equation (11.13) indicates that as W_{1s} increases, the gas velocity behind the traveling shock decreases. Such a result helps to explain the results of Examples 11.1 and 11.2. Example 11.2 has a larger W_{1s} and yields a smaller V_b; it also leads to a smaller M_b, as expected, because the acoustic speed is a function of static temperature, which does not depend on whether or not the shock is traveling. The Mach number of the induced mass action of gas behind the shock observed by the laboratory (X-coordinate) observer is

$$M_b = \frac{V_b}{a_b} = \frac{V_b}{a_2} - \frac{2}{\gamma + 1}\left(M_{1s} - \frac{1}{M_{1s}}\right)\frac{a_1}{a_2} \tag{11.16}$$

because $a = \sqrt{\gamma RT}$, $a_1/a_2 = \sqrt{T_1/T_2}$ and therefore

$$M_b = \frac{2}{\gamma + 1}\left(M_{1s} - \frac{1}{M_{1s}}\right)\sqrt{\frac{T_1}{T_2}} \text{ (for } V_a = 0) \tag{11.17}$$

Attention should be paid to the use of Eqn. (11.13), because values of V_b for $V_a = 0$ and $V_a \neq 0$ are different for a given ratio, p_2/p_1. Therefore, an expression like Eqn. (11.17) is not readily available for $V_a \neq 0$, because $W_{1s} \neq V_s$ and Eqn. (11.15) cannot be used. Example 11.3 illustrates such a difference between values of V_b for $V_a = 0$ and $V_a \neq 0$. For a very strong normal shock ($p_2/p_1 \to \infty$) traveling into either a quiescent or flowing gas, it can be shown that (left as an end-of-chapter problem)

$$\lim_{p_2/p_1 \to \infty} M_b = \sqrt{\frac{2}{\gamma(\gamma - 1)}} \tag{11.18}$$

For comparison, for very a strong stationary normal shock,

$$M_2 = \sqrt{\frac{\gamma - 1}{2\gamma}} \text{ (} M_2 = M_{2s} \text{ in Fig. 11–2b)} \tag{4.21b}$$

Once again, as in results in Example 11.1, $M_b \neq M_{2s}$. For $\gamma = 1.4$, $M_b = 1.890$ and $M_2 = 0.378$. (But as shown in Example 11.2, if $V_a = -W_{1s}$ then $M_{2s} = M_b = 0.378$, i.e., the result of a stationary shock with $M_1 = M_{1s}$.) It is interesting to note from these results that the induced gas motion behind the traveling shock can be supersonic for sufficient shock strength!

EXAMPLE 11.3 Use Eqns. (11.13) and (11.17) to find out the induced velocities and Mach numbers for the conditions in Examples 11.1 and 11.2.

Solutions –
For Example 11.1, $V_a = 0$ and

$$a_1 = \sqrt{\gamma RT_1} = \sqrt{(1.4)(288 \ J/kg \cdot K)(216 \ K)} = 295.1 \text{ m/s}$$

$$V_s = W_{1s} = M_{1s}\sqrt{\gamma R T_1} = 590.2 \text{ m/s}$$

$$V_b = \frac{a_1}{\gamma}\left(\frac{p_2}{p_1} - 1\right)\sqrt{\frac{\frac{2\gamma}{\gamma+1}}{\frac{p_2}{p_1} + \frac{\gamma-1}{\gamma+1}}} = \frac{\sqrt{(1.4)(288J/kg\cdot K)(216\ K)}}{1.4} \times 3.5 \times 0.5$$

$$= 368.9\ m/s$$

$$M_b = \frac{V_b}{a_2} = \frac{1}{\gamma}\frac{a_1}{a_2}\left(\frac{p_2}{p_1} - 1\right)\sqrt{\frac{\frac{2\gamma}{\gamma+1}}{\frac{p_2}{p_1} + \frac{\gamma-1}{\gamma+1}}} = \frac{1}{\gamma}\sqrt{\frac{T_1}{T_2}}\left(\frac{p_2}{p_1} - 1\right)\sqrt{\frac{\frac{2\gamma}{\gamma+1}}{\frac{p_2}{p_1} + \frac{\gamma-1}{\gamma+1}}}$$

$$= \frac{1}{\gamma}\left(\frac{p_2}{p_1} - 1\right)\sqrt{\frac{\frac{2\gamma}{\gamma+1}}{\frac{p_2}{p_1} + \frac{\gamma-1}{\gamma+1}}}\sqrt{\frac{1 + \frac{\gamma+1}{\gamma-1}\left(\frac{p_2}{p_1}\right)}{\frac{\gamma+1}{\gamma-1}\left(\frac{p_2}{p_1}\right) + \left(\frac{p_2}{p_1}\right)^2}}$$

where Eqn. (11.7) is used for T_1/T_2 or because $M_{1s} = M_a = 2$, $T_2/T_1 = 1.6875$ from Appendix D can be used directly. Therefore,

$$M_b = \frac{1}{1.4} \times 3.5 \times \sqrt{\frac{\frac{2.8}{2.4}}{4.5 + \frac{0.4}{2.4}}}\sqrt{\frac{1 + \frac{2.4}{0.4} \times 4.5}{\frac{2.4}{0.4}(4.5) + (4.5)^2}} = 0.962$$

Alternatively, Eqn. (11.16) can be used to determine the value of M_b –

$$M_b = \frac{2}{\gamma+1}\left(M_{1s} - \frac{1}{M_{1s}}\right)\frac{a_1}{a_2} = \frac{2}{\gamma+1}\left(M_{1s} - \frac{1}{M_{1s}}\right)\sqrt{\frac{T_1}{T_2}}$$

$$= \frac{2}{2.4} \times \left(2 - \frac{1}{2}\right) \times \sqrt{\frac{1}{1.6875}} = 0.962$$

For Example 11.2, the contribution of $V_a = -100$ m/s to V_b is calculated, using Eqn. (X-13). In this case, $W_{1s} = (V_s - V_a) = 590.2$ m/s and $V_s = 490.2$ m/s

$$V_b = \left(1 - \frac{\rho_1}{\rho_2}\right)V_s + \frac{\rho_1}{\rho_2}V_a = \left(1 - \frac{p_1}{p_2}\frac{T_2}{T_1}\right)V_s + \frac{\rho_1}{\rho_2}V_a$$

$$V_b = \left(1 - \frac{1.6875}{4.5}\right) \times 490.2\ \text{m/s} + \frac{1.6875}{4.5}(-100\ \text{m/s}) = 268.8\ \text{m/s}$$

$$M_b = \frac{V_b}{a_2} = \frac{V_b}{a_2} = \frac{269.9\ \text{m/s}}{\sqrt{\gamma R T_2}} = \frac{268.8\ \text{m/s}}{\sqrt{(1.4)(288J/kg\cdot K)(1.6875 \times 216\ K)}} = 0.701$$

Comments –
The reader can find that for $V_a = -590.2$ m/s, $M_b = M_{2s} = 0.577$ (stationary shock result).   □

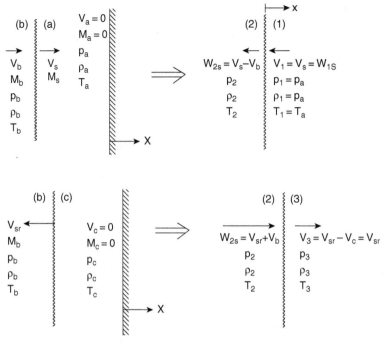

Figure 11.3 (a) A normal shock propagating toward a wall, with the flow field seen by an observer riding on the wave (the x-coordinate); (b) the reflected normal shock wave from the wall, with the flow field seen by an observer riding on the wave.

11.4 Reflected Normal Shock Waves

The reflected normal shock wave depicted in Fig. 11.1b is now further described in Fig. 11.3 with the details of the shock movements and relevant thermodynamic and flow variables. Fig. 11.3a shows the incident wave, with state ahead of the incident wave having quiescent gas while carrying along with it gas behind it at state b. The reflected wave, shown in Fig. 11.3b, travels into the gas at state b leaving gas behind it at state c. To satisfy the non-moving wall condition, $V_c = 0$, which then determines the strength of the shock wave, which is such that it brought the gas behind it to rest and is expected to be different from the strength of the incident wave. The following example illustrates the intuitive approach in finding state c, behind the reflected shock by using the stationary shock relations with coordinate transformation.

EXAMPLE 11.4 An incident normal shock wave travels in quiescent air (at 100 kPa and 300 K) in a direction perpendicular to a plane wall. The incident wave causes a pressure ratio of 2.60. Find the temperature, pressure, velocity, and Mach number in state c behind the reflected shock wave.

Solution –
Referring to Fig. 11.3a, one finds the same results for this moving shock problem for the following quantities. Note that for quiescent air ahead of the shock, $V_a = 0$, $W_{1s} = V_s$, and $M_{1s} = M_a$.

For $p_2/p_1 = 2.6$

$M_{si} = M_{1s} = M_a = 1.54$ (The subscript si denotes the incident wave.)

$V_{si} = W_{1s} = M_{si}\sqrt{\gamma R T_1} = 1.54\sqrt{(1.4)(288J/kg \cdot K)(300\ K)} = 535.6$ m/s

Therefore,

$$M_2 = 0.687$$

$$T_2/T_1 = 1.347; \quad T_2 = 404.1\text{K}$$

$$p_2/p_1 = p_b/p_a = 2.60 \text{ as given.}$$

$$W_{2s} = (V_s - V_b) = M_2\sqrt{\gamma R T_2} = (0.687)\sqrt{(1.4)(288J/kg \cdot K)(404.1\ K)}$$
$$= 277.3 \text{ m/s.}$$

$$V_b = 258.3 \text{ m/s}$$

For the reflected shock, the wall condition requires that $V_c = 0$. Therefore,

$$V_2 = V_{sr} + V_b \text{ (The subscript } sr \text{ denotes the reflected wave.)}$$

$$V_3 = V_{sr}$$

The continuity equation, Eqn. (11.3), for the reflected wave becomes

$$\rho_2(V_{sr} + V_b) = \rho_3 V_{sr}$$

This expression is rearranged to give

$$\frac{\rho_3}{\rho_2} = \frac{V_{sr} + V_b}{V_{sr}}$$

Note that

$$M_{sr} = \frac{V_{sr} + V_b}{\sqrt{\gamma R T_2}}$$

Equation (4.18d) is used to relate ρ_3/ρ_2 to M_{sr}:

$$\frac{\rho_3}{\rho_2} = \frac{(\gamma+1)M_{sr}^2}{(\gamma-1)M_{sr}^2 + 2} = \frac{(\gamma+1)\frac{(V_{sr}+V_b)^2}{\gamma R T_2}}{(\gamma-1)\frac{(V_{sr}+V_b)^2}{\gamma R T_2} + 2} = \frac{V_{sr} + V_b}{V_{sr}}$$

Rearranging the above expression yields

$$(\gamma + 1)(V_{sr} + V_b)^2 V_{sr} = (\gamma - 1)(V_{sr} + V_b)^2 V_{sr} + (\gamma - 1)(V_{sr} + V_b)^2 V_b$$
$$+ 2(\gamma RT_2)V_{sr} + 2(\gamma RT_2)V_b$$

With the values of V_b and T_2, the above equation becomes

$$2V_{sr}^3 + 930V_{sr}^2 - 245,713V_{sr} - 91,042,585 = 0$$

$$V_{sr} = 329.0 \text{ m/s}$$

$$M_2 = M_{sr} = \frac{V_{sr} + V_b}{\sqrt{\gamma RT_2}} = \frac{329.0 \text{ m/s} + 258.3 \text{ m/s}}{\sqrt{(1.4)(288 \ J/kg \cdot K)(404.1 \ K)}} \approx 1.455 \text{ and } M_3 \approx 0.72$$

It is noted that $(V_{sr} + V_b) = V_2$ and

$$V_2 = M_2\sqrt{\gamma RT_2} = 1.455 \times \sqrt{1.4 \times 287 \times 404} \approx 587.0 \text{ m/s same as } (V_{sr} + V_b).$$

$$p_3/p_2 = p_c/p_b \approx 2.30; \ p_3 = p_c = 2.30 \times 2.6 \times 100 \text{ kPa} = 598.0 \text{ kPa}$$

$$T_3/T_2 = T_c/T_b \approx 1.29; \ T_3 = T_c = 1.29 \times 1.347 \times 300 \text{ K} = 521.3 \text{ K}$$

$$V_3 = M_3\sqrt{\gamma RT_3} = 0.72 \times \sqrt{1.4 \times 287 \times 521.3} = 329.0 \text{ m/s}$$

This value of V_3 is the same as $(V_{sr} - V_c)$, where $V_c = 0$ as required by the boundary condition due to reflection. □

11.5 Isentropic One-Dimensional Waves of Finite Amplitude and Characteristics

Small perturbation theory was applied to one-dimensional acoustic wave propagation in Chapter 9, where the perturbation is negligibly small so that the entire induced flow field is isentropic. In this section, waves causing finite, but weak, changes are of concern. Examples of finite wave include the moving shock wave discussed in the previous section. A rarefaction wave causing significant density decrease is also a finite wave. Due to the finite change in temperature, the acoustic speed changes and the wave propagation and interaction become nonlinear, unlike in Chapter 9, where a constant acoustic speed is used throughout the entire flow field and the phenomenon is linear.

On the other hand, the complex wave interaction makes tracking local acoustic speed and particle velocity a complicated and tedious task; for example, if one follows the methods described in Sections 11.2 and 11.3, accounting for the relative speeds due to multiple passing of waves by using the stationary shock theory (i.e., moving waves with coordinate transformation) might be daunting. Here the finite changes are assumed to be still small so that the flow field is considered isentropic and the method of characteristics (MOC) can be extended for solving complex nonlinear propagation and interactions of waves (of the same of opposite families). The following describes the development of MOC for nonlinear wave propagation, providing an algorithmic approach to the problem. The extent to which the isentropic assumption is suitable will also be discussed.

In Section 9.6, acoustic wave propagation equation and its general solution are, respectively,

$$\frac{\partial^2 \phi}{\partial t^2} - a_1^2 \nabla^2 \phi = 0 \qquad \text{(9.49) (11.19)}$$

$$\phi = F(x - a_1 t) + G(x + a_1 t) \qquad \text{(9.50) (11.20)}$$

where ϕ is the perturbation velocity potential. The functions $F(x - a_1 t)$ and $G(x + a_1 t)$ are constant for constant values of $(x - a_1 t)$ and $(x + a_1 t)$, respectively. Again, lines corresponding to $(x - a_1 t) = $ constant and $(x + a_1 t) = $ constant are called characteristic lines or simply *characteristics* on the *x-t* plane described by $dx/dt = \pm a_1$. Equation (11.19) is derived using small-perturbation theory and is thus linear and, as a consequence, both $F(x - a_1 t)$ and $G(x + a_1 t)$ are its solutions and so are their linear combinations shown by Eqn. (11.20) is an example. It is also noted from Eqns. (11.19) and (11.20) that due to the small amplitudes (of ρ', p', T' and $\vec{v}'$) and linearization, the acoustic wave possesses the following features: (1) it travels at a constant wave velocity a_1, (2) the wave shape is permanent, and (3) they are isentropic and linear waves.

The continuity and momentum equations for finite waves can be derived from the more general form given in Chapter 8:

$$\frac{\partial \rho}{\partial t} + \nabla \cdot \left(\rho \vec{V} \right) = \frac{\partial \rho}{\partial t} + \rho \nabla \cdot \vec{V} + \left(\vec{V} \cdot \nabla \right) \rho = 0 \qquad \text{(8.1d) (11.21)}$$

$$\rho \frac{\partial \vec{V}}{\partial t} + \left(\rho \vec{V} \cdot \nabla \right) \vec{V} = -\nabla p \qquad \text{(8.2e) (11.22)}$$

These equations are nonlinear and finite waves are also nonlinear. The technique for solving these equations is the method of characteristics, as detailed in the following.

For one-dimensional flow, Eqns. (11.21) and (11.22) are reduced, respectively, to

$$\frac{\partial \rho}{\partial t} + u \frac{\partial \rho}{\partial x} + \rho \frac{\partial u}{\partial x} = 0 \qquad \text{(11.23)}$$

$$\frac{\partial u}{\partial t} + u\frac{\partial u}{\partial x} + \frac{1}{\rho}\frac{\partial p}{\partial x} = 0 \tag{11.24}$$

Assuming the medium ahead of the wave is quiescent, $\nabla \cdot s = 0$ and $dp = a^2 d\rho$. (Note that entropy changes due to the passing of finite waves.) A flow in a region with $\nabla \cdot s = 0$ is *homentropic*. Then Eqn. (11.23) becomes

$$\frac{1}{a^2}\left(\frac{\partial p}{\partial t} + u\frac{\partial p}{\partial x}\right) + \rho\frac{\partial u}{\partial x} = 0 \tag{11.25}$$

Multiplying Eqn. (11.25) by a and dividing by ρ and then by adding/subtracting the result to/from Eqn. (11.24) yields

$$\left(\frac{\partial u}{\partial t} \pm \frac{1}{\rho a}\frac{\partial p}{\partial t}\right) + (u \pm a)\left(\frac{\partial u}{\partial x} \pm \frac{1}{\rho a}\frac{\partial p}{\partial x}\right) = 0 \tag{11.26}$$

The general solution to the nonlinear Eqn. (11.26) is difficult to obtain. However, for the observer "riding" on the particle, i.e., moving with $(u + a)$,

$$dx = (u + a)dt \text{ or } \frac{dx}{dt} = u + a \tag{11.27}$$

Substituting this expression into the exact differential of u leads to

$$du \equiv \frac{\partial u}{\partial t}dt + \frac{\partial u}{\partial x}dx = \frac{\partial u}{\partial t}dt + (u + a)\frac{\partial u}{\partial x}dt = \left[\frac{\partial u}{\partial t} + (u + a)\frac{\partial u}{\partial x}\right]dt \tag{11.28}$$

Similarly,

$$dp = \left[\frac{\partial p}{\partial t} + (u + a)\frac{\partial p}{\partial x}\right]dt \tag{11.29}$$

By multiplying Eqn. (11.26) by dt, followed by invoking Eqns. (11.28) and (11.29), one finds

$$du + \frac{dp}{\rho a} = 0 \tag{11.30}$$

Similarly for the observer moving with $(u - a)$,

$$du - \frac{dp}{\rho a} = 0 \tag{11.31}$$

The reader is reminded of the resemblance that Eqns. (11.30) and (11.31) bear with the results of Problem 9.8 for the linear acoustic wave, where

$$u' = \pm\frac{p'}{\rho_1 a_1} \text{ (the } \pm \text{ signs are for right- and left-running linear acoustic waves,}$$

respectively)

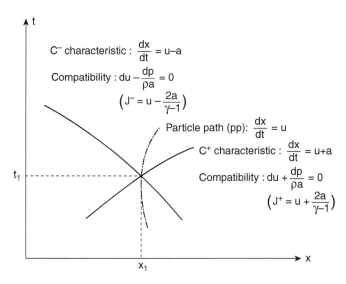

Figure 11.4 Schematic showing the characteristic lines, with their slopes, compatibility conditions, and particle path affected by propagation of a finite-amplitude, nonlinear wave.

The above derivation reduces the governing partial differential equations, Eqn. (11.23) and (11.24) and their combined form, Eqn. (11.26), to ordinary differential equations, Eqns. (11.30) and (11.31). The requirement for such a reduction is that *the rate of changes are observed by riding on particles moving with velocities*, $(u \pm a)$. The trajectory or these particles on an *x-t* plot is shown in Fig. 11.4.

Equations (11.30) and (11.31) hold only along the following characteristic lines:

$$C^+ \text{ characteristic line } : \quad \frac{dx}{dt} = u + a$$

$$C^- \text{ characteristic line } : \quad \frac{dx}{dt} = u - a$$

Equations (11.30) and (11.31) are called the *compatibility equations* along the C^+ and C^- characteristic lines, respectively. It is noted that the characteristic lines shown in Fig. 11.4 are in general not straight lines, owing to the interaction between the finite-amplitude waves passing each other, resulting in changes in both u and a. In contrast, characteristics of linear acoustic waves described in Chapter 9 are straight lines with $dx/dt = a_1 = $ constant.

Integrating Eqn. (11.30) along a C^+ characteristic yields

$$u + \int \frac{dp}{\rho a} = u + F = \text{const.} = J^+ \tag{11.32}$$

The $\nabla \cdot s = 0$ condition allows that $a^2 = \gamma R T = \gamma p / \rho$ and $\rho = \gamma p / a^2$. It can be seen that in Eqn. (11.32) $F = f(p, \rho)$ is itself a thermodynamic property. Following Eqn. (4.4),

$$p = c_1 T^{\gamma/(\gamma-1)} = c_2 a^{2\gamma/(\gamma-1)} \tag{11.33}$$

where c_1 and c_2 are constants. Therefore,

$$dp = c_2 \left(\frac{2\gamma}{\gamma - 1} \right) a^{(\gamma+1)/(\gamma-1)} da \tag{11.34}$$

and

$$\rho = c_2 \gamma a^{2/(\gamma-1)} \tag{11.35}$$

Substituting Eqns. (11.34) and (11.35) into Eqn. (11.32) and integrating leads to

$$u + \frac{2a}{\gamma - 1} = \text{const.} = J^+ \ (\text{along a } C^+ \text{ characteristic}) \tag{11.36a}$$

Similarly

$$u - \frac{2a}{\gamma - 1} = \text{const.} = J^- \ (\text{along a } C^- \text{ characteristic}) \tag{11.36b}$$

J^+ and J^- in Eqns. (11.36a) and (11.36b) are called the *Riemann invariants* for constant values of γ (γ is assumed to be constant during the integration process). Therefore, a characteristic *carries on it* information that relates not only u and a but also changes in them. It is left as an exercise to show that

$$F \equiv \int \frac{dp}{\rho a} = \frac{2a}{\gamma - 1} \tag{11.37}$$

Therefore, along the a C^+ and a C^- characteristics (i.e., for an observer riding with the wave propagation speeds $(u + a)$ and $(u - a)$), respectively,

$$\frac{D^+}{Dt}(u + F) \equiv \left[\frac{\partial}{\partial t} + (u + a) \frac{\partial}{\partial x} \right] (u + F) = 0 \ (\text{along a } C^+ \text{ characteristic}) \tag{11.38a}$$

$$\frac{D^-}{Dt}(u + F) \equiv \left[\frac{\partial}{\partial t} + (u - a) \frac{\partial}{\partial x} \right] (u - F) = 0 \ (\text{along a } C^- \text{ characteristic}) \tag{11.38b}$$

where $D^\pm/Dt$ represent the rate of change to the observer riding on the wave, or traveling with a velocity of $(u + a)$ and $(u - a)$, respectively. Such rates of change are analogous to the material or substantial derivatives in general fluid flow, i.e., $D/Dt \equiv [\partial/\partial t + u(\partial/\partial x)]$. Equations (11.38a) and (11.38b) indicate that $(u \pm F)$ remains unchanged on each of $C^\pm$ characteristics, respectively. In the limit of waves with negligible amplitude ($u \ll a$), i.e., linear acoustic waves,

$$\frac{D^\pm}{Dt} = \frac{\partial}{\partial t} \pm a \frac{\partial}{\partial x} \tag{11.39}$$

with general solutions having the functional forms of $F(x - a_1 t)$ or $G(x + a_1 t)$, as Eqn. (9.50) has demonstrated. Nonlinear (finite-amplitude) waves thus carry information at a velocity of $(u \pm a)$, as opposed to the acoustic velocity a of linear waves. It is now possible to solve Eqns. (11.35) and (11.36) for u and a:

$$a = \frac{\gamma - 1}{4}(J^+ - J^-) \tag{11.40}$$

$$u = \frac{1}{2}(J^+ + J^-) \tag{11.41}$$

Thus at any given point on the x-t plane, u and a can be solved in Eqns. (11.40) and (11.41) given known values of J^+ and J^-. The solution is schematically shown in Fig. 11.4, where the solution leading to the particle path is also shown by the curve with $dx/dt = u$. The particle velocity u is due to the induced mass motion, similar to V_b in Sections 11.2 and 11.3.

11.6 Types of Waves and Interactions

Several observations of the current unsteady nonlinear wave propagation can be made. First, the calculation procedure is similar to that using the method of characteristics for two-dimensional isentropic (and irrotational) flows discussed in Chapter 10. While the Riemann invariants in Chapter 10 are $(\nu + \delta)$ and $(\nu - \delta)$, they are now J^+ and J^-. The *data curve* or the *initial value line* (as in Chapter 10) will necessarily provide the initial data for calculations further along the characteristics and into the t-x space, as demonstrated later by Example 11.6.

Secondly, the simple (wave) region exists where there is no interaction between waves, and a simple wave is one that propagates into a region of uniform flow (or no flow at all). In the simple wave region the flow is determined by either J^+ or J^-, depending on the direction of propagation. As a consequence, both u and a are constant along a given characteristic and the characteristic is a straight line. An example of a simple wave region is given in Fig. 11.5a, which shows disturbances generated at some point in a uniform flow, with a speed of u_o, in a long tube. To the laboratory observer the disturbances are carried by the wave traveling at a speed of $(u + a)$ with the leading wave front at a speed of $(u_o + a_o)$. Note that the acoustic speed may vary with time (e.g., a sinusoidal disturbance creates alternatively compression and expansion waves causing an increase or decrease in the acoustic speed). resulting in different slopes for the C^+ characteristics on the t-x plane, shown in Fig. 11.5b. Along each of the C^+ characteristics (propagating in the $+x$ direction, or right-running) is a respective constant $J^+ = u + 2a/(\gamma - 1) = $ constant, and the constant varies from one to the other characteristic. On the other hand, all the C^- characteristics (propagating in the $-x$ direction, or left-running) have the same constant for $J^- (= u_o - 2a_o/(\gamma - 1) = u - 2a/(\gamma - 1))$ as they originate from the uniform flow region where both u and a are constants. An infinite number of C^-

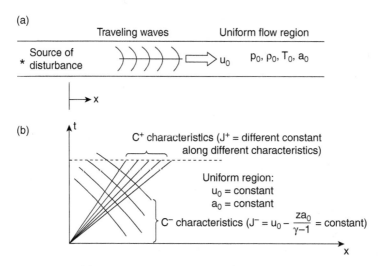

Figure 11.5 (a) Generation and propagation of a simple wave; (b) characteristics on the *t-x* plane with their Riemann invariants.

characteristics can be drawn on the *t-x* plane. Thus one can conclude that $J^- = u_o - 2a_o/(\gamma - 1) =$ constant *everywhere* and that the flow is determined by J^+).

Third, in a non-simple (wave) region, wave interaction (passing each other) takes place. An example is shown in Fig. 11.6a, where a diaphragm divides a high- and low-pressure zones in a tube with both ends closed, and the compression and expansion waves are generated after the rupture of the diaphragm (Fig. 11.6b), Sometime later the reflected waves from the end walls (Fig. 11.6c) pass through each other (Fig. 11.6d). Figure 11.6d shows the non-simple wave region, where the right- and left-running waves (corresponding to the reflected expansion and compression waves, respectively) intersect. At the point of intersection, the flow is determined by both J^+ or J^-. One can further see the complexity of interaction by extending the event of Fig. 11.6d to a later moment when the left-running compression wave is reflected from the left wall to become right-running and meet with the left-running expansion wave reflected from the right end wall (Fig. 11.6e). By this time, the compression wave has been weakened while the expansion wave also has a diminished rarefying capability, resulting from their encounter (Fig. 11.6d) with the other.

For a more quantitative description of wave interactions, consider the example shown in Fig. 11.7a. In Fig. 11.7, compression and expansion waves interact due to the action of piston 1 continuously accelerating to the right to reach a constant speedu_p, generating a series of compression waves that travels to the right and piston 2 also continuously accelerating to the right to reach a constant speed $u_p/2$, generating a series a series of expansion waves that travels to the left. For illustration, assume that no reflected wave from either piston surface has yet occurred and that the two pistons reach their respective steady-state speed at the same time. Consider three characteristics generated by each piston action at three different instants: at

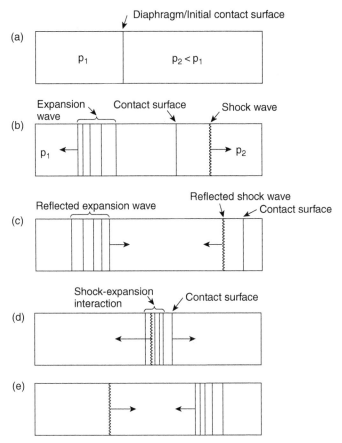

Figure 11.6 (a) A diaphragm separating high- and low-pressure gases in a tube; (b) after the rupture of the diaphragm, a normal shock wave forms and propagates into the low-pressure gas while an expansion forms and propagates into the high-pressure gas, with the contact surface being carried along by the shock wave (or being repulsed away from the expansion wave); (c) reflected normal shock and expansion waves; (d) interacting shock and expansion waves, forming non-simple wave region; (e) impending second interaction of shock and expansion waves.

the beginning, in the middle, and at the end of acceleration. The three right-running characteristics generated by piston 1 emanate from points, A, B, and C on the *t-x* plane, and the three left-running characteristics by piston from point D, E, and F. Several qualitative features of Fig. 11.7b are summarized in the following, which also help to point out the need of repeatedly using the method of characteristics for quantitative analysis of the flow field and wave propagation.

(1) Because of the compression action of piston 1, the slopes of the three characteristics prior to interaction are such that $(dx/dt)_{C-3} > (dx/dt)_{B-2} > (dx/dt)_{A-1}$. The slopes of these right-running characteristics are $(dx/dt) = (u + a)$. Equation (11.33), $p = c_1 T^{\gamma/(\gamma-1)} = c_2 a^{2\gamma/(\gamma-1)}$, indicates increases in T and thus in a on the characteristic generated following characteristic A-1. The induced

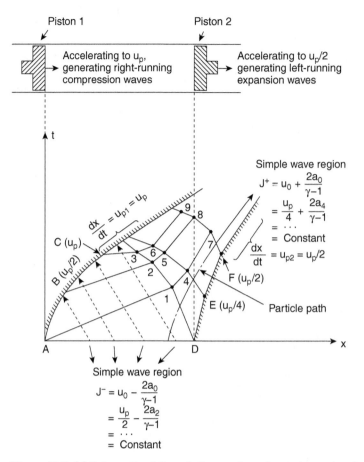

Figure 11.7 (a) Schematic of the relative motion of two pistons in a long tube; (b) characteristic waves formed due to piston 1 moving with a constant speed u_p and piston 2 with $u_\mathrm{p}/2$, both to the right; also included is an example of particle path. The grid numbering is used in Example 11.5.

      mass motion also increases from A-1 to C-3 due to the increased piston speed. Therefore, the slope increases from A-1 to C-3 due to (1) an increase in u by the increased piston speed and (2) an increase in a resulting from compression.

(2) Similar reasoning can be made for the left-running characteristics. These characteristics have slopes $(dx/dt) = (u - a)$ with magnitudes in the order of $|dx/dt|_{D-1} > |dx/dt|_{E-4} > |dx/dt|_{F-7}$. Such an order arises due to two factors: (1) an increase in u in the positive x direction by the piston motion and (2) decreases in T and thus in a on characteristics from D-1 to F-7 caused by the passing expansion/rarefaction waves.

(3) It is noted that the regions A-1–2-3-C-B-A and D-E-F-7-4-1-D on the t-x plane are simple wave regions, where the qualitative trend of the slope of a characteristic can be easily explained as the slope changes due to either compression or expansion/rarefaction. The region A-D-1-A is the undisturbed region. The qualitative trend of the slopes of the rest of the characteristic line

segments (1–4, 1–2, 2–3, 3–6, 4–5, . . ., etc.) cannot be easily determined as they are now in the non-simple wave region, where C^+ and C^- characteristics intersect. They will have to be determined by repeatedly using the method of characteristics. In the case depicted by Fig. 11.7, the relative magnitude of their slopes depend on the relative piston speeds of $(dx/dt)_{p1}$ and $(dx/dt)_{p2}$.

(4) Whether in simple or non-simple wave regions, the local magnitudes of slopes, $(u + a)$ and $(u - a)$, (in other words, u and a) are needed to determine the location of wave intersection and to construct the network of characteristics. The following example illustrates the successive use of the method of characteristics for quantitative results of the flow field and wave propagation.

EXAMPLE 11.5 Find the slopes of line segments A-1, B-2, D-1, E-4, 1–2, and 1–4 in Fig. 11.7b (i.e., find the gas and acoustic velocities at each of the intersection locations of characteristics). Assume that both pistons are continuously moved to reach their respective steady-state speed over the same period of time that is sufficiently short so that no reflected waves from each piston have occurred.

Solutions –

The gas between the two piston is initially at rest; therefore $u_o = 0$ and $a = a_o$.

Lines	Riemann Invariants	Known	Solutions	Slope
A-1	$J^+ = u_o + \frac{2a_o}{\gamma-1} = u_1 + \frac{2a_1}{\gamma-1}$	$u_o = 0$	$u_1 = u_o = 0$	$\left(\frac{dx}{dt}\right)_{A-1} = u_1 + a_1 = a_o$
D-1	$J^- = u_o - \frac{2a_o}{\gamma-1} = u_1 - \frac{2a_1}{\gamma-1}$	a_o	$a_1 = a_o$	$\left(\frac{dx}{dt}\right)_{D-1} = u_1 - a_1 = -a_o$
A-B	$J^- = u_A - \frac{2a_A}{\gamma-1} = u_B - \frac{2a_B}{\gamma-1}$	$u_A = 0$	$a_B = a_o + \frac{\gamma-1}{4} u_p$	
B-2	$J^+ = u_B + \frac{2a_B}{\gamma-1} = u_2 + \frac{2a_2}{\gamma-1}$	$u_B = \frac{u_p}{2}$	$u_2 = \frac{1}{2} u_p$	$\left(\frac{dx}{dt}\right)_{B-2} = u_2 + a_2 = a_o + \frac{\gamma+1}{4} u_p$
		$a_A = a_o$	$a_2 = a_o + \frac{\gamma-1}{4} u_p$	
1–2	$J^- = u_1 - \frac{2a_1}{\gamma-1} = u_2 - \frac{2a_2}{\gamma-1}$			$\left(\frac{dx}{dt}\right)_{1-2} = u_2 - a_2 = -a_o + \frac{3-\gamma}{4} u_p$
B-C	$J^- = u_B - \frac{2a_B}{\gamma-1} = u_C - \frac{2a_C}{\gamma-1}$	$u_C = u_p$	$a_C = a_o + \frac{\gamma-1}{2} u_p$	
C-3	$J^+ = u_C + \frac{2a_C}{\gamma-1} = u_3 + \frac{2a_3}{\gamma-1}$		$u_3 = u_p$	$\left(\frac{dx}{dt}\right)_{C-3} = u_3 + a_3 = a_o + \frac{1+\gamma}{2} u_p$
			$a_3 = a_o + \frac{\gamma-1}{2} u_p$	
2–3	$J^- = u_2 - \frac{2a_2}{\gamma-1} = u_3 - \frac{2a_3}{\gamma-1}$			$\left(\frac{dx}{dt}\right)_{2-3} = u_2 - a_3 = -a_o + \frac{3-\gamma}{2} u_p$
D-E	$J^+ = u_D + \frac{2a_D}{\gamma-1} = u_E + \frac{2a_E}{\gamma-1}$	$u_D = 0$	$a_E = a_o - \frac{\gamma-1}{8} u_p$	
1–4	$J^+ = u_1 + \frac{2a_1}{\gamma-1} = u_4 + \frac{2a_4}{\gamma-1}$	$u_E = \frac{u_p}{4}$	$u_4 = \frac{1}{4} u_p$	$\left(\frac{dx}{dt}\right)_{1-4} = u_4 + a_4 = a_o + \frac{3-\gamma}{8} u_p$
		$a_D = a_o$	$a_4 = a_o - \frac{\gamma-1}{8} u_p$	
E-4	$J^- = u_E - \frac{2a_E}{\gamma-1} = u_4 - \frac{2a_4}{\gamma-1}$			$\left(\frac{dx}{dt}\right)_{E-4} = u_4 - a_4 = -a_o + \frac{\gamma+1}{8} u_p$
2–5	$J^+ = u_2 + \frac{2a_2}{\gamma-1} = u_5 + \frac{2a_5}{\gamma-1}$		$u_5 = \frac{3}{4} u_p$	$\left(\frac{dx}{dt}\right)_{2-5} = u_5 + a_5 = a_o + \frac{\gamma+5}{8} u_p$
4–5	$J^- = u_4 - \frac{2a_4}{\gamma-1} = u_5 - \frac{2a_5}{\gamma-1}$		$a_5 = a_o + \frac{\gamma-1}{8} u_p$	$\left(\frac{dx}{dt}\right)_{4-5} = u_5 - a_5 = -a_o + \frac{7-\gamma}{8} u_p$
3–6	$J^+ = u_3 + \frac{2a_3}{\gamma-1} = u_6 + \frac{2a_6}{\gamma-1}$		$u_6 = \frac{5}{4} u_p$	$\left(\frac{dx}{dt}\right)_{3-6} = u_6 + a_6$
			$a_6 = a_o + \frac{3(\gamma-1)}{8} u_p$	$= a_o + \left[\frac{3\gamma+7}{8}\right] u_p$

(continued)

(*continued*)

Lines	Riemann Invariants	Known	Solutions	Slope
5–6	$J^- = u_5 - \frac{2a_5}{\gamma-1} = u_6 - \frac{2a_6}{\gamma-1}$			$\left(\frac{dx}{dt}\right)_{5-6} = u_6 - a_6$ $= -a_o + \left[\frac{13-3\gamma}{8}\right]u_p$
F-F	$J^+ = u_E + \frac{2a_F}{\gamma-1} = u_F + \frac{2a_F}{\gamma-1}$	$u_F = \frac{u_p}{2}$	$a_F = a_o - \frac{\gamma-1}{4}u_p$	
4–7	$J^+ = u_4 + \frac{2a_4}{\gamma-1} = u_7 + \frac{2a_7}{\gamma-1}$		$u_7 = \frac{1}{2}u_p$ $a_7 = a_o - \frac{\gamma-1}{4}u_p$	$\left(\frac{dx}{dt}\right)_{4-7} = u_7 + a_7$ $= a_o + \left[\frac{3-\gamma}{4}\right]u_p$
F-7	$J^- = u_F - \frac{2a_F}{\gamma-1} = u_7 - \frac{2a_7}{\gamma-1}$			$\left(\frac{dx}{dt}\right)_{4-7} = u_7 + a_7$ $= a_o + \left[\frac{\gamma+1}{4}\right]u_p$
5–8	$J^+ = u_5 + \frac{2a_5}{\gamma-1} = u_8 + \frac{2a_8}{\gamma-1}$		$u_8 = u_p$ $a_8 = a_o$	$\left(\frac{dx}{dt}\right)_{5-8} = u_8 + a_8$ $= a_o + u_p$
7–8	$J^- = u_7 - \frac{2a_7}{\gamma-1} = u_8 - \frac{2a_8}{\gamma-1}$			$\left(\frac{dx}{dt}\right)_{7-8} = u_8 - a_8$ $= -a_o + u_p$
6–9	$J^+ = u_6 + \frac{2a_6}{\gamma-1} = u_9 + \frac{2a_9}{\gamma-1}$		$u_9 = \frac{3}{2}u_p$ $a_9 = a_o + \frac{(\gamma-1)}{4}u_p$	$\left(\frac{dx}{dt}\right)_{6-9} = u_9 + a_9$ $= a_o + \left[\frac{7-\gamma}{4}\right]u_p$
8–9	$J^- = u_8 - \frac{2a_8}{\gamma-1} = u_9 - \frac{2a_9}{\gamma-1}$			$\left(\frac{dx}{dt}\right)_{6-9} = u_9 - a_9$ $= -a_o + \left[\frac{\gamma+5}{4}\right]u_p$

Comments –
It is interesting to note that $u_8 = u_p$ and $a_8 = a_o$. At point 8, the intersecting J^+ and J^- characteristics originate from points B and F, respectively, where the amount of piston advancing and withdrawing are equal in magnitude ($= u_p/2$) and cancel out each other. Because the difference in the piston advancing speed for the points of interest has the unit of $u_p/8$, one can see that variations in the acoustic and gas velocity are multiples of $(\gamma - 1)u_p/8$ and $u_p/8$, respectively. The increment in the acoustic velocity is smaller than that in the gas velocity by a factor of $(\gamma - 1)$.   □

It is useful to revisit the qualitative trend regarding the slope of characteristics, especially those in the non-simple wave region, based on the quantitative results of Example 11.5. The following observation can be made.

(1) A left-running characteristic (i.e., of the C^- family) passing through successive C^+ characteristics (here the compression waves) experiences increases in both u and a. Consider for example the characteristic D-1-2-3 that experiences three compression waves emanating from Points A, B, and C. As shown in Example 11.5, $u_1 < u_2 < u_3$ and $a_1 < a_2 < a_3$. The increase in u from point 1 to point 3 is due to the action of piston 1. As comments of Example 11.5 indicate, the increase in a from point 1 to point 3 is smaller than that in gas velocity u. Consequently, the absolute values of slopes of these characteristic line segment are such that $|dx/dt|_{D-1} < |dx/dt|_{1-2} < |dx/dt|_{2-3}$. Similar qualitative observations can be made for the other two characteristics of the C^- family: E-4-5-6 and F-7-8-9. Therefore, the left-running (C^-) characteristics accelerate as they propagate into the gas being compressed by the motion of piston 1.

(2) A right-running characteristic (i.e., of the C^+ family) passing through successive C^- characteristics experiences increases in u and decreases in a. Consider for example the characteristic A-1-4-7 that experiences three rarefaction waves emanating from Points D, E, and F. As shown in Example 11.5, $u_1 < u_4 < u_7$ and $a_1 > a_4 > a_7$. Therefore, $|dx/dt|_{A-1} < |dx/dt|_{1-4} < |dx/dt|_{4-7}$. Similar qualitative observations can be made for the other two characteristics of the C^+ family: B-2-5-8 and C-3-6-9. One can conclude that the right-running (C^+) characteristics decelerate as they propagate into the gas being rarefied by the motion of piston 2.

(3) To improve the accuracy and resolution of the flow region, more characteristics can be added. The number of both C^+ and C^- characteristics can be increased to a number n by, for example, choosing characteristics with velocity increments of u_p/n and $u_p/2n$, respectively, on the advancing piston 1 and withdrawing piston 2 (so that there are $(n+1)$ characteristics for each of the C^+ and C^- families).

(4) The reader is encouraged to make observations for the scenario where the advancing motion is twice as weak as the withdrawing motion ($u_p/2$ vs. u_p), which is left as an exercise (Problem 11.5).

It is again noted that for the simple wave region A-1-2-3-C-B-A shown in Fig. 11.7b, the flow is determined by the Riemann invariant J^+, because J^- is uniformly constant in the region it originates from the uniform region A-D-1-A where both the flow and acoustic velocity are constant (u_o and a_o, respectively). The following example demonstrates how the flow can be determined in the simple wave regions.

EXAMPLE 11.6 Determine the acoustic velocities at points B, C, E, and F in Fig. 11.7b.

Solutions –
Consider a C^- characteristic that originates from the uniform region and intersects the surface of piston 1 at point B and C. Then

$$J^- = u_o - \frac{2a_o}{\gamma - 1} = \text{const.} = u_B - \frac{2a_B}{\gamma - 1}$$

Because $u_o = 0$ and $u_B = \frac{u_p}{2}$,

$$a_B = a_o + \frac{\gamma - 1}{4} u_p$$

Similarly, for point C

$$J^- = u_o - \frac{2a_o}{\gamma - 1} = \text{const.} = u_C - \frac{2a_C}{\gamma - 1}$$

The fact that $u_C = u_p$ leads to

$$a_C = a_o + \frac{\gamma - 1}{2} u_p$$

Now consider a C^+ characteristic that originates from the uniform (undisturbed) region and intersects the surface of piston 2 at point E and F. Then

$$J^+ = u_o + \frac{2a_o}{\gamma - 1} = \text{const.} = u_E + \frac{2a_E}{\gamma - 1}$$

Because $u_o = 0$ and $u_E = \frac{u_p}{4}$

$$a_E = a_o - \frac{\gamma - 1}{8} u_p$$

For point F,

$$J^+ = u_o + \frac{2a_o}{\gamma - 1} = \text{const.} = u_F + \frac{2a_F}{\gamma - 1}$$

With $u_o = 0$ and $u_F = \frac{u_p}{2}$

$$a_F = a_o - \frac{\gamma - 1}{4} u_p$$

Comments –

These acoustic velocities determined in Example 11.6 using the simple wave theory are identical with those obtained in Example 11.5 assuming that the piston path line on the *t-x* plane is a characteristic. For each of the simple wave regions, A-1-2-3-C-B-A and D-E-F-7-4-1-D, one can draw multiple characteristics originating from the uniform flow region. These characteristics reach the surface of the two pistons. Each of the Riemann invariants is therefore constant, J^- and J^+, respectively, for A-1-2-3-C-B-A and D-E-F-7-4-1-D. Thus, *the piston path line is a characteristic*, when it is still in the simple wave region.   □

11.7 Centered Expansion

In Fig. 11.8a, the expansion/rarefaction is caused by a continuously withdrawn piston in a long tube filled with initially quiescent gas (similar to Piston 2 described in Fig. 11.7), where the C^+ characteristics emanate from the curved *t-x* path line of the piston. Imagine that the piston motion is impulsively achieved, the curvature of the piston path line is zero and all the characteristics emanate (or rather for this case, radiate) from a single point, as shown in Fig. 11.8b. This idealized wave system is used to illustrate the variation of the flow and gas properties throughout the field.

For the impulsively withdrawn piston motion described in Fig. 11.8b, let its location be $X_p(t)$ such that

$$\frac{dX_p}{dt} = u_p \ (< \ 0 \text{ in Fig. 11.8})$$

The piston motion depicted in Fig. 11.8 generates right-running, C^+-family characteristics that form a centered expansion fan. By setting the velocity increment

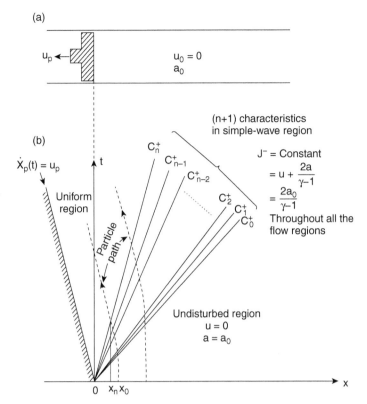

Figure 11.8 (a) Schematic of piston motion with a constant speed away from the initially quiescent gas in a long tube, generating centered expansion/rarefaction waves into the gas; (b) characteristic waves formed due to the piston moving with a constant speed u_p; also included are two examples of particle path.

between any two adjacent characteristics to be u_p/n, there are $(n+1)$ characteristics with the leading and trailing ones denoted by C_0^+ and C_n^+, respectively. Ahead of the C_0^+ characteristic (i.e., $x/a_ot \geq 1$) is the undisturbed region where $u=0$ and $a=a_o$. Downstream of the C_n^+ characteristic (i.e., $x/a_ot \geq u_p/a_o$) is the uniform region where the gas has attained the velocity of the piston, that is, on and downstream of the C_n^+ characteristic $u=u_p$. Between these two regions lies the expansion fan comprised of simple waves, whose fronts are represented by C_0^+, C_1^+, ... C_n^+ in Fig. 11.8b. Recall that on each characteristic, the local gas velocity $u=$ constant and the local acoustic speed $a=$ constant.

Therefore, each of these C^+ characteristics is a straight line, as $dx/dt = (u+a) =$ constant. Because of the simple-wave nature,

$$ J^- = u - \frac{2a}{\gamma - 1} = \text{constant} = u_o - \frac{2a_o}{\gamma - 1} \quad \text{(throughout the entire flow region)} $$

With this constant value of J^-, the acoustic speed on each characteristic can be calculated. For example on the C_n^+ characteristic, $u_n = u_p$ leads to $a_n = [a_o + (\gamma - 1)u_p/2]$. In general, on the ith characteristic

$$u_i = \frac{i u_p}{n} \tag{11.42a}$$

Upon substituting Eqn. (11.42a) into the above equation for J^-, one obtains

$$a_i = a_o + \frac{i(\gamma - 1)}{2n} u_p \quad (i = 0, 1, 2, \ldots, n) \tag{11.42b}$$

$$\left(\frac{dx}{dt} \right)_i = u_i + a_i = a_o + \frac{i(\gamma + 1)}{2n} u_p$$

Eqn. (11.42) provides flow condition *along characteristics*. For a given n the flow field is determined with the associated resolution and accuracy. One the other hand, it is desirable to know the flow and acoustic velocities *at an arbitrary point* (x, t) on the t-x plane. The procedure briefly discussed so far in this section leads to:

$$\frac{u}{a_o} = 0$$

$$\frac{a}{a_o} = 1 \text{ for } \frac{x}{a_o t} \geq 1 \text{ (i.e., the undisturbed region)} \tag{11.43}$$

The C_n^+ characteristic provides the boundary for the uniform region, so that

$$\frac{dx}{dt} = \frac{x}{t} = u_p + a = u_p + \left[a_o + \frac{(\gamma - 1)}{2} u_p \right] = a_o + \frac{\gamma + 1}{2} u_p = \left(\frac{dx}{dt} \right)_n$$

The simple wave (i.e., the fan) region is therefore defined by the following expression:

$$1 \geq \frac{x}{a_o t} \geq \frac{u_p}{a_o} + \frac{\gamma + 1}{2} \frac{u_p}{a_o} \text{ and } a_i = a_o + \frac{i(\gamma - 1)}{2n} u_p$$

In the uniform region, Eqn. (11.42) yields

$$\frac{u}{a_o} = \frac{u_p}{a_o}$$

$$\frac{a}{a_o} = 1 + \frac{\gamma - 1}{2} \frac{u_p}{a_o} \text{ for } 1 + \frac{\gamma + 1}{2} \frac{u_p}{a_o} \geq \frac{x}{a_o t} \geq \frac{u_p}{a_o} \text{ (i.e., the uniform region)} \tag{11.43}$$

In the fan (i.e., simple wave) region,

$$\frac{dx}{dt} = \frac{x}{t} = u_p + a = u_p + \left[a_o + \frac{(\gamma - 1)}{2} u_p \right] = a_o + \frac{(\gamma + 1)}{2} u_p$$

Solving for u in terms of x and t and then substituting the result in solution for a yields

$$\frac{u}{a_o} = -\frac{2}{\gamma+1}\left(1 - \frac{x}{a_o t}\right) \tag{11.44a}$$

$$\frac{a}{a_o} = 1 - \frac{\gamma-1}{\gamma+1}\left(1 - \frac{x}{a_o t}\right) \text{ for } 1 \geq \frac{x}{a_o t} \geq 1 + \frac{\gamma+1}{2}\frac{u_p}{a_o} \text{ (i.e., the fan region)} \tag{11.44b}$$

The derivation of these two expressions is left as an exercise (Problem 2.27). The velocity thus obtained is now used to determine the *particle trajectory (PT)*, X_{PT}, as a function of time. Attention is paid to the fan region, as the particle trajectory in the other two regions is readily known due to the uniform and undisturbed conditions there. Because the characteristic is a straight line, Eqn. (11.44a) can be written as

$$u = \frac{x}{t} \left(\text{or } \frac{dx}{dt}\right) = -\frac{2a_o}{\gamma+1} + \frac{2}{\gamma+1}\frac{x}{t} \tag{11.45}$$

which is a linear differential solution. One may choose to find a particular solution for

$$\frac{dx}{dt} = \frac{x}{t} = -\frac{2a_o}{\gamma+1} + \frac{2}{\gamma+1}\frac{x}{t} \tag{*}$$

and a general solution for

$$\frac{dx}{dt} = \frac{2}{\gamma+1}\frac{x}{t}$$

By solving the above expression (*) for x/t, the particular solution in the fan region is

$$\frac{x}{t} = -\frac{2a_o}{\gamma-1} \text{ for } 1 \geq \frac{x}{a_o t} \geq 1 + \frac{\gamma+1}{2}\frac{u_p}{a_o} \text{ (i.e., the fan region)}$$

while the general solution is

$$x = Kt^{2/(\gamma+1)}$$

with K being a constant to be determined. Combining the particular and general solutions yields

$$x = -\frac{2}{\gamma-1}a_o t + Kt^{2/(\gamma+1)} \tag{11.46}$$

For the initial condition, assume that the expansion wave passes a particle located at $x_o (=a_o t_o)$ with $u = 0$ at $t = t_o$ for $t \leq t_o$. Differentiating both sides of Eqn. (11.46) and setting the result equal to zero gives

$$u(x_0, \ t_0) = \left(\frac{dx}{dt}\right)_{x_0, \ t_0} = 0 = -\frac{2}{\gamma-1}a_o + K\frac{2}{\gamma+1}t_o^{(1-\gamma)/(\gamma+1)}$$

Therefore,

$$K = \frac{\gamma+1}{\gamma-1} a_o t_o^{(\gamma+1)/(1-\gamma)}$$

Substituting this expression into Eqn. (11.46) and rearranging the result yields

$$x = -\frac{2}{\gamma-1} a_o t + \frac{\gamma+1}{\gamma-1} x_o \left(\frac{t}{t_o}\right)^{2/(\gamma+1)} \tag{11.47}$$

$$\frac{x}{a_o t} = -\frac{2}{\gamma-1} + \frac{\gamma+1}{\gamma-1} \frac{x_o}{a_o t_o} \left(\frac{t}{t_o}\right)^{(1-\gamma)/(\gamma+1)} \tag{11.48}$$

Define non-dimensional parameters $\tilde{x}$ and $\tilde{t}$ as

$$\tilde{x} \equiv \frac{x}{a_o t}; \quad \tilde{t} \equiv \frac{t}{t_o} \tag{11.49}$$

Then Eqn. (11.47) becomes

$$\tilde{x} = -\frac{2}{\gamma-1} + \frac{\gamma+1}{\gamma-1} \tilde{x}_o \tilde{t}^{(1-\gamma)/(\gamma+1)} = -\frac{2}{\gamma-1} + \frac{\gamma+1}{\gamma-1} \tilde{t}^{(1-\gamma)/(\gamma+1)} \tag{11.50}$$

where $\tilde{x}_o = x_o/a_o t_o = 1$. Eqn. (11.44) can be rewritten as

$$\tilde{u} = -\frac{2}{\gamma+1}(1-\tilde{x})$$

$$\tilde{a} = 1 - \frac{\gamma-1}{\gamma+1}(1-\tilde{x}) \text{ for } 1 \le \tilde{x} \le 1 + \frac{\gamma+1}{2}\tilde{u}_p \text{ (i.e., the fan region)} \tag{11.51}$$

where

$$\tilde{u} \equiv \frac{u}{a_o}; \quad \tilde{u}_p \equiv \frac{u_p}{a_o}; \quad \tilde{a} \equiv \frac{a}{a_o} \tag{11.52}$$

Examining Eqns. (11.50) and (11.51) suggests that $\tilde{x} = x/a_o t$ and $\tilde{t} = t/t_o$ are the *similarity variables* for the one-dimensional unsteady flow induced by the centered expansion wave. The self-similar nature of the gas motion is clearly seen in the wave diagram of Fig. 11.8. The relevant time scale is t_o and the lack of a geometric dimension in the direction of motion necessitates the use of $a_o t$ as the unique and relevant length scale. Such a lack is also seen in that as t increases, the fan occupies an increasingly large extent while keeping a fixed fan angle. The spread of the fan arises from the decrease in acoustic speed of successive waves as time passes.

It is of interest to check the validity of Eqn. (11.50). Consider the beginning of the expansion at $t = t_o$, when $\tilde{t} = 1$ and $x = x_o = a_o t_o$. That is, $\tilde{x} = \tilde{x}_o = 1$ for $\tilde{t} = 1$ and such a condition is indeed satisfied by Eqn. (11.50).

Referring to Fig. 11.8b, several results of the gas particle motion within the expansion fan can be observed and summarized in the following. These results are left as exercise problems:

(1) As successive expansion wave fronts pass the particle, the location changes from $\tilde{x} = \tilde{x}_o = 1$ upon encountering C_0^+ characteristic to $\tilde{x} = \tilde{x}_n = 1 + \frac{\gamma+1}{2}\tilde{u}_p$ at the passing of the C_n^+ characteristic. Therefore, the total displacement of the particle is

$$\Delta\tilde{x} = \tilde{x}_n - \tilde{x}_o = \frac{\gamma+1}{2}\tilde{u}_p \qquad (11.53)$$

That is, the faster the piston motion, the larger the particle displacement. Because $\tilde{u}_p < 0$, the particle moves in the $-x$ direction, which is consistent with the previous finding that an expansion wave (right-running in the $+x$ direction, as in Fig. 11.8b) displaces particles in the direction opposite to its propagation.

(2) There exists a piston speed that is equal to the local sonic speed, for which Eqn. (11.42b) yields

$$\left(\frac{dx}{dt}\right)_{C^+} = u + a = a_o + \frac{\gamma+1}{2}u_p = 0$$

That is, the C_n^+ characteristic becomes vertical in the t-x plane when

$$|u_p|_{M=1} = \frac{2}{\gamma+1}a_o \qquad (11.54)$$

(3) To be carried out as an end-of-chapter exercise (Problem 11.6), the particle at x_o spends a duration of

$$\Delta\tilde{t} = \tilde{t}_n - \tilde{t}_o = \left[1 + \frac{\gamma-1}{2}\tilde{u}_p\right]^{(1+\gamma)/(1-\gamma)} - 1 \qquad (11.55)$$

in the expansion fan. Because $\tilde{u}_p < 0$ and $(1+\gamma)/(1-\gamma) < 0$, as the piston speed increases, the time the particle spends inside the expansion fan increases. Such an increase is accompanied by the fact that the fan angle increases as the slope of the C_n^+ characteristic decreases (dx/dt approaching zero in Fig. 11.8b). In fact, when the piston becomes supersonic ($|u_p| > \frac{2}{\gamma+1}a_o$) the slope of the C_n^+ characteristic becomes negative on the t-x plane, which is not possible as discussed below in (4).

(4) Because $(1+\gamma)/(1-\gamma) < 0$ and the value in the bracket of Eqn. (11.55) must be greater than zero for $\Delta\tilde{t} > 0$, therefore, the following condition must be satisfied:

$$|\tilde{u}_p| \le \frac{2}{\gamma-1} \quad \text{or} \quad |u_p| \le \frac{2a_o}{\gamma-1} \qquad (11.56)$$

The less-than-or-equal sign in Eqn. (11.56) holds for the escape velocity defined as

$$|u_p|_{\text{escape}} = \frac{2a_o}{\gamma-1} \qquad (11.57)$$

with which the piston leaves a vacuum where no sound propagation is possible (and the temperature is absolute zero) and the gas can no longer sense the piston motion, and it *separates* or *escapes* from the piston surface. Therefore, it is not possible for the piston to be withdrawn at a supersonic speed.

(5) Along the C^+ family of characteristics, which form the centered expansion fan in Fig. 11.8, Eqn. (11.30) indicates that $du + dp/\rho a = 0$ on each characteristic. Because on a given characteristic, $u = $ constant p is also constant. The changes in u and p across the fan region are given by the changes on a C^- characteristic originating in the undisturbed region. For example, the change in u from the C_{i-1}^+ to C_i^+ characteristic is equal to Δu_i:

$$\Delta u_i \equiv u_i - u_{i-1} \tag{11.58}$$

Similarly, because along a C^- characteristic $du = dp/\rho a$,

$$\Delta p_i \equiv p_i - p_{i-1} = \int_{i-1}^{i} \rho a \, du \tag{11.59}$$

The pressure change due to the fan is thus

$$\Delta p = \int_0^n \rho a \, du = \sum_{i=1}^n \Delta p_i = \sum_{i=1}^n \left(\int_{i-1}^i \rho a \, du \right) \tag{11.60}$$

Because both ρ and a vary through the fan, piecewise integration is needed, for infinitesimal changes from the leading characteristic (C_0^+) to the trailing characteristic (C_n^+), $\Delta p = \rho a_o u'$.

11.8 Approximation for Small-Amplitude Waves

For the center expansion wave shown in Fig. 11.8, the limiting case where $|\tilde{u}_p| \rightarrow 0$ or $|u_p| \ll a_o$ constitutes the acoustic wave. The slope of the wave is

$$\left(\frac{dx}{dt} \right)_{C^+} = \left[a_o + \frac{\gamma+1}{2} u_p \right] \approx a_o \tag{11.61}$$

for *all* characteristics (as for acoustic waves) and the expansion fan angle is negligible. (In Chapter 9, the acoustic speed in the undisturbed gas medium is denoted by a_1.) The duration for a particle under the influence of the expansion wave (or the time it takes for the particle to attain the piston speed after t_o) is

$$\Delta \tilde{t} = \left[1 + \frac{\gamma-1}{2} \tilde{u}_p \right]^{(1+\gamma)/(1-\gamma)} - 1 \approx -\frac{\gamma+1}{2} \tilde{u}_p \text{ or } \Delta t = -\frac{\gamma+1}{2} \frac{x_o u_p}{a_o^2} \tag{11.62}$$

while the last wave, the C_n^+ characteristic, passes at

$$t = t_n = t_o + \Delta t = \left(1 - \frac{\gamma + 1}{2}\frac{u_p}{a_o}\right)\frac{x_o}{a_o} \approx \frac{x_o}{a_o} = t_o \qquad (11.63)$$

That is, the particle spends no (or negligible) time in the expansion fan (i.e., very thin wave). Equation (11.53) is used to obtain the displacement of a particle, which is

$$\Delta x \approx (\Delta\tilde{x})a_o t_o = -\left(\frac{\gamma + 1}{2}\frac{u_p}{a_o}\right)^2 x_o \qquad (11.64)$$

Equation (11.64) suggests that Δx due to the acoustic wave is of the order of $(u_p/a_o)^2$ and is thus negligible, unless for very large x_o. Due to the negligible expansion fan angle, the particle must be very far away from the piston to experience an appreciable displacement.

11.9 Compression Wave and Shock Formation

An advancing piston will lead to the collapse of compression waves, eventually forming a shock wave in a sufficiently long cylinder. Two types of piston motion are considered: the impulsively advanced and the gradually advanced, with the former being the limiting case of the latter. An impulsively advanced piston with $dX_p/dt = u_p > 0$, shown in Fig. 11.9, will necessarily cause a shock wave at the origin of motion. In Fig. 11.9, the region ahead of the compression wave is designated as region I, while that downstream of the wave is region II. Assuming that the gas in region I is initially quiescent, the following can be stated:

$$\text{Along the } C_I^+ \text{ characteristic}: \left(\frac{dx}{dt}\right)_{C_I^+} = a_I \text{ and } J_I^+ = \frac{2a_I}{\gamma - 1} \qquad (11.65)$$

$$\text{Along the } C_{II}^+ \text{ characteristic}: \left(\frac{dx}{dt}\right)_{C_{II}^+} = u_p + a_{II} \text{ and } J_{II}^+ = u_p + \frac{2a_{II}}{\gamma - 1} \qquad (11.66)$$

As a result of compression $u_p > 0$ and $a_{II} > a_I$. Thus,

$$\left(\frac{dx}{dt}\right)_{C_{II}^+} > \left(\frac{dx}{dt}\right)_{C_I^+} \qquad (11.67)$$

Therefore the C_{II}^+ characteristics are steeper on the *t-x* plane and will intersect (i.e., catch up) with the C_I^+ characteristics. Thompson (1972) describes details that lead to the following relationship

$$\frac{u_p}{a_I} = \frac{2}{\gamma - 1}\left(M_{In} - \frac{1}{M_{In}}\right) \qquad (11.68)$$

where M_{In} is the Mach number of the shock propagating into region I. With M_{In} thus determined, normal shock relationships can be used to determine ratios of

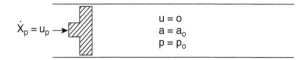

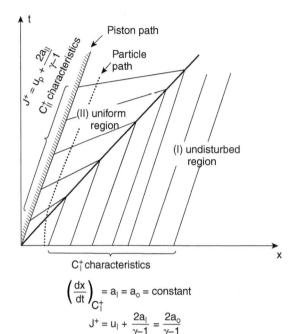

Figure 11.9 (a) Schematic of piston moving into the initially quiescent gas in a long tube, generating compression waves into the gas; (b) characteristic waves formed due to the piston moving with a constant speed u_p; also included is an example of particle path.

properties across the shock: p_{II}/p_I, T_{II}/T_I, and so on. It is noted that due to the non-isentropic and non-homentropic nature of the piston-induced shock flow field, the concept of Riemanian invariants is no longer valid. For example, there is no similar result for Δp as 24)shown by Eqn. (11.60) for expansion waves.

Consider the general continuous piston motion prescribed by $X_p = X_p(t)$, depicted in Fig. 11.10 (adopted from Thompson, 1972) With acceleration of the piston motion, the later compression waves catch up with the earlier ones to cause a finite pressure increase across a collapsed, thin wave, thus forming a shock wave (this has been described in Section 4.3). The location of shock formation is designated as point *a* in Fig. 11.10., where the first two compression waves collapse. As the acceleration is increased, point *a* moves closer to the origin of motion, approaching the limiting case depicted in Fig. 11.9. Once the shock forms, the condition of homentropy ceases to be valid and the method of characteristics is no longer applicable. The Riemann invariant, J^-, is only uniform on the *t-x* plane in the region bounded by the limiting C^- characteristic and the piston path line schematically shown in Fig. 11.10. The limiting C^- characteristic is the one in Fig. 11.10 that passes point *a* and still possesses the equality, $J_I^- = J_{II}^-$. The C^- characteristic passing point *b* in Fig. 11.10 illustrates that downstream of the limiting C^- characteristic, the Riemann invariant J^-assumes different values as it crosses from region (I) into

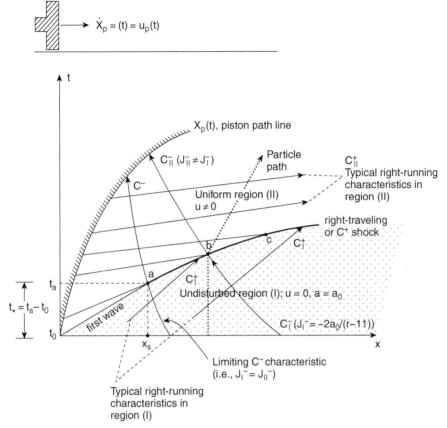

Figure 11.10 (a) (a) Schematic of piston moving with varying speeds into the initially quiescent gas in a long tube, generating compression waves into the gas; (b) characteristic waves formed due to the piston moving with a speed $u_p(t)$ that increases with time. Note that C_I^- is not possible downstream of x_* (corresponding to a time $t_* = t_a - t_o$) as characteristics do not exist if $\nabla s \neq 0$.

region (II), with point b separating the two segments of the two characteristics C_I^- and C_{II}^-. In region (I), $J_I^- = -2a_o/(\gamma - 1)$ while in region (II), $J_{II}^- = u_{II} - 2a_{II}/(\gamma - 1) = u_p - 2a_{II}/(\gamma - 1)$.

The location of shock formation (point a in Fig. 11.10, with location denoted by x_s) is of interest, and so is the associated time it takes for shock formation (t_s). It is instructive to consider a finite-amplitude compression waveform shown in Fig. 11.11 (adopted from Zeldovich and Raizer, 1966) at its initiation at $t = t_o$. The waveforms for pressure, acoustic speed, and density are similar in shape. For compression waves, all three non-dimensional parameters—$(p - p_0)/p_0$, $(a - a_0)/a_0$, and $(\rho - \rho_0)/\rho_0$—are all positive at all times. The increase in $(a - a_0)/a_0$, or simply in a, is the direct result of increase in temperature due to compression. In Fig. 11.11, the similarity variable $(x - a_o t)$ is used as the horizontal coordinate, so that the waveform spans over the same wavelength λ at all times. Points on the waves having the same amplitude travel with the same velocity and thus maintain a fixed separation

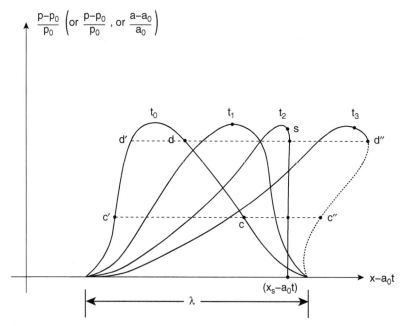

Figure 11.11 Evolution of the shape (or waveform) of a finite-amplitude compression wave. Note that the horizontal axis is similarity variable is $(x - a_o t)$.

between them (such as points c and c' in on the initial waveform). The finite-amplitude, nonlinear nature of the wave causes the waveform to begin to distort for $t > t_o$, as shown in Fig. 11.10. Recall that compression results in increases in temperature and acoustic speed, with larger compression ratios causing larger increases. For example, point *d* on the initial waveform thus travels faster than, and will at some later time catch up with, point *c*, causing *steepening* of the leading edge and *flattening* of the trailing edge.

Prior to shock formation, the steepening (or distortion) of the wave form can be understood as follows. Across the C^+ characteristics generated by the piston shown in Fig. 11.10, J^- remains constant:

$$J^- = u - \frac{2a}{\gamma - 1} = \text{constant} = -\frac{2a_o}{\gamma - 1}$$

which in turn gives

$$u = \frac{2}{\gamma - 1}(a - a_o)$$

The slope of a C^+ characteristic is

$$\left(\frac{dx}{dt}\right)_{C^+} = u + a = \frac{\gamma + 1}{\gamma - 1}a - \frac{2a_o}{\gamma - 1}$$

The speed of leading edge of the right-running nonlinear wave has an increasing $(p - p_0)/p_0$ and $(a - a_0)/a_0$. Therefore $(dx/dt)_{C^+}$ increases monotonically with a. As a result the later characteristics/waves catch up with the earlier ones, causing the steepening of the gradient and the increasingly distorted waveform, as shown in Fig. 11.11; for example, point d is catching up with point c. On the other hand, the trailing edge becomes increasingly flattened, resulting from the decreasing $(p - p_0)/p_0$ and $(a - a_0)/a_0$. On the $(x - a_0 t)$ coordinate, the trailing edge lags and flattens behind the leading edge while the leading steepens. At $t = t_2$, a portion of the leading front in the neighborhood of point s becomes vertical with points c and d being at the same location, x_s, and infinite gradients in pressure and density. In Fig. 11.11, this time is designated as t_2, and it is the time of formation of the first shock. At $t = t_3$, point d moves ahead of point c and the waveform becomes triple-valued for a given $(x - a_0 t)$. The multivalued waveform is mathematically impossible; it is also physically absurd in that in a one-dimensional flow no physical mechanism exists for such overtaking (note that in multidimensions, overtaking can occur in rotational flows) and some physical mechanism must be formed at $t = t_2$ ($t_1 < t_2 < t_3$) to prevent this from occurring. It is physically plausible for two values to exist within a few mean free paths (Vincenti and Kruger, 1975). Mathematically, one then obtains an infinitely steepened wave that is a shock wave (Zeldovich and Raizer, 1966), with an infinite gradient at point s in Fig. 11.11. It is said that at t_2, the first shock already forms. The impossible waveform for $t = t_3$ is shown by the dashed line in Fig. 11.11.

The location of point s in Fig. 11.11 can be expressed in terms of the similarity variable $(x_s - a_0 t_s)$, which in turn requires specifying both x_s and t_s. The following discussion is based on Thompson [1972]. To express x_s in terms of a_0 and $X_p(t)$, consider this location in a t-x plot shown in Fig. 11.12. Assume that the two intersecting characteristics C_i^+ and C_{i+1}^+ to form the first shock emanate from the piston path, at two *arbitrary* times t and $t + dt$ with corresponding piston locations of $x_{p,i}$ and $x_{p,i} + dx$, respectively. The two characteristics intersect after a time period of $t_s - t$ at x_s. The C_{i+1}^+ characteristic can be imagined to have emanated at the same time as the C_i^+ characteristic. Then its point of origination will be at $(x_{p,i} - dl)$. Geometrical consideration of the diagram of Fig. 11.12 leads to

$$t_s - t = \frac{dl}{d(u + a)} \tag{11.69}$$

That is the extra distance, dl, that the C_{i+1}^+ characteristic has to travel to meet the C_i^+ characteristic, due to the change in the slope of the right-running characteristic over the time period from t and $t + dt$. Once dl is found, then t_s and x_s can be determined. Before the intersection occurs, J^- remains constant through the flow field. Thus

$$a = a_o + \frac{\gamma - 1}{2} u$$

Applying this result to the slope of the C^+ characteristic leads to

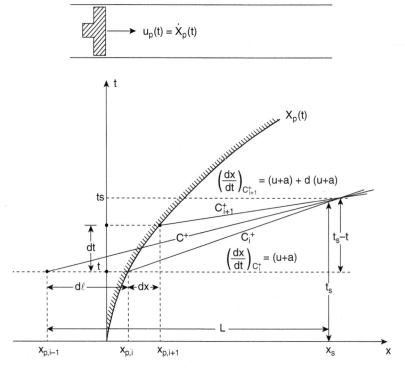

Figure 11.12 The *t-x* plot showing the time (t_s) and location (x_s) where the shock wave forms due to a piston moving with a speed $u_p(t)$, increasing with time, into an initially quiescent gas in a long tube.

$$\left(\frac{dx}{dt}\right)_{C^+} = u + a = a_o + \frac{\gamma+1}{2}u = a_o + \frac{\gamma+1}{2}u_p = a_o + \frac{\gamma+1}{2}\frac{dX_p}{dt} \qquad (11.70)$$

Therefore,

$$d(u+a) = \frac{\gamma+1}{2}\frac{d^2X_p}{dt^2}dt \qquad (11.71)$$

Considering a general C^+ characteristic in Fig. 11.12, one arrives at

$$dl + dx = \left(\frac{dx}{dt}\right)_{C^+}dt = \left(a_o + \frac{\gamma+1}{2}\frac{dX_p}{dt}\right)dt$$

With $dx = (dX_p/dt)dt$

$$dl = \left(a_o + \frac{\gamma-1}{2}\frac{dX_p}{dt}\right)dt \qquad (11.72)$$

Substituting Eqns. (11.70) and (11.72) into Eqn. (11.69) yields

$$t_s = t + \frac{\left[a_o + \left(\frac{\gamma-1}{2}\right)\frac{dX_p}{dt}\right]dt}{\left(\frac{\gamma+1}{2}\right)\frac{d^2X_p}{dt^2}dt} = t + \frac{1}{\ddot{X}_p}\left(\frac{2}{\gamma+1}a_o + \frac{\gamma-1}{\gamma+1}\dot{X}_p\right) \tag{11.73}$$

where $\dot{X}_p = dX_p/dt (= u_p)$ and $\ddot{X}_p = d^2X_p/dt^2$.

Note that Eqn. (11.73) is derived for an arbitrary time when two characteristics intersect. For the earliest moment of shock formation, t_s takes up a minimum value $t_{s,min}$:

$$\frac{dt_s}{dt} = 0 = 1 - \left(\ddot{X}_p\right)^{-2}\left(\frac{2}{\gamma+1}a_o + \frac{\gamma-1}{\gamma+1}\dot{X}_p\right)\dddot{X}_p + \left(\ddot{X}_p\right)^{-1}\left(\frac{\gamma-1}{\gamma+1}\ddot{X}_p\right)$$

The above expression leads to an implicit equation for $t_{s,min}$:

$$\left(\ddot{X}_p\right)^2 = \frac{1}{\gamma}\left(a_o + \frac{\gamma-1}{2}\dot{X}_p\right)\dddot{X}_p \tag{11.74}$$

The corresponding location where the first shock forms is

$$x_{s,min} = X_p + (u+a)\left(t_{s,min} - t\right) = X_p + \left(a_o + \frac{\gamma+1}{2}\dot{X}_p\right)\frac{1}{\ddot{X}_p}\left(\frac{2}{\gamma+1}a_o + \frac{\gamma-1}{\gamma+1}\dot{X}_p\right) \tag{11.75}$$

The traveling shock formed by intersection of C^+ characteristics, such as those shown in Figs. 11.9 and 11.10 are call the right-traveling or C^+ shocks. The flow field after shock formation clearly has become *piecewise homentropic*, as entropy is only constant (equal to different values) in uniform regions. In Fig. 11.9, regions I and II are both homentropic, with region II having higher entropy than region I. In Fig. 11.10, the undisturbed region I (shaded) is a homentropic region, while in region II (non-shaded), only that part of the t-x plane bounded by the piston path line and the limiting C^- characteristic is homentropic.

EXAMPLE 11.7 Consider $X_p = A(1 - \cos \omega t)$ in a long tube containing quiescent gas at $t = 0$. Assume that $A = a_1/2\omega$. Find the time and the location of the first shock formation.

Solutions –
Equation (11.74) provides

$$A^2\omega^4 \cos^2\omega t = \frac{1}{\gamma}\left(a_o + \frac{\gamma-1}{2}A\omega \sin \omega t\right)\left(-A\omega^3 \sin \omega t\right)$$

Substituting $A = a_o/2\omega$ and $\cos^2\omega t = (1 - \sin^2\omega t)$ into the above expression and rearranging leads to

$$\frac{-(\gamma+1)}{8\gamma}\sin^2\omega t + \frac{1}{2\gamma}\sin \omega t + \frac{1}{4} = 0$$

The solution for this quadratic equation for $\gamma = 7/5$ (for diatomic gases) is

$$\sin \omega t = \left(-5 \pm \sqrt{67}\right)/6$$

Because $|\sin \omega t| \leq 1$, the above solutions yield only meaningful results of $\sin \omega t \approx 0.531$

$$t_{s,min} \approx \frac{0.560}{\omega}$$

If $\omega = 1 \ s^{-1}$, then $t_{s,min} = 0.56$ s.

The value of $t_{s,min}$ depends on both the amplitude, A, and ω. In this example, $A = a_1/2\omega$ is a function of ω, leaving $t_{s,min}$ solely dependent on ω. The value of A is chosen so that $\dot{X}_p = (a_1/2) \sin \omega t$, ensuring that the amplitude of the piston velocity is not too small compared to the acoustic speed in the quiescent gas for the nonlinear theory to apply.

The corresponding location where the first shock forms is

$$x_{s,min} = X_p + (u + a)(t_{s,min} - t) = X_p + \left(a_1 + \frac{\gamma+1}{2}\dot{X}_p\right)$$

$$\frac{1}{\ddot{X}_p}\left(\frac{2}{\gamma+1}a_1 + \frac{\gamma-1}{\gamma+1}\dot{X}_p\right)$$

With $X_p = A(1 - \cos \omega t)$ and $A = a_1/2\omega$ and $\gamma = 7/5$

$$x_{s,min} \approx 2.881 \frac{a_1}{\omega}$$

$\square$

11.10 Flow with Weak Traveling Shock

The relationship between jumps in entropy and pressure across a weak shock wave has been shown in Chapter 4 as

$$\frac{s_2 - s_1}{R} \approx \frac{2\gamma}{3(\gamma+1)^2} \left(M_1^2 - 1\right)^3 = \frac{\gamma+1}{12\gamma^2}\beta^3 \tag{4.20f}$$

where the shock strength is given by Eqn. (4.18d):

$$\beta \equiv \frac{p_2 - p_1}{p_1} = \frac{2\gamma}{\gamma+1}\left(M_1^2 - 1\right) \tag{4.16e}$$

For a weak shock, $\beta \ll 1$. The entropy change across a weak shock wave is of the order of β^3, which is negligibly small. Therefore, the flow field with weak shocks can be approximated as homentropic, and the above results from the method of characteristics that can be used for flow calculations, with negligible changes in the Riemann invariants. Figure 11.13 schematically shows the invariants and the particle

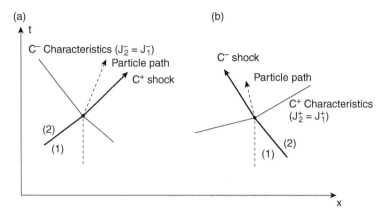

Figure 11.13 Schematics showing the characteristics and associated invariants and the particle trajectory across a C^+ and C^- shocks traveling into a quiescent gas.

trajectory across a C^+ and a C^- shocks traveling into a quiescent gas, respectively. The following derivation shows the negligible changes in the invariants across a weak shock, which in turns requires showing negligible changes in both the flow velocity, u, and the acoustic speed, a.

Take a C^+ shock as an example. The change in u across the weak C^+ shock can be obtained by requiring that $J_2^- = J_1^-$ along a C^- characteristic, shown in Fig. 11.13a. Therefore, by combining the result of Problem 11.5,

$$u_2 - \int_{p_0}^{p_2} \left(\frac{dp}{\rho a} \right)_s = u_1 - \int_{p_0}^{p_1} \left(\frac{dp}{\rho a} \right)_s$$

and

$$u_2 - u_1 = \int_{p_1}^{p_2} \left(\frac{dp}{\rho a} \right)_s \tag{11.76}$$

The following expression is useful for the integration to be carried out for Eqn. (11.76)

$$\left(\frac{\partial v}{\partial p} \right)_s = \frac{dv}{d\rho} \left(\frac{\partial \rho}{\partial p} \right)_s = -\frac{1}{\rho^2 a^2} \tag{11.77}$$

For small increases in p from p_1, a Taylor series in $(p - p_1)$ is useful by considering the state ahead of the shock wave (state 1 in Fig. 11.13) as the reference state and expanding:

$$\left(\frac{\partial v}{\partial p} \right)_s = \left(\frac{\partial v}{\partial p} \right)_{s,1} + \left(\frac{\partial^2 v}{\partial p^2} \right)_{s,1} (p - p_1) + \frac{1}{2} \left(\frac{\partial^3 v}{\partial p^3} \right)_{s,1} (p - p_1)^2 + \dots \tag{11.78}$$

By repeating the similar steps shown in Eqn. (11.77),

$$\left(\frac{\partial^2 v}{\partial p^2}\right)_{s,1} = \left(\frac{d}{dp}\left(\frac{\partial v}{\partial p}\right)\frac{\partial \rho}{\partial p}\right)_{s,1} = \left(\frac{d}{dp}\left(-\frac{1}{\rho^2 a^2}\right)\frac{\partial \rho}{\partial p}\right)_{s,1} = \frac{2}{\rho_1^3 a_1^4}$$

and

$$\left(\frac{\partial^3 v}{\partial p^3}\right)_{s,1} = \left(\frac{d}{dp}\left(\frac{\partial^2 v}{\partial p^2}\right)\frac{\partial \rho}{\partial p}\right)_{s,1} = \left(\frac{d}{dp}\left(\frac{2}{\rho_1^3 a_1^4}\right)\frac{\partial \rho}{\partial p}\right)_{s,1} = \frac{-6}{\rho_1^4 a_1^6}$$

With these two expressions, Eqn. (11.78) becomes

$$\left(\frac{\partial v}{\partial p}\right)_s = -\frac{1}{\rho_1^2 a_1^2} + \frac{2}{\rho_1^3 a_1^4}(p - p_1) - \frac{3}{\rho_1^4 a_1^6}(p - p_1)^2 + \dots \qquad (11.79)$$

Substituting Eqn. (11.77) into Eqn. (11.76) yields

$$u_2 - u_1 = \int_{p_1}^{p_2}\left(\frac{dp}{\rho a}\right)_s = \int_{p_1}^{p_2}\sqrt{-\left(\frac{\partial v}{\partial p}\right)_s}\,dp$$

Using Eqn. (11.79), one obtains

$$u_2 - u_1 = \frac{1}{\rho_1 a_1}\int_{p_1}^{p_2}\sqrt{1 - \left[\frac{2}{\rho_1 a_1^2}(p - p_1) - \frac{3}{\rho_1^2 a_1^4}(p - p_1)^2\right]}\,dp$$

Performing a binomial expansion, for small $(p - p_1)$, and integration for this expression leads to

$$u_2 - u_1 = \frac{1}{\rho_1 a_1}\left[(p_2 - p_1) - \frac{1}{2}\frac{1}{\rho_1 a_1^2}(p_2 - p_1)^2 + \frac{1}{2}\frac{1}{\rho_1^2 a_1^4}(p_2 - p_1)^3\right] \qquad (11.80)$$

The relative change in gas velocity compared to the acoustic velocity in state 1 is, after using $\beta = (p_2 - p_1)/p_1 = \gamma(p_2 - p_1)/\rho_1\gamma RT_1 = \gamma(p_2 - p_1)/\rho_1 a_1^2$,

$$\frac{u_2 - u_1}{a_1} = \frac{1}{\gamma}\beta - \frac{1}{2\gamma^2}\beta^2 + \frac{1}{2\gamma^3}\beta^3 \text{ (for a } C^+\text{shock with } \beta \ll 1) \qquad (11.81)$$

For comparison, the velocity change across a stationary normal shock wave is, following Eqns. (4.20c) and (4.20d),

$$\frac{V_2 - V_1}{a_1} \approx -\frac{2M_1}{\gamma + 1}(M_1^2 - 1) \approx -\frac{\beta}{\gamma} \text{ (for a stationary normal shock with } M_1 \to 1$$

$$\text{and } \beta \ll 1) \qquad (11.82)$$

The difference of velocity change across a weak stationary and a weak moving shock is of $O(\beta^2)$, which is even smaller than the pressure change. It is noted that the signs of the leading terms are opposite, for the following reasons. Moving shock waves carry gas along with them and thus *increase* the gas velocity (from zero). Across a stationary shock, the increase in pressure is accompanied by a *decrease* in gas velocity.

The change in acoustic speed across the C^+ shock is related to the change in static temperature. For both a stationary or a moving shock

$$\frac{T_2}{T_1} \approx 1 + \frac{2\gamma(\gamma - 1)}{(\gamma + 1)^2} (M_1^2 - 1) \tag{4.22b}$$

Therefore,

$$\frac{a_2 - a_1}{a_1} = \sqrt{\frac{T_2}{T_1}} - 1 = \sqrt{1 + \frac{2\gamma(\gamma - 1)}{(\gamma + 1)^2} (M_1^2 - 1)} - 1$$

By using Eqn. (4.16e) and binomial expansion for small β, the above expression becomes

$$\frac{a_2 - a_1}{a_1} = \frac{\gamma - 1}{2(\gamma + 1)} \beta \tag{11.83}$$

Subtracting Eqn. (11.83) from Eqn. (11.81), the change in J^- across the weak C^+ shock is

$$\frac{J_2^- - J_1^-}{a_1} = \frac{1}{\gamma(\gamma + 1)} \beta - \frac{1}{2\gamma^2} \beta^2 + \frac{1}{2\gamma^3} \beta^3 \tag{11.84}$$

which is $\ll 1$, validating the requirement of $J_2^- = J_1^-$ and allowing applicability of the method of characteristics for weak shocks.

EXAMPLE 11.8 A long piston-cylinder device contains air having a temperature of 300 K ($a_1 = 347.8 \ m/s$) and a pressure of 1 atmosphere. The air is initially at rest and then the piston is impulsively moved to compress the air at a speed of 100 m/s. Find the speed of the resultant shock before it reaches the end of the cylinder and the associated pressure and temperature increases.

Solutions –
Consider the shock to be a right-running (C^+) shock. Although $(u_2 - u_1)/a_1 \approx 0.29$ is not much smaller than unity, one might make an initial attempt to solve the problem using the method of characteristics.

In the simple wave region, the method of characteristics requires $J_1^- = J_2^-$. Therefore,

$$u_2 - \frac{2}{\gamma - 1} a_2 = -\frac{2}{\gamma - 1} a_1$$

$$a_2 = a_1 + \frac{\gamma - 1}{2} u_2 = 347.8 \ m/s + 0.2 \times 100 \ m/s = 367.8 \ m/s$$

With the weak shock assumption, it is reasonable to calculate the wave speed, V_s, as

$$V_s = \frac{1}{2}\left[\left(\frac{dx}{dt}\right)_1 + \left(\frac{dx}{dt}\right)_2\right] = \frac{1}{2}[(a_1 + u_1) + (a_2 + u_2)] = 407.8 \ m/s$$

With the isentropic assumption for the method of characteristics,

$$\frac{p_2}{p_1} = \left(\frac{T_2}{T_1}\right)^{\gamma/(\gamma-1)} = \left(\frac{a_2}{a_1}\right)^{2\gamma/(\gamma-1)} = 1.48$$

$$\frac{T_2}{T_1} = \left(\frac{a_2}{a_1}\right)^2 = 1.12$$

Comments –
(1) For a traveling shock with a speed of 407.8 m/s into the quiescent air, the Mach number is calculated using Eqn. (4.22d):

$$\frac{V_2}{V_1} \approx 1 - \frac{2}{\gamma+1}(M_1^2 - 1) \ \text{or} \ \frac{V_1 - V_2}{V_1} \approx \frac{2}{\gamma+1}(M_1^2 - 1)$$

For an observer riding on the wave, $V_1 = V_s = 407.8 \ m/s$ and $V_2 = V_s - u_2 = 307.8 \ m/s$. Therefore,

$$M_s = M_1 = 1.16$$

for which,

$$\frac{p_2}{p_1} = 1.40$$

$$\frac{T_2}{T_1} = 1.10$$

Therefore, solutions using the method of characteristics and shock relationship yield very similar results, although both $\beta \approx 0.4$ and $\dot{X}_p/a_1 = u_2/a_1 \approx 0.29$ are not much smaller than unity.

(2) If the shock is left-running (C^-), one would obtain the same results – this is left as an exercise problem. □

11.11 Wave Interaction

A non-simple wave region is where waves interact. Example 11.5 illustrates the interaction between compression and expansion waves. There are other types of interaction. They generally fall into two categories. The first category is that of

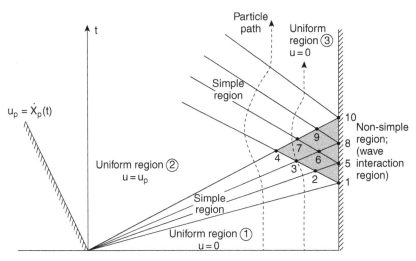

Figure 11.14 Characteristics and particle path resulting from a piston moving with a constant speed u_p in a long tube away from the gas bounded by the piston and the closed end of a long tube.

characteristics (i.e., isentropic waves and weak shocks) and the second of shock waves.

Consider the situation where centered expansion waves are generated by a piston in a long tube with a rigid end wall, as shown in Fig. 11.14. The right-running rarefaction, or expansion, wave (C^+ characteristics) will be turned back by the wall to form left-running rarefaction waves (C^- characteristics). These C^+ and C^- characteristics interact in the near-wall region to create the non-simple region, shown as the shaded region in Fig. 11.14. The regions outside the simple and non-simple regions are uniform regions such as regions 1, 2, and 3. In both uniform regions 1 and 3, the gas velocity is zero, while in uniform region 2 the gas velocity is equal to that of the piston impulsively set in motion with a constant velocity $u_p = \dot{X}_p$, at $t = 0$. Gas velocities in these uniform regions are required to satisfy the stationary and moving boundaries of the end wall and the moving piston, respectively. Two representative particle trajectories are given in Fig. 11.14, with one going through only the simple wave and uniform regions and the other through both simple- and non-simple wave regions. It is noted that an expansion wave pushes particles in the direction opposite to its propagation, as qualitatively described by these two trajectories.

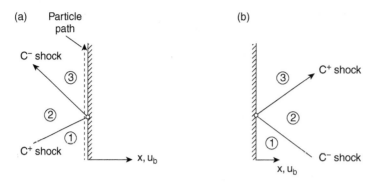

Figure 11.15 Incident and reflected shock waves from a wall that is either stationary ($u_b = 0$) or moving ($u_b \neq 0$); $J_i^+ = -J_r^-$ at a stationary wall, and $J_r^+ = -J_i^- + 2u_b$ at a moving wall.

In Fig. 11.15a the incident C^+ and reflected C^- shocks carry invariant values of J_i^+ and J_r^-, respectively:

$$J_i^+ = u + \int_b \frac{dp}{\rho a} \tag{11.85}$$

$$J_r^- = u - \int_b \frac{dp}{\rho a} \tag{11.86}$$

where subscripts i and r denote incident and reflected waves, respectively. Recall that in general the integration in Eqns. (11.85) and (11.86) results in different values from one to another characteristic. However, since both the C^+ and C^- characteristics share the common wall condition, the integration represent a thermodynamic condition at the wall (designated by the subscript, b). At the stationary wall surface, $u = 0$, and thus for a pair of incident and reflected characteristics at the wall

$$J_i^+ = -J_r^- \quad \text{(at a stationary wall)} \tag{11.87}$$

In general, the wall boundary might be moving with a velocity $u = u_b$, then

$$J_r^- = -J_i^+ + 2u_b \quad \text{(at a moving wall)} \tag{11.88}$$

If the reflection is from a left boundary, then the incident wave is a C^- characteristic (Fig. 11.15b) and

$$J_r^+ = -J_i^- + 2u_b$$

It is seen in Fig. 11.15 that whether or not $u_b = 0$, the particle trajectory at the boundary is that of the moving wall as required by the boundary conditions. Due to thin shock waves, regions 1, 2, and 3 in Fig. 11.15 are each a uniform region.

EXAMPLE 11.9 The wall boundary in Fig. 11.15a is moving with a constant speed $u_b = a_1/8$ to the right while the gas behind the incident wave is traveling at a speed equal to $a_1/4$. Find the pressure in each of regions 1, 2, and 3, given that the pressure in region 1 is p_1. Also find the pressures for $u_b = 0$

Solution –

$$u_1 = u_b = a_1/8$$

$$u_2 = a_1/4$$

$$u_3 = u_b = a_1/8$$

In the simple wave region across the C^+ characteristic

$$J^- = u_1 - \frac{2a_1}{\gamma - 1} = u_2 - \frac{2a_2}{\gamma - 1}; \text{ that is } \frac{1}{8}a_1 - \frac{2a_1}{\gamma - 1} = \frac{1}{4}a_1 - \frac{2a_2}{\gamma - 1}$$

Across the C^- characteristic

$$J^+ = u_2 + \frac{2a_2}{\gamma - 1} = u_3 + \frac{2a_3}{\gamma - 1}; \text{ that is } \frac{1}{4}a_1 + \frac{2a_2}{\gamma - 1} = \frac{1}{8}a_1 + \frac{2a_3}{\gamma - 1}$$

Therefore,

$$a_2 = a_1 + \frac{\gamma - 1}{16}a_1$$

$$a_3 = a_1 + \frac{\gamma - 1}{8}a_1$$

Applying isentropic relations for MOC yields:

$$\frac{p_2}{p_1} = \left(\frac{T_2}{T_1}\right)^{\gamma/(\gamma-1)} = \left(\frac{a_2}{a_1}\right)^{2\gamma/(\gamma-1)} = \left(\frac{\gamma + 15}{16}\right)^{2\gamma/(\gamma-1)}$$

$$\frac{p_3}{p_2} = \left(\frac{T_3}{T_2}\right)^{\gamma/(\gamma-1)} = \left(\frac{a_3}{a_2}\right)^{2\gamma/(\gamma-1)} = \left[\frac{2(\gamma + 7)}{\gamma + 15}\right]^{2\gamma/(\gamma-1)}$$

For $\gamma = 1.4$, $p_2/p_1 = 1.19$ and $p_3/p_2 = 1.18$.

For $u_b = 0$,

$$J^- = u_1 - \frac{2a_1}{\gamma - 1} = u_2 - \frac{2a_2}{\gamma - 1}; \quad -\frac{2a_1}{\gamma - 1} = \frac{1}{4}a_1 - \frac{2a_2}{\gamma - 1} \text{ and } a_2 = a_1 + \frac{\gamma - 1}{8}a_1$$

$$J^+ = u_2 + \frac{2a_2}{\gamma - 1} = u_3 + \frac{2a_3}{\gamma - 1}; \quad \frac{1}{4}a_1 + \frac{2a_2}{\gamma - 1} = \frac{2a_3}{\gamma - 1} \text{ and } a_3 = a_1 + \frac{\gamma - 1}{4}a_1$$

$$\frac{p_2}{p_1} = \left(\frac{a_2}{a_1}\right)^{2\gamma/(\gamma-1)} = \left(\frac{\gamma + 7}{8}\right)^{2\gamma/(\gamma-1)}$$

$$\frac{p_3}{p_2} = \left(\frac{a_3}{a_2}\right)^{2\gamma/(\gamma-1)} = \left[\frac{2(\gamma + 3)}{\gamma + 7}\right]^{2\gamma/(\gamma-1)}$$

For $\gamma = 1.4$, $p_2/p_1 = 1.41$ and $p_3/p_2 = 1.38$.

Comments –

Because the gas is traveling at a higher speed ($a_0/4$) after the wave passes, one expects the wave to be a shock wave. Therefore, it is instructive to compare the above results with shock solutions. For simplicity, consider the case of $u_b = 0$; this is left as an end-of-chapter problem, where one result is that $u_b = 0$ leads to a harder hit of the wave on the wall than that in this example. □

EXAMPLE 11.10 Consider a long tube, as described in Fig. 11.14, with the right end closed and the left end being a movable piston. The gas in the tube is air and is initially at rest. A centered expansion is then generated by withdrawing the piston impulsively to a velocity u_p. Find the flow conditions in the non-simple wave region.

Solutions –

For convenience, choose characteristics in the expansion fan that have velocities equal to 0, $-u_p/3$, $-2u_p/3$, and $-u_p$. The non-simple wave region is shown by the shaded area in Fig. 11.14. A table similar to that of Example 11.5 helps to summarize the calculation results, as shown below, where the subscript "o" designates the undisturbed region:

Comments –

For point 2, J^- from point 1 can be used so that $u_1 - \frac{2a_1}{\gamma-1} = u_2 - \frac{2a_2}{\gamma-1}$. Knowing $u_1 = 0$ $u_2 = -u_p/3$ and $a_1 = a_0$, the same results for a_2 and u_2 are found. Similarly, for point 3, J^- from point 2 can be used so that $u_2 - \frac{2a_2}{\gamma-1} = u_3 - \frac{2a_3}{\gamma-1}$ leading to the same results for a_3 and u_3 as shown in the table.

To increase the accuracy of the flow conditions (i.e., u, a, and the slopes of the characteristics, $u + a$ and $u - a$), more characteristics emanating from point O can be chosen or the slopes of characteristics can be taken as the average value of those at the two points connected by the given characteristic. For example the slope of the characteristic segment connecting points 6 and 7 is now

Point	Invariants	Known	Found	$u + a$	$u - a$
1	$J^+ = u_o + \frac{2a_o}{\gamma-1} = u_1 + \frac{2a_1}{\gamma-1}$	$u_o = 0$ $u_1 = 0$	$a_1 = a_0$	$u_1 + a_1 = a_0$	$\left(\frac{dx}{dt}\right)_{1-2} = u_1 - a_1 = -a_0$
2	$J^- = u_0 - \frac{2a_0}{\gamma-1} = u_2 - \frac{2a_2}{\gamma-1}$	$u_2 = -u_p/3$	$a_2 = a_0 - \frac{\gamma-1}{6} u_p$	$u_2 + a_2 = a_0 - \left(\frac{1}{3} + \frac{\gamma-1}{6}\right) u_p$	$u_2 - a_2 = -a_0 - \left(\frac{1}{3} - \frac{\gamma-1}{6}\right) u_p$
3	$J^- = u_0 - \frac{2a_0}{\gamma-1} = u_3 - \frac{2a_3}{\gamma-1}$	$u_3 = -\frac{2}{3} u_p$	$a_3 = a_0 - \frac{\gamma-1}{3} u_p$	$u_3 + a_3 = a_0 - \left(\frac{2}{3} + \frac{\gamma-1}{3}\right) u_p$	$u_3 - a_3 = -a_0 - \left(\frac{2}{3} - \frac{\gamma-1}{3}\right) u_p$
4	$J^- = u_0 - \frac{2a_0}{\gamma-1} = u_4 - \frac{2a_4}{\gamma-1}$	$u_4 = -u_p$	$a_4 = a_0 - \frac{\gamma-1}{2} u_p$	$u_4 + a_4 = a_0 - \left(1 + \frac{\gamma-1}{2}\right) u_p$	$u_4 - a_4 = -a_0 - \left(1 - \frac{\gamma-1}{2}\right) u_p$
5	$J^+ = u_2 + \frac{2a_2}{\gamma-1} = u_5 + \frac{2a_5}{\gamma-1}$	$u_5 = 0$	$a_5 = a_0 - \frac{\gamma-1}{3} u_p$	$u_5 + a_5 = a_0 - \frac{\gamma-1}{3} u_p$	$u_5 - a_5 = -a_0 + \frac{\gamma-1}{3} u_p$
6	$J^+ = u_3 + \frac{2a_3}{\gamma-1} = u_6 + \frac{2a_6}{\gamma-1}$ $J^- = u_5 - \frac{2a_5}{\gamma-1} = u_6 - \frac{2a_6}{\gamma-1}$		$u_6 = -\frac{1}{3} u_p$ $a_6 = a_0 - \frac{\gamma-1}{2} u_p$	$u_6 + a_6 = a_0 - \left(\frac{1}{3} + \frac{\gamma-1}{2}\right) u_p$	$u_6 - a_6 = -a_0 - \left(\frac{1}{3} - \frac{\gamma-1}{2}\right) u_p$
7	$J^+ = u_4 + \frac{2a_4}{\gamma-1} = u_7 + \frac{2a_7}{\gamma-1}$ $J^- = u_6 - \frac{2a_6}{\gamma-1} = u_7 - \frac{2a_7}{\gamma-1}$		$u_7 = -\frac{2}{3} u_p$ $a_7 = a_0 - \frac{2(\gamma-1)}{3} u_p$	$u_7 + a_7 = a_0 - \left(\frac{2}{3} + \frac{2(\gamma-1)}{3}\right) u_p$	$u_7 - a_7 = -a_0 - \left(\frac{2}{3} - \frac{2(\gamma-1)}{3}\right) u_p$
8	$J^+ = u_6 + \frac{2a_6}{\gamma-1} = u_8 + \frac{2a_8}{\gamma-1}$	$u_8 = 0$	$a_8 = a_0 - \frac{2(\gamma-1)}{3} u_p$	$u_8 + a_8 = a_0 - \frac{2(\gamma-1)}{3} u_p$	$u_8 - a_8 = -a_0 + \frac{2(\gamma-1)}{3} u_p$
9	$J^+ = u_7 + \frac{2a_7}{\gamma-1} = u_9 + \frac{2a_9}{\gamma-1}$ $J^- = u_8 - \frac{2a_8}{\gamma-1} = u_9 - \frac{2a_9}{\gamma-1}$		$u_9 = -\frac{1}{3} u_p$ $a_9 = a_0 - \frac{5(\gamma-1)}{6} u_p$	$u_9 + a_9 = a_0 - \left(\frac{1}{3} + \frac{5(\gamma-1)}{6}\right) u_p$	$u_9 - a_9 = -a_0 - \left(\frac{1}{3} - \frac{5(\gamma-1)}{6}\right) u_p$
10	$J^+ = u_9 + \frac{2a_9}{\gamma-1} = u_{10} + \frac{2a_{10}}{\gamma-1}$	$u_{10} = 0$	$a_{10} = a_0 - (\gamma-1) u_p$	$u_{10} + a_{10} = a_0 - (\gamma-1) u_p$	$u_{10} - a_{10} = -a_0 + (\gamma-1) u_p$

$$(u - a) = -a_0 - \left[\frac{1}{2} - \frac{7(\gamma - 1)}{12}\right]u_p.$$

Once the slopes are determined, they are used to determine the locations of the intersections of these characteristics.   □

11.12 The Shock Tube

The capability of achieving rapid and steep temperature increases finds important applications in gas-phase chemical kinetics – for example, determining the temperature for ignition and the ignition delay time under prescribed pressures. Rapid temperature rises can be generated by passing a shock wave through the gas medium. In practice, the shock tube provides a means for such a purpose.

The schematic of a simple shock tube of finite length is shown in Fig. 11.16a. The diaphragm separates the low-pressure (driven) gas from the high-pressure (driver) gas. For chemical kinetics and ignition studies, the low-pressure, driven gas is the gas to be tested (the test gas that is a reactive mixture). For convenience, the pressure-temperature pairs of the driving and driven gases are (p_4, T_4) and (p_1, T_1), respectively. The diaphragm only has a limited mechanical strength designed to rupture, resulting from a sufficient pressure difference across it or by other mechanical and/or electrical means. The initial pressure ratio (p_4/p_1) sets up the condition for the shock strength and the associated temperature rise. The diaphragm would ideally disappear upon its rupture, to avoid flow tripping that causes the flow to cease to be one-dimensional. After the rupture takes place at $t = 0$, the following observations are made.

(1) The pressure discontinuity propagates in two directions: as a shock into the driven/test gas 1, and as a centered rarefaction/expansion wave into the driving gas. A snapshot of the flow regimes at a particular (though arbitrary) time $t = t_1$, prior to occurrences of wave reflection from end walls, is shown in Fig. 11.16b. Figures 11.16c and 11.16d show for $t = t_1$ the pressure and temperature distributions, respectively, throughout the shock tube. At $t = t_1$, the flow regimes within the shock tube are either of the uniform or the simple-wave types, as shown in the wave diagram shown in Fig. 11.16e. The wave pattern as time proceeds, for example, $t_4 > t_3 > t_2 > t_1$, is also illustrated in Fig. 11.16e.

(2) The contact surface (shown in Fig. 11.16b), which is a materials surface and behaves like a gas particle, follows the shock wave in the direction toward the driven gas, as a compression wave carries gas particles. Reasoning based on the expansion wave propagation leads to the same result, as the left-propagating expansion wave pushes gas and the contact surface in the opposite direction, that is, toward the driven gas. In region 2, the particle path, represented by $X_{PP}(t)$, in Fig. 11.16e, is parallel to that of the contact surface, just like the gas near the surface of a moving piston.

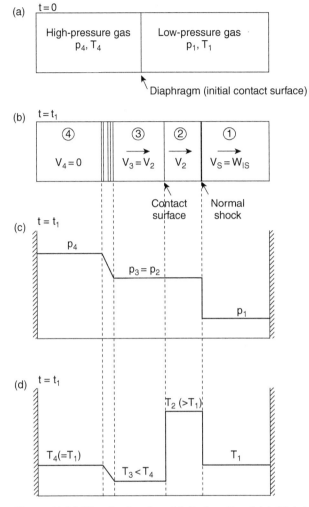

Figure 11.16 The shock tube of finite length – (a) initial state (e.g., before the diaphragm is ruptured); (b) the waves and the states of gases at $t = t_1$ shortly after the rupture; (c) pressure distribution at $t = t_1$; (d) temperature distribution at $t = t_1$; (e) characteristics and wave patterns in the t-x plot.

(3) At $t = t_1$, pressures and velocities across the contact surface match, so that $p_2 = p_3$ (Fig. 11.6c) and $u_2 = u_3$. However, in general $T_2 \neq T_3$ (Fig. 11.16d), as T_2 is raised from T_1 by shock and T_3 decreases from T_4 by expansion.

(4) The pressure ratio p_2/p_1 determines the Mach number of the shock wave and the temperature ratio T_2/T_1. However, p_2/p_1 and, consequently, T_2/T_1 are related to the initial pressure ratio p_4/p_1.

(5) At a later time, expansion and the shock waves reflect from the end walls. For example, the incident shock reflects from the wall at $t = t_2$ (see Fig. 11.16e). The reflected shock meets the contact surface at point a at $t = t_3$. Since $p_2 = p_3$ and $u_2 = u_3$, the reflected shock would propagate into region 3 without

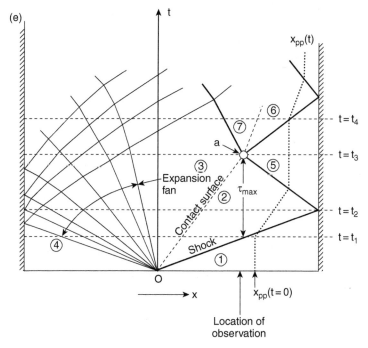

Figure 11.16 (cont.)

interference. In practice, reflection may occur at the contact surface due to several factors, among which are the acoustic impedance resulting from the driver, the driven gases being different, and the non-ideal conditions such as boundary layer development in the shock-induced flow. In Fig. 11.16e, the reflected shock by the contact surface is shown, separating regions 5 and 6. In region 5, $u_5 = 0$ and the gas particle is motionless as depicted by the $X_{PP}(t)$ in the wave diagram.

At $t = t_2$ the shock wave is reflected from the end wall to propagate into region 2. The reflected shock meets the contact surface at $t = t_3$ (i.e., point a, as shown in Fig. 11.16e). For chemical kinetics studies, T_2 is the temperature of interest as well as the time (between the passing of incident shock and t_3) the test gas spends at this temperature. To maximize the test time, the location for observation is chosen for the duration τ_{max} shown in Fig. 11.16e. It is of interest to find T_2/T_1 and τ_{max} as related to the initial pressure ratio, p_4/p_1, as follows.

Assume the gases (or gas mixtures) in Fig. 11.16a are ideal gases. Across the left-running expansion wave (i.e., along C^+ characteristics: also refer to Fig. 11.16b or 11.16e for subscripts),

$$J^+ = u_3 + \frac{2a_3}{\gamma - 1} = u_4 + \frac{2a_4}{\gamma - 1}$$

For expansion, isentropic relation applies. By rearranging the above expression, the pressure ratio across the expansion wave is

$$\frac{p_3}{p_4} = \left(\frac{T_3}{T_4}\right)^{\gamma_4/(\gamma_4-1)} = \left(\frac{a_3}{a_4}\right)^{2\gamma_4/(\gamma_4-1)} = \left(1 - \frac{\gamma_4-1}{2}\frac{u_3}{a_4}\right)^{2\gamma_4/(\gamma_4-1)} \tag{11.89}$$

where $\gamma_3 = \gamma_4$ and $u_4 = 0$. This expression suggests that p_3 and therefore $p_2 \ (= p_3)$ are determined once u_3 is known. Because $u_3 = u_2$ (at the contact surface), the shock relations can be used to find u_3. By combining the following equations

$$\beta \equiv \frac{p_2 - p_1}{p_1} = \frac{2\gamma_1}{\gamma_1+1}\left(M_{1s}^2 - 1\right) \tag{4.16e} \text{ or } (11.9)$$

with M_{1s} being the Mach number with which the shock propagates into the quiescent gas region 1 ($u_1 = 0$). From Eqn. (4.20d) for stationary shock:

$$\frac{u_2 - u_1}{a_1} = -\frac{1}{M_1}\frac{\beta}{\rho_1 a_1^2/p_1} = -\frac{2}{\gamma_1+1}\left(M_1 - \frac{1}{M_1}\right)$$

one finds

$$u_2 = \frac{2}{\gamma_1+1}a_1\left(M_{1s} - \frac{1}{M_{1s}}\right) = u_3 \tag{11.90a}$$

for the moving shock. Equation (11.17) can be used for M_2:

$$M_2 = \frac{2}{\gamma_1+1}\left(M_{1s} - \frac{1}{M_{1s}}\right)\sqrt{\frac{T_1}{T_2}} \tag{11.90b}$$

Combining Eqns. (11.89) and (11.90) yields

$$\frac{p_3}{p_4} = \left[1 - \frac{\gamma_4-1}{\gamma_1+1}\frac{a_1}{a_4}\left(M_{1s} - \frac{1}{M_{1s}}\right)\right]^{2\gamma_4/(\gamma_4-1)} \tag{11.91}$$

The propagation Mach number M_{1s} is found by again using Eqn. (4.16b):

$$\frac{p_2}{p_1} = \frac{2\gamma_1 M_{1s}^2 - (\gamma_1-1)}{\gamma_1+1} \tag{11.92a}$$

$$\frac{T_2}{T_1} = \left(\frac{p_2}{p_1}\right)^{(\gamma_1-1)/\gamma_1} = \frac{\left(1 + \frac{\gamma_1-1}{2}M_{1s}^2\right)\left(\frac{2\gamma_1}{\gamma_1-1}M_{1s}^2 - 1\right)}{\left[\frac{(\gamma_1+1)^2}{2(\gamma_1-1)}\right]M_{1s}^2} \tag{11.92b}$$

The matching condition $p_2 = p_3$ leads to M_{1s} as a function of the initial pressure ratio p_4/p_1, as shown in the following implicit equation for M_{1s}.

$$\frac{p_4}{p_1} = \frac{p_4 p_3 p_2}{p_3 p_2 p_1} = \left[1 - \frac{\gamma_4-1}{\gamma_1+1}\frac{a_1}{a_4}\left(M_{1s} - \frac{1}{M_{1s}}\right)\right]^{-2\gamma_4/(\gamma_4-1)}\left[\frac{2\gamma_1 M_{1s}^2 - (\gamma_1-1)}{\gamma_1+1}\right] \tag{11.93}$$

For strong shock applications (for large T_2/T_1, to initiate chemical reaction), $p_4/p_1 \rightarrow \infty$, which with $\gamma_4/(\gamma_4 - 1) > 0$ and the second bracket is > 0, requiring the expression in the first bracket in Eqn. (11.93) to approach zero. Therefore,

$$M_{1s} - \frac{1}{M_{1s}} \rightarrow \frac{\gamma_1 + 1}{\gamma_4 - 1} \frac{a_4}{a_1} \tag{11.94}$$

For strong shock application, $M_{1s} \gg 1$, yielding

$$M_{1s} \approx \frac{\gamma_1 + 1}{\gamma_4 - 1} \frac{a_4}{a_1} \tag{11.95}$$

It is desirable to complete the flow conditions at $t = t_1$ and those in region 5. By using $u_2 = u_3$, $M_3\sqrt{\gamma_3 R_3 T_3} = M_2\sqrt{\gamma_2 R_2 T_2}$. For ideal gases $\gamma_2 = \gamma_1$, $\gamma_3 = \gamma_4$, $R_2 = R_1$ and $R_3 = R_4$. Therefore, the velocity matching at the contact surface becomes

$$M_3\sqrt{\gamma_4 R_4 T_3} = M_2\sqrt{\gamma_1 R_1 T_2}$$

Across the simple wave regime, $J^+ = u + 2a/(\gamma - 1) = \text{constant}$. With $u_4 = 0$,

$$\frac{2}{\gamma_4 - 1} a_4 = u_3 + \frac{2}{\gamma_4 - 1} a_3 = u_2 + \frac{2}{\gamma_4 - 1} a_3$$

Rearranging the above expression while using Eqn. (11.90), one obtains

$$a_4 = a_3 + \frac{\gamma_4 - 1}{\gamma_1 + 1} a_1 \left(M_{1s} - \frac{1}{M_{1s}} \right) \tag{11.96}$$

With $a \equiv \sqrt{\gamma R T}$, Eqn. (11.96) can be rewritten to obtain the temperature in region 3 as

$$\frac{T_3}{T_4} = \left[1 - \left(\frac{\gamma_4 - 1}{\gamma_1 + 1} \right) \sqrt{\frac{\gamma_1 R_1 T_1}{\gamma_4 R_4 T_4}} \left(M_{1s} - \frac{1}{M_{1s}} \right) \right]^2 \tag{11.97}$$

The pressure in region 3 is given by

$$\frac{p_3}{p_1} = \frac{p_2}{p_1} = \frac{2\gamma_1 M_{1s}^2 - (\gamma_1 - 1)}{\gamma_1 + 1} \tag{11.98}$$

It is noted that M_{1s} is found is using the implicit equation, Eqn. (11.93).

EXAMPLE 11.11 Assume that the test gas in a shock tube experiment consists of a very dilute fuel-air mixture, the driving gas is nitrogen, and that the initial temperature is uniform throughout the entire tube. Find for the strong shock condition, the initial pressure ratio required, and the temperature and the pressure under which the experiment is being conducted.

Solution –

For both nitrogen and the dilute mixture, $\gamma_1 = \gamma_4 = 1.4$ and Eqns. (11.94) and (11.95) yield $M_{1s} = 6.16$ and 6, respectively. For convenience, take $M_{1s} = 6$. Thus

$$\frac{p_2}{p_1} = \frac{2\gamma_1 M_{1s}^2 - (\gamma_1 - 1)}{\gamma_1 + 1} = 41.8$$

The shock relationship gives (for $\gamma = 1.4$; or from the shock table)

$$\frac{T_2}{T_1} = \frac{\left(1 + \frac{\gamma-1}{2} M_{1s}^2\right)\left(\frac{2\gamma}{\gamma-1} M_{1s}^2 - 1\right)}{\left[\frac{(\gamma+1)^2}{2(\gamma-1)}\right] M_{1s}^2} = 7.94$$

To generate the above temperature rise, the required initial pressure ratio is

$$\frac{p_4}{p_1} = \left[1 - \frac{\gamma_4 - 1}{\gamma_1 + 1}\frac{a_1}{a_4}\left(M_{1s} - \frac{1}{M_{1s}}\right)\right]^{-2\gamma_4/(\gamma_4 - 1)}\left[\frac{2\gamma_1 M_{1s}^2 - (\gamma_1 - 1)}{\gamma_1 + 1}\right] = 3.3 \times 10^{12}$$

Comments –

The initial pressure for the given conditions is too enormous to be realistic. Equation (11.95) suggests that it is desirable to have a combination of (1) a large value of a_4/a_1 and (2) a small specific ratio γ_4 (i.e., approaching 1). For example, raising T_4 helps to simultaneously increase a_4 and γ_4. Using polyatomic gas as the driver helps to decrease γ_4. For example, one may use propane (C_3H_8) at 290 K as the driver gas, so that $\gamma_4 = 1.13$. Assume also $T_1 = 290$ K. Thus $p_2/p_1 = 209.3$ and $T_2/T_1 = 4.6$. The initial pressure ratio is then $p_4/p_1 = 4.5 \times 10^5$, which is more manageable, especially for small p_1. When a desired value of T_2/T_1 is known, Eqns. (11.92) and (11.93) are then used to find M_{1s} p_4/p_1, respectively. □

The result of Example 11.11 suggests that to lessen the requirement for large p_4/p_1, observations can be made in region 5 in the wave diagram, because $T_5 > T_2$. Region 5 offers the added benefit that observation is made of the same motionless gas particle. Further benefits can be derived if the acoustic impedances are matched across the contact interface – under this condition there is no reflected shock from the contact and the duration of the observation time is until the reflected expansion wave reaches region 5. It is then of interest to find T_5 and p_5 as functions of p_4/p_1. First of the velocity and Mach number of the reflected shock should be found, as described in the following.

Let the velocity of the reflected shock be V_{sr} in the x direction, with the gas velocity in region 2 relative to the reflected shock equal to $(V_{sr} + u_2)$, while the gas velocity downstream of and relative to the reflected shock (i.e., region 5) is V_{sr}. Then

$$M_{sr} = \frac{V_{sr} + u_2}{a_2} \tag{11.99}$$

Equation (11.92a) can be rewritten with M_{1s} replaced by M_{sr} and subscripts 2 and 1 by 5 and 2, respectively. Further assuming that no reaction has been initiated yet, so that $\gamma_2 = \gamma_1$, one arrives at

$$M_{sr}^2 = \frac{\gamma_1 + 1}{2\gamma_1}\left(\frac{p_5}{p_2}\right) + \frac{\gamma_1 - 1}{2\gamma_1} \tag{11.100}$$

Rewriting Eqn. (4.18) with the aid of Eqns. (11.9) and (11.100) for the reflected shock and assuming that $\gamma_5 = \gamma_1$, one obtains the following

$$\frac{\rho_5}{\rho_2} = \frac{(\gamma_1 + 1)M_{sr}^2}{(\gamma_1 - 1)M_{sr}^2 + 2} = \frac{(\gamma_1 + 1)\left(\frac{p_5}{p_2}\right) + (\gamma_1 - 1)}{(\gamma_1 - 1)\left(\frac{p_5}{p_2}\right) + (\gamma_1 + 1)} \tag{11.101}$$

$$\frac{T_5}{T_2} = \frac{\left(1 + \frac{\gamma_1 - 1}{2}M_{sr}^2\right)\left(\frac{2\gamma_1}{\gamma_1 - 1}M_{sr}^2 - 1\right)}{\left[\frac{(\gamma_1 + 1)^2}{2(\gamma_1 - 1)}\right]M_{sr}^2} = \left(\frac{p_5}{p_2}\right)\frac{\left(\frac{\gamma_1 + 1}{\gamma_1 - 1}\right) + \left(\frac{p_5}{p_2}\right)}{1 + \left(\frac{\gamma_1 + 1}{\gamma_1 - 1}\right)\left(\frac{p_5}{p_2}\right)} \tag{11.102}$$

$$M_{sr} = \frac{V_{sr} + u_2}{a_2} = \sqrt{\frac{\gamma_1 + 1}{2\gamma_1}\left(\frac{p_5}{p_2}\right) + \frac{\gamma_1 - 1}{2\gamma_1}} \tag{11.103}$$

Equations (11.101) through (11.103) calls for the knowledge of p_5. The following derivation is needed for this purpose. First, look at the relationship between M_{sr} and M_{1s}. Continuity across the incident shock requires

$$\frac{V_{sr} + u_2}{V_{sr}} = \frac{\rho_5}{\rho_2}$$

Because $V_{sr} + u_2 = M_{sr}a_2$,

$$u_2 = M_{sr}a_2\left(1 - \frac{\rho_2}{\rho_5}\right)$$

By using Eqn. (11.101) and combining Eqn. (11.90a), one finds

$$u_2 = \frac{2}{\gamma_1 + 1}a_2\left(M_{sr} - \frac{1}{M_{sr}}\right) = \frac{2}{\gamma_1 + 1}a_1\left(M_{1s} - \frac{1}{M_{1s}}\right) \tag{11.104}$$

By using the Rankine-Hugoniot relationship, Eqn. (4.18b), the relation between a_1 and a_2 can be written as

$$\left(\frac{a_2}{a_1}\right)^2 = \frac{T_2}{T_1} = \frac{p_2\rho_1}{p_1\rho_2} = \frac{p_2}{p_1}\left[\frac{\gamma_1 + 1}{\gamma_1 - 1} + \frac{p_2}{p_1}\right]\left[\left(\frac{\gamma_1 + 1}{\gamma_1 - 1}\right)\frac{p_2}{p_1} + 1\right]^{-1} \tag{11.105}$$

In a similar fashion for the incident shock, Eqn. (11.92a) can be rewritten for the reflected shock:

$$\frac{p_5}{p_2} = \frac{2\gamma_1 M_{sr}^2 - (\gamma_1 - 1)}{\gamma_1 + 1} \tag{11.106}$$

Inserting Eqns. (11.92a), (11.105), and (11.106) into Eqn. (11.104), one finds

$$\frac{p_5}{p_2} = \frac{\frac{\gamma_1+1}{\gamma_1-1} + 2 - \frac{p_1}{p_2}}{1 + \left(\frac{\gamma_1+1}{\gamma_1-1}\right)\frac{p_1}{p_2}} \tag{11.107}$$

By using Eqn. (11.98) for p_2/p_1 as a function of M_{1s} and using the relationship $T_5/T_2 = (p_5/p_2)(\rho_2/\rho_5)$, respectively, leads to

$$\frac{p_5}{p_1} = \left[\frac{2\gamma_1 M_{1s}^2 - (\gamma_1 - 1)}{\gamma_1 + 1}\right]\left[\frac{(3\gamma_1 - 1)M_{1s}^2 - 2(\gamma_1 - 1)}{(\gamma_1 - 1)M_{1s}^2 + 2}\right] \tag{11.108}$$

$$\frac{T_5}{T_1} = \frac{[2\gamma_1 M_{1s}^2 + (3 - \gamma_1)][(3\gamma_1 - 1)M_{1s}^2 - 2(\gamma_1 - 1)]}{(\gamma_1 + 1)^2 M_{1s}^2} \tag{11.109}$$

Equations (11.108) and (11.109) allow determining p_5 and T_5, respectively, where M_{1s} has to be first calculated using the implicit relationship, Eqn. (11.93). Thus, by choosing the initial pressure ratio in the driver and the driven chambers and initial pressure p_1 and the initial temperature T_1, the conditions (p_5 and T_5) under which the shock-induced chemical reaction takes place is determined.

EXAMPLE 11.12 A simple shock tube as shown in Fig. 11.16 is used so that the chemical reaction in region 5 is to be observed. For the desired temperature ratio $T_5/T_1 = 7.94$, what is the pressure ratio of the driver to driven gas is needed? Assume that $\gamma = 1.4$ for all regions of gases.

Solution –

$$\frac{T_5}{T_1} = \frac{[2\gamma_1 M_{1s}^2 + (3 - \gamma_1)][(3\gamma_1 - 1)M_{1s}^2 - 2(\gamma_1 - 1)]}{(\gamma_1 + 1)^2 M_{1s}^2} = 7.94$$

One finds

$$M_{1s} = 2.33$$

So that

$$\frac{p_5}{p_1} = \left[\frac{2\gamma_1 M_{1s}^2 - (\gamma_1 - 1)}{\gamma_1 + 1}\right]\left[\frac{(3\gamma_1 - 1)M_{1s}^2 - 2(\gamma_1 - 1)}{(\gamma_1 - 1)M_{1s}^2 + 2}\right] = 89.12$$

Comments –

If observation is to be made in region 2 for the same temperature ratio T_5/T_1, an exceedingly high p_5/p_1(on the order of hundreds of thousands) is needed, as shown in Example 11.11.   □

Problems

Problem 11.1 Show all steps leading to Eqn. (11.15).

Problem 11.2 Derive Eqns. (11.97) and (11.102).

Problem 11.3 Equation (11.26) can be further simplified by introducing a thermodynamic function $F = F(p, s)$ such that

$$F \equiv \int_{p_0}^{p} \left(\frac{dp}{\rho a}\right)_s$$

where p_0 is a reference pressure and the subscript s indicates that entropy is being held constant. The $s = $ const. requirement is consistent with the earlier assumption $\nabla \cdot s = 0$. Show that

$$\left[\frac{\partial}{\partial t} + (u \pm a)\frac{\partial}{\partial x}\right](u \pm F) = 0$$

By comparing Eqn. (11.36),

$$F = \frac{2a}{\gamma - 1}$$

Problem 11.4 Consider a piston in a long tube where the gas is initially at rest. If the piston is withdrawn to the left following the path line $X_p(t)$, find (1) the slopes of the (right-running) characteristics originating from the piston surface, (2) the sonic velocity that the piston can attain, and (3) the piston velocity that would cause a vacuum. Compare these results with those in Problem 4.15 for steady flows.

Problem 11.5 Repeat Example 11.5 with the advancing motion is twice as weak as the withdrawing motion, that is, $u_B = u_p/4, u_c = u_p/2, u_E = u_p$, and $u_p = u_p$.

Problem 11.6 For the piston motion described in Fig. 11.8b, show

(1) that the particle position at the passing of the ith (i.e., C_i^+) characteristic is

$$\tilde{x}_i = 1 + \frac{i(\gamma + 1)}{2n}\tilde{u}_p;$$

(2) that the time the particle spends in the fan is

$$\Delta \tilde{t} = \tilde{t}_n - \tilde{t}_o = \left[1 + \frac{\gamma - 1}{2} \tilde{u}_p \right]^{(1+\gamma)/(1-\gamma)} - 1;$$

(3) the total particle displacement through the expansion fan is

$$\Delta \tilde{x} = \tilde{x}_n - \tilde{x}_o = \frac{(\gamma + 1)}{2} \tilde{u}_p \text{ or } \Delta x = \frac{(\gamma + 1)}{2} \tilde{u}_p$$

Problem 11.7 Discuss the usefulness of $\Delta p = \rho a \Delta u$ for (1) simple acoustic waves, (2) simple nonlinear continuous waves, and (3) shock waves.

Problem 11.8 Consider the shock to be a left-running (C^-) shock in Example 11.8. Show that the results are the same.

Problem 11.9 In Example 11.5, compare the pressures at locations given by the following two sets of points: (1) $2, 5$, and 8, and (2) $4, 5$, and 6, and observe pattern of changes in their sound speeds.

Problem 11.10 The wall boundary in Fig. 11.15 is moving with a constant speed $u_b = a_1/4$ to the right while the gas behind the incident wave is traveling at a speed equal to $a_1/2$. Find the pressure in each of regions $1, 2$, and 3, given that the pressure in region 1 is p_1. Compare these results with those of Example 11.9.

Problem 11.11 Compare the results of the incident wave in Example 11.9 for $u_b = 0$ using exact solutions for traveling shock.

Problem 11.12 For the results of Example 11.10, show that the slopes of characteristics at points 4, 7, 9, and 10 are in the following order: $(u_{10} - a_{10}) > (u_9 - a_9) > (u_7 - a_7) > (u_{14} - a_4)$. Explain this result.

Problem 11.13 Expand the results of Example 11.10 to find the speed of the piston so that the reflected wave does not reach the piston. Compare the result with

$$|u_p|_{\text{escape}} = \frac{2a_o}{\gamma - 1}$$

and explain the finding.

Problem 11.14 The shock tube test as described in Fig. 11.16 has the following properties: $T_1 = 300$ K and $T_4 = 500$ K, $\gamma_4 = 1.1$, $\gamma_1 = 1.4$. To achieve $T_2/T_1 = 5$, find the needed p_4/p_1

Problem 11.15 Derive Eqn. (11.18).

Problem 11.16 For a general expansion wave, consider a piston motion in a cylinder containing initially quiescent gas. The piston trajectory is described by

$$x_p = \frac{1}{2} a t_p^2$$

where t_p is the time counted from the instant when the piston is set in motion. The *t-x* plot of the piston movement is as shown.

(a) Show that the particle location in the initially quiescent gas can be described by

$$(x - x_p) = (u - u_p)(t - t_p) = \left(\frac{\gamma + 1}{2}\alpha\, t_p - a_1\right)(t - t_p),$$

where u_p is the speed of the piston and the subscript 1 designates the initially quiescent gas.

(b) By defining the dimensionless parameters: $\tilde{t}_p \equiv \alpha t_p/a_1$, $\tilde{t} \equiv \alpha t/a_1$, and $\tilde{x} \equiv \alpha x/a_1^2$, show that

$$\tilde{x}_p = \frac{1}{2}\tilde{t}_p^2$$

and

$$\tilde{t}_p = \frac{1}{\gamma}\left\{\left(1 + \frac{\gamma + 1}{2}\tilde{t}\right) + \left[1 - (\gamma - 1)\tilde{t} + \left(\frac{\gamma + 1}{2}\right)^2\tilde{t}^2 - 2\gamma\tilde{x}\right]^{\frac{1}{2}}\right\}$$

(c) Show that the speed of sound and the pressure in the gas, respectively, are

$$\frac{a}{a_1} = 1 - \frac{\gamma - 1}{2}\tilde{t}_p$$

$$\frac{p}{p_1} = \left[1 - \frac{\gamma - 1}{2}\tilde{t}_p\right]^{\frac{2\gamma}{\gamma - 1}}$$

(d) Show that the piston reaches the sonic and escape speeds, respectively, at

$$\tilde{t}_p^* = \frac{2}{\gamma + 1}$$

$$\tilde{t}_{p,\ escape} = \frac{2}{\gamma - 1}$$

And that the escape speed is reached at the location

$$\tilde{x}_p = \frac{2}{(\gamma - 1)^2}$$

Problem 11.17 Consider $X_p = A\sin\omega t$ in a long tube containing quiescent gas at $t = 0$. Assume that $A = a_1/2\omega$, a_1 being the speed of sound in the undisturbed gas.

Find the time and the location of the first shock formation, if the shock formation takes place at all. Explain your results by examining the slopes of the characteristics and compare the finding with those of Example 11.7.

Problem 11.18 In a long tube, a moving normal shock wave is propagating against an $M = 2$ air flow (at 1 atm and 400 K), bringing it to a full stop relative to a laboratory observer. Determine (1) the wave speed as observed by the laboratory observer, (2) the pressure ratio across the shock wave, and (3) the temperature in the stagnant air behind the wave and, by comparing this with the stagnation temperature in the incoming flow, explain what is found.

Problem 11.19 A one-dimensional shock wave is moving normal to itself into stagnant air (at 300 K) causing a pressure ratio $p_2/p_1 = 6$, with 1 and 2 denoting the upstream and downstream regions of the shock, respectively. Find the Mach number behind the shock, T_{t2}, p_{t2} and compare these results with those of a stationary normal shock with the same p_2/p_1.

Problem 11.20 The piston in a long piston-cylinder device is suddenly withdrawn with a constant speed u_p ($\ll a_1$), as shown in the following figure, where some representative incident characteristics (which form an expansion fan) are also shown. Region 1 is the undisturbed, uniform region, while region 2 is the uniform region downstream of the expansion fan. How would the first reflected characteristic, denoted by AB, curve in the non-simple region?

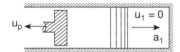

Problem 11.21 The piston in a long piston-cylinder device is suddenly advanced with a constant speed u_p ($\ll a_1$) to generate compression waves leading to a normal shock, as shown in the following figure. For the condition given in the figure, find the values of (a) u_p and (b) V_{rs}, the speed of the reflected shock.

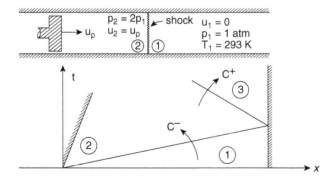

Problem 11.22 A long cylinder has two pistons that are suddenly set into relative motion at $t = 0^+$ with constant speeds, with u_L and u_R being the speed of the left and

right pistons, respectively. At $t = 0$, the gas between the two pistons is at rest (designated as the state 1) and at $t = t_i$ the waves generated by the piston motion begin to interact and various flow regimes are as shown in the figure, where the dash lines are either compression or expansion waves depending on whether the piston motion is advancing or withdrawing. Find u_4, a_4, and p_4/p_1 for the following scenarios, where $u_p \ll a_1$ –

(1) $u_L = u_R = u_p > 0$
(2) $u_R = u_p > 0$ and $u_L = -u_p < 0$
(3) $u_R = 2u_p > 0$ and $u_L = u_p > 0$
(4) $u_R = 2u_p > 0$ and $u_L = 0$

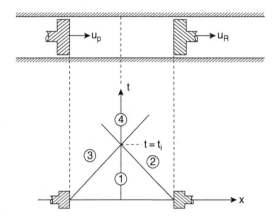

Problem 11.23 A piston is set into an advancing motion in a very long tube where the initial state (denoted as state 1) of the gas is quiescent. The piston speed $u_p(t)$ is such that a normal shock wave (speed $= W_{1s}$) forms ahead of it. For $u_p(t) \ll a_1$

(1) Show that $\frac{W_{1s}}{a_1} = \left(\frac{\gamma+1}{4}\right)\left(\frac{u_p}{a_1}\right) + \sqrt{1 + \left[\left(\frac{\gamma+1}{4}\right)\left(\frac{u_p}{a_1}\right)\right]^2}$

(2) Show that $W_{1s} = \frac{1}{2}\left(\sum \text{speeds of the two characteristics that intersect to form the shock}\right)$

(3) Find W_{1s} for the piston motion described in Example 11.7.

(4) The pressure rise when the shock is formed.

Problem 11.24 A diaphragm in a long tube enables separation of two regions of unequal pressures. After the diaphragm is ruptured, a shock and an expansion waves form and propagate in opposite direction, as shown in the figure, where the flow regimes are also denoted by 1, 2, 3, and 4. The original contact surface is also set in motion. Find the difference between the shock wave and the contact surface speeds, $(W_{1s} - W_{cs})$ for the following scenarios –

(1) Very weak shock
(2) Very strong shock

(3) Use the method of characteristics to show that $u_2 > 0$.

(4) Determine u_2/a_1 and find its strong shock value.

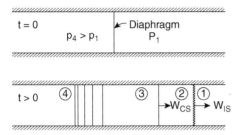

Problem 11.25 A piston in a long tube is suddenly withdrawn at a constant speed, u_p, as shown in the figure. (a) Find the pressure on the piston face using the method of characteristics. (b) For $u_p \ll a_1$, using binomial expansion to show that this pressure is same as that found using the momentum equation.

Problem 11.26 Consider the same device as in Problem 11.25, except that the piston is advancing with a speed $u_p = at$, where a is the instant acoustic speed. Find the pressure at the piston face and determine when it becomes infinity.

Problem 11.27 Show steps leading to Eqns. (11.44a) and (11.44b).

12 Introduction to Inviscid Hypersonic Flows

Hypersonic flows are characterized by high Mach numbers and high temperatures resulting from strong shocks. They are characterized by the associated oblique shocks with small wave angles generated at the leading edge of the body, as typically occurs in the case of reentry vehicles. It is expected that due to the small shock wave angle, the interaction between the shock wave and the boundary can be significant. Due to the strong shocks, the temperature rises significantly to cause dissociation of gas molecules and reactions among them. The large increase in temperature due to shock or aerodynamic heating leads to challenges in the design of space vehicle. The gas mixture behind the shock wave cannot be assumed to be a perfect gas. Without treating the detailed aspects of the shock-boundary interaction and chemical reactions in such flows, this chapter focuses on the key effect of a large Mach number on the behaviors of shock and expansion waves in inviscid flows.

12.1 Relationships for Shock Waves in Hypersonic Flow

A hypersonic flow is one in which Mach number is much greater than unity, and therefore the free-stream Mach number M_1 is such that

$$M_1 \equiv \frac{V_\infty}{a_\infty} \gg 1 \tag{12.1}$$

The shock relations obtained in Chapter 4 can be used by letting $M_1 \to \infty$ and $M_{1n}^2 = M_1^2 \sin^2\theta \gg 1$. Thus,

$$M_{2n}^2 = \frac{M_{1n}^2 + \frac{2}{\gamma-1}}{\frac{2\gamma}{\gamma-1}M_{1n}^2 - 1} \to \frac{\gamma-1}{2\gamma} \text{ and } M_{2n} \to \sqrt{\frac{\gamma-1}{2\gamma}} \tag{12.2}$$

$$\frac{p_2}{p_1} = \left(\frac{2\gamma M_{1n}^2}{\gamma+1} - \frac{\gamma-1}{\gamma+1}\right) \to \frac{2\gamma}{\gamma+1}M_1^2\sin^2\theta \tag{12.3}$$

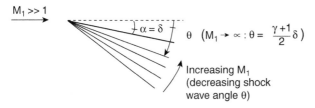

Figure 12.1 A flat plate at an angle of attack to free-stream hypersonic flow ($M_1 \gg 1$); variation of the shock wave angle, θ, with M_1 and the angle of attack, α ($= \delta$), shown schematically and mathematically.

$$\frac{T_2}{T_1} = \frac{\left(1 + \frac{\gamma-1}{2}M_{1n}^2\right)\left(\frac{2\gamma}{\gamma-1}M_{1n}^2 - 1\right)}{\left[\frac{(\gamma+1)^2}{2(\gamma-1)}\right]M_{1n}^2} \rightarrow \frac{2\gamma(\gamma-1)}{(\gamma+1)^2}M_{1n}^2 = \frac{2\gamma(\gamma-1)}{(\gamma+1)^2}M_1^2\sin^2\theta \quad (12.4)$$

$$\frac{\rho_2}{\rho_1} = \frac{V_{1n}}{V_{2n}} = \frac{M_{1n}\sqrt{\gamma RT_1}}{M_{2n}\sqrt{\gamma RT_2}} = \left[\frac{(\gamma+1)M_{1n}^2}{(\gamma-1)M_{1n}^2 + 2}\right] \rightarrow \frac{\gamma+1}{\gamma-1} \quad (12.5)$$

$$\beta \equiv \frac{p_2 - p_1}{p_1} = \left[\frac{2\gamma}{\gamma+1}\left(M_{1n}^2 - 1\right)\right] \rightarrow \frac{2\gamma}{\gamma+1}M_1^2\sin^2\theta \quad (12.6)$$

To reduce the shock strength, hypersonic vehicles are designed to produce a small shock wave angle θ, which in turn requires a small deflection angle δ (i.e., slim body shapes). For small deflection angles ($\delta \ll 1$) and $M_1^2\sin^2\theta \gg 1$ Eqn. (4.33c) continues to hold. With $\sin\theta \approx \theta$, $\cos 2\theta \approx 1$, and $\tan\delta \approx \sin\delta \approx \delta$, Eqn. (4.33c) simplifies to

$$\theta = \frac{\gamma+1}{2}\delta \quad (12.7)$$

This result indicates a linear relationship between the shock wave angle and the deflection (or wedge) angles. Inspection of Fig. 4.12 (for $\gamma = 1.4$ and $\theta = 1.2\delta$) provides support for the linearity expressed in Eqn. (12.7) as $M_1 \rightarrow \infty$ over a range of δ, up to approximately $\delta = 35°$ (i.e., $\delta \approx 0.611$), which is surprisingly larger than permitted by the assumption of $\delta \ll 1$. The result of Eqn. (12.7) is schematically shown in Fig. 12.1, where a flat plate is at an angle of attack to the free-stream flow at $M_1 \gg 1$.

EXAMPLE 12.1

Access the validity of Eqn. (12.7) for $\delta = 5°$ and $35°$ for an infinite free-stream Mach number $M_1 \rightarrow \infty$, and for $\delta = 30°$ and $M_1 = 10$.

Solution –
For $\delta = 5° \approx 0.0873$, $\theta \approx 6° = 1.2\delta$. In fact for $\delta = 35° \approx 0.611$, $\theta \approx 42.08° \approx 1.202\delta$, quite a pleasantly surprising result considering the $35°$ deflection is not quite a small angle. For other finite, but still large values of M_1, the linear relation

between θ and δ is limited o a smaller range of δ, as can be seen from Fig. 4.12. For finite values of M_1, the deviations from the prediction using Eqn. (12.7) increase with decreasing M_1. Such a result is expected, because of the underlying assumption that M_1 be infinity, and can be clearly visualized as the intercept on the θ axis increases with decreasing value of M_1. For example for $M_1 = 10$ and $\delta = 30°$, Fig. 4.12 gives, $\theta \approx 39.25° \approx 1.308\delta$, with Eqn. (12.7) under-predicting by approximately 9%. □

For the pressure coefficient across an oblique shock, Eqn. (4.57) results in

$$C_p \equiv \frac{2}{\gamma M_1^2} \frac{p_2 - p_1}{p_1} = \frac{2}{\gamma M_1^2} \beta = \frac{4}{\gamma + 1} \sin^2\theta \tag{12.8}$$

For small deflection angles ($\delta \ll 1$ and $\sin\theta \approx \theta = [\gamma + 1]\delta/2 \ll 1$),

$$C_p \approx (\gamma + 1)\delta^2 \tag{12.9}$$

Recalling for small δ and moderate values of M_1, Eqn. (4.35)

$$\beta = \frac{p_2 - p_1}{p_1} = \frac{2\gamma}{\gamma + 1}(M_1^2 \sin^2\theta - 1) = \left(\frac{\gamma M_1^2}{\sqrt{M_1^2 - 1}}\right) \cdot \delta \tag{4.35}$$

can be used to obtain

$$C_p = \frac{\pm 2\delta}{\sqrt{M_1^2 - 1}} \tag{4.59c}$$

where the positive and the negative signs denote, respectively, deflections toward the free stream (i.e., compression) and away from the free-stream flow (i.e., expansion). Comparing Eqns. (4.59c) and (12.9) reveals that for small values of δ: As $M_1 \to \infty$, C_p is independent of M_1 and $C_p \propto \delta$ for small δ and, as opposed to that for moderate and small supersonic values of M_1, C_p is a function of both δ and M_1.

12.2 Practical Considerations of Hypersonic Flow of Non-Perfect Gases

Consider a hypersonic flow around a slender body of revolution, with a blunt nose/leading edge, shown in Fig. 12.2. A slender body is of practical importance, as a thick body incurs tremendous pressure losses and is not at all appealing for use. In the nose region, the shock wave is detached and has the form of a normal shock. Away from the nose region, a bow shock forms, because the shock strength weakens in the direction away from the nose region (i.e., $\beta \neq$ constant) along the bow shock, while further away the shock wave takes up the form of an oblique shock. Therefore, an entropy layer forms along the surfaces (see the discussion of Fig. 4.13). Since $M_1 \gg 1$, Eqn. (12.2) indicates that $M_2 = 0.378$, which causes the temperature in the nose

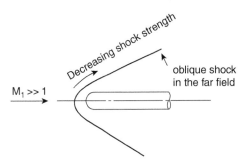

Figure 12.2 A hypersonic flow around a slender body of revolution, with a blunt nose/leading edge, with the shock transitioning from a bow shock in the nose region to an oblique shock in the far field.

region to increase to nearly stagnation value. For $M_1 = 20$, $T_2 = 77.78\, T_1$ (from Eqn. (12.4)) which at an altitude of 30 km based on the ideal gas properties and $\gamma = 1.4$, T_2 is approximately 17,000 K, compared to the stagnation value of 17,600 K on the surface of the nose. Such high temperatures have several effects, including dissociation and ionization of gases (air in the atmosphere) and chemical reactions. For example, the dissociation of O_2 and N_2 in air leads to formation of NO (nitric oxide), which is a non-equilibrium process. Ablation of the surface materials may also occur under such harsh environments. High temperatures lead to strong thermal radiation from both gases and surfaces. Because ionization and formation of new species require energy, and radiation emits heat, the temperature of the newly established equilibrium mixture further downstream of the shock wave would be lower, although still high.

As a result of ionization and dissociation, the assumption for ideal gases is no longer valid. Furthermore, determining the gas composition in this region (and in the hypersonic boundary layer where the low gas velocity causes a high "recovery" temperature near the stagnation value) requires non-equilibrium thermodynamic and chemical models. Validation of these models is as challenging as obtaining experimental data for gaseous species in such environments (traditional sampling probes or non-intrusive optical methods) and is not a trivial matter. Consequently, thermal and physical properties (such as c_p, c_v, and γ) of the gases are not easily determined because of the change in gas composition across the shock wave, which also varies along the shock wave with non-uniform strength along its surface.

Knowledge of how internal energy storage depends on temperature can help shed light on the value of c_v. The dissociation results from the increasingly vigorous vibration between atoms that breaks down their bonds. The high temperature therefore has significant effects on the heat capacity c_v in the following manner. Combining the result for c_v at moderate temperatures from Chapter 1 (i.e., $c_v = \xi R/2$, where ξ is the number of degrees of freedom of the molecular translation and rotation) and the contribution from vibration leads to

$$c_v = \frac{\xi}{2}R + c_{v,vib} \tag{12.11}$$

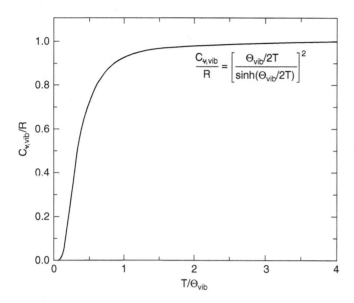

Figure 12.3 Variation of $c_{v,vib}$, the constant-volume heat capacity due to vibration of the molecule, with temperature.

where $c_{v,vib}$ represent the constant-volume heat capacity due to vibration of the molecule.

It is useful to note the "vibrational temperature," Θ_{vib}, which is the characteristic temperature above which the contribution from the vibrational motion to the total heat capacity c_v is not negligible. At high temperatures the electrons are also "excited"; however, at very high temperatures, the energy level of electrons varies only very slightly and does not contribute significantly to c_v. The values of Θ_{vib} for O_2, N_2, and NO are 2,270 K, 3,390 K, and 2,740 K, respectively (Vincent and Kruger, 1975). The temperature dependence of $c_{v,vib}$ on temperature is

$$c_{v,vib} = R\left[\frac{\Theta_{vib}/T}{\sinh(\Theta_{vib}/2T)}\right]^2 \tag{12.12}$$

Results of Eqn. (12.12) are shown in Fig. 12.3. It can be seen that as the temperature approaches 17,000 K, the mixture of the diatomic gases of interest (O_2, N_2, and NO; $\zeta = 5$) possesses a constant-volume specific heat equal to

$$c_v = \frac{\zeta}{2}R + R = \frac{\zeta + 2}{2}R \tag{12.13}$$

Assuming that the gas behind the shock wave comprises of mainly O_2, N_2, and NO, the specific heat ratio is thus

$$\gamma = \frac{c_p}{c_v} = \frac{\frac{\zeta+2}{2}R}{\frac{\zeta+2}{2}R} = 1 \tag{12.14}$$

At moderate temperatures, values of c_v vary and increase with temperature, resulting in a temperature regime of variable γ. In the flow around the body shown in

Fig. 12.2, the value of γ also varies from one region to another because of the non-uniform shock strength and temperature jump across the shock wave system (for the same reason for the entropy layer). Therefore it is not unusual to choose an effective value of γ (γ_{eff}) for the convenience of solving the problem, bypassing the difficulty of ascertaining the detailed and spatially varying composition and temperature of the gas undergoing dissociation reactions. For diatomic gases (of which air is one), the effective value lies between unity and 1.4. In the limit of $M_1 = \infty$, $\gamma = \gamma_{\text{eff}} = 1$. Under these conditions, Eqn. (12.7) becomes

$$\theta = \delta \tag{12.15}$$

This result has several implications. First, the shock wave is inclined at the same angle as the two-dimensional wedge. In reality, θ is slightly larger than δ, making the interaction between the boundary layer and the shock wave significant. For the qualitative analysis, further consider the results for $M_1 \to \infty$ shown in Fig. 4.11 (where a constant $\gamma = 1.4$ is used and in the limit of M_1 ∞, $\theta \approx 1.2\delta$). Figure 4.11 indicates that $\theta\theta$ decreases with increasing M_1 and that for the constant value γ the shock wave moves toward the surface as M_1 is increased. Equation (12.15) suggests that in hypersonic flows, such a movement with increasing M_1 brings the shock wave to the surface; such a movement is illustrated in Fig. 12.1, over a thin flat plate. The shock wave-boundary layer interaction is outside the scope of this chapter; for this topic one is referred to, for example, Anderson (1989) and Park (1990).

With $M_1 = \infty$ and $\gamma = \gamma_{\text{eff}} = 1$, Eqns. (12.9a) and (12.9b) for a flat plate at an angle of attack α becomes

$$C_p = 2\sin^2\theta = 2\sin^2\alpha \tag{11.16a}$$

Equation (11.16a) is the so-called *sine-squared law* for hypersonic flow. Because $\alpha = \delta \ll 1$ and $\sin\theta \approx \theta = [\gamma_{\text{eff}} + 1]\delta/2 \ll 1$, it becomes

$$C_p \approx (\gamma + 1)\delta^2 = 2\theta^2 = 2\alpha^2 \tag{11.16b}$$

It is of interest to be able to obtain values of C_p on the surface of a blunt body in hypersonic flow (Fig. 12.4). For the portion of the surface of the blunt body that is normal to the streamline, a normal shock wave forms. The flow between the normal shock wave and the body decelerate to zero velocity, forming a stagnation point at the surface, where the pressure is the stagnation pressure ($p_2 = p_{t2}$) behind the shock and the surface pressure coefficient is the maximum on the body, denoted as C_{pmax}. Let the subscript 2 designate the state behind the shock. Thus

$$C_{p,\text{max}} = \frac{(p_{t2} - p_1)}{\frac{1}{2}\rho_1 V_1^2} \tag{12.17}$$

The pressure coefficient elsewhere can be expressed as (Thompson, 1972)

$$C_p = C_{p,\text{max}}\sin^2\theta \tag{12.18a}$$

or

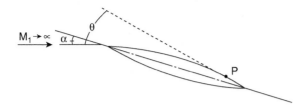

Figure 12.4 Schematic showing the local angle of attack, θ, formed by the tangent to the airfoil surface and the flow direction, and the angle of attack formed by the chord line, α, in hypersonic flow.

$$\frac{C_p}{C_{p,max}} = \sin^2\theta \qquad (12.18b)$$

where θ is now taken as the local angle of attack, or the inclination angle to the free-stream flow, of the surface location away from the stagnation. Such a local angle of θ (e.g., of point P) is illustrated in Fig. 12.4.

EXAMPLE 12.2

Estimate the Mach number and static temperature of the post-shock region for the flat-plate configuration shown in Fig. 8.3 with $\alpha = 35°$. The given flow conditions are $M_1 = 20$ and $T_1 = 220$ K.

Solution –

For the free stream, the isentropic relationship Eqn. (4.6c) can be used and for $\gamma = 1.4$

$$\frac{T_{1t}}{T_1} = 1 + \frac{\gamma-1}{2}M_1^2 = 1 + 0.2 \times 20^2 = 81; \ T_{1t} = 81 \times 220K = 17,820 \ K$$

Oblique shock theory

Immediately downstream of the shock wave where dissociation/ionization/reaction have not occurred yet, $\gamma = 1.4$. From Fig. 4.12 $\theta = 45°$, then $M_{1n} = M_1 \sin\theta = 14.14$. The shock relationship Eqn. (4.22d) provides

$$\frac{T_2}{T_1} = \frac{\left(1 + \frac{\gamma-1}{2}M_1^2\sin^2\theta\right)\left(\frac{2\gamma}{\gamma-1}M_1^2\sin^2\theta - 1\right)}{\left[\frac{(\gamma+1)^2}{2(\gamma-1)}\right]M_1^2\sin^2\theta} = 39.82$$

Or using $M_1^2\sin^2\theta \gg 1$

$$\frac{T_2}{T_1} = \frac{2\gamma(\gamma-1)}{(\gamma+1)^2}M_1^2\sin^2\theta = 38.89$$

Therefore $T_2 \approx 8,556 - 8,760 \ K$. The shock relationship, Eqn. (4.22a), gives M_{2n} as

$$M_{2n}^2 = \frac{M_{1n}^2 + \frac{2}{\gamma-1}}{\frac{2\gamma}{\gamma-1}M_{1n}^2 - 1}$$

$$M_{2t} = 0.3828$$

$$M_{2t} = M_{1t}\sqrt{\frac{T_1}{T_2}} = M_1\cos\theta\sqrt{\frac{T_1}{T_2}} = 20 \times \cos 45° \times \sqrt{\frac{1}{39.82}} = 2.241$$

$$M_2 = 2.2735$$

It should be noted that the estimation of T_2 is based on $\gamma = 1.4$, which is itself contradictory as $8,556\ K$ exceeds the vibrational temperature for typical species present behind the shock wave $\Theta_{vib} = 2{,}270$ K, $3{,}390$ K, and $2{,}740$ K for O_2, N_2, and NO, respectively.

Hypersonic theory

For $M_1 = 20$, assume that $M_1 \to \infty$, then $\gamma_{eff} \to 1$ in this region. Further downstream of the shock wave, a new equilibrium exists and $\gamma_{eff} \to 1$. As a first approximation, one may still assume an ideal gas in this region and that the gas velocity does not change due to a new equilibrium. Also $\theta = \delta = \alpha = 35°$. One readily find $M_{1n} = M_1\sin\theta = 20 \times \sin 35° = 11.47$. However, both

$$\frac{T_{2t}}{T_2} = 1 + \frac{\gamma_{eff} - 1}{2}M_2^2 = 1 \quad \text{and}$$

$$M_{2n} \to \sqrt{\frac{\gamma_{eff} - 1}{2\gamma_{eff}}} \to 0$$

are apparently not the case because $M_{2n} = M_2\sin\theta$ would imply that $\theta = 0$ and a stagnation flow downstream of the shock, violating the given condition $\theta = 35°$. One can use the l'Hospital rule for the following result

$$M_{2n}^2 = \lim_{\gamma \to 1}\left[\frac{M_{1n}^2 + \frac{2}{\gamma-1}}{\frac{2\gamma}{\gamma-1}M_{1n}^2 - 1}\right] = \frac{1}{M_{1n}^2} \quad \text{or}\ M_{2n} = \frac{1}{M_{1n}} = \frac{1}{11.47} = 0.08718$$

$$\frac{M_{2t}}{M_{1t}} = \sqrt{\frac{T_1}{T_2}}$$

Also $\theta = \delta = \alpha = 35°$.

Assume that the momentum flux in the tangential direction along the shock wave remains the same. Then following Eqn. (4.21d),

$$V_{2t} = V_{1t} \quad \text{and}$$

$$\frac{T_2}{T_1} = \frac{\left(1 + \frac{\gamma-1}{2} M_1^2 \sin^2\theta\right)\left(\frac{2\gamma}{\gamma-1} M_1^2 \sin^2\theta - 1\right)}{\left[\frac{(\gamma+1)^2}{2(\gamma-1)}\right] M_1^2 \sin^2\theta} = 26.5227$$

$$M_{2t} = M_{1t}\sqrt{\frac{T_1}{T_2}} = M_1 \cos\theta\sqrt{\frac{T_1}{T_2}} = 20 \times \cos 35° \times \sqrt{\frac{1}{26.5227}} = 3.1812$$

$$M_2 - 3.1824$$

One can see that the hypersonic theory ($\gamma_{\text{eff}} \to 1$) predicts a smaller T_2 and a larger M_2 than the oblique shock theory assuming $\gamma = 1.4$.   □

EXAMPLE 12.3 Fine the pressure coefficients for a flat plate in a hypersonic air flow at angles of attack of $5°$ and $35°$, respectively.

Solutions –
For $\alpha = 5° = 0.08731$, $C_p \approx 2\alpha^2 = 0.0152$ (also $C_p = 2\sin^2\alpha = 0.0152$).

For $\alpha = 35° = 0.611$, which is not much smaller than unity, $C_p = 2\sin^2\alpha = 0.658$ ($2\alpha^2$ would give a value of 0.746).   □

EXAMPLE 12.4 Find an expression for C_{pmax} and C_p of a blunt body in hypersonic flow. The angle θ is as shown in the figure.

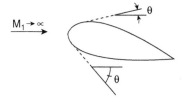

Solution –
The stagnation pressure behind the normal shock wave in Eqn. (11.17), p_{t2}, is related to the free-stream stagnation, p_{t1}, by Eqn. (4.19d):

$$\frac{p_{t2}}{p_{t1}} = \left[\frac{\frac{\gamma+1}{2} M_1^2}{1 + \frac{\gamma-1}{2} M_1^2}\right]^{\gamma/(\gamma-1)} \left[\frac{1}{\frac{2\gamma}{\gamma+1} M_1^2 - \frac{\gamma-1}{\gamma+1}}\right]^{1/(\gamma-1)}$$

$$\approx = \left[\frac{\frac{\gamma+1}{2} M_1^2}{1 + \frac{\gamma-1}{2} M_1^2}\right]^{\gamma/(\gamma-1)} \left[\frac{1}{\frac{2\gamma}{\gamma+1} M_1^2}\right]^{1/(\gamma-1)} \quad (*)$$

By using the ideal gas law, $p = \rho RT$, $C_{p,max}$ can be rewritten as

$$C_{p,max} = \frac{(p_{t2} - p_1)}{\frac{1}{2}\rho_1 V_1^2} = \frac{(p_{t2} - p_1)}{\frac{\gamma p_1}{2} \frac{V_1^2}{\gamma R T_1}} = \left(\frac{p_{t2}}{p_1}\right)\left(\frac{2}{\gamma M_1^2}\right) - \frac{2}{\gamma M_1} \approx \left(\frac{p_{t2}}{p_1}\right)\left(\frac{2}{\gamma M_1^2}\right)$$

$$= \left(\frac{p_{t2}}{p_{t1}}\right)\left(\frac{p_{t1}}{p_1}\right)\left(\frac{2}{\gamma M_1^2}\right)$$

Substituting (*) and the isentropic relation for p_{t1}/p_1, with $M_1 \gg 1$, into this expression yields

$$C_{p,max} = \left[\frac{\frac{\gamma+1}{2} M_1^2}{1 + \frac{\gamma-1}{2} M_1^2}\right]^{\gamma/(\gamma-1)} \left[\frac{2\gamma}{\gamma+1} M_1^2\right]^{-1/(\gamma-1)} \left[\frac{\gamma-1}{2} M_1^2\right]^{\gamma/(\gamma-1)} \left[\frac{2}{\gamma M_1^2}\right]$$

$$= \left[\frac{\gamma+1}{\gamma}\right]^{\frac{\gamma}{\gamma-1}} \left[\frac{\gamma+1}{4}\right]^{\frac{1}{\gamma-1}}$$

which is independent of M_1 for $M_1 \gg 1$.

$$C_p = C_{p,max} \sin^2\theta$$

For $\gamma = 1.4$,

$$C_p = 1.839 \sin^2\theta$$

For $\gamma = \gamma_{\text{eff}} = 1$,

$$C_p = 2 \sin^2\theta$$

which recovers the result shown by Eqn. (12.16a) for a flat plate; for the correct result, θ is the local angle relative to the free stream. □

12.3 Lift and Drag in Hypersonic Flow

An engineering method of calculating pressure on surfaces in hypersonic flows is the *Newtonian impact theory*. The theory requires that fluid particles in the free stream only impact the frontal surface area of the object and that they cannot flow around the object. The latter requirement appears to be reasonable as high-speed flows have large inertia that prevents turning. For illustration, consider an infinitely thin flat plate at an angle of attack α to a hypersonic stream, as shown in Fig. 12.5. It is noted that $\theta = \alpha$ everywhere on the lower surface of a flat plate. The control volume in

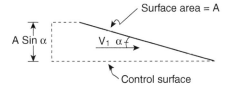

Figure 12.5 Schematic of an infinitely thin flat plate at an angle of attack α to a hypersonic stream, showing relevant parameters for the Newtonian impact theory.

Fig. 12.5 is drawn such that there is no flow over the upper surface, as there is zero inlet area over the upper surface. As a consequence, $p_U = p_1$ and $C_{pU} = 0$. In contrast, the flow impacting the lower surface enters the control surface through an area equal to $A \sin \alpha$. Applying the Reynolds transport theorem, Eqn. (2.8), for the inviscid flow under steady state condition leads to

$$\sum \vec{F} = -\int_{CS} (p_L - p_1) \cdot d\vec{A} = \int_{CS} \vec{V} \left(\rho \vec{V} \cdot d\vec{A} \right)$$

For the pressure force acting normal to the flat plate,

$$p_L - p_1 = \rho_1 V_1^2 \sin^2 \alpha \qquad (12.19)$$

which in turn leads to the sine-squared law for the lower surface:

$$C_{pL} = 2 \sin^2 \alpha \qquad (12.20)$$

In arriving at Eqn. (11.20), conditions of $M_1 \rightarrow \infty$, the presence of a shock wave and $\gamma = \gamma_{\text{eff}}$ (i.e., high-temperature thermodynamics) are not invoked. Furthermore, Eqn. (12.20) suggests the independence of pressure coefficient on M_1. The *modified Newtonian impact theory* can be obtained by considering shock relations, where γ affects the pressure ratio across a shock wave. Thus, Eqn. (12.8) essentially represents the modified theory – for a flat plate, $\theta = \alpha$ and $C_p = [4/(\gamma + 1)] \sin^2 \alpha < 2 \sin^2 \alpha$ for $\gamma > \gamma_{\text{eff}}$ as already demonstrated in Example 11.4.

For a body of an arbitrary shape, θ varies from one surface location to another and thus $\theta \neq \alpha$. By using the local angle of attack θ in place of α in Eqn. (12.20), C_{pL} is independent of both γ and M_1 for $M_1 \rightarrow \infty$. The independence of γ is advantageous because the composition arising from dissociation of gases need not be known. Equation (12.20) can thus be extended to different gas species in the free stream.

Together with the assumptions and results of Eqns. (12.16a) and (12.16b), one expects the accuracy of Newtonian impact theory to improve as M_1 and θ (α in the case of a flat plate) increase. Such improvement of the Newtonian impact theory might be expected since, as M_1 is increased toward ∞, the very large flow inertia renders going around the object increasingly difficult, consistent with the assumption of the theory. Increasing both M_1 and θ causes the backside of the object not to be reached by the flow, thus becoming a shadow region. According to shock theory, increasing both M_1 and θ increases shock strength and the temperature behind the shock, leading to the larger degrees of dissociation and pushing the value of γ toward 1.

The coefficient 2 in Eqn. (11.20) is in strikingly good agreement with that of Eqn. (11.16a). This agreement for $M_1 \rightarrow \infty$ is worth noting as the Newtonian impact theory does not consider the high-temperature effects on gas properties shown by Eqns. (12.11) through (12.14), as the Reynolds transport theorem is concerned with the rate of momentum change that does not depend on composition of the gas.

Following similar procedures for the thin airfoils in Section 4.10, the lift coefficient for the thin flat-plate airfoil (where the angle of attack is equal to the flow deflection angle, $\alpha = \delta$) would generate lift and drag coefficients as

$$C_L = \left(C_{pL} - C_{pU}\right)\cos\alpha = 2\sin^2\alpha\cos\alpha \ (\approx 2\alpha^2 \text{ for small } \alpha) \tag{12.21}$$

$$C_D = \left(C_{pL} - C_{pU}\right)\sin\alpha = 2\sin^3\alpha \ (\approx 2\alpha^3 \text{ for small } \alpha) \tag{12.22}$$

Comparison with lift and drag coefficients for thin airfoil at small angles of attack in supersonic flows, Eqns. (4.59a) and (4.59b), reveals (1) the lift and drag coefficients for thin supersonic airfoil depend on the free-stream Mach number and vary with α and α^2, respectively, while (2) the lift and drag coefficients in hypersonic flows are independent of the free-stream Mach number and are proportional to α^2 and α^3, respectively. It is left as an end-of-chapter problem (Problem 12.1) to show that C_L for a flat plate in hypersonic flow reaches a maximum of approximately 0.770 at $\alpha \approx 55°$ and that the C_L/C_D ratio deteriorates rapidly as α is increased.

It can also be shown that Newtonian theory predicts the drag coefficient for a large-aspect ratio circular cylinder, with the span placed perpendicular to the flow, to be 4/3. Similarly, for a sphere, the drag coefficient is 1. The process of arriving at these drag coefficients is left as an end-of-chapter problem (Problem 12.2).

12.4 Hypersonic Similarity

Recalling the similarity variables for subsonic and supersonic flows described in Chapter 9, one might ask for the existence of similarity variables for hypersonic flow over slender bodies (characterized by, for example, small deflection angle δ). The following steps demonstrate how such a similarity variable is obtained. First, consider the oblique shock relation governing the deflection and shock wave angles applies for hypersonic flows:

$$\tan\delta = 2\cot\theta \frac{M_1^2\sin^2\theta - 1}{M_1^2(\gamma + \cos 2\theta) + 2} \tag{4.27}$$

By taking advantage of smaller values of δ and θ, for which $\tan\delta \approx \delta$, $\sin\theta \approx \theta$, $\cos 2\theta \approx 1$, and $\cos\theta \approx 1$, Eqn. (4.27) can be rearranged to become

$$\delta = \frac{2}{\theta}\left[\frac{M_1^2\theta^2 - 1}{M_1^2(\gamma + 1) + 2}\right] \tag{12.23}$$

Rearranging Eqn. (11.23) leads to

$$M_1^2\theta^2 - 1 = \left[\frac{M_1^2(\gamma + 1)}{2} + 1\right]\delta\theta \approx \left[\frac{M_1^2(\gamma + 1)}{2}\right]\delta\theta \tag{12.24}$$

where $M_1^2(\gamma+1)/2 \gg 1$ is valid but $M_1^2\theta^2 \gg 1$ cannot be assumed because θ is small. Note that if constants 1 and 2 are neglected in Eqn. (12.23), Eqn. (12.7) is recovered: $\theta = \delta(\gamma+1)/2$. Equation (11.24) is then cast in the form of the quadratic equation:

$$\left(\frac{\theta}{\delta}\right)^2 - \frac{\gamma+1}{2}\left(\frac{\theta}{\delta}\right) - \frac{1}{M_1^2\delta^2} = 0 \tag{12.25}$$

which has the solution:

$$\frac{\theta}{\delta} = \frac{\gamma+1}{4} + \sqrt{\left(\frac{\gamma+1}{4}\right)^2 + \frac{1}{M_1^2\delta^2}} \tag{12.26}$$

Equation (12.26) indicates that the ratio of the shock wave angle to the deflection angle is only a function of the product M_1 and δ; that is, $\theta/\delta = f(M_1\delta)$. In the explicit form, $\theta = f(M_1, \delta)$, which is not similar to the weak solution for oblique shock waves described in Chapter 4 by Eqn. (4.34).

One now can relate the pressure ratios shown in Eqns. (12.3) with θ given by Eqn. (12.26) to the $M_1\delta$ as the following

$$\frac{p_2}{p_1} = \frac{2\gamma}{\gamma+1}M_1^2\sin^2\theta \approx \frac{2\gamma}{\gamma+1}M_1^2\theta^2 = \frac{2\gamma}{\gamma+1} + \frac{\gamma(\gamma+1)}{4}M_1^2\delta^2$$

$$+\gamma M_1^2\delta^2\sqrt{\left(\frac{\gamma+1}{4}\right)^2 + \frac{1}{M_1^2\delta^2}} \tag{12.27a}$$

Similarly,

$$\frac{T_2}{T_1} = \frac{2\gamma(\gamma-1)}{(\gamma+1)^2} + \frac{\gamma(\gamma-1)}{4}M_1^2\delta^2 + \frac{\gamma(\gamma-1)}{(\gamma+1)}M_1^2\delta^2\sqrt{\left(\frac{\gamma+1}{4}\right)^2 + \frac{1}{M_1^2\delta^2}} \tag{12.27b}$$

Equation (12.5) remains unchanged:

$$\frac{\rho_2}{\rho_1} = \frac{V_{1n}}{V_{2n}} = \left[\frac{(\gamma+1)M_1^2\sin^2\theta}{(\gamma-1)M_1^2\sin^2\theta + 2}\right] \approx \frac{\gamma+1}{\gamma-1} \tag{12.27c}$$

while Eqn. (12.6), the shock strength, becomes

$$\beta \equiv \frac{p_2 - p_1}{p_1} = \frac{\gamma-1}{\gamma+1} + \frac{\gamma(\gamma+1)}{4}M_1^2\delta^2 + \gamma M_1^2\delta^2\sqrt{\left(\frac{\gamma+1}{4}\right)^2 + \frac{1}{M_1^2\delta^2}} \tag{12.27d}$$

It is of significance to note that in Eqns. (11.28a) through (11.28d), the common factor $M_1\delta$ determines the values of these ratios. Therefore

$$K \equiv M_1\delta \tag{12.28}$$

is the *hypersonic similarity parameter*.

Equations (11.27a) through (11.27d) can now be rewritten as

$$\frac{p_2}{p_1} = \frac{2\gamma}{\gamma+1} + \frac{\gamma(\gamma+1)}{4} K^2 + \gamma K^2 \sqrt{\left(\frac{\gamma+1}{4}\right)^2 + \frac{1}{K^2}} \tag{12.29a}$$

$$\frac{T_2}{T_1} = \frac{2\gamma(\gamma-1)}{(\gamma+1)^2} + \frac{\gamma(\gamma-1)}{4} K^2 + \frac{\gamma(\gamma-1)}{(\gamma+1)} K^2 \sqrt{\left(\frac{\gamma+1}{4}\right)^2 + \frac{1}{K^2}} \tag{12.29b}$$

$$\beta = \frac{\gamma-1}{\gamma+1} + \frac{\gamma(\gamma+1)}{4} K^2 + \gamma K^2 \sqrt{\left(\frac{\gamma+1}{4}\right)^2 + \frac{1}{K^2}} \tag{12.29c}$$

The pressure coefficient, shown in Eqn. (11.9a), can be expressed as

$$C_p \equiv \frac{2}{\gamma M_1^2} \frac{p_2 - p_1}{p_1} = \frac{2}{\gamma M_1^2} \left[\frac{\gamma-1}{\gamma+1} + \frac{\gamma(\gamma+1)}{4} M_1^2 \delta^2 + \gamma M_1^2 \delta^2 \sqrt{\left(\frac{\gamma+1}{4}\right)^2 + \frac{1}{M_1^2 \delta^2}} \right]$$

which can be further reduced to

$$C_p = 2\delta^2 \left[\frac{\gamma+1}{4} + \sqrt{\left(\frac{\gamma+1}{4}\right)^2 + \frac{1}{K^2}} \right] \tag{12.29d}$$

Equation (12.29d) shows that the pressure coefficient on a slender body in hypersonic flow is proportional to δ^2 for a given value of $K \equiv M_1\delta$, or

$$\frac{C_p}{\delta^2} = f(K, \gamma) \tag{11.30}$$

The result of Eqn. (11.30) is in contrast to the pressure coefficient for moderately supersonic values of M_1, expressed by Eqn. (4.59c):

$$C_p = \frac{\pm 2\delta}{\sqrt{M_1^2 - 1}} \quad (+ \text{ for oblique shocks; for moderately supersonic values of } M_1)$$

$$\tag{4.59c}$$

12.5 Hypersonic Expansion

Consider the centered Prandtl-Meyer expansion of a hypersonic flow ($M_1 \to \infty$) around a corner through an angle of δ, as shown in Fig. 12.6. Adopting the same notations as for supersonic expansion, (M) is the Prandtl-Meyer function:

$$\nu(M) = \sqrt{\frac{\gamma+1}{\gamma-1}} \tan^{-1} \sqrt{\frac{\gamma-1}{\gamma+1}(M^2 - 1)} - \tan^{-1} \sqrt{M^2 - 1} \tag{4.47c}$$

As $M_1 \to \infty$, Eqn. (4.47c) becomes

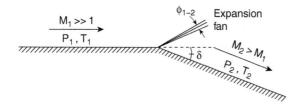

Figure 12.6 Hypersonic flow expansion around a corner.

$$\nu(M) = \sqrt{\frac{\gamma + 1}{\gamma - 1}} \tan^{-1} \sqrt{\frac{\gamma - 1}{\gamma + 1}} M - \tan^{-1} M \qquad (12.31)$$

The trigonometric identity for the $\tan^{-1}$ function provides

$$\tan^{-1} M = \frac{\pi}{2} - \tan^{-1} \frac{1}{M} \qquad (12.32a)$$

Similarly,

$$\tan^{-1} \sqrt{\frac{\gamma - 1}{\gamma + 1}} M = \frac{\pi}{2} - \tan^{-1} \sqrt{\frac{\gamma + 1}{\gamma - 1}} \frac{1}{M} \qquad (12.32b)$$

For $\varepsilon \ll 1$, the series expansion

$$\tan^{-1} \varepsilon = \varepsilon - \frac{1}{3} \varepsilon^3 + \frac{1}{5} \varepsilon^5 - \frac{1}{7} \varepsilon^7 + \dots \qquad (12.32c)$$

can be used for $1/M \ll 1$, while $\sqrt{(\gamma + 1)/(\gamma - 1)}$ is of order unity for expansion, in Eqns. (12.32a) and (12.32b). By manipulating Eqns. (11.33a) and (11.33b), substituting the results into Eqn. (12.31), and then retaining the first-order terms yields:

$$\nu(M) = \left[\sqrt{\frac{\gamma + 1}{\gamma - 1}} - 1 \right] \frac{\pi}{2} - \frac{2}{\gamma - 1} \frac{1}{M} \qquad (12.33)$$

Recalling that the turning angle δ is related to the Prandtl-Meyer function by

$$\delta = \Delta \nu_{1-2} = \nu(M_2) - \nu(M_1) = \frac{2}{\gamma - 1} \left[\frac{1}{M_1} - \frac{1}{M_2} \right] \qquad (12.34)$$

For Prandtl-Meyer expansion, $M_2 > M_1$ and $\delta > 0$.

The expansion fan angle, ϕ, according to Eqn. (4.49) is

$$\phi_{1-2} = \Delta \nu_{1-2} + \sin^{-1} \frac{1}{M_1} - \sin^{-1} \frac{1}{M_2} == \nu(M_2) - \nu(M_1) + \sin^{-1} \frac{1}{M_1} - \sin^{-1} \frac{1}{M_2}$$

Because $M_2 > M_1$ and $M_1 \to \infty$, both δ (for no flow separation) and ϕ_{1-2} are expected to be small. For $M_1 = 13$ and $M_2 = 15$, $\delta = 2.94°$ and $\phi_{1-2} = 3.53°$.

Assuming no flow separation due to the hypersonic turning around the angle δ, the isentropic pressure relation holds. For both M_2 and $M_1 \to \infty$, Eqn. (4.4) becomes

$$\frac{p_2}{p_1} = \left(\frac{1 + \frac{\gamma-1}{2} M_1^2}{1 + \frac{\gamma-1}{2} M_2^2}\right)^{\gamma/(\gamma-1)} = \left(\frac{M_1}{M_2}\right)^{2\gamma/(\gamma-1)} \tag{12.35}$$

By rewriting Eqn. (12.34) as

$$\frac{M_1}{M_2} = 1 - \frac{\gamma-1}{2} M_1 \delta \tag{12.36}$$

and substituting this result into Eqn. (12.35) yields

$$\frac{p_2}{p_1} = \left(1 - \frac{\gamma-1}{2} M_1 \delta\right)^{2\gamma/(\gamma-1)} \tag{11.37}$$

The variable $M_1 \delta$ is defined as the hypersonic similarity parameter for expansion: $K \equiv M_1 \delta$. Thus

$$\frac{p_2}{p_1} = \left(1 - \frac{\gamma-1}{2} K\right)^{2\gamma/(\gamma-1)} \tag{12.38a}$$

Other ratios can be similarly found to be

$$\frac{T_2}{T_1} = \frac{1 + \frac{\gamma-1}{2} M_1^2}{1 + \frac{\gamma-1}{2} M_2^2} = \left(1 - \frac{\gamma-1}{2} K\right)^{2\gamma} \tag{12.38b}$$

$$\beta = \frac{p_2 - p_1}{p_1} = \left(1 - \frac{\gamma-1}{2} K\right)^{2\gamma/(\gamma-1)} - 1 \tag{12.38c}$$

$$C_p = \frac{2}{\gamma M_1^2} \frac{p_2 - p_1}{p_1} = \frac{2}{\gamma M_1^2} \left[\left(1 - \frac{\gamma-1}{2} K\right)^{2\gamma/(\gamma-1)} - 1\right]$$

$$= \frac{2\delta^2}{\gamma K^2} \left[\left(1 - \frac{\gamma-1}{2} K\right)^{2\gamma/(\gamma-1)} - 1\right] \tag{12.38d}$$

Similar to the pressure coefficient found for oblique shock, Eqn. (12.38d) can be cast in the functional form as

$$\frac{C_p}{\delta^2} = f(K, \gamma) \tag{12.39}$$

Problems

Problem 12.1 Use the Newtonian impact theory to find the angle of attack for which the lift coefficient is maximum for a thin flat plate in a hypersonic free stream. What are the values of C_L/C_D and C_D when C_L is maximum? To visualize these results, plot these parameters as a function of α.

Problem 12.2 Use the Newtonian impact theory to find the drag coefficient for a long cylinder (with span placed perpendicular to the flow) and a sphere in hypersonic flow.

Problem 12.3 Show that the pressure coefficient at the stagnation point of a blunt body in hypersonic air flow, $C_{p,max}$ is equal to 2 by using the Newtonian impact theory and that

$$\frac{C_p}{C_{p,max}} = \sin^2\theta$$

Problem 12.4 By using the normal shock theory to find the pressure at the stagnation point of a blunt body in hypersonic air low, show that

$$C_{p,max} = 4/(\gamma + 1)$$

$$C_p = \frac{4}{\gamma + 1} \sin^2\theta$$

When this result is used for Problem 11.3, a *modified Newtonian impact theory* is obtained, where the value of γ is a function of temperature behind the normal shock wave and $C_{p,max} = 1.667$ for $\gamma = 1.4$ (no dissociation of gases due to shock heating) and 2 for $\gamma = 1$ (vibration modes of gas molecules fully excited); the realistic value of $C_{p,max}$ should fall between 1.667 and 2. A modified theory assuming $\gamma = 1.4$ predicts that $C_p = 1.839 \sin^2\theta$.

Problem 12.5 Use the modified Newtonian theory shown in Problem 12.4 to obtain the drag force per diameter on a sphere flying at hypersonic speed. Also show that the drag coefficient is

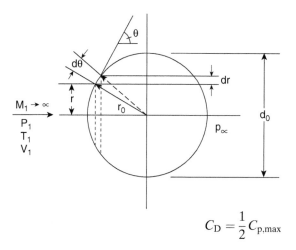

$$C_D = \frac{1}{2} C_{p,max}$$

Problem 12.6 Show that a hypersonic flow over a two-dimensional wedge with a small turning angle, δ, results in the downstream Mach number

$$M_2 \approx \sqrt{\frac{2}{\gamma(\gamma - 1)} \frac{1}{\delta}}$$

(Note that this result is not valid for $\gamma = \gamma_{\text{eff}} \to 1$. The value of M_2 is strongly dependent on γ, pointing to the importance of knowing the values of c_p, c_v, and γ for hypersonic approximations.) What can be said about the Mach number variation in a hypersonic flow over a blunt body, where a bow shock forms and the turning angle diminishes in the downstream direction away from the stagnation point region?

APPENDIX A

Universal Physical Constants

Table A.1 *Universal Physical Constants*

h	= Planck's constant	= 6.26192×10^{-34} J·s
k	= Boltzmann's constant	= 1.38054×10^{-23} J/K
N_A	= Avogadro's number	= 6.02252×10^{23} 1/mol
$\hat{R}$	= universal gas constant	= 8.3143 kJ/kmol · K

APPENDIX B

Properties of Some Ideal Gases

Table B.1 *Thermodynamic Constant for Common Gases**

	Chemical Symbol	Molecular Weight	$c_p(\text{J/kg} \cdot \text{K})$ at 298.15 K	$c_v(\text{J/kg} \cdot \text{K})$ at 298.15 K	γ	$R(\text{J/kg} \cdot \text{K})$
Air		28.96	1004.0	716.5	1.400	287.06
Argon	Ar	39.95	520.3	312.2	1.667	208.13
Carbon Dioxide	CO_2	44.01	841.8	652.9	1.289	188.92
Carbon Monoxide	CO	28.01	1041.3	744.5	1.398	296.83
Helium	He	4.00	5192.6	3115.6	1.667	2077.03
Hydrogen	H_2	2.02	14209.1	10084.9	1.409	4124.18
Methane	CH_4	16.04	2253.7	1735.4	1.299	518.35
Nitrogen	N_2	28.01	1041.6	744.8	1.400	296.80
Oxygen	O_2	32.00	921.6	661.8	1.393	259.83
Water	H_2O	18.02	1872.3	1410.8	1.327	461.5

* *Sources* – (1) *Handbook of Chemistry and Physics*, 74th Edition, CRC Press (1993); (2) *Introduction to Thermodynamics, Classical and Statistical*, 3rd Edition, R.E. Sonntag and G.J. Van Wylen, John Wiley & Sons (1991).

APPENDIX C

Tables for Isentropic and Prandtl-Meyer Expansion Flows

Note – The numerical values of the table can be directly obtained from the following equations. They are tabulated for $\gamma = 1.4$ convenience. (The subscript t denotes stagnation properties; the superscript * denotes critical condition for $M = 1$.)

$$\frac{T}{T_t} = \left(1 + \frac{\gamma - 1}{2} M^2\right)^{-1} \tag{4.6c}$$

$$\frac{p}{p_t} = \left(1 + \frac{\gamma - 1}{2} M^2\right)^{-\gamma/(\gamma-1)} \tag{4.7}$$

$$\frac{\rho}{\rho_t} = \left(1 + \frac{\gamma - 1}{2} M^2\right)^{-1/(\gamma-1)} \tag{4.8}$$

$$\nu(M) = \sqrt{\frac{\gamma + 1}{\gamma - 1}} \tan^{-1} \sqrt{\frac{\gamma - 1}{\gamma + 1}(M^2 - 1)} - \tan^{-1}\sqrt{M^2 - 1} \quad \text{(Prandtl-Meyer Function)} \tag{4.47f}$$

$$\frac{A}{A^*} = \frac{1}{M}\left[\left(\frac{2}{\gamma + 1}\right)\left(1 + \frac{\gamma - 1}{2} M^2\right)\right]^{(\gamma+1)/2(\gamma-1)} \tag{5.14}$$

γ = 1.4

Mach Number	Isentropic Temp Ratio T/T_t	Isentropic Pressure Ratio p/p_t	Isentropic Density Ratio ρ/ρ_t	Isentropic Area Ratio A/A^*	Prandtl-Meyer Function v (degrees)
0	1	1	1		–
0.02	0.999920006	0.99972005	0.999800028	28.94213019	–
0.04	0.999680102	0.998880806	0.999200448	14.48148593	–
0.06	0.999280518	0.997484077	0.998202266	9.665910065	–
0.08	0.998721636	0.995532872	0.996807154	7.261609645	–
0.1	0.998003992	0.993031385	0.995017448	5.82182875	–
0.12	0.997128271	0.989984975	0.992836132	4.864317646	–
0.14	0.996095306	0.986400146	0.990266835	4.182399799	–
0.16	0.994906081	0.982284517	0.987313813	3.672738634	–
0.18	0.99356172	0.977646787	0.983981938	3.277926451	–
0.2	0.992063492	0.972496703	0.980276677	2.96352	–
0.22	0.990412804	0.966845014	0.976204074	2.7076021	–
0.24	0.988611199	0.960703428	0.971770732	2.495562452	–
0.26	0.986660352	0.954084563	0.966983786	2.31728731	–
0.28	0.984562067	0.947001894	0.961850883	2.165553576	–
0.3	0.982318271	0.939469698	0.956380153	2.035065262	–
0.32	0.979931013	0.931503001	0.950580183	1.921851275	–
0.34	0.977402455	0.923117512	0.944459989	1.822875752	–
0.36	0.974734872	0.914329565	0.938028987	1.735778282	–
0.38	0.971930643	0.905156057	0.931296964	1.658696166	–
0.4	0.968992248	0.895614383	0.924274043	1.59014	–
0.42	0.965922263	0.885722374	0.916970659	1.5289048	–
0.44	0.962723352	0.875498229	0.90939752	1.474005374	–
0.46	0.959398265	0.864960453	0.901565579	1.42462855	–
0.48	0.955949832	0.854127793	0.893486002	1.380097348	–
0.5	0.952380952	0.843019175	0.885170134	1.33984375	–
0.52	0.948694596	0.831653642	0.876629471	1.303387758	–
0.54	0.944893794	0.820050293	0.867875626	1.270321102	–
0.56	0.940981632	0.808228222	0.858920296	1.240294439	–
0.58	0.936961247	0.796206467	0.849775239	1.213007209	–
0.6	0.932835821	0.784003951	0.840452235	1.188199506	–
0.62	0.928608573	0.771639428	0.830963068	1.165645539	–
0.64	0.924282757	0.75913144	0.821319487	1.145148306	–
0.66	0.919861653	0.746498261	0.811533189	1.126535244	–
0.68	0.915348565	0.733757861	0.801615788	1.109654642	–
0.7	0.910746812	0.720927861	0.791578791	1.094372679	–
0.72	0.906059727	0.708025494	0.781433577	1.080570939	–
0.74	0.901290648	0.695067572	0.771191372	1.06814435	–
0.76	0.896442915	0.682070453	0.760863231	1.056999429	–
0.78	0.891519863	0.669050012	0.750460018	1.047052813	–
0.8	0.886524823	0.656021618	0.739992386	1.03823	–
0.82	0.88146111	0.643000108	0.729470763	1.030464279	–
0.84	0.876332025	0.629999769	0.718905336	1.023695814	–
0.86	0.871140846	0.617034321	0.708306037	1.017870853	–
0.88	0.865890828	0.604116905	0.697682531	1.012941048	–
0.9	0.860585198	0.591260072	0.687044203	1.008862865	–

(continued)

γ = 1.4

Mach Number	Isentropic Temp Ratio T/T_t	Isentropic Pressure Ratio p/p_t	Isentropic Density Ratio ρ/ρ_t	Isentropic Area Ratio $A/A*$	Prandtl-Meyer Function v (degrees)
0.92	0.855227148	0.578475773	0.676400152	1.005597068	–
0.94	0.849819838	0.565775357	0.665759179	1.003108275	–
0.96	0.844366387	0.553169566	0.65512978	1.001364565	–
0.98	0.838869875	0.540668533	0.644520144	1.000337135	–
1	0.833333333	0.528281788	0.633938145	1	0
1.02	0.827759751	0.516018259	0.623391338	1.000329724	0.125688396
1.04	0.822152065	0.503886279	0.612886959	1.001305188	0.350982534
1.06	0.816513162	0.491893592	0.60243192	1.002907379	0.636686713
1.08	0.810845874	0.480047365	0.592032814	1.005119207	0.968039969
1.1	0.805152979	0.468354195	0.58169591	1.007925341	1.336200924
1.12	0.799437196	0.456820126	0.571427159	1.011312058	1.735039347
1.14	0.793701187	0.445450658	0.561232193	1.015267118	2.159960005
1.16	0.787947554	0.434250764	0.55111633	1.01977964	2.607346221
1.18	0.782178837	0.423224904	0.541084575	1.024840005	3.074255071
1.2	0.776397516	0.412377042	0.53114163	1.030439753	3.558233358
1.22	0.770606005	0.401710664	0.521291894	1.036571503	4.057198466
1.24	0.764806657	0.391228791	0.511539469	1.043228873	4.569357035
1.26	0.759001761	0.380934004	0.50188817	1.050406417	5.093147097
1.28	0.753193541	0.370828457	0.492341526	1.058099555	5.627195529
1.3	0.747384155	0.360913895	0.482902792	1.066304519	6.170285886
1.32	0.7415757	0.351191678	0.473574954	1.075018305	6.721333523
1.34	0.735770204	0.341662794	0.464360737	1.084238622	7.279365987
1.36	0.729969633	0.332327881	0.455262611	1.093963852	7.843507289
1.38	0.724175888	0.323187244	0.446282802	1.104193011	8.412965121
1.4	0.718390805	0.314240876	0.437423299	1.114925714	8.987020318
1.42	0.712616156	0.305488472	0.428685862	1.126162143	9.565018091
1.44	0.706853653	0.29692945	0.420072031	1.137903018	10.14636065
1.46	0.701104941	0.288562969	0.411583134	1.150149572	10.73050094
1.48	0.695371607	0.280387945	0.403220296	1.162903522	11.3169373
1.5	0.689655172	0.272403066	0.394984446	1.176167052	11.90520883
1.52	0.683957102	0.264606813	0.38687633	1.189942793	12.49489142
1.54	0.6782788	0.256997471	0.378896512	1.204233798	13.08559425
1.56	0.67262161	0.249573148	0.37104539	1.219043533	13.67695673
1.58	0.66698682	0.242331786	0.3633232	1.234375854	14.26864583
1.6	0.661375661	0.23527118	0.355730024	1.250235	14.86035366
1.62	0.655789308	0.228388988	0.348265801	1.266625574	15.45179536
1.64	0.650228881	0.221682746	0.340930329	1.283552533	16.04270725
1.66	0.644695446	0.21514988	0.333723282	1.30102118	16.63284507
1.68	0.639190018	0.208787717	0.326644207	1.319037148	17.22198252
1.7	0.633713561	0.202593498	0.31969254	1.337606397	17.80990982
1.72	0.628266988	0.19656439	0.312867608	1.3567352	18.39643254
1.74	0.622851163	0.190697491	0.306168637	1.376430139	18.9813704
1.76	0.617466904	0.184989848	0.299594758	1.396698095	19.56455633
1.78	0.61211498	0.179438457	0.293145019	1.417546247	20.14583544
1.8	0.606796117	0.17404028	0.286818382	1.438982058	20.72506425

(continued)

$\gamma = 1.4$

Mach Number	Isentropic Temp Ratio T/T_t	Isentropic Pressure Ratio p/p_t	Isentropic Density Ratio ρ/ρ_t	Isentropic Area Ratio A/A^*	Prandtl-Meyer Function ν (degrees)
1.82	0.601510996	0.168792248	0.280613736	1.461013275	21.30210986
1.84	0.596260256	0.16369127	0.274529902	1.483647923	21.87684925
1.86	0.591044494	0.158734239	0.268565634	1.5068943	22.44916864
1.88	0.585864267	0.153918044	0.262719631	1.530760974	23.01896282
1.9	0.580720093	0.149239569	0.256990537	1.555256776	23.58613467
1.92	0.575612452	0.144695702	0.251376949	1.580390801	24.15059458
1.94	0.570541786	0.140283341	0.245877418	1.606172401	24.71226003
1.96	0.565508505	0.1359994	0.24049046	1.632611184	25.27105507
1.98	0.560512981	0.13184081	0.235214553	1.659717013	25.82690997
2	0.555555556	0.127804525	0.230048146	1.6875	26.37976081
2.02	0.550636536	0.123887527	0.224989661	1.715970508	26.92954915
2.04	0.5457562	0.120086828	0.220037496	1.745139146	27.47622164
2.06	0.540914795	0.116399472	0.215190032	1.775016769	28.01972978
2.08	0.536112541	0.112822542	0.210445632	1.805614476	28.56002957
2.1	0.531349628	0.109353159	0.205802645	1.836943609	29.0970813
2.12	0.526626222	0.105988484	0.201259412	1.869015753	29.63084922
2.14	0.521942461	0.102725722	0.196814265	1.901842733	30.16130138
2.16	0.517298461	0.099562123	0.192465532	1.935436615	30.68840936
2.18	0.512694311	0.096494985	0.188211539	1.969809704	31.21214807
2.2	0.508130081	0.093521652	0.184050612	2.004974545	31.73249557
2.22	0.503605818	0.090639517	0.179981077	2.040943923	32.24943286
2.24	0.499121546	0.087846025	0.176001267	2.077730858	32.76294373
2.26	0.494677273	0.085138668	0.17210952	2.115348612	33.27301459
2.28	0.490272984	0.082514993	0.168304182	2.153810685	33.77963431
2.3	0.485908649	0.079972598	0.164583606	2.193130815	34.28279408
2.32	0.481584219	0.077509131	0.16094616	2.233322979	34.78248726
2.34	0.47729963	0.075122294	0.15739022	2.274401391	35.2787093
2.36	0.473054799	0.072809841	0.153914179	2.316380509	35.77145756
2.38	0.468849631	0.070569578	0.150516443	2.359275026	36.26073121
2.4	0.464684015	0.068399364	0.147195432	2.403099877	36.74653115
2.42	0.460557828	0.066297109	0.143949586	2.447870237	37.22885988
2.44	0.456470932	0.064260773	0.14077736	2.493601525	37.70772142
2.46	0.452423179	0.06228837	0.137677229	2.540309398	38.1831212
2.48	0.448414407	0.060377962	0.134647686	2.588009758	38.65506599
2.5	0.444444444	0.058527663	0.131687243	2.63671875	39.12356383
2.52	0.44051311	0.056735636	0.128794434	2.686452762	39.5886239
2.54	0.43662021	0.055000093	0.125967812	2.737228428	40.05025652
2.56	0.432765545	0.053319291	0.123205953	2.789062629	40.50847302
2.58	0.428948904	0.05169154	0.120507452	2.841972489	40.9632857
2.6	0.425170068	0.050115191	0.117870929	2.895975385	41.41470778
2.62	0.421428812	0.048588645	0.115295023	2.951088938	41.86275331
2.64	0.417724903	0.047110345	0.112778397	3.007331021	42.30743714
2.66	0.414058101	0.04567878	0.110319736	3.06471976	42.74877486
2.68	0.410428159	0.044292482	0.107917748	3.123273528	43.18678274
2.7	0.406834825	0.042950025	0.105571162	3.183010955	43.62147769

(continued)

$\gamma = 1.4$

Mach Number	Isentropic Temp Ratio T/T_t	Isentropic Pressure Ratio p/p_t	Isentropic Density Ratio ρ/ρ_t	Isentropic Area Ratio A/A^*	Prandtl-Meyer Function v (degrees)
2.72	0.403277842	0.041650025	0.103278733	3.243950925	44.05287723
2.74	0.399756948	0.040391136	0.101039235	3.306112575	44.48099942
2.76	0.396271874	0.039172056	0.098851468	3.369515302	44.90586283
2.78	0.39282235	0.037991519	0.096714251	3.434178758	45.32748652
2.8	0.3894081	0.036848298	0.094626429	3.500122857	45.74588996
2.82	0.386028844	0.035741201	0.092586868	3.567367771	46.16109305
2.84	0.382684301	0.034669076	0.090594455	3.635933935	46.57311605
2.86	0.379374184	0.033630801	0.088648101	3.705842048	46.98197956
2.88	0.376098207	0.032625293	0.086746739	3.777113071	47.38770447
2.9	0.372856078	0.0316515	0.084889322	3.849768233	47.79031199
2.92	0.369647504	0.030708402	0.083074827	3.923829029	48.18982355
2.94	0.366472192	0.029795014	0.08130225	3.999317224	48.58626084
2.96	0.363329845	0.028910377	0.07957061	4.076254852	48.97964573
2.98	0.360220167	0.028053566	0.077878944	4.154664218	49.37000029
3	0.357142857	0.027223684	0.076226314	4.234567901	49.75734674
3.02	0.354097618	0.02641986	0.074611799	4.315988754	50.14170747
3.04	0.351084148	0.025641255	0.073034499	4.398949905	50.52310495
3.06	0.348102147	0.024887052	0.071493532	4.483474761	50.9015618
3.08	0.345151314	0.024156464	0.06998804	4.569587006	51.27710069
3.1	0.342231348	0.023448726	0.068517179	4.657310605	51.64974438
3.12	0.339341948	0.022763101	0.067080126	4.746669805	52.0195157
3.14	0.336482812	0.022098872	0.065676079	4.837689136	52.3864375
3.16	0.333653641	0.021455347	0.064304249	4.930393413	52.75053266
3.18	0.330854133	0.020831856	0.06296387	5.024807738	53.11182409
3.2	0.32808399	0.020227752	0.06165419	5.1209575	53.47033468
3.22	0.325342911	0.019642408	0.060374476	5.218868377	53.82608734
3.24	0.322630601	0.019075215	0.059124011	5.318566341	54.17910494
3.26	0.319946761	0.018525588	0.057902097	5.420077652	54.52941032
3.28	0.317291096	0.017992959	0.05670805	5.523428868	54.87702629
3.3	0.31466331	0.017476778	0.055541201	5.628646842	55.22197561
3.32	0.312063112	0.016976514	0.0544009	5.735758723	55.56428098
3.34	0.309490208	0.016491653	0.053286511	5.844791961	55.90396503
3.36	0.306944308	0.016021698	0.052197411	5.955774305	56.24105033
3.38	0.304425124	0.015566168	0.051132995	6.068733808	56.57555935
3.4	0.301932367	0.015124598	0.05009267	6.183698824	56.90751449
3.42	0.299465753	0.014696539	0.049075859	6.300698014	57.23693806
3.44	0.297024998	0.014281555	0.048081998	6.419760349	57.56385225
3.46	0.294609819	0.013879226	0.047110536	6.540915103	57.88827915
3.48	0.292219936	0.013489146	0.046160936	6.664191866	58.21024077
3.5	0.289855072	0.01311092	0.045232674	6.789620536	58.52975896
3.52	0.287514951	0.012744169	0.044325239	6.917231326	58.84685549
3.54	0.285199297	0.012388524	0.043438131	7.047054766	59.16155198
3.56	0.28290784	0.012043631	0.042570863	7.1791217	59.47386994
3.58	0.280640309	0.011709145	0.041722961	7.313463293	59.78383073
3.6	0.278396437	0.011384733	0.040893961	7.450111029	60.09145559

(continued)

$\gamma = 1.4$

Mach Number	Isentropic Temp Ratio T/T_t	Isentropic Pressure Ratio p/p_t	Isentropic Density Ratio ρ/ρ_t	Isentropic Area Ratio A/A^*	Prandtl-Meyer Function v (degrees)
3.62	0.276175957	0.011070074	0.040083411	7.589096715	60.39676563
3.64	0.273978608	0.010764858	0.039290869	7.730452481	60.6997818
3.66	0.271804127	0.010468782	0.038515906	7.874210783	61.00052491
3.68	0.269652256	0.010181557	0.037758101	8.020404403	61.29901565
3.7	0.267522739	0.009902902	0.037017046	8.169066453	61.59527452
3.72	0.265415322	0.009632543	0.03629234	8.320230374	61.88932191
3.74	0.263329752	0.009370219	0.035583595	8.47392994	62.18117804
3.76	0.26126578	0.009115676	0.034890431	8.630199258	62.47086297
3.78	0.25922316	0.008868666	0.034212476	8.789072773	62.75839662
3.8	0.257201646	0.008628953	0.033549369	8.950585263	63.04379875
3.82	0.255200996	0.008396306	0.032900759	9.114771849	63.32708895
3.84	0.253220971	0.008170504	0.0322663	9.281667989	63.60828668
3.86	0.251261332	0.00795133	0.031645659	9.451309487	63.88741122
3.88	0.249321845	0.007738578	0.031038507	9.623732487	64.16448169
3.9	0.247402276	0.007532045	0.030444527	9.798973482	64.43951705
3.92	0.245502396	0.007331538	0.029863406	9.97706931	64.71253612
3.94	0.243621977	0.007136867	0.029294842	10.15805716	64.98355752
3.96	0.241760792	0.006947852	0.028738538	10.34197457	65.25259975
3.98	0.23991862	0.006764315	0.028194207	10.52885944	65.51968113
4	0.238095238	0.006586087	0.027661567	10.71875	65.7848198
4.02	0.236290429	0.006413003	0.027140343	10.91168486	66.04803376
4.04	0.234503977	0.006244904	0.026630268	11.10770299	66.30934086
4.06	0.232735668	0.006081634	0.02613108	11.30684369	66.56875876
4.08	0.230985291	0.005923046	0.025642526	11.50914665	66.82630496
4.1	0.229252636	0.005768995	0.025164358	11.71465192	67.08199684
4.12	0.227537498	0.005619341	0.024696332	11.9233999	67.33585156
4.14	0.225839672	0.00547395	0.024238212	12.13543136	67.58788616
4.16	0.224158956	0.00533269	0.02378977	12.35078745	67.83811751
4.18	0.22249515	0.005195435	0.023350779	12.56950968	68.08656232
4.2	0.220848057	0.005062063	0.022921021	12.79163993	68.33323714
4.22	0.219217481	0.004932455	0.022500282	13.01722046	68.57815837
4.24	0.217603231	0.004806497	0.022088354	13.24629391	68.82134223
4.26	0.216005115	0.004684078	0.021685033	13.47890327	69.06280481
4.28	0.214422945	0.004565091	0.021290122	13.71509194	69.30256204
4.3	0.212856535	0.004449431	0.020903426	13.95490369	69.54062968
4.32	0.2113057	0.004336998	0.020524757	14.19838267	69.77702335
4.34	0.20977026	0.004227696	0.020153932	14.4455734	70.01175851
4.36	0.208250033	0.004121429	0.01979077	14.69652081	70.24485046
4.38	0.206744844	0.004018106	0.019435097	14.9512702	70.47631437
4.4	0.205254516	0.00391764	0.019086742	15.20986727	70.70616525
4.42	0.203778875	0.003819945	0.018745538	15.47235811	70.93441794
4.44	0.202317752	0.003724937	0.018411322	15.73878919	71.16108717
4.46	0.200870977	0.003632538	0.018083937	16.00920738	71.3861875
4.48	0.199438382	0.003542669	0.017763226	16.28365995	71.60973333
4.5	0.198019802	0.003455256	0.017449041	16.56219457	71.83173895

(continued)

$\gamma = 1.4$

Mach Number	Isentropic Temp Ratio T/T_t	Isentropic Pressure Ratio p/p_t	Isentropic Density Ratio ρ/ρ_t	Isentropic Area Ratio A/A^*	Prandtl-Meyer Function v (degrees)
4.52	0.196615075	0.003370225	0.017141232	16.84485931	72.05221848
4.54	0.195224039	0.003287506	0.016839657	17.13170263	72.2711859
4.56	0.193846536	0.003207031	0.016544175	17.42277339	72.48865506
4.58	0.192482407	0.003128734	0.016254649	17.71812089	72.70463966
4.6	0.191131498	0.003052551	0.015970947	18.01779478	72.91915326
4.62	0.189793656	0.00297842	0.015692937	18.32184517	73.13220928
4.64	0.188468729	0.002906281	0.015420493	18.63032256	73.34382102
4.66	0.187156568	0.002836075	0.01515349	18.94327785	73.55400161
4.68	0.185857024	0.002767747	0.014891809	19.26076236	73.76276409
4.7	0.184569952	0.002701242	0.014635329	19.58282785	73.97012132
4.72	0.183295208	0.002636507	0.014383937	19.90952645	74.17608606
4.74	0.182032649	0.00257349	0.01413752	20.24091076	74.38067093
4.76	0.180782136	0.002512143	0.013895967	20.57703376	74.58388842
4.78	0.179543529	0.002452416	0.013659173	20.91794887	74.78575088
4.8	0.17831669	0.002394264	0.013427031	21.26370994	74.98627054
4.82	0.177101486	0.00233764	0.01319944	21.61437123	75.18545952
4.84	0.175897782	0.002282503	0.012976301	21.96998743	75.38332979
4.86	0.174705447	0.002228807	0.012757516	22.33061368	75.5798932
4.88	0.173524349	0.002176514	0.01254299	22.69630552	75.77516149
4.9	0.172354361	0.002125582	0.01233263	23.06711895	75.96914627
4.92	0.171195354	0.002075974	0.012126346	23.44311039	76.16185903
4.94	0.170047205	0.002027651	0.011924049	23.8243367	76.35331114
4.96	0.168909789	0.001980578	0.011725653	24.21085517	76.54351385
4.98	0.167782983	0.001934718	0.011531074	24.60272354	76.7324783
5	0.166666667	0.001890038	0.01134023	25	76.92021551
5.5	0.141843972	0.001074826	0.007577522	36.86896307	81.24479016
6	0.12195122	0.000633361	0.005193563	53.17978395	84.95549818
6.5	0.105820106	0.000385468	0.003642675	75.1343149	88.16816076
7	0.092592593	0.000241555	0.002608799	104.1428571	90.97273233
7.5	0.081632653	0.000155426	0.001903969	141.8414834	93.43966839
8	0.072463768	0.000102429	0.001413521	190.109375	95.62467171
8.5	0.064724919	6.89843E-05	0.001065808	251.0861673	97.57220575
9	0.058139535	4.7386E-05	0.00081504	327.1893004	99.31809865
9.5	0.052493438	3.31411E-05	0.000631339	421.1313734	100.8914832
10	0.047619048	2.35631E-05	0.000494825	535.9375	102.3162532
12	0.033557047	6.92218E-06	0.000206281	1276.214892	106.8786261
14	0.024875622	2.42778E-06	9.75966E-05	2685.383929	110.1798378
16	0.019157088	9.73091E-07	5.07953E-05	5144.554688	112.6758959
18	0.015197568	4.32722E-07	2.84731E-05	9159.28215	114.6278183
20	0.012345679	2.09075E-07	1.69351E-05	15377.34375	116.1952976

$\gamma = 5/3$

Mach Number	Isentropic Temp Ratio T/T_t	Isentropic Pressure Ratio p/p_t	Isentropic Density Ratio ρ/ρ_t	Isentropic Area Ratio A/A^*	Prandtl-Meyer Function ν (degrees)
0	1	1	1		–
0.02	0.999866684	0.999666744	0.999800033	28.1325005	–
0.04	0.999466951	0.99866791	0.999200533	14.077504	–
0.06	0.998801438	0.997006289	0.998202696	9.3975135	–
0.08	0.997871208	0.994686514	0.996808512	7.061282	–
0.1	0.996677741	0.991715036	0.995020753	5.6625625	–
0.12	0.99522293	0.988100079	0.992842959	4.732608	–
0.14	0.993509074	0.983851597	0.990279428	4.070528643	–
0.16	0.991538868	0.978981214	0.987335187	3.575881	–
0.18	0.989315394	0.973502154	0.984015977	3.1928645	–
0.2	0.986842105	0.967429169	0.980328225	2.888	–
0.22	0.984122819	0.960778452	0.976279011	2.639983682	–
0.24	0.981161695	0.953567548	0.971876045	2.434614	–
0.26	0.977963229	0.945815254	0.967127625	2.262060038	–
0.28	0.974532225	0.937541522	0.962042607	2.115300571	–
0.3	0.970873786	0.928767347	0.956630367	1.9891875	–
0.32	0.966993296	0.91951466	0.95090076	1.8798605	–
0.34	0.962896392	0.909806214	0.94486408	1.784368265	–
0.36	0.958588957	0.899665474	0.938531022	1.700416	–
0.38	0.954077089	0.889116496	0.931912636	1.626192658	–
0.4	0.949367089	0.878183818	0.925020288	1.56025	–
0.42	0.944465433	0.866892346	0.917865616	1.501416214	–
0.44	0.939378758	0.855267241	0.910460487	1.448733091	–
0.46	0.934113837	0.843333813	0.902816958	1.401409587	–
0.48	0.928677563	0.831117416	0.894947233	1.358787	–
0.5	0.923076923	0.818643343	0.886863621	1.3203125	–
0.52	0.917318982	0.805936735	0.878578499	1.285518769	–
0.54	0.911410864	0.793022488	0.870104273	1.254008167	–
0.56	0.90535973	0.779925164	0.861453341	1.225440286	–
0.58	0.899172761	0.766668916	0.852638057	1.199522086	–
0.6	0.892857143	0.75327741	0.843670699	1.176	–
0.62	0.886420045	0.739773757	0.834563434	1.154653565	–
0.64	0.879868606	0.726180455	0.825328293	1.13529025	–
0.66	0.87320992	0.712519331	0.815977138	1.117741227	–
0.68	0.866451017	0.698811493	0.806521638	1.101857882	–
0.7	0.859598854	0.685077287	0.796973244	1.087508929	–
0.72	0.8526603	0.671336263	0.787343169	1.074578	–
0.74	0.845642124	0.657607142	0.777642366	1.062961635	–
0.76	0.838550984	0.643907794	0.767881508	1.052567579	–
0.78	0.831393415	0.63025522	0.758070979	1.043313346	–
0.8	0.824175824	0.616665536	0.748220851	1.035125	–
0.82	0.816904477	0.603153969	0.738340879	1.02793611	–
0.84	0.809585492	0.58973485	0.728440486	1.021686857	–
0.86	0.802224837	0.576421617	0.718528759	1.016323267	–
0.88	0.794828317	0.563226821	0.708614438	1.011796545	–
0.9	0.787401575	0.550162136	0.698705913	1.0080625	–

(continued)

$\gamma = 5/3$

Mach Number	Isentropic Temp Ratio T/T_t	Isentropic Pressure Ratio p/p_t	Isentropic Density Ratio ρ/ρ_t	Isentropic Area Ratio A/A^*	Prandtl-Meyer Function v (degrees)
0.92	0.779950083	0.537238369	0.688811221	1.005081043	–
0.94	0.772479143	0.524465477	0.678938042	1.002815755	–
0.96	0.76499388	0.511852586	0.669093701	1.0012335	–
0.98	0.757499243	0.499408014	0.659285166	1.000304092	–
1	0.75	0.48713929	0.649519053	1	0
1.02	0.742500743	0.475053181	0.639801624	1.000296088	0.112893887
1.04	0.73500588	0.463155722	0.630138798	1.001169385	0.314627205
1.06	0.727519643	0.451452237	0.620536148	1.002598877	0.569606709
1.08	0.720046083	0.439947375	0.610998914	1.004565333	0.864339518
1.1	0.712589074	0.428645133	0.601532004	1.007051136	1.190715632
1.12	0.705152313	0.417548893	0.592140004	1.010040143	1.543101358
1.14	0.697739325	0.406661447	0.582827185	1.013517553	1.917270958
1.16	0.690353461	0.395985028	0.573597513	1.017469793	2.309899831
1.18	0.682997905	0.385521346	0.564454653	1.021884415	2.71828616
1.2	0.675675676	0.375271611	0.555401984	1.02675	3.140182376
1.22	0.668389627	0.365236567	0.546442603	1.032056074	3.573685743
1.24	0.661142454	0.355416523	0.537579338	1.037793032	4.017163459
1.26	0.653936699	0.345811377	0.528814757	1.043952071	4.469199208
1.28	0.64677475	0.336420647	0.520151176	1.050525125	4.928553752
1.3	0.639658849	0.327243501	0.511590673	1.057504808	5.394135038
1.32	0.632591093	0.318278778	0.503135092	1.064884364	5.864975027
1.34	0.625573442	0.309525018	0.494786059	1.072657619	6.340211359
1.36	0.61860772	0.300980486	0.486544988	1.080818941	6.819072621
1.38	0.61169562	0.292643193	0.478413093	1.089363196	7.300866322
1.4	0.60483871	0.284510924	0.470391394	1.098285714	7.784968967
1.42	0.598038434	0.276581252	0.462480732	1.107582261	8.270817765
1.44	0.591296121	0.268851569	0.454681773	1.117249	8.757903635
1.46	0.584612986	0.261319095	0.446995023	1.127282473	9.245765258
1.48	0.577990136	0.253980905	0.43942083	1.137679568	9.733983977
1.5	0.571428571	0.246833942	0.431959398	1.1484375	10.22217939
1.52	0.564929196	0.239875034	0.424610793	1.159553789	10.71000553
1.54	0.558492814	0.233100913	0.417374954	1.17102624	11.19714752
1.56	0.552120141	0.226508225	0.410251697	1.182852923	11.68331861
1.58	0.545811804	0.220093547	0.403240724	1.195032158	12.16825763
1.6	0.539568345	0.213853399	0.396341633	1.2075625	12.65172665
1.62	0.533390228	0.207784256	0.389553923	1.220442722	13.13350895
1.64	0.52727784	0.201882558	0.382877001	1.233671805	13.6134072
1.66	0.521231496	0.196144722	0.376310188	1.247248922	14.09124177
1.68	0.515251443	0.190567151	0.369852727	1.261173429	14.56684934
1.7	0.509337861	0.185146241	0.363503787	1.275444853	15.04008148
1.72	0.50349087	0.179878393	0.357262473	1.290062884	15.51080351
1.74	0.497710532	0.174760016	0.351127825	1.305027362	15.97889337
1.76	0.491996851	0.169787537	0.345098828	1.320338273	16.44424066
1.78	0.486349783	0.164957405	0.339174419	1.335995736	16.90674571
1.8	0.480769231	0.160266098	0.333353483	1.352	17.36631874

(continued)

$\gamma = 5/3$

Mach Number	Isentropic Temp Ratio T/T_t	Isentropic Pressure Ratio p/p_t	Isentropic Density Ratio ρ/ρ_t	Isentropic Area Ratio A/A^*	Prandtl-Meyer Function ν (degrees)
1.82	0.475255054	0.155710127	0.327634869	1.368351434	17.82287917
1.84	0.469807066	0.151286042	0.322017383	1.385050522	18.27635486
1.86	0.464425042	0.146990434	0.316499802	1.402097855	18.7266815
1.88	0.459108717	0.14281994	0.311080872	1.419494128	19.17380205
1.9	0.453857791	0.138771246	0.305759312	1.437240132	19.61766618
1.92	0.448671931	0.134841089	0.30053382	1.45533675	20.05822975
1.94	0.443550772	0.131026261	0.295403073	1.473784954	20.49545443
1.96	0.43849392	0.127323609	0.290365733	1.492585796	20.92930719
1.98	0.433500954	0.123730037	0.28542045	1.511740409	21.35975997
2	0.428571429	0.120242511	0.280565859	1.53125	21.7867893
2.2	0.382653061	0.090575967	0.236705194	1.746181818	25.86630898
2.4	0.342465753	0.068634563	0.200412923	1.998375	29.60144219
2.6	0.307377049	0.05238158	0.17041474	2.289846154	33.00872276
2.8	0.276752768	0.04029303	0.14559215	2.622892857	36.1131625
3	0.25	0.03125	0.125	3	38.94244127
3.2	0.226586103	0.024438986	0.107857389	3.42378125	41.52398701
3.4	0.206043956	0.019270798	0.093527606	3.896941176	43.88357677
3.6	0.187969925	0.015318665	0.081495298	4.42225	46.04474819
3.8	0.172018349	0.01227261	0.071344773	5.002526316	48.02863993
4	0.157894737	0.009906475	0.062741006	5.640625	49.85405234
4.5	0.129032258	0.005980614	0.046349755	7.5078125	53.82857743
5	0.107142857	0.003757578	0.035070732	9.8	57.12165044
5.5	0.090225564	0.002445254	0.027101567	12.56321023	59.886881
6	0.076923077	0.001641125	0.021334623	15.84375	62.23721625
6.5	0.066298343	0.001131766	0.017070805	19.68810096	64.25675012
7	0.057692308	0.000799456	0.013857244	24.14285714	66.0089832
7.5	0.050632911	0.000576876	0.011393295	29.2546875	67.54255199
8	0.044776119	0.000424244	0.009474792	35.0703125	68.89522082
8.5	0.03986711	0.000317349	0.007960166	41.63648897	70.09669137
9	0.035714286	0.000241049	0.006749366	49	71.17060491
9.5	0.032171582	0.000185644	0.005770436	57.20764803	72.13598787
10	0.029126214	0.000144781	0.004970797	66.30625	73.00830955
12	0.020408163	5.9499E-05	0.002915452	112.546875	75.79069327
14	0.015075377	2.79043E-05	0.001850982	176.7901786	77.79475833
16	0.011583012	1.44396E-05	0.001246615	262.0351563	79.30566282
18	0.009174312	8.06183E-06	0.00087874	371.28125	80.48490123
20	0.007444169	4.78124E-06	0.00064228	507.528125	81.43058911

$\gamma = 1.3$

Mach Number	Isentropic Temp Ratio T/T_t	Isentropic Pressure Ratio p/p_t	Isentropic Density Ratio ρ/ρ_t	Isentropic Area Ratio A/A^*	Prandtl-Meyer Function v (degrees)
0	1	1	1		–
0.02	0.999940004	0.999740042	0.999800026	29.2681205	–
0.04	0.999760058	0.998960665	0.999200416	14.64415972	–
0.06	0.999460291	0.997663366	0.998202104	9.774002412	–
0.08	0.999040921	0.995850628	0.996806645	7.342304564	–
0.1	0.998502247	0.993525918	0.995016207	5.886000133	–
0.12	0.997844656	0.990693669	0.992833567	4.917402747	–
0.14	0.997068618	0.987359265	0.990262101	4.227505951	–
0.16	0.996174689	0.983529022	0.987305773	3.711808266	–
0.18	0.995163505	0.979210162	0.983969123	3.312254883	–
0.2	0.994035785	0.974410788	0.980257253	2.994014338	–
0.22	0.992792328	0.969139853	0.976175809	2.734922458	–
0.24	0.99143401	0.963407129	0.971730967	2.520204335	–
0.26	0.989961787	0.957223168	0.966929411	2.339632315	–
0.28	0.98837669	0.950599269	0.961778316	2.185901989	–
0.3	0.986679822	0.943547432	0.956285322	2.053657791	–
0.32	0.984872361	0.93608032	0.950458514	1.938884001	–
0.34	0.982955551	0.928211214	0.944306397	1.838510679	–
0.36	0.980930707	0.919953966	0.937837872	1.750150971	–
0.38	0.978799209	0.911322954	0.931062209	1.671921354	–
0.4	0.9765625	0.902333029	0.923989022	1.602315822	–
0.42	0.974222084	0.892999474	0.91662824	1.540116015	–
0.44	0.971779523	0.883337948	0.908990082	1.484325856	–
0.46	0.969236436	0.873364438	0.901085026	1.434123245	–
0.48	0.966594494	0.863095212	0.892923782	1.388823828	–
0.5	0.963855422	0.852546763	0.884517267	1.347853461	–
0.52	0.961020989	0.841735768	0.87587657	1.310727031	–
0.54	0.958093012	0.83067903	0.867012931	1.277031984	–
0.56	0.95507335	0.819393439	0.857937707	1.24641538	–
0.58	0.951963902	0.807895919	0.848662347	1.218573635	–
0.6	0.948766603	0.796203385	0.839198368	1.193244304	–
0.62	0.945483426	0.784332697	0.829557321	1.170199466	–
0.64	0.94211637	0.772300621	0.819750771	1.149240341	–
0.66	0.938667468	0.760123783	0.809790271	1.130192893	–
0.68	0.935138775	0.747818632	0.799687332	1.112904206	–
0.7	0.931532371	0.735401405	0.789453408	1.09723948	–
0.72	0.927850356	0.722888088	0.779099866	1.083079535	–
0.74	0.924094849	0.710294384	0.768637965	1.070318715	–
0.76	0.920267982	0.697635684	0.75807884	1.058863121	–
0.78	0.9163719	0.684927035	0.747433476	1.048629125	–
0.8	0.912408759	0.672183115	0.736712694	1.039542101	–
0.82	0.908380721	0.659418209	0.72592713	1.031535341	–
0.84	0.904289952	0.646646188	0.71508722	1.024549133	–
0.86	0.900138621	0.633880486	0.704203187	1.018529963	–
0.88	0.895928899	0.621134087	0.693285023	1.013429827	–
0.9	0.891662951	0.608419506	0.682342476	1.009205638	–

(continued)

γ = 1.3

Mach Number	Isentropic Temp Ratio T/T_t	Isentropic Pressure Ratio p/p_t	Isentropic Density Ratio ρ/ρ_t	Isentropic Area Ratio A/A^*	Prandtl-Meyer Function ν (degrees)
0.92	0.88734294	0.595748777	0.671385042	1.005818704	–
0.94	0.882971021	0.583133446	0.660421953	1.003234285	–
0.96	0.878549339	0.570584555	0.649462164	1.001421195	–
0.98	0.874080031	0.558112642	0.638514349	1.000351457	–
1	0.869565217	0.545727734	0.627586894	1	0
1.02	0.865007007	0.533439342	0.616687886	1.000344396	0.131267258
1.04	0.860407489	0.521256465	0.605825114	1.001364619	0.36688068
1.06	0.855768737	0.509187585	0.595006061	1.00304284	0.666106121
1.08	0.851092803	0.497240673	0.584237901	1.005363245	1.013653043
1.1	0.846381718	0.485423191	0.5735275	1.008311862	1.40038103
1.12	0.84163749	0.4737421	0.562881413	1.011876426	1.819961129
1.14	0.836862102	0.462203862	0.552305883	1.016046242	2.267653326
1.16	0.832057512	0.450814454	0.541806843	1.020812072	2.73972878
1.18	0.827225651	0.439579369	0.531389916	1.026166029	3.233153684
1.2	0.822368421	0.428503633	0.521060418	1.032101486	3.745398608
1.22	0.817487697	0.417591813	0.510823362	1.038612993	4.274315179
1.24	0.812585321	0.406848029	0.500683458	1.045696202	4.818052017
1.26	0.807663108	0.396275964	0.490645122	1.053347799	5.374994979
1.28	0.802722836	0.385878882	0.480712476	1.061565445	5.943723273
1.3	0.797766254	0.375659637	0.470889354	1.070347724	6.522976317
1.32	0.792795078	0.36562069	0.461179313	1.079694088	7.111628141
1.34	0.787810988	0.355764121	0.45158563	1.089604822	7.708667227
1.36	0.782815631	0.346091648	0.442111315	1.100080996	8.313180366
1.38	0.777810619	0.336604637	0.432759117	1.111124436	8.924339548
1.4	0.772797527	0.327304118	0.423531528	1.122737691	9.541391177
1.42	0.767777897	0.318190804	0.414430794	1.134924	10.16364709
1.44	0.762753234	0.309265102	0.40545892	1.147687272	10.79047705
1.46	0.757725006	0.300527132	0.396617677	1.161032061	11.42130231
1.48	0.752694647	0.291976738	0.387908615	1.174963544	12.05559022
1.5	0.747663551	0.283613505	0.379333063	1.189487504	12.69284948
1.52	0.74263308	0.275436776	0.370892145	1.204610313	13.33262613
1.54	0.737604555	0.267445662	0.362586782	1.220338919	13.97449997
1.56	0.732579265	0.259639061	0.354417704	1.236680829	14.61808149
1.58	0.727558459	0.252015669	0.346385456	1.253644103	15.26300913
1.6	0.722543353	0.244573994	0.338490408	1.271237342	15.90894688
1.62	0.717535123	0.237312372	0.33073276	1.289469675	16.55558215
1.64	0.712534914	0.230228976	0.323112554	1.308350758	17.20262382
1.66	0.707543832	0.223321832	0.315629678	1.327890765	17.84980059
1.68	0.70256295	0.216588829	0.308283876	1.348100382	18.49685942
1.7	0.697593303	0.210027733	0.301074756	1.368990802	19.14356415
1.72	0.692635895	0.203636197	0.294001795	1.390573724	19.78969427
1.74	0.687691694	0.19741177	0.287064351	1.412861348	20.43504378
1.76	0.682761634	0.191351912	0.280261665	1.435866375	21.07942014
1.78	0.677846617	0.185454002	0.273592871	1.459602002	21.72264341
1.8	0.67294751	0.179715344	0.267057002	1.484081927	22.36454532

(continued)

$\gamma = 1.3$

Mach Number	Isentropic Temp Ratio T/T_t	Isentropic Pressure Ratio p/p_t	Isentropic Density Ratio ρ/ρ_t	Isentropic Area Ratio A/A^*	Prandtl-Meyer Function v (degrees)
1.82	0.66806515	0.174133184	0.260652997	1.509320343	23.00496853
1.84	0.66320034	0.168704709	0.254379709	1.535331944	23.6437659
1.86	0.658353852	0.163427066	0.248235908	1.562131923	24.28079986
1.88	0.653526429	0.15829736	0.242220289	1.589735972	24.91594177
1.9	0.64871878	0.153312668	0.236331477	1.618160289	25.5490714
1.92	0.643931589	0.148470042	0.230568037	1.647421578	26.18007636
1.94	0.639165506	0.143766519	0.22492847	1.677537051	26.80885169
1.96	0.634421154	0.139199126	0.21941123	1.708524433	27.43529939
1.98	0.62969913	0.134764882	0.214014719	1.740401964	28.05932797
2	0.625	0.130460811	0.208737298	1.773188407	28.68085215
2.2	0.579374276	0.093934091	0.162130241	2.155552632	34.74334953
2.4	0.536480687	0.067308162	0.125462414	2.653523986	40.49622775
2.6	0.49652433	0.048129237	0.096932284	3.295436021	45.91676103
2.8	0.459558824	0.034420048	0.074898025	4.116479872	51.00152922
3	0.425531915	0.024662623	0.057957164	5.159771816	55.7584169
3.2	0.394321767	0.017729096	0.044960987	6.477594472	60.20174997
3.4	0.365764448	0.012799896	0.034994917	8.132804216	64.34926578
3.6	0.339673913	0.009288299	0.027344752	10.20040384	68.22020898
3.8	0.31585597	0.006778278	0.021460029	12.76928281	71.83414037
4	0.294117647	0.004976508	0.016920127	15.94412912	75.21020605
4.5	0.247678019	0.002363265	0.009541681	27.38695934	82.72711082
5	0.210526316	0.001168587	0.005550788	45.9565175	89.12342651
5.5	0.180586907	0.000601137	0.003328799	75.21969871	94.60781975
6	0.15625	0.000321036	0.002054629	120.0964754	99.34724887
6.5	0.136286201	0.000177538	0.001302686	187.2173211	103.4742577
7	0.119760479	0.000101397	0.000846665	285.3371996	107.0940706
7.5	0.105960265	5.96513E-05	0.000562959	425.8095325	110.2905606
8	0.094339623	3.60585E-05	0.00038222	623.1234688	113.1309908
8.5	0.084477297	2.23464E-05	0.000264525	895.5076955	115.6696866
9	0.076045627	1.41684E-05	0.000186315	1265.60395	117.9508535
9.5	0.068787618	9.1737E-06	0.000133363	1761.213327	120.0107342
10	0.0625	6.05545E-06	9.68873E-05	2416.118397	121.8792625
12	0.044247788	1.3558E-06	3.06412E-05	7566.459573	127.8816248
14	0.032894737	3.75158E-07	1.14048E-05	20209.03829	132.2409423
16	0.025380711	1.2195E-07	4.80484E-06	47783.04401	135.5448414
18	0.02016129	4.49689E-08	2.23046E-06	102659.4343	138.1326204
20	0.016393443	1.83472E-08	1.11918E-06	204201.683	140.2130595

APPENDIX D

Tables for Normal Shock Waves

Note – The numerical values of the table can be directly obtained from the following equations. They are tabulated for $\gamma = 1.4$ for convenience. They are applicable for oblique shock waves whose normal and tangential components of Mach number are denoted, respectively, by M_n and M_t. For normal shock waves, $M_{1n} = M_1$ and $M_{2n} = M_2$. (The subscripts t for thermodynamic properties denote stagnation values; the subscripts 1 and 2 denote upstream and downstream of the shock, respectively. The superscript * denotes critical condition for $M = 1$.)

$$M_{2n}^2 = \frac{M_{1n}^2 + \frac{2}{\gamma-1}}{\frac{2\gamma}{\gamma-1} M_{1n}^2 - 1} \tag{4.22a}$$

$$\frac{p_{t2}}{p_{t1}} = \frac{A_1^*}{A_2^*} = \left[\frac{\frac{\gamma+1}{2} M_{1n}^2}{1 + \frac{\gamma-1}{2} M_{1n}^2}\right]^{\gamma/(\gamma-1)} \left[\frac{1}{\frac{2\gamma}{\gamma+1} M_{1n}^2 - \frac{\gamma-1}{\gamma+1}}\right]^{1/(\gamma-1)} \tag{4.22b}$$

$$\frac{p_2}{p_1} = \frac{2\gamma M_{1n}^2}{\gamma+1} - \frac{\gamma-1}{\gamma+1} \tag{4.22c}$$

$$\frac{T_2}{T_1} = \left(\frac{M_{1t}}{M_{2t}}\right)^2 = \frac{\left(1 + \frac{\gamma-1}{2} M_{1n}^2\right)\left(\frac{2\gamma}{\gamma-1} M_{1n}^2 - 1\right)}{\left[\frac{(\gamma+1)^2}{2(\gamma-1)}\right] M_{1n}^2} \tag{4.22d}$$

$$\frac{\rho_2}{\rho_1} = \frac{V_{1n}}{V_{2n}} = \frac{M_{1n}\sqrt{\gamma R T_1}}{M_{2n}\sqrt{\gamma R T_2}} = \frac{(\gamma+1)M_{1n}^2}{(\gamma-1)M_{1n}^2 + 2} \tag{4.22e}$$

$$\frac{T_{t2}}{T_{t1}} = 1 \tag{4.22f}$$

$\gamma = 1.4$

M_1	M_2	$P_{t1}/P_{t2} = A_2^*/A_1^*$	P_2/P_1	T_2/T_1	$\rho_2/\rho_1 = V_{1n}/V_{2n}$	T_{t2}/T_{t1}
1	1	1	1	1	1	1
1.02	0.980519493	1.000009968	1.047133333	1.01324878	1.033441494	1
1.04	0.962025499	1.000076723	1.0952	1.02634497	1.067087609	1
1.06	0.944445301	1.000249345	1.1442	1.039311606	1.100921027	1
1.08	0.927713359	1.000569625	1.194133333	1.052169608	1.134924754	1
1.1	0.911770421	1.001073135	1.245	1.064938017	1.169082126	1
1.12	0.896562769	1.001790146	1.2968	1.077634184	1.203376823	1
1.14	0.882041571	1.002746407	1.349533333	1.090273954	1.237792876	1
1.16	0.868162321	1.003963821	1.4032	1.102871819	1.272314675	1
1.18	0.854884363	1.005461017	1.4578	1.115441051	1.306926976	1
1.2	0.84217047	1.007253842	1.513333333	1.127993827	1.341614907	1
1.22	0.829986484	1.009355783	1.5698	1.140541333	1.376363973	1
1.24	0.818300995	1.011778331	1.6272	1.153093861	1.411160059	1
1.26	0.807085068	1.014531287	1.685533333	1.165660891	1.445989435	1
1.28	0.796311994	1.017623034	1.7448	1.178251172	1.480838756	1
1.3	0.78595708	1.021060758	1.805	1.190872781	1.515695067	1
1.32	0.775997454	1.024850648	1.866133333	1.20353319	1.5505458	1
1.34	0.7664119	1.028998065	1.9282	1.216239318	1.585378775	1
1.36	0.757180705	1.033507686	1.9912	1.228997578	1.6201822	1
1.38	0.748285526	1.038383625	2.055133333	1.241813921	1.654944673	1
1.4	0.739709275	1.043629544	2.12	1.254693878	1.689655172	1
1.42	0.731436004	1.049248741	2.1858	1.267642591	1.724303061	1
1.44	0.723450815	1.055244231	2.252533333	1.280664849	1.758878082	1
1.46	0.715739774	1.061618814	2.3202	1.293765115	1.793370352	1
1.48	0.708289828	1.068375129	2.3888	1.306947553	1.82777036	1
1.5	0.701088742	1.075515709	2.458333333	1.320216049	1.862068966	1
1.52	0.694125028	1.08304302	2.5288	1.333574238	1.896257387	1
1.54	0.687387893	1.090959498	2.6002	1.347025519	1.930327202	1
1.56	0.680867185	1.099267581	2.672533333	1.360573073	1.96427034	1
1.58	0.674553345	1.107969738	2.7458	1.374219885	1.998079078	1
1.6	0.668437365	1.117068486	2.82	1.38796875	2.031746032	1
1.62	0.662510745	1.126566416	2.895133333	1.401822295	2.065264152	1
1.64	0.656765462	1.136466204	2.9712	1.415782986	2.098626717	1
1.66	0.651193934	1.146770629	3.0482	1.429853143	2.131827325	1
1.68	0.645788989	1.157482582	3.126133333	1.444034946	2.16485989	1
1.7	0.640543841	1.16860508	3.205	1.45833045	2.197718631	1
1.72	0.635452058	1.18014127	3.2848	1.47274159	2.23039807	1
1.74	0.630507546	1.192094437	3.365533333	1.487270191	2.262893019	1
1.76	0.625704523	1.204468015	3.4472	1.501917975	2.295198577	1
1.78	0.621037498	1.217265582	3.5298	1.516686567	2.327310122	1
1.8	0.616501258	1.230490875	3.613333333	1.531577503	2.359223301	1
1.82	0.612090843	1.244147786	3.6978	1.546592235	2.390934026	1
1.84	0.607801539	1.258240365	3.7832	1.561732136	2.422438466	1
1.86	0.603628856	1.272772826	3.869533333	1.576998506	2.453733037	1
1.88	0.59956852	1.287749546	3.9568	1.592392576	2.484814398	1
1.9	0.595616455	1.303175066	4.045	1.607915512	2.515679443	1
1.92	0.591768777	1.319054094	4.134133333	1.623568422	2.54632529	1
1.94	0.58802178	1.335391502	4.2242	1.639352354	2.576749281	1

(continued)

$\gamma = 1.4$

M_1	M_2	$P_{t1}/P_{t2} = A_2^*/A_1^*$	P_2/P_1	T_2/T_1	$\rho_2/\rho_1 = V_{1n}/V_{2n}$	T_{t2}/T_{t1}
1.96	0.584371924	1.35219233	4.3152	1.655268305	2.606948969	1
1.98	0.58081583	1.369461786	4.407133333	1.67131722	2.636922111	1
2	0.577350269	1.387205243	4.5	1.6875	2.666666667	1
2.02	0.573972154	1.405428241	4.5938	1.703817498	2.696180785	1
2.04	0.57067853	1.424136488	4.688533333	1.720270528	2.725462801	1
2.06	0.567466572	1.443335859	4.7842	1.736859864	2.754511229	1
2.08	0.564333573	1.463032392	4.8808	1.753586243	2.783324756	1
2.1	0.561276941	1.483232295	4.978333333	1.770450365	2.811902232	1
2.12	0.558294192	1.503941939	5.0768	1.787452901	2.840242669	1
2.14	0.555382944	1.52516786	5.1762	1.804594489	2.868345234	1
2.16	0.552540913	1.546916761	5.276533333	1.821875735	2.896209237	1
2.18	0.549765907	1.569195506	5.3778	1.839297222	2.923834133	1
2.2	0.547055823	1.592011126	5.48	1.856859504	2.951219512	1
2.22	0.544408639	1.615370814	5.583133333	1.87456311	2.978365094	1
2.24	0.541822415	1.639281927	5.6872	1.892408546	3.005270724	1
2.26	0.539295285	1.663751984	5.7922	1.910396296	3.031936365	1
2.28	0.536825457	1.688788667	5.898133333	1.928526822	3.058362096	1
2.3	0.534411206	1.71439982	6.005	1.946800567	3.084548105	1
2.32	0.532050873	1.74059345	6.1128	1.965217955	3.110494683	1
2.34	0.52974286	1.767377725	6.221533333	1.983779391	3.136202222	1
2.36	0.527485632	1.794760973	6.3312	2.002485263	3.161671208	1
2.38	0.525277707	1.822751686	6.4418	2.021335944	3.186902217	1
2.4	0.523117659	1.851358517	6.553333333	2.04033179	3.211895911	1
2.42	0.521004115	1.880590278	6.6658	2.059473144	3.236653034	1
2.44	0.518935749	1.910455944	6.7792	2.078760333	3.261174408	1
2.46	0.516911285	1.94096465	6.893533333	2.098193673	3.285460929	1
2.48	0.51492949	1.972125694	7.0088	2.117773465	3.30951356	1
2.5	0.512989176	2.003948532	7.125	2.1375	3.333333333	1
2.52	0.511089196	2.036442784	7.242133333	2.157373556	3.356921342	1
2.54	0.509228441	2.069618229	7.3602	2.177394401	3.380278738	1
2.56	0.507405844	2.10348481	7.4792	2.197562793	3.40340673	1
2.58	0.505620372	2.138052629	7.599133333	2.217878978	3.426306578	1
2.6	0.503871026	2.173331951	7.72	2.238343195	3.448979592	1
2.62	0.502156842	2.209333203	7.8418	2.258955673	3.471427127	1
2.64	0.500476888	2.246066974	7.964533333	2.279716631	3.493650581	1
2.66	0.498830263	2.283544015	8.0882	2.300626282	3.515651396	1
2.68	0.497216096	2.321775241	8.2128	2.32168483	3.537431048	1
2.7	0.495633542	2.36077173	8.338333333	2.342892471	3.55899105	1
2.72	0.494081788	2.400544722	8.4648	2.364249394	3.580332946	1
2.74	0.492560042	2.441105623	8.5922	2.385755783	3.601458313	1
2.76	0.491067542	2.482466001	8.720533333	2.407411814	3.622368755	1
2.78	0.489603546	2.524637591	8.8498	2.429217654	3.6430659	1
2.8	0.488167338	2.567632292	8.98	2.451173469	3.663551402	1
2.82	0.486758224	2.611462169	9.111133333	2.473279416	3.683826936	1
2.84	0.485375531	2.656139453	9.2432	2.495535648	3.703894195	1
2.86	0.484018608	2.701676542	9.3762	2.51794231	3.723754894	1
2.88	0.482686823	2.748086	9.510133333	2.540499546	3.743410759	1

(continued)

γ = 1.4

M_1	M_2	$P_{t1}/P_{t2} = A_2^*/A_1^*$	P_2/P_1	T_2/T_1	$\rho_2/\rho_1 = V_{1n}/V_{2n}$	T_{t2}/T_{t1}
2.9	0.481379562	2.795380561	9.645	2.563207491	3.762863535	1
2.92	0.480096233	2.843573126	9.7808	2.586066279	3.782114975	1
2.94	0.478836259	2.892676765	9.917533333	2.609076037	3.801166847	1
2.96	0.477599082	2.942704717	10.0552	2.632236888	3.820020928	1
2.98	0.476384159	2.993670393	10.1938	2.655548953	3.838679001	1
3	0.475190963	3.045587373	10.33333333	2.679012346	3.857142857	1
3.02	0.474018985	3.09846941	10.4738	2.702627179	3.875414294	1
3.04	0.472867729	3.152330428	10.6152	2.72639356	3.893495113	1
3.06	0.471736712	3.207184524	10.75753333	2.750311593	3.911387117	1
3.08	0.470625468	3.263045968	10.9008	2.77438138	3.929092114	1
3.1	0.469533542	3.319929207	11.045	2.798603018	3.94661191	1
3.12	0.468460494	3.37784886	11.19013333	2.822976602	3.963948311	1
3.14	0.467405894	3.436819723	11.3362	2.847502223	3.981103125	1
3.16	0.466369326	3.496856768	11.4832	2.872179971	3.998078155	1
3.18	0.465350385	3.557975145	11.63113333	2.897009931	4.014875202	1
3.2	0.464348678	3.620190181	11.78	2.921992188	4.031496063	1
3.22	0.463363821	3.683517382	11.9298	2.94712682	4.047942531	1
3.24	0.462395442	3.747972434	12.08053333	2.972413907	4.064216395	1
3.26	0.46144318	3.813571204	12.2322	2.997853525	4.080319435	1
3.28	0.460506682	3.880329738	12.3848	3.023445747	4.096253427	1
3.3	0.459585605	3.948264267	12.53833333	3.049190644	4.112020138	1
3.32	0.458679617	4.017391202	12.6928	3.075088286	4.12762133	1
3.34	0.457788392	4.087727139	12.8482	3.101138739	4.143058754	1
3.36	0.456911616	4.159288859	13.00453333	3.12734207	4.158334152	1
3.38	0.456048979	4.232093327	13.1618	3.15369834	4.173449258	1
3.4	0.455200184	4.306157695	13.32	3.180207612	4.188405797	1
3.42	0.454364938	4.381499303	13.47913333	3.206869946	4.203205481	1
3.44	0.453542957	4.458135677	13.6392	3.233685398	4.217850014	1
3.46	0.452733966	4.536084532	13.8002	3.260654025	4.232341087	1
3.48	0.451937694	4.615363776	13.96213333	3.287775881	4.246680382	1
3.5	0.45115388	4.695991502	14.125	3.31505102	4.260869565	1
3.52	0.450382266	4.777986	14.2888	3.342479494	4.274910295	1
3.54	0.449622605	4.861365749	14.45353333	3.370061351	4.288804216	1
3.56	0.448874652	4.946149422	14.6192	3.397796642	4.30255296	1
3.58	0.448138172	5.032355887	14.7858	3.425685412	4.316158146	1
3.6	0.447412933	5.120004206	14.95333333	3.453727709	4.329621381	1
3.62	0.44669871	5.209113639	15.1218	3.481923577	4.342944257	1
3.64	0.445995283	5.299703639	15.2912	3.510273059	4.356128353	1
3.66	0.445302438	5.391793863	15.46153333	3.538776197	4.369175238	1
3.68	0.444619967	5.48540416	15.6328	3.567433034	4.382086461	1
3.7	0.443947664	5.580554585	15.805	3.596243608	4.394863563	1
3.72	0.443285332	5.677265389	15.97813333	3.62520796	4.407508069	1
3.74	0.442632777	5.775557028	16.1522	3.654326126	4.420021488	1
3.76	0.441989808	5.875450158	16.3272	3.683598144	4.432405317	1
3.78	0.441356241	5.976965641	16.50313333	3.71302405	4.44466104	1
3.8	0.440731896	6.080124541	16.68	3.742603878	4.456790123	1
3.82	0.440116596	6.18494813	16.8578	3.772337663	4.468794022	1

(continued)

γ = 1.4

M_1	M_2	$P_{t1}/P_{t2} = A_2^*/A_1^*$	P_2/P_1	T_2/T_1	$\rho_2/\rho_1 = V_{1n}/V_{2n}$	T_{t2}/T_{t1}
3.84	0.43951017	6.291457885	17.03653333	3.802225439	4.480674176	1
3.86	0.438912449	6.399675491	17.2162	3.832267237	4.492432009	1
3.88	0.438323269	6.50962284	17.3968	3.862463089	4.504068933	1
3.9	0.43774247	6.621322036	17.57833333	3.892813025	4.515586343	1
3.92	0.437169895	6.734795392	17.7608	3.923317076	4.526985623	1
3.94	0.436605392	6.850065431	17.9442	3.953975271	4.53826814	1
3.96	0.436048812	6.96715489	18.12853333	3.984787638	4.549435247	1
3.98	0.435500007	7.08608672	18.3138	4.015754206	4.560488282	1
4	0.434958836	7.206884085	18.5	4.046875	4.571428571	1
4.02	0.434425159	7.329570363	18.68713333	4.078150048	4.582257424	1
4.04	0.43389884	7.454169151	18.8752	4.109579375	4.592976137	1
4.06	0.433379746	7.580704263	19.0642	4.141163006	4.603585991	1
4.08	0.432867746	7.709199728	19.25413333	4.172900965	4.614088255	1
4.1	0.432362714	7.839679798	19.445	4.204793278	4.624484182	1
4.12	0.431864523	7.972168945	19.6368	4.236839966	4.634775011	1
4.14	0.431373054	8.106691859	19.82953333	4.269041053	4.644961969	1
4.16	0.430888186	8.243273455	20.0232	4.301396561	4.655046266	1
4.18	0.430409804	8.381938872	20.2178	4.333906511	4.665029102	1
4.2	0.429937793	8.522713471	20.41333333	4.366570925	4.674911661	1
4.22	0.429472041	8.66562284	20.6098	4.399389823	4.684695112	1
4.24	0.429012441	8.810692791	20.8072	4.432363225	4.694380614	1
4.26	0.428558884	8.957949367	21.00553333	4.465491152	4.70396931	1
4.28	0.428111268	9.107418836	21.2048	4.498773622	4.71346233	1
4.3	0.427669488	9.259127698	21.405	4.532210654	4.722860792	1
4.32	0.427233447	9.413102681	21.60613333	4.565802267	4.732165799	1
4.34	0.426803044	9.569370745	21.8082	4.599548478	4.741378442	1
4.36	0.426378186	9.727959085	22.0112	4.633449306	4.7504998	1
4.38	0.425958777	9.888895125	22.21513333	4.667504766	4.759530937	1
4.4	0.425544726	10.05220653	22.42	4.701714876	4.768472906	1
4.42	0.425135944	10.21792119	22.6258	4.736079652	4.777326747	1
4.44	0.424732341	10.38606724	22.83253333	4.770599111	4.786093487	1
4.46	0.424333832	10.55667305	23.0402	4.805273267	4.794774141	1
4.48	0.423940332	10.72976723	23.2488	4.840102136	4.803369711	1
4.5	0.423551759	10.90537863	23.45833333	4.875085734	4.811881188	1
4.52	0.423168031	11.08353634	23.6688	4.910224074	4.820309551	1
4.54	0.422789069	11.26426968	23.8802	4.945517171	4.828655765	1
4.56	0.422414795	11.44760824	24.09253333	4.980965039	4.836920787	1
4.58	0.422045133	11.63358182	24.3058	5.016567691	4.845105557	1
4.6	0.421680008	11.8222205	24.52	5.052325142	4.853211009	1
4.62	0.421319347	12.01355458	24.73513333	5.088237403	4.861238062	1
4.64	0.420963079	12.2076146	24.9512	5.124304489	4.869187624	1
4.66	0.420611132	12.40443138	25.1682	5.160526411	4.877060594	1
4.68	0.420263438	12.60403597	25.38613333	5.196903181	4.884857857	1
4.7	0.41991993	12.80645967	25.605	5.233434812	4.892580288	1
4.72	0.419580541	13.01173403	25.8248	5.270121316	4.900228752	1
4.74	0.419245205	13.21989084	26.04553333	5.306962703	4.907804104	1
4.76	0.418913861	13.43096218	26.2672	5.343958986	4.915307185	1

(continued)

γ = 1.4

M_1	M_2	$P_{t1}/P_{t2} = A_2^*/A_1^*$	P_2/P_1	T_2/T_1	$\rho_2/\rho_1 = V_{1n}/V_{2n}$	T_{t2}/T_{t1}
4.78	0.418586444	13.64498036	26.4898	5.381110175	4.922738829	1
4.8	0.418262894	13.86197793	26.71333333	5.418416281	4.930099857	1
4.82	0.41794315	14.08198772	26.9378	5.455877314	4.937391083	1
4.84	0.417627155	14.30504281	27.1632	5.493493286	4.944613306	1
4.86	0.417314849	14.53117654	27.38953333	5.531264206	4.95176732	1
4.88	0.417006177	14.76042251	27.6168	5.569190083	4.958853906	1
4.9	0.416701082	14.99281457	27.845	5.607270929	4.965873837	1
4.92	0.41639951	15.22838685	28.07413333	5.645506752	4.972827873	1
4.94	0.416101408	15.46717372	28.3042	5.683897561	4.979716769	1
4.96	0.415806723	15.70920983	28.5352	5.722443366	4.986541268	1
4.98	0.415515404	15.95453008	28.76713333	5.761144176	4.993302103	1
5	0.415227399	16.20316967	29	5.8	5	1
5.2	0.412519192	18.88005879	31.38	6.197085799	5.063670412	1
5.4	0.410093218	21.92958513	33.85333333	6.609681451	5.121779859	1
5.6	0.407911868	25.39147028	36.42	7.037793367	5.174917492	1
5.8	0.40594356	29.30832295	39.08	7.481426873	5.223602484	1
6	0.404161618	33.72574463	41.83333333	7.94058642	5.268292683	1
6.2	0.402543391	38.69243541	44.68	8.415275754	5.309392265	1
6.4	0.401069549	44.26030002	47.62	8.905498047	5.347258486	1
6.6	0.399723516	50.48455396	50.65333333	9.411255994	5.382207578	1
6.8	0.398491018	57.42382972	53.78	9.932551903	5.414519906	1
7	0.397359707	65.14028317	57	10.46938776	5.444444444	1
7.2	0.396318863	73.69969986	60.31333333	11.02176526	5.472202674	1
7.4	0.395359142	83.17160152	63.72	11.5896859	5.497991968	1
7.6	0.394472369	93.62935255	67.22	12.17315097	5.521988528	1
7.8	0.39365137	105.1502665	70.81333333	12.77216159	5.544349939	1
8	0.392889827	117.8157125	74.5	13.38671875	5.565217391	1
8.2	0.392182161	131.7112223	78.28	14.01682332	5.584717608	1
8.4	0.391523428	146.9265961	82.15333333	14.66247606	5.602964531	1
8.6	0.390909238	163.5560098	86.12	15.32367766	5.62006079	1
8.8	0.39033568	181.6981213	90.18	16.00042872	5.636098981	1
9	0.38979926	201.4561774	94.33333333	16.69272977	5.651162791	1
9.2	0.389296852	222.9381199	98.58	17.40058129	5.665327979	1
9.4	0.388825651	246.256693	102.92	18.1239837	5.678663239	1
9.6	0.388383133	271.5295492	107.3533333	18.8629374	5.691230959	1
9.8	0.387967025	298.8793565	111.88	19.61744273	5.703087886	1
10	0.387575273	328.4339049	116.5	20.3875	5.714285714	1
11	0.385922492	514.1223892	141	24.47107438	5.761904762	1
12	0.384661215	777.2297061	167.8333333	28.94347994	5.798657718	1
13	0.383677106	1140.131886	197	33.80473373	5.827586207	1
14	0.382894653	1629.206319	228.5	39.05484694	5.850746269	1
15	0.382262375	2275.165169	262.3333333	44.69382716	5.869565217	1
20	0.380387347	9278.597922	466.5	78.721875	5.925925926	1
1.00E+07	0.377964473	2.77867E+32	1.16667E+14	1.94444E+13	6	1

		$\gamma = 5/3$				
M_1	M_2	$P_{t1}/P_{t2} = A_2^*/A_1^*$	P_2/P_1	T_2/T_1	$\rho_2/\rho_1 = V_{1n}/V_{2n}$	T_{t2}/T_{t1}
1	1	1	1	1	1	1
1.02	0.980582543	1.000009565	1.0505	1.019905854	1.02999703	1
1.04	0.96226442	1.000073271	1.102	1.03964571	1.05997648	1
1.06	0.944955378	1.00023703	1.1545	1.059250667	1.089921428	1
1.08	0.928575067	1.00053907	1.208	1.078748971	1.119815668	1
1.1	0.913051678	1.001011152	1.2625	1.098166322	1.149643705	1
1.12	0.898320807	1.001679587	1.318	1.117526148	1.179390748	1
1.14	0.884324504	1.002566103	1.3745	1.136849838	1.209042702	1
1.16	0.871010458	1.003688558	1.432	1.156156956	1.238586156	1
1.18	0.858331311	1.00506156	1.4905	1.175465419	1.268008378	1
1.2	0.846244074	1.006696976	1.55	1.194791667	1.297297297	1
1.22	0.834709616	1.008604372	1.6105	1.2141508	1.326441494	1
1.24	0.823692236	1.010791377	1.672	1.233556712	1.355430183	1
1.26	0.813159283	1.013263999	1.7345	1.253022203	1.384253204	1
1.28	0.803080831	1.016026884	1.798	1.272559082	1.412901	1
1.3	0.793429393	1.019083545	1.8625	1.292178254	1.441364606	1
1.32	0.784179673	1.022436547	1.928	1.311889807	1.469635628	1
1.34	0.775308347	1.026087672	1.9945	1.33170308	1.497706231	1
1.36	0.766793872	1.030038052	2.062	1.35162673	1.525569119	1
1.38	0.758616312	1.034288288	2.1305	1.371668793	1.553217519	1
1.4	0.750757194	1.038838546	2.2	1.391836735	1.580645161	1
1.42	0.743199367	1.043688642	2.2705	1.412137498	1.607846264	1
1.44	0.735926889	1.048838109	2.342	1.432577546	1.634815516	1
1.46	0.72892492	1.054286265	2.4145	1.453162906	1.661548055	1
1.48	0.722179623	1.060032252	2.488	1.473899196	1.688039457	1
1.5	0.715678085	1.066075091	2.5625	1.494791667	1.714285714	1
1.52	0.709408236	1.072413707	2.638	1.515845222	1.740283218	1
1.54	0.703358781	1.07904697	2.7145	1.53706445	1.766028744	1
1.56	0.69751914	1.085973709	2.792	1.558453649	1.791519435	1
1.58	0.691879391	1.093192746	2.8705	1.580016844	1.816752784	1
1.6	0.686430215	1.100702902	2.95	1.601757813	1.841726619	1
1.62	0.681162859	1.108503019	3.0305	1.623680098	1.866439087	1
1.64	0.676069086	1.116591973	3.112	1.645787032	1.890888639	1
1.66	0.671141138	1.124968676	3.1945	1.668081743	1.915074015	1
1.68	0.666371706	1.133632094	3.278	1.690567177	1.938994229	1
1.7	0.661753894	1.142581246	3.3625	1.713246107	1.962648557	1
1.72	0.657281189	1.151815213	3.448	1.736121147	1.98603652	1
1.74	0.65294744	1.161333141	3.5345	1.759194758	2.009157874	1
1.76	0.648746827	1.171134245	3.622	1.782469267	2.032012595	1
1.78	0.644673845	1.181217807	3.7105	1.805946866	2.054600869	1
1.8	0.640723276	1.191583183	3.8	1.82962963	2.076923077	1
1.82	0.636890176	1.202229801	3.8905	1.853519518	2.098979786	1
1.84	0.633169857	1.213157161	3.982	1.877618384	2.120771736	1
1.86	0.629557868	1.224364837	4.0745	1.901927983	2.142299833	1
1.88	0.626049982	1.235852473	4.168	1.926449977	2.163565132	1
1.9	0.62264218	1.247619786	4.2625	1.951185942	2.184568835	1
1.92	0.619330643	1.259666563	4.358	1.97613737	2.205312276	1
1.94	0.616111734	1.271992661	4.4545	2.001305678	2.225796913	1

(continued)

$\gamma = 5/3$

M_1	M_2	$P_{t1}/P_{t2} = A_2^*/A_1^*$	P_2/P_1	T_2/T_1	$\rho_2/\rho_1 = V_{1n}/V_{2n}$	T_{t2}/T_{t1}
1.96	0.612981994	1.284598004	4.552	2.026692212	2.246024322	1
1.98	0.609938123	1.297482584	4.6505	2.052298248	2.265996185	1
2	0.606976979	1.310646457	4.75	2.078125	2.285714286	1
2.02	0.604095564	1.324089743	4.8505	2.104173623	2.305180498	1
2.04	0.601291018	1.337812624	4.952	2.130445213	2.324396783	1
2.06	0.598560611	1.351815343	5.0545	2.156940817	2.343365178	1
2.08	0.595901733	1.3660982	5.158	2.183661428	2.362087792	1
2.1	0.593311893	1.380661553	5.2625	2.210607993	2.380566802	1
2.12	0.590788707	1.395505815	5.368	2.237781417	2.398804441	1
2.14	0.588329896	1.410631452	5.4745	2.26518256	2.416802998	1
2.16	0.585933279	1.426038983	5.582	2.292812243	2.434564809	1
2.18	0.583596766	1.441728977	5.6905	2.32067125	2.452092255	1
2.2	0.581318359	1.45770205	5.8	2.348760331	2.469387755	1
2.22	0.57909614	1.473958867	5.9105	2.377080198	2.486453761	1
2.24	0.576928272	1.490500137	6.022	2.405631537	2.503292756	1
2.26	0.574812993	1.507326614	6.1345	2.434414999	2.519907248	1
2.28	0.572748612	1.524439093	6.248	2.46343121	2.536299766	1
2.3	0.570733509	1.541838412	6.3625	2.492680766	2.552472859	1
2.32	0.568766126	1.559525447	6.478	2.522164239	2.56842909	1
2.34	0.566844966	1.577501113	6.5945	2.551882177	2.584171032	1
2.36	0.564968593	1.595766362	6.712	2.581835105	2.59970127	1
2.38	0.563135626	1.614322181	6.8305	2.612023524	2.61502239	1
2.4	0.561344737	1.633169593	6.95	2.642447917	2.630136986	1
2.42	0.559594649	1.652309652	7.0705	2.673108744	2.645047649	1
2.44	0.557884132	1.671743446	7.192	2.70400645	2.659756969	1
2.46	0.556212006	1.691472093	7.3145	2.735141458	2.674267533	1
2.48	0.554577131	1.711496741	7.438	2.766514178	2.68858192	1
2.5	0.552978412	1.731818568	7.5625	2.798125	2.702702703	1
2.52	0.551414792	1.752438777	7.688	2.829974301	2.716632444	1
2.54	0.549885254	1.773358602	7.8145	2.862062442	2.730373693	1
2.56	0.548388817	1.794579299	7.942	2.894389771	2.74392899	1
2.58	0.546924536	1.816102152	8.0705	2.926956621	2.757300857	1
2.6	0.545491499	1.837928468	8.2	2.959763314	2.770491803	1
2.62	0.544088826	1.860059576	8.3305	2.992810158	2.783504319	1
2.64	0.542715669	1.88249683	8.462	3.026097452	2.796340876	1
2.66	0.541371209	1.905241605	8.5945	3.059625481	2.80900393	1
2.68	0.540054653	1.928295297	8.728	3.09339452	2.821495915	1
2.7	0.538765239	1.951659321	8.8625	3.127404835	2.833819242	1
2.72	0.537502229	1.975335114	8.998	3.161656683	2.845976304	1
2.74	0.53626491	1.999324129	9.1345	3.196150308	2.85796947	1
2.76	0.535052593	2.023627841	9.272	3.230885948	2.869801085	1
2.78	0.533864611	2.04824774	9.4105	3.265863833	2.881473472	1
2.8	0.532700322	2.073185334	9.55	3.301084184	2.89298893	1
2.82	0.531559103	2.098442147	9.6905	3.336547212	2.904349732	1
2.84	0.53044035	2.124019719	9.832	3.372253124	2.915558126	1
2.86	0.529343482	2.149919608	9.9745	3.408202119	2.926616337	1
2.88	0.528267933	2.176143382	10.118	3.444394387	2.937526562	1

(continued)

$\gamma = 5/3$

M_1	M_2	$P_{t1}/P_{t2} = A_2^*/A_1^*$	P_2/P_1	T_2/T_1	$\rho_2/\rho_1 = V_{1n}/V_{2n}$	T_{t2}/T_{t1}
2.9	0.527213159	2.202692629	10.2625	3.480830113	2.948290973	1
2.92	0.52617863	2.229568946	10.408	3.517509476	2.958911716	1
2.94	0.525163835	2.256773947	10.5545	3.55443265	2.96939091	1
2.96	0.524168277	2.284309258	10.702	3.591599799	2.979730649	1
2.98	0.523191477	2.312176517	10.8505	3.629011086	2.989932999	1
3	0.522232968	2.340377375	11	3.666666667	3	1
3.02	0.5212923	2.368913494	11.1505	3.704566691	3.009933666	1
3.04	0.520369034	2.397786549	11.302	3.742711305	3.019735982	1
3.06	0.519462748	2.426998226	11.4545	3.78110065	3.02940891	1
3.08	0.51857303	2.456550219	11.608	3.819734863	3.038954382	1
3.1	0.517699481	2.486444236	11.7625	3.858614074	3.048374306	1
3.12	0.516841713	2.516681992	11.918	3.897738412	3.057670562	1
3.14	0.515999352	2.547265215	12.0745	3.937108001	3.066845003	1
3.16	0.515172033	2.578195641	12.232	3.976722961	3.075899458	1
3.18	0.514359402	2.609475013	12.3905	4.016583407	3.084835728	1
3.2	0.513561116	2.641105086	12.55	4.056689453	3.093655589	1
3.22	0.512776841	2.673087623	12.7105	4.097041207	3.102360791	1
3.24	0.512006254	2.705424394	12.872	4.137638775	3.110953058	1
3.26	0.511249039	2.738117178	13.0345	4.178482258	3.11943409	1
3.28	0.510504892	2.771167763	13.198	4.219571758	3.127805559	1
3.3	0.509773515	2.804577942	13.3625	4.260907369	3.136069114	1
3.32	0.509054619	2.838349517	13.528	4.302489186	3.144226381	1
3.34	0.508347925	2.872484299	13.6945	4.344317298	3.152278957	1
3.36	0.507653159	2.906984101	13.862	4.386391794	3.160228418	1
3.38	0.506970056	2.941850748	14.0305	4.42871276	3.168076315	1
3.4	0.506298359	2.977086069	14.2	4.471280277	3.175824176	1
3.42	0.505637816	3.012691899	14.3705	4.514094426	3.183473504	1
3.44	0.504988184	3.04867008	14.542	4.557155287	3.191025779	1
3.46	0.504349225	3.085022461	14.7145	4.600462933	3.19848246	1
3.48	0.503720708	3.121750894	14.888	4.64401744	3.205844981	1
3.5	0.503102408	3.15885724	15.0625	4.687818878	3.213114754	1
3.52	0.502494108	3.196343363	15.238	4.731867317	3.22029317	1
3.54	0.501895593	3.234211133	15.4145	4.776162824	3.227381596	1
3.56	0.501306656	3.272462426	15.592	4.820705466	3.23438138	1
3.58	0.500727096	3.311099123	15.7705	4.865495307	3.241293847	1
3.6	0.500156715	3.350123108	15.95	4.910532407	3.248120301	1
3.62	0.499595323	3.389536272	16.1305	4.955816829	3.254862025	1
3.64	0.499042732	3.42934051	16.312	5.001348629	3.261520284	1
3.66	0.498498762	3.469537721	16.4945	5.047127867	3.268096319	1
3.68	0.497963234	3.510129809	16.678	5.093154596	3.274591353	1
3.7	0.497435976	3.551118682	16.8625	5.139428871	3.281006591	1
3.72	0.49691682	3.592506251	17.048	5.185950746	3.287343216	1
3.74	0.496405601	3.634294434	17.2345	5.23272027	3.293602392	1
3.76	0.495902161	3.67648515	17.422	5.279737494	3.299785267	1
3.78	0.495406342	3.719080322	17.6105	5.327002467	3.305892969	1
3.8	0.494917993	3.76208188	17.8	5.374515235	3.311926606	1
3.82	0.494436966	3.805491754	17.9905	5.422275846	3.31788727	1

(continued)

$\gamma = 5/3$

M_1	M_2	$P_{t1}/P_{t2} = A_2^*/A_1^*$	P_2/P_1	T_2/T_1	$\rho_2/\rho_1 = V_{1n}/V_{2n}$	T_{t2}/T_{t1}
3.84	0.493963116	3.849311878	18.182	5.470284342	3.323776035	1
3.86	0.493496302	3.893544192	18.3745	5.51854077	3.329593957	1
3.88	0.493036385	3.938190636	18.568	5.56704517	3.335342077	1
3.9	0.492583233	3.983253156	18.7625	5.615797584	3.341021417	1
3.92	0.492136713	4.028733699	18.958	5.664798053	3.346632982	1
3.94	0.491696698	4.074634217	19.1545	5.714046616	3.352177762	1
3.96	0.491263062	4.120956664	19.352	5.763543312	3.357656732	1
3.98	0.490835685	4.167702997	19.5505	5.813288178	3.363070848	1
4	0.490414446	4.214875175	19.75	5.86328125	3.368421053	1
4.02	0.48999923	4.262475162	19.9505	5.913522564	3.373708273	1
4.04	0.489589923	4.310504922	20.152	5.964012156	3.378933422	1
4.06	0.489186415	4.358966425	20.3545	6.014750058	3.384097395	1
4.08	0.488788596	4.40786164	20.558	6.065736303	3.389201075	1
4.1	0.488396361	4.457192542	20.7625	6.116970925	3.394245331	1
4.12	0.488009607	4.506961104	20.968	6.168453954	3.399231016	1
4.14	0.487628233	4.557169306	21.1745	6.220185421	3.40415897	1
4.16	0.48725214	4.607819128	21.382	6.272165357	3.409030021	1
4.18	0.486881232	4.658912553	21.5905	6.32439379	3.413844982	1
4.2	0.486515414	4.710451564	21.8	6.376870748	3.418604651	1
4.22	0.486154594	4.762438151	22.0105	6.429596261	3.423309817	1
4.24	0.485798683	4.814874301	22.222	6.482570354	3.427961254	1
4.26	0.485447591	4.867762006	22.4345	6.535793055	3.432559723	1
4.28	0.485101234	4.921103258	22.648	6.58926439	3.437105974	1
4.3	0.484759527	4.974900055	22.8625	6.642984383	3.441600745	1
4.32	0.484422387	5.029154392	23.078	6.696953061	3.44604476	1
4.34	0.484089734	5.083868268	23.2945	6.751170446	3.450438733	1
4.36	0.483761489	5.139043685	23.512	6.805636563	3.454783367	1
4.38	0.483437576	5.194682645	23.7305	6.860351434	3.459079353	1
4.4	0.483117918	5.250787152	23.95	6.915315083	3.46332737	1
4.42	0.482802442	5.307359213	24.1705	6.970527531	3.467528088	1
4.44	0.482491076	5.364400835	24.392	7.0259888	3.471682164	1
4.46	0.482183749	5.421914029	24.6145	7.081698911	3.475790246	1
4.48	0.481880392	5.479900804	24.838	7.137657884	3.479852972	1
4.5	0.481580938	5.538363173	25.0625	7.193865741	3.483870968	1
4.52	0.481285319	5.597303151	25.288	7.2503225	3.487844851	1
4.54	0.480993472	5.656722753	25.5145	7.307028181	3.491775229	1
4.56	0.480705332	5.716623997	25.742	7.363982802	3.495662699	1
4.58	0.480420838	5.7770089	25.9705	7.421186383	3.499507849	1
4.6	0.480139929	5.837879482	26.2	7.478638941	3.503311258	1
4.62	0.479862544	5.899237765	26.4305	7.536340494	3.507073495	1
4.64	0.479588627	5.961085772	26.662	7.59429106	3.510795121	1
4.66	0.479318119	6.023425526	26.8945	7.652490654	3.514476687	1
4.68	0.479050964	6.086259051	27.128	7.710939294	3.518118736	1
4.7	0.478787108	6.149588376	27.3625	7.769636996	3.521721802	1
4.72	0.478526496	6.213415527	27.598	7.828583776	3.525286411	1
4.74	0.478269077	6.277742532	27.8345	7.887779649	3.52881308	1
4.76	0.478014799	6.342571423	28.072	7.947224631	3.53230232	1

(continued)

M_1	M_2	$P_{t1}/P_{t2} = A_2^*/A_1^*$	P_2/P_1	T_2/T_1	$\rho_2/\rho_1 = V_{1n}/V_{2n}$	T_{t2}/T_{t1}
4.78	0.47776361	6.40790423	28.3105	8.006918736	3.535754631	1
4.8	0.477515461	6.473742985	28.55	8.066861979	3.539170507	1
4.82	0.477270305	6.540089721	28.7905	8.127054374	3.542550434	1
4.84	0.477028092	6.606946474	29.032	8.187495936	3.54589489	1
4.86	0.476788777	6.674315278	29.2745	8.248186678	3.549204346	1
4.88	0.476552313	6.742198169	29.518	8.309126612	3.552479265	1
4.9	0.476318656	6.810597186	29.7625	8.370315754	3.555720104	1
4.92	0.476087762	6.879514367	30.008	8.431754115	3.558927311	1
4.94	0.475859588	6.94895175	30.2545	8.493441707	3.56210133	1
4.96	0.475634091	7.018911377	30.502	8.555378544	3.565242595	1
4.98	0.47541123	7.089395289	30.7505	8.617564638	3.568351535	1
5	0.475190963	7.160405527	31	8.68	3.571428571	1
5.2	0.473122614	7.899905771	33.55	9.318065828	3.600532623	1
5.4	0.471274281	8.69429113	36.2	9.981069959	3.626865672	1
5.6	0.469615963	9.545616174	38.95	10.66902105	3.650756694	1
5.8	0.468122586	10.45594136	41.8	11.38192628	3.672489083	1
6	0.466773066	11.42733204	44.75	12.11979167	3.692307692	1
6.2	0.46554957	12.46185765	47.8	12.88262227	3.71042471	1
6.4	0.464436937	13.56159104	50.95	13.67042236	3.727024568	1
6.6	0.463422213	14.72860797	54.2	14.48319559	3.742268041	1
6.8	0.462494276	15.96498669	57.55	15.32094507	3.756295695	1
7	0.461643536	17.27280751	61	16.18367347	3.769230769	1
7.2	0.460861686	18.65415258	64.55	17.0713831	3.781181619	1
7.4	0.460141506	20.1111056	68.2	17.98407597	3.792243767	1
7.6	0.459476694	21.64575162	71.95	18.92175381	3.802501646	1
7.8	0.458861733	23.26017686	75.8	19.88441815	3.812030075	1
8	0.458291772	24.95646857	79.75	20.87207031	3.820895522	1
8.2	0.457762536	26.73671489	83.8	21.88471148	3.829157175	1
8.4	0.457270245	28.60300477	87.95	22.92234269	3.836867863	1
8.6	0.456811547	30.55742782	92.2	23.98496485	3.844074844	1
8.8	0.45638346	32.60207431	96.55	25.07257877	3.850820487	1
9	0.455983324	34.73903501	101	26.18518519	3.857142857	1
9.2	0.455608764	36.97040121	105.55	27.32278474	3.863076221	1
9.4	0.45525765	39.29826462	110.2	28.485378	3.868651489	1
9.6	0.454928069	41.72471734	114.95	29.67296549	3.873896595	1
9.8	0.454618297	44.25185182	119.8	30.88554769	3.878836834	1
10	0.454326783	46.88176082	124.75	32.123125	3.883495146	1
11	0.453098238	61.64618847	151	38.68595041	3.903225806	1
12	0.452162168	79.29402747	179.75	45.87369792	3.918367347	1
13	0.451432685	100.0870895	211	53.68639053	3.930232558	1
14	0.450853237	124.2872518	244.75	62.12404337	3.939698492	1
15	0.450385363	152.1564348	281	71.18666667	3.947368421	1
20	0.448999778	355.7065504	499.75	125.8745313	3.970223325	1
1.00E+07	0.447213595	4.36732E+19	1.25E+14	3.125E+13	4	1

$\gamma = 1.3$

M_1	M_2	$P_{t1}/P_{t2} = A_2^*/A_1^*$	P_2/P_1	T_2/T_1	$\rho_2/\rho_1 = V_{1n}/V_{2n}$	T_{t2}/T_{t1}
1	1	1	1	1	1	1
1.02	0.980491952	1.0000101	1.045669565	1.010361197	1.034946283	1
1.04	0.961920674	1.000077901	1.092243478	1.020588709	1.070209251	1
1.06	0.944220566	1.000253686	1.139721739	1.030701349	1.105773016	1
1.08	0.927332174	1.000580688	1.188104348	1.040716202	1.141621842	1
1.1	0.911201473	1.001096091	1.237391304	1.050648815	1.177740161	1
1.12	0.895779238	1.001831897	1.287582609	1.06051336	1.214112577	1
1.14	0.88102052	1.002815672	1.338678261	1.070322776	1.250723886	1
1.16	0.866884173	1.004071191	1.390678261	1.080088896	1.287559076	1
1.18	0.853332454	1.005618988	1.443582609	1.08982256	1.324603345	1
1.2	0.840330669	1.007476845	1.497391304	1.099533711	1.361842105	1
1.22	0.827846867	1.009660195	1.552104348	1.109231486	1.399260991	1
1.24	0.815851566	1.012182486	1.607721739	1.118924287	1.436845869	1
1.26	0.804317513	1.015055485	1.664243478	1.128619858	1.474582842	1
1.28	0.793219477	1.018289545	1.721669565	1.138325343	1.512458258	1
1.3	0.782534057	1.021893834	1.78	1.148047337	1.550458716	1
1.32	0.772239517	1.025876535	1.839234783	1.157791944	1.588571066	1
1.34	0.762315637	1.030245014	1.899373913	1.167564812	1.626782422	1
1.36	0.752743583	1.035005971	1.960417391	1.177371179	1.66508016	1
1.38	0.743505785	1.040165567	2.022365217	1.187215905	1.703451924	1
1.4	0.734585834	1.045729536	2.085217391	1.197103507	1.741885626	1
1.42	0.72596838	1.05170328	2.148973913	1.207038184	1.780369455	1
1.44	0.717639055	1.058091951	2.213634783	1.217023847	1.818891872	1
1.46	0.709584388	1.064900527	2.2792	1.22706414	1.857441617	1
1.48	0.701791736	1.072133871	2.345669565	1.237162463	1.896007708	1
1.5	0.694249222	1.079796785	2.413043478	1.247321991	1.934579439	1
1.52	0.686945676	1.08789406	2.481321739	1.257545692	1.973146388	1
1.54	0.679870583	1.096430515	2.550504348	1.26783634	2.011698408	1
1.56	0.673014035	1.105411033	2.620591304	1.278196536	2.050225634	1
1.58	0.666366684	1.114840592	2.691582609	1.288628715	2.088718479	1
1.6	0.659919708	1.12472429	2.763478261	1.299135161	2.12716763	1
1.62	0.653664771	1.135067372	2.836278261	1.309718018	2.165564054	1
1.64	0.647593988	1.145875248	2.909982609	1.320379301	2.203898991	1
1.66	0.641699898	1.157153514	2.984591304	1.331120903	2.242163952	1
1.68	0.635975434	1.168907961	3.060104348	1.341944606	2.280350719	1
1.7	0.630413896	1.181144598	3.136521739	1.352852088	2.318451343	1
1.72	0.625008929	1.193869657	3.213843478	1.363844928	2.356458137	1
1.74	0.619754499	1.207089606	3.292069565	1.374924617	2.394363679	1
1.76	0.614644877	1.220811162	3.3712	1.386092562	2.432160804	1
1.78	0.609674613	1.235041294	3.451234783	1.397350089	2.469842604	1
1.8	0.604838524	1.249787234	3.532173913	1.408698453	2.507402423	1
1.82	0.600131679	1.265056482	3.614017391	1.420138839	2.544833852	1
1.84	0.595549378	1.280856816	3.696765217	1.431672368	2.58213073	1
1.86	0.591087143	1.297196294	3.780417391	1.443300103	2.619287134	1
1.88	0.586740706	1.314083259	3.864973913	1.455023049	2.656297381	1
1.9	0.582505991	1.331526347	3.950434783	1.466842158	2.693156017	1
1.92	0.578379109	1.349534491	4.0368	1.478758333	2.72985782	1
1.94	0.574356346	1.368116923	4.124069565	1.490772433	2.766397791	1

(continued)

M_1	M_2	$P_{t1}/P_{t2} = A_2^*/A_1^*$	P_2/P_1	T_2/T_1	$\rho_2/\rho_1 = V_{1n}/V_{2n}$	T_{t2}/T_{t1}
$\gamma = 1.3$						
1.96	0.570434152	1.38728318	4.212243478	1.50288527	2.802771152	1
1.98	0.566609131	1.407043107	4.301321739	1.515097617	2.838973339	1
2	0.562878036	1.427406862	4.391304348	1.527410208	2.875	1
2.02	0.559237759	1.448384919	4.482191304	1.539823742	2.910846991	1
2.04	0.555685323	1.469988073	4.573982609	1.552338881	2.946510368	1
2.06	0.552217877	1.492227441	4.666678261	1.56495626	2.981986386	1
2.08	0.548832689	1.515114469	4.760278261	1.577676478	3.017271492	1
2.1	0.545527137	1.538660932	4.854782609	1.590500109	3.052362323	1
2.12	0.542298707	1.562878941	4.950191304	1.6034277	3.087255698	1
2.14	0.539144988	1.587780944	5.046504348	1.616459771	3.121948617	1
2.16	0.536063662	1.613379732	5.143721739	1.62959682	3.156438253	1
2.18	0.533052506	1.639688443	5.241843478	1.64283932	3.19072195	1
2.2	0.530109381	1.666720562	5.340869565	1.656187724	3.224797219	1
2.22	0.527232233	1.694489929	5.4408	1.669642464	3.25866173	1
2.24	0.524419086	1.723010741	5.541634783	1.683203954	3.29231331	1
2.26	0.521668039	1.752297558	5.643373913	1.696872589	3.325749941	1
2.28	0.518977265	1.782365303	5.746017391	1.710648747	3.358969749	1
2.3	0.516345001	1.813229273	5.849565217	1.724532788	3.391971006	1
2.32	0.513769553	1.844905136	5.954017391	1.73852506	3.424752125	1
2.34	0.511249288	1.877408939	6.059373913	1.752625892	3.45731165	1
2.36	0.508782632	1.910757114	6.165634783	1.766835602	3.489648259	1
2.38	0.506368069	1.944966479	6.2728	1.781154495	3.521760756	1
2.4	0.504004136	1.980054247	6.380869565	1.795582861	3.553648069	1
2.42	0.501689422	2.016038024	6.489843478	1.81012098	3.585309243	1
2.44	0.499422567	2.052935824	6.599721739	1.824769119	3.616743439	1
2.46	0.497202257	2.090766063	6.710504348	1.839527536	3.64794993	1
2.48	0.495027225	2.129547573	6.822191304	1.854396478	3.678928096	1
2.5	0.492896245	2.169299601	6.934782609	1.869376181	3.709677419	1
2.52	0.490808136	2.210041819	7.048278261	1.884466874	3.740197484	1
2.54	0.488761754	2.251794326	7.162678261	1.899668774	3.770487971	1
2.56	0.486755996	2.294577655	7.277982609	1.914982092	3.800548653	1
2.58	0.484789794	2.338412779	7.394191304	1.93040703	3.830379392	1
2.6	0.482862116	2.383321116	7.511304348	1.945943781	3.859980139	1
2.62	0.480971963	2.429324535	7.629321739	1.961592534	3.889350926	1
2.64	0.47911837	2.47644536	7.748243478	1.977353468	3.918491865	1
2.66	0.477300401	2.524706382	7.868069565	1.993226756	3.947403146	1
2.68	0.475517153	2.574130857	7.9888	2.009212564	3.976085031	1
2.7	0.47376775	2.624742518	8.110434783	2.025311054	4.004537855	1
2.72	0.472051343	2.67656558	8.232973913	2.041522379	4.03276202	1
2.74	0.47036711	2.729624746	8.356417391	2.057846689	4.060757993	1
2.76	0.468714257	2.783945213	8.480765217	2.074284128	4.088526304	1
2.78	0.467092012	2.839552679	8.606017391	2.090834833	4.116067542	1
2.8	0.465499627	2.896473349	8.732173913	2.107498939	4.143382353	1
2.82	0.463936377	2.954733947	8.859234783	2.124276574	4.170471439	1
2.84	0.462401561	3.014361713	8.9872	2.141167864	4.197335554	1
2.86	0.460894497	3.07538442	9.116069565	2.158172926	4.2239755	1
2.88	0.459414523	3.137830374	9.245843478	2.175291878	4.250392129	1

(continued)

$\gamma = 1.3$

M_1	M_2	$P_{t1}/P_{t2} = A_2^*/A_1^*$	P_2/P_1	T_2/T_1	$\rho_2/\rho_1 = V_{1n}/V_{2n}$	T_{t2}/T_{t1}
2.9	0.457960998	3.201728426	9.376521739	2.192524832	4.276586337	1
2.92	0.456533301	3.267107977	9.508104348	2.209871895	4.302559062	1
2.94	0.455130826	3.333998985	9.640591304	2.227333172	4.328311286	1
2.96	0.453752987	3.402431973	9.773982609	2.244908763	4.353844027	1
2.98	0.452399215	3.472438038	9.908278261	2.262598767	4.379158341	1
3	0.451068956	3.544048857	10.04347826	2.280403277	4.404255319	1
3.02	0.449761675	3.617296696	10.17958261	2.298322384	4.429136086	1
3.04	0.448476847	3.692214417	10.3165913	2.316356177	4.453801797	1
3.06	0.447213967	3.768835486	10.45450435	2.334504741	4.478253637	1
3.08	0.445972542	3.847193981	10.59332174	2.352768159	4.502492819	1
3.1	0.444752092	3.927324601	10.73304348	2.37114651	4.526520582	1
3.12	0.443552152	4.009262675	10.87366957	2.389639871	4.550338189	1
3.14	0.442372268	4.093044168	11.0152	2.408248318	4.573946929	1
3.16	0.441212001	4.17870569	11.15763478	2.426971924	4.597348109	1
3.18	0.440070922	4.266284508	11.30097391	2.445810757	4.620543058	1
3.2	0.438948613	4.355818549	11.44521739	2.464764887	4.643533123	1
3.22	0.43784467	4.447346413	11.59036522	2.483834378	4.66631967	1
3.24	0.436758698	4.540907381	11.73641739	2.503019297	4.68890408	1
3.26	0.435690313	4.636541424	11.88337391	2.522319703	4.711287749	1
3.28	0.43463914	4.734289209	12.03123478	2.541735657	4.733472086	1
3.3	0.433604816	4.834192114	12.18	2.561267218	4.755458515	1
3.32	0.432586986	4.936292231	12.32966957	2.580914441	4.77724847	1
3.34	0.431585305	5.040632382	12.48024348	2.600677383	4.798843394	1
3.36	0.430599437	5.147256121	12.63172174	2.620556095	4.820244743	1
3.38	0.429629054	5.25620775	12.78410435	2.64055063	4.841453977	1
3.4	0.428673837	5.367532324	12.9373913	2.660661037	4.862472568	1
3.42	0.427733475	5.481275665	13.09158261	2.680887366	4.88330199	1
3.44	0.426807664	5.597484367	13.24667826	2.701229663	4.903943727	1
3.46	0.42589611	5.71620581	13.40267826	2.721687975	4.924399265	1
3.48	0.424998525	5.837488168	13.55958261	2.742262345	4.944670094	1
3.5	0.424114627	5.96138042	13.7173913	2.762952818	4.964757709	1
3.52	0.423244144	6.08793236	13.87610435	2.783759435	4.984663607	1
3.54	0.422386807	6.217194605	14.03572174	2.804682238	5.004389285	1
3.56	0.421542356	6.349218609	14.19624348	2.825721265	5.023936244	1
3.58	0.420710538	6.484056674	14.35766957	2.846876556	5.043305982	1
3.6	0.419891105	6.621761954	14.52	2.868148148	5.0625	1
3.62	0.419083814	6.762388475	14.68323478	2.889536078	5.081519797	1
3.64	0.418288431	6.905991139	14.84737391	2.911040381	5.100366869	1
3.66	0.417504724	7.052625735	15.01241739	2.932661091	5.119042714	1
3.68	0.416732468	7.202348957	15.17836522	2.954398243	5.137548823	1
3.7	0.415971445	7.355218405	15.34521739	2.976251869	5.155886687	1
3.72	0.415221441	7.511292604	15.51297391	2.998222001	5.174057794	1
3.74	0.414482245	7.670631012	15.68163478	3.02030867	5.192063625	1
3.76	0.413753654	7.833294032	15.8512	3.042511906	5.20990566	1
3.78	0.413035469	7.999343025	16.02166957	3.064831738	5.227585373	1
3.8	0.412327494	8.168840316	16.19304348	3.087268195	5.245104232	1
3.82	0.411629539	8.341849215	16.36532174	3.109821305	5.262463702	1

(continued)

$\gamma = 1.3$

M_1	M_2	$P_{t1}/P_{t2} = A_2^*/A_1^*$	P_2/P_1	T_2/T_1	$\rho_2/\rho_1 = V_{1n}/V_{2n}$	T_{t2}/T_{t1}
3.84	0.410941418	8.518434019	16.53850435	3.132491096	5.279665239	1
3.86	0.41026295	8.698660032	16.7125913	3.155277592	5.296710294	1
3.88	0.409593957	8.882593571	16.88758261	3.178180821	5.313600314	1
3.9	0.408934265	9.070301982	17.06347826	3.201200807	5.330336736	1
3.92	0.408283706	9.261853649	17.24027826	3.224337575	5.346920991	1
3.94	0.407642113	9.457318011	17.41798261	3.247591148	5.363354504	1
3.96	0.407009324	9.656765568	17.5965913	3.27096155	5.379638689	1
3.98	0.406385181	9.8602679	17.77610435	3.294448803	5.395774957	1
4	0.405769529	10.06789767	17.95652174	3.31805293	5.411764706	1
4.02	0.405162216	10.27972866	18.13784348	3.341773952	5.427609329	1
4.04	0.404563095	10.49583575	18.32006957	3.36561189	5.44331021	1
4.06	0.403972021	10.71629495	18.5032	3.389566765	5.458868724	1
4.08	0.403388851	10.94118342	18.68723478	3.413638596	5.474286237	1
4.1	0.402813448	11.17057946	18.87217391	3.437827403	5.489564106	1
4.12	0.402245675	11.40456256	19.05801739	3.462133205	5.504703679	1
4.14	0.401685401	11.64321336	19.24476522	3.486556021	5.519706296	1
4.16	0.401132495	11.88661373	19.43241739	3.511095868	5.534573285	1
4.18	0.400586829	12.13484672	19.62097391	3.535752765	5.549305966	1
4.2	0.400048281	12.3879966	19.81043478	3.560526729	5.56390565	1
4.22	0.399516727	12.64614891	20.0008	3.585417776	5.578373637	1
4.24	0.39899205	12.90939039	20.19206957	3.610425923	5.592711219	1
4.26	0.398474132	13.17780908	20.38424348	3.635551187	5.606919675	1
4.28	0.397962858	13.45149428	20.57732174	3.660793582	5.621000277	1
4.3	0.397458119	13.73053658	20.77130435	3.686153124	5.634954286	1
4.32	0.396959802	14.01502789	20.9661913	3.711629829	5.648782953	1
4.34	0.396467803	14.30506144	21.16198261	3.73722371	5.662487517	1
4.36	0.395982015	14.60073176	21.35867826	3.762934783	5.67606921	1
4.38	0.395502336	14.90213477	21.55627826	3.78876306	5.689529252	1
4.4	0.395028664	15.20936774	21.75478261	3.814708557	5.702868852	1
4.42	0.394560902	15.52252931	21.9541913	3.840771285	5.716089211	1
4.44	0.394098953	15.84171952	22.15450435	3.866951259	5.729191517	1
4.46	0.393642721	16.16703982	22.35572174	3.89324849	5.742176949	1
4.48	0.393192114	16.49859309	22.55784348	3.919662992	5.755046677	1
4.5	0.392747042	16.83648362	22.76086957	3.946194777	5.767801858	1
4.52	0.392307414	17.1808172	22.9648	3.972843856	5.78044364	1
4.54	0.391873143	17.53170104	23.16963478	3.999610242	5.792973161	1
4.56	0.391444145	17.88924388	23.37537391	4.026493945	5.805391548	1
4.58	0.391020334	18.25355594	23.58201739	4.053494977	5.817699918	1
4.6	0.390601629	18.62474894	23.78956522	4.080613348	5.829899377	1
4.62	0.390187948	19.00293617	23.99801739	4.10784907	5.841991023	1
4.64	0.389779213	19.38823245	24.20737391	4.135202153	5.85397594	1
4.66	0.389375347	19.78075415	24.41763478	4.162672607	5.865855205	1
4.68	0.388976272	20.18061925	24.6288	4.190260443	5.877629884	1
4.7	0.388581915	20.58794731	24.84086957	4.217965669	5.889301032	1
4.72	0.388192202	21.0028595	25.05384348	4.245788296	5.900869693	1
4.74	0.387807061	21.42547865	25.26772174	4.273728332	5.912336905	1
4.76	0.387426422	21.85592919	25.48250435	4.301785788	5.92370369	1

(continued)

M_1	M_2	$P_{t1}/P_{t2} = A_2^*/A_1^*$	P_2/P_1	T_2/T_1	$\rho_2/\rho_1 = V_{1n}/V_{2n}$	T_{t2}/T_{t1}
γ = 1.3						
4.78	0.387050216	22.29433726	25.6981913	4.329960672	5.934971066	1
4.8	0.386678376	22.74083065	25.91478261	4.358252993	5.946140036	1
4.82	0.386310834	23.19553887	26.13227826	4.38666276	5.957211596	1
4.84	0.385947526	23.65859312	26.35067826	4.41518998	5.968186732	1
4.86	0.385588387	24.13012637	26.56998261	4.443834663	5.97906642	1
4.88	0.385233354	24.6102733	26.7901913	4.472596817	5.989851624	1
4.9	0.384882367	25.09917039	27.01130435	4.501476448	6.000543301	1
4.92	0.384535364	25.59695588	27.23332174	4.530473566	6.011142398	1
4.94	0.384192286	26.10376984	27.45624348	4.559588178	6.021649852	1
4.96	0.383853075	26.61975413	27.68006957	4.588820292	6.032066589	1
4.98	0.383517674	27.14505247	27.9048	4.618169913	6.042393529	1
5	0.383186026	27.67981043	28.13043478	4.647637051	6.052631579	1
5.2	0.380064322	33.5810597	30.43652174	4.94877328	6.150316456	1
5.4	0.377263109	40.60290575	32.83304348	5.261668028	6.240044659	1
5.6	0.374740378	48.92453205	35.32	5.586326531	6.322580645	1
5.8	0.372460733	58.74800005	37.8973913	5.922753136	6.398610652	1
6	0.370394188	70.30039548	40.56521739	6.270951481	6.46875	1
6.2	0.36851521	83.8361021	43.32347826	6.630924624	6.533550103	1
6.4	0.366801955	99.63920616	46.17217391	7.002675154	6.593505039	1
6.6	0.365235643	118.0260343	49.11130435	7.386205273	6.649057606	1
6.8	0.363800067	139.347828	52.14086957	7.781516866	6.700604839	1
7	0.362481178	163.9935576	55.26086957	8.18861155	6.748502994	1
7.2	0.361266753	192.3928785	58.47130435	8.607490723	6.793072015	1
7.4	0.360146121	225.019233	61.77217391	9.038155595	6.834599522	1
7.6	0.359109928	262.3930998	65.16347826	9.480607219	6.873344371	1
7.8	0.358149953	305.0853948	68.64521739	9.934846516	6.909539799	1
8	0.357258947	353.7210255	72.2173913	10.40087429	6.943396226	1
8.2	0.356430497	408.982602	75.88	10.87869126	6.975103734	1
8.4	0.355658913	471.6143073	79.63304348	11.36829803	7.004834254	1
8.6	0.354939137	542.4259299	83.47652174	11.86969518	7.032743509	1
8.8	0.354266657	622.2970602	87.41043478	12.38288319	7.058972733	1
9	0.35363744	712.1814559	91.43478261	12.90786249	7.08365019	1
9.2	0.35304787	813.1115759	95.54956522	13.44463349	7.106892523	1
9.4	0.352494702	926.2032879	99.75478261	13.99319652	7.128805949	1
9.6	0.351975013	1052.660751	104.0504348	14.55355191	7.149487318	1
9.8	0.351486167	1193.781477	108.4365217	15.12569992	7.169025055	1
10	0.351025781	1350.961571	112.9130435	15.70964083	7.1875	1
11	0.349081792	2434.216918	136.6521739	18.806246	7.266318538	1
12	0.347596475	4195.090842	162.6521739	22.19770006	7.327433628	1
13	0.346436457	6955.072479	190.9130435	25.88401696	7.375711575	1
14	0.345513445	11146.48781	221.4347826	29.86520582	7.414473684	1
15	0.344767131	17338.31191	254.2173913	34.14127284	7.446043165	1
20	0.342551516	111792.5101	452.0434783	59.94489603	7.540983607	1
1.00E+07	0.33968311	1.02531E+43	1.13043E+14	1.47448E+13	7.666666667	1

APPENDIX E

Oblique-Shock Chart

Note: The oblique-shock chart for $\gamma = 1.4$ is taken from NACA Report 1135 (Equations, Tables, and Charts for Compressible Flow, by Ames Research Staff, 1953).

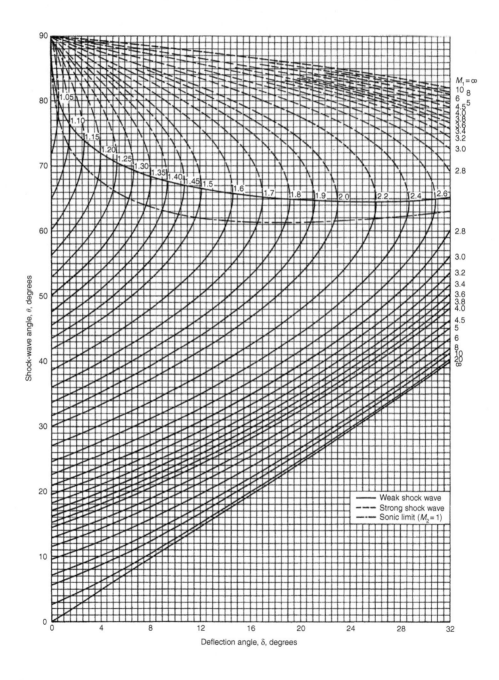

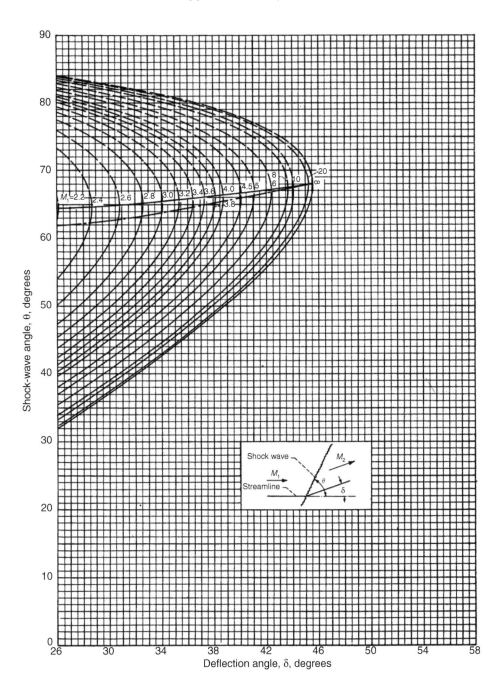

APPENDIX F

Table for Fanno Line Flow

Note – The numerical values of the table can be directly obtained from the following equations. They are tabulated for $\gamma = 1.4$ for convenience. (The subscripts t for thermodynamic properties denote stagnation values; the superscript * denotes critical condition for $M = 1$.)

$$\frac{fL^*}{D_h} = \left(\frac{\gamma+1}{2\gamma}\right)\left(\ln\frac{\frac{\gamma+1}{2}M^2}{1+\frac{\gamma-1}{2}M^2}\right) - \frac{M^2-1}{\gamma M^2} \tag{6.21a}$$

$$\frac{p}{p^*} = \left(\frac{1}{M}\right)\left[\frac{(\gamma+1)/2}{1+\frac{\gamma-1}{2}M^2}\right]^{1/2} \tag{6.26}$$

$$\frac{p_t}{p_t^*} = \left(\frac{1}{M}\right)\left[\frac{2}{\gamma+1}\left(1+\frac{\gamma-1}{2}M^2\right)\right]^{\frac{\gamma+1}{2(\gamma-1)}} \tag{6.27}$$

$$\frac{T}{T^*} = \frac{(\gamma+1)/2}{1+\frac{\gamma-1}{2}M^2} \tag{6.28}$$

$$\frac{\rho}{\rho^*} = \frac{V^*}{V} = \left(\frac{1}{M}\right)\left[\frac{2}{\gamma+1}\left(1+\frac{(\gamma-1)}{2}M^2\right)\right]^{1/2} \tag{6.29}$$

$\gamma = 1.4$					
Mach Number	fL^*/D_h	p/p^*	p_t/p_t^*	T/T^*	ρ/ρ^*
0				1.2	
0.02	1778.44988	54.770065	28.9421302	1.19990401	45.6453722
0.04	440.352214	27.3817471	14.4814859	1.19961612	22.8254244
0.06	193.031082	18.2508495	9.66591007	1.19913662	15.2199917
0.08	106.718216	13.6843088	7.26160964	1.19846596	11.4181872
0.1	66.92156	10.9435131	5.82182875	1.19760479	9.13783344
0.12	45.407962	9.11559228	4.86431765	1.19655392	7.61820432
0.14	32.511306	7.80931667	4.1823998	1.19531437	6.53327433
0.16	24.19783	6.82907187	3.67273863	1.1938873	5.72003059
0.18	18.5426545	6.0661835	3.27792645	1.19227406	5.08791031
0.2	14.5332665	5.45544726	2.96352	1.19047619	4.58257569
0.22	11.5960544	4.95536975	2.7076021	1.18849536	4.16944811
0.24	9.38648053	4.53828896	2.49556245	1.18633344	3.82547504
0.26	7.68756658	4.18505446	2.31728731	1.18399242	3.53469699
0.28	6.35721449	3.88198758	2.16555358	1.18147448	3.28571429
0.3	5.29925311	3.61905747	2.03506526	1.17878193	3.07016708
0.32	4.44674347	3.38874115	1.92185128	1.17591722	2.88178547
0.34	3.75195255	3.1852859	1.82287575	1.17288295	2.71577476
0.36	3.18011752	3.00421752	1.73577828	1.16968185	2.5684057
0.38	2.70544783	2.84200386	1.65869617	1.16631677	2.43673411
0.4	2.30849265	2.69581933	1.59014	1.1627907	2.31840462
0.42	1.97436603	2.56337663	1.5289048	1.15910672	2.21151046
0.44	1.69152484	2.44280443	1.47400537	1.15526802	2.11449152
0.46	1.45091135	2.33255695	1.42462855	1.15127792	2.02605897
0.48	1.24534129	2.23134609	1.38009735	1.1471398	1.94513876
0.5	1.06906031	2.13808994	1.33984375	1.14285714	1.87082869
0.52	0.91741799	2.05187308	1.30338776	1.13843352	1.80236531
0.54	0.78662509	1.97191578	1.2703211	1.13387255	1.73909825
0.56	0.67357071	1.89754974	1.24029444	1.12917796	1.68047005
0.58	0.57568302	1.8281989	1.21300721	1.1243535	1.6260001
0.6	0.49082205	1.76336404	1.18819951	1.11940299	1.57527188
0.62	0.41719657	1.70261038	1.16564554	1.11433029	1.52792256
0.64	0.35329885	1.64555755	1.14514831	1.10913931	1.48363468
0.66	0.29785323	1.59187127	1.12653524	1.10383398	1.44212925
0.68	0.24977519	1.54125665	1.10965464	1.09841828	1.40316005
0.7	0.20813851	1.4934525	1.09437268	1.09289617	1.36650903
0.72	0.17214885	1.44822667	1.08057094	1.08727167	1.33198234
0.74	0.14112224	1.40537213	1.06814435	1.08154878	1.29940707
0.76	0.11446756	1.36470364	1.05699943	1.0757315	1.2686285
0.78	0.09167216	1.32605498	1.04705281	1.06982384	1.2395078
0.8	0.07228997	1.28927656	1.03823	1.06382979	1.21191996
0.82	0.05593167	1.25423336	1.03046428	1.05775333	1.18575222
0.84	0.04225644	1.22080325	1.02369581	1.05159843	1.16090251
0.86	0.03096515	1.18887545	1.01787085	1.04536902	1.13727826
0.88	0.02179454	1.15834925	1.01294105	1.03906899	1.11479532
0.9	0.01451239	1.12913287	1.00886287	1.03270224	1.09337699
0.92	0.00891334	1.10114253	1.00559707	1.02627258	1.07295328
0.94	0.00481545	1.07430155	1.00310828	1.01978381	1.0534601

(continued)

$\gamma = 1.4$

Mach Number	fL^*/D_h	p/p^*	p_t/p_t^*	T/T^*	ρ/ρ^*
0.96	0.00205714	1.04853965	1.00136456	1.01323966	1.03483873
0.98	0.0004947	1.02379227	1.00033714	1.00664385	1.01703524
1	0	1	1	1	1
1.02	0.00045869	0.97710808	1.00032972	0.9933117	0.98368727
1.04	0.0017685	0.95506595	1.00130519	0.98658248	0.96805484
1.06	0.00383785	0.93382684	1.00290738	0.97981579	0.95306367
1.08	0.0065846	0.91334746	1.00511921	0.97301505	0.93867762
1.1	0.009935	0.89358761	1.00792534	0.96618357	0.92486318
1.12	0.01382273	0.87450999	1.01131206	0.95932464	0.91158922
1.14	0.01818811	0.85607987	1.01526712	0.95244142	0.89882679
1.16	0.02297735	0.83826492	1.01977964	0.94553707	0.88654897
1.18	0.02814193	0.82103495	1.02484001	0.9386146	0.87473064
1.2	0.03363807	0.80436182	1.03043975	0.93167702	0.86334835
1.22	0.03942619	0.78821916	1.0365715	0.92472721	0.8523802
1.24	0.04547052	0.77258232	1.04322887	0.91776799	0.8418057
1.26	0.05173869	0.7574282	1.05040642	0.91080211	0.83160567
1.28	0.05820139	0.7427351	1.05809955	0.90383225	0.82176211
1.3	0.06483209	0.72848265	1.06630452	0.89686099	0.81225816
1.32	0.07160673	0.71465171	1.07501831	0.88989084	0.80307795
1.34	0.07850352	0.70122424	1.08423862	0.88292425	0.79420658
1.36	0.0855027	0.68818327	1.09396385	0.87596356	0.78563002
1.38	0.09258633	0.67551278	1.10419301	0.86901107	0.77733507
1.4	0.09973817	0.66319764	1.11492571	0.86206897	0.76930926
1.42	0.10694349	0.65122357	1.12616214	0.85513939	0.76154084
1.44	0.11418891	0.63957707	1.13790302	0.84822438	0.75401873
1.46	0.12146232	0.62824535	1.15014957	0.84132593	0.74673242
1.48	0.12875273	0.6172163	1.16290352	0.83444593	0.73967201
1.5	0.13605022	0.60647843	1.17616705	0.82758621	0.73282811
1.52	0.14334576	0.59602087	1.18994279	0.82074852	0.72619183
1.54	0.15063121	0.58583326	1.2042338	0.81393456	0.71975474
1.56	0.15789921	0.57590577	1.21904353	0.80714593	0.71350886
1.58	0.1651431	0.56622907	1.23437585	0.80038418	0.7074466
1.6	0.17235689	0.55679425	1.250235	0.79365079	0.70156076
1.62	0.17953517	0.54759286	1.26662557	0.78694717	0.6958445
1.64	0.18667309	0.53861681	1.28355253	0.78027466	0.6902913
1.66	0.19376628	0.52985842	1.30102118	0.77363454	0.68489499
1.68	0.20081083	0.52131033	1.31903715	0.76702802	0.67964966
1.7	0.20780326	0.51296555	1.3376064	0.76045627	0.6745497
1.72	0.21474046	0.50481738	1.3567352	0.75392039	0.66958977
1.74	0.22161967	0.4968594	1.37643014	0.7474214	0.66476476
1.76	0.22843844	0.48908551	1.3966981	0.74096028	0.66006981
1.78	0.23519464	0.48148984	1.41754625	0.73453798	0.65550026
1.8	0.24188637	0.47406677	1.43898206	0.72815534	0.6510517
1.82	0.24851201	0.46681093	1.46101327	0.72181319	0.64671987
1.84	0.25507015	0.45971718	1.48364792	0.71551231	0.64250073
1.86	0.26155959	0.45278055	1.5068943	0.70925339	0.6383904
1.88	0.2679793	0.44599632	1.53076097	0.70303712	0.63438517

(continued)

$\gamma = 1.4$

Mach Number	fL^*/D_h	p/p^*	p_t/p_t^*	T/T^*	ρ/ρ^*
1.9	0.27432844	0.43935993	1.55525678	0.69686411	0.6304815
1.92	0.28060634	0.43286701	1.5803908	0.69073494	0.62667599
1.94	0.28681244	0.42651334	1.6061724	0.68465014	0.62296539
1.96	0.29294634	0.4202949	1.63261118	0.67861021	0.61934657
1.98	0.29900775	0.4142078	1.65971701	0.67261558	0.61581654
2	0.3049965	0.40824829	1.6875	0.66666667	0.61237244
2.02	0.31091251	0.40241278	1.71597051	0.66076384	0.6090115
2.04	0.31675578	0.3966978	1.74513915	0.65490744	0.6057311
2.06	0.32252642	0.39110002	1.77501677	0.64909775	0.60252868
2.08	0.32822461	0.3856162	1.80561448	0.64333505	0.59940182
2.1	0.33385058	0.38024325	1.83694361	0.63761955	0.59634817
2.12	0.33940465	0.37497818	1.86901575	0.63195147	0.59336548
2.14	0.34488716	0.3698181	1.90184273	0.62633095	0.59045158
2.16	0.35029855	0.36476022	1.93543661	0.62075815	0.58760439
2.18	0.35563927	0.35980184	1.9698097	0.61523317	0.58482191
2.2	0.36090982	0.35494037	2.00497455	0.6097561	0.5821022
2.22	0.36611074	0.35017329	2.04094392	0.60432698	0.57944342
2.24	0.37124261	0.34549818	2.07773086	0.59894586	0.57684376
2.26	0.37630603	0.34091269	2.11534861	0.59361273	0.57430151
2.28	0.38130164	0.33641454	2.15381069	0.58832758	0.57181501
2.3	0.38623008	0.33200154	2.19313082	0.58309038	0.56938265
2.32	0.39109204	0.32767157	2.23332298	0.57790106	0.56700288
2.34	0.39588821	0.32342256	2.27440139	0.57275956	0.56467423
2.36	0.40061929	0.31925253	2.31638051	0.56766576	0.56239526
2.38	0.40528602	0.31515954	2.35927503	0.56261956	0.56016456
2.4	0.40988913	0.31114172	2.40309988	0.55762082	0.55798082
2.42	0.41442936	0.30719726	2.44787024	0.55266939	0.55584273
2.44	0.41890747	0.30332441	2.49360153	0.54776512	0.55374904
2.46	0.42332422	0.29952146	2.5403094	0.54290781	0.55169856
2.48	0.42768037	0.29578675	2.58800976	0.53809729	0.5496901
2.5	0.43197669	0.2921187	2.63671875	0.53333333	0.54772256
2.52	0.43621396	0.28851573	2.68645276	0.52861573	0.54579483
2.54	0.44039294	0.28497635	2.73722843	0.52394425	0.54390587
2.56	0.44451442	0.28149909	2.78906263	0.51931865	0.54205465
2.58	0.44857915	0.27808252	2.84197249	0.51473868	0.54024019
2.6	0.45258792	0.27472527	2.89597538	0.51020408	0.53846154
2.62	0.4565415	0.271426	2.95108894	0.50571457	0.53671777
2.64	0.46044064	0.26818339	3.00733102	0.50126988	0.53500799
2.66	0.46428612	0.26499619	3.06471976	0.49686972	0.53333133
2.68	0.46807868	0.26186316	3.12327353	0.49251379	0.53168696
2.7	0.47181909	0.25878311	3.18301096	0.48820179	0.53007406
2.72	0.47550808	0.25575486	3.24395092	0.48393341	0.52849185
2.74	0.47914641	0.2527773	3.30611258	0.47970834	0.52693956
2.76	0.48273481	0.24984932	3.3695153	0.47552625	0.52541646
2.78	0.48627401	0.24696984	3.43417876	0.47138682	0.52392182
2.8	0.48976473	0.24413783	3.50012286	0.46728972	0.52245496
2.82	0.49320769	0.24135227	3.56736777	0.46323461	0.5210152

(continued)

$\gamma = 1.4$

Mach Number	fL^*/D_h	p/p^*	p_t/p_t^*	T/T^*	ρ/ρ^*
2.84	0.4966036	0.23861218	3.63593394	0.45922116	0.51960188
2.86	0.49995315	0.23591659	3.70584205	0.45524902	0.51821438
2.88	0.50325705	0.23326456	3.77711307	0.45131785	0.51685207
2.9	0.50651597	0.23065519	3.84976823	0.44742729	0.51551436
2.92	0.5097306	0.22808759	3.92382903	0.443577	0.51420067
2.94	0.5129016	0.22556089	3.99931722	0.43976663	0.51291043
2.96	0.51602963	0.22307425	4.07625485	0.43599581	0.51164311
2.98	0.51911535	0.22062685	4.15466422	0.4322642	0.51039816
3	0.52215941	0.21821789	4.2345679	0.42857143	0.50917508
3.02	0.52516243	0.21584659	4.31598875	0.42491714	0.50797335
3.04	0.52812504	0.21351218	4.39894991	0.42130098	0.5067925
3.06	0.53104787	0.21121393	4.48347476	0.41772258	0.50563206
3.08	0.53393152	0.20895111	4.56958701	0.41418158	0.50449155
3.1	0.53677659	0.20672301	4.6573106	0.41067762	0.50337053
3.12	0.53958369	0.20452896	4.7466698	0.40721034	0.50226858
3.14	0.54235338	0.20236827	4.83768914	0.40377937	0.50118526
3.16	0.54508625	0.20024029	4.93039341	0.40038437	0.50012016
3.18	0.54778286	0.19814439	5.02480774	0.39702496	0.49907288
3.2	0.55044378	0.19607994	5.1209575	0.39370079	0.49804305
3.22	0.55306956	0.19404633	5.21886838	0.39041149	0.49703026
3.24	0.55566073	0.19204296	5.31856634	0.38715672	0.49603416
3.26	0.55821783	0.19006926	5.42007765	0.38393611	0.49505439
3.28	0.56074138	0.18812466	5.52342887	0.38074931	0.4940906
3.3	0.56323191	0.1862086	5.62864684	0.37759597	0.49314244
3.32	0.56568992	0.18432055	5.73575872	0.37447573	0.4922096
3.34	0.56811592	0.18245998	5.84479196	0.37138825	0.49129173
3.36	0.5705104	0.18062636	5.95577431	0.36833317	0.49038853
3.38	0.57287384	0.17881921	6.06873381	0.36531015	0.4894997
3.4	0.57520672	0.17703802	6.18369882	0.36231884	0.48862493
3.42	0.57750952	0.17528231	6.30069801	0.3593589	0.48776394
3.44	0.57978269	0.17355162	6.41976035	0.35643	0.48691644
3.46	0.58202669	0.17184549	6.5409151	0.35353178	0.48608215
3.48	0.58424197	0.17016346	6.66419187	0.35066392	0.4852608
3.5	0.58642898	0.16850509	6.78962054	0.34782609	0.48445214
3.52	0.58858813	0.16686997	6.91723133	0.34501794	0.48365591
3.54	0.59071987	0.16525766	7.04705477	0.34223916	0.48287185
3.56	0.59282461	0.16366775	7.1791217	0.33948941	0.48209973
3.58	0.59490275	0.16209985	7.31346329	0.33676837	0.4813393
3.6	0.59695472	0.16055357	7.45011103	0.33407572	0.48059034
3.62	0.59898091	0.15902851	7.58909671	0.33141115	0.47985262
3.64	0.6009817	0.1575243	7.73045248	0.32877433	0.47912592
3.66	0.60295749	0.15604058	7.87421078	0.32616495	0.47841002
3.68	0.60490866	0.15457699	8.0204044	0.32358271	0.47770472
3.7	0.60683557	0.15313316	8.16906645	0.32102729	0.4770098
3.72	0.60873861	0.15170877	8.32023037	0.31849839	0.47632507
3.74	0.61061812	0.15030346	8.47392994	0.3159957	0.47565034
3.76	0.61247446	0.14891692	8.63019926	0.31351894	0.47498541

(continued)

$\gamma = 1.4$

Mach Number	fL^*/D_h	p/p^*	p_t/p_t^*	T/T^*	ρ/ρ^*
3.78	0.61430798	0.14754881	8.78907277	0.31106779	0.47433009
3.8	0.61611903	0.14619883	8.95058526	0.30864198	0.47368421
3.82	0.61790794	0.14486666	9.11477185	0.3062412	0.47304759
3.84	0.61967505	0.143552	9.28166799	0.30386516	0.47242005
3.86	0.62142067	0.14225454	9.45130949	0.3015136	0.47180142
3.88	0.62314514	0.14097401	9.62373249	0.29918621	0.47119154
3.9	0.62484877	0.13971012	9.79897348	0.29688273	0.47059025
3.92	0.62653186	0.13846258	9.97706931	0.29460288	0.46999739
3.94	0.62819472	0.13723113	10.1580572	0.29234637	0.4694128
3.96	0.62983766	0.13601549	10.3419746	0.29011295	0.46883633
3.98	0.63146097	0.13481541	10.5288594	0.28790234	0.46826784
4	0.63306493	0.13363062	10.71875	0.28571429	0.46770717
4.02	0.63464984	0.13246088	10.9116849	0.28354852	0.4671542
4.04	0.63621597	0.13130594	11.107703	0.28140477	0.46660878
4.06	0.6377636	0.13016555	11.3068437	0.2792828	0.46607077
4.08	0.639293	0.12903949	11.5091467	0.27718235	0.46554005
4.1	0.64080443	0.12792751	11.7146519	0.27510316	0.46501649
4.12	0.64229817	0.12682939	11.9233999	0.273045	0.46449995
4.14	0.64377446	0.12574491	12.1354314	0.27100761	0.46399032
4.16	0.64523356	0.12467384	12.3507874	0.26899075	0.46348747
4.18	0.64667572	0.12361598	12.5695097	0.26699418	0.46299129
4.2	0.64810119	0.12257111	12.7916399	0.26501767	0.46250166
4.22	0.6495102	0.12153903	13.0172205	0.26306098	0.46201847
4.24	0.65090299	0.12051953	13.2462939	0.26112388	0.4615416
4.26	0.65227981	0.11951242	13.4789033	0.25920614	0.46107095
4.28	0.65364086	0.1185175	13.7150919	0.25730753	0.46060641
4.3	0.65498639	0.11753458	13.9549037	0.25542784	0.46014788
4.32	0.65631661	0.11656347	14.1983827	0.25356684	0.45969526
4.34	0.65763174	0.115604	14.4455734	0.25172431	0.45924844
4.36	0.658932	0.11465597	14.6965208	0.24990004	0.45880733
4.38	0.66021759	0.11371922	14.9512702	0.24809381	0.45837183
4.4	0.66148873	0.11279356	15.2098673	0.24630542	0.45794186
4.42	0.66274562	0.11187883	15.4723581	0.24453465	0.45751731
4.44	0.66398845	0.11097487	15.7387892	0.2427813	0.45709809
4.46	0.66521743	0.11008151	16.0092074	0.24104517	0.45668413
4.48	0.66643275	0.10919858	16.28366	0.23932606	0.45627533
4.5	0.6676346	0.10832593	16.5621946	0.23762376	0.45587162
4.52	0.66882318	0.1074634	16.8448593	0.23593809	0.45547289
4.54	0.66999865	0.10661085	17.1317026	0.23426885	0.45507909
4.56	0.67116122	0.10576812	17.4227734	0.23261584	0.45469011
4.58	0.67231104	0.10493507	17.7181209	0.23097889	0.4543059
4.6	0.67344831	0.10411155	18.0177948	0.2293578	0.45392637
4.62	0.6745732	0.10329742	18.3218452	0.22775239	0.45355145
4.64	0.67568587	0.10249255	18.6303226	0.22616248	0.45318105
4.66	0.67678649	0.10169679	18.9432778	0.22458788	0.45281512
4.68	0.67787523	0.10091001	19.2607624	0.22302843	0.45245359
4.7	0.67895225	0.10013209	19.5828278	0.22148394	0.45209637

(continued)

$\gamma = 1.4$

Mach Number	fL^*/D_h	p/p^*	p_t/p_t^*	T/T^*	ρ/ρ^*
4.72	0.68001771	0.09936288	19.9095265	0.21995425	0.45174341
4.74	0.68107177	0.09860227	20.2409108	0.21843918	0.45139463
4.76	0.68211457	0.09785013	20.5770338	0.21693856	0.45104998
4.78	0.68314628	0.09710634	20.9179489	0.21545223	0.45070939
4.8	0.68416705	0.09637078	21.2637099	0.21398003	0.45037279
4.82	0.68517701	0.09564333	21.6143712	0.21252178	0.45004012
4.84	0.68617633	0.09492387	21.9699874	0.21107734	0.44971133
4.86	0.68716513	0.09421229	22.3306137	0.20964654	0.44938636
4.88	0.68814355	0.09350848	22.6963055	0.20822922	0.44906514
4.9	0.68911175	0.09281233	23.067119	0.20682523	0.44874762
4.92	0.69006985	0.09212373	23.4431104	0.20543443	0.44843375
4.94	0.69101799	0.09144257	23.8243367	0.20405665	0.44812346
4.96	0.69195629	0.09076875	24.2108552	0.20269175	0.44781671
4.98	0.69288489	0.09010217	24.6027235	0.20133958	0.44751344
5	0.69380392	0.08944272	25	0.2	0.4472136
5.5	0.71400429	0.07501245	36.8689631	0.17021277	0.44069817
6	0.72987528	0.06375767	53.179784	0.14634146	0.43567742
6.5	0.74254432	0.05482282	75.1343149	0.12698413	0.4317297
7	0.75280216	0.04761905	104.142857	0.11111111	0.42857143
7.5	0.76121412	0.04173124	141.841483	0.09795918	0.42600643
8	0.76819161	0.03686049	190.109375	0.08695652	0.42389562
8.5	0.77403906	0.03278744	251.086167	0.0776699	0.42213824
9	0.77898519	0.02934836	327.1893	0.06976744	0.42065988
9.5	0.78320428	0.02641919	421.131373	0.06299213	0.41940467
10	0.78683083	0.02390457	535.9375	0.05714286	0.41833001
12	0.79721157	0.0167225	1276.21489	0.04026846	0.41527546
14	0.80356079	0.01234098	2685.38393	0.02985075	0.41342275
16	0.80771861	0.00947623	5144.55469	0.02298851	0.41221581
18	0.81058621	0.00750249	9159.28215	0.01823708	0.41138629
20	0.81264596	0.00608581	15377.3438	0.01481481	0.41079192

APPENDIX G

Table for Rayleigh Line Flow

Note – The numerical values of the table can be directly obtained from the following equations. They are tabulated for $\gamma = 1.4$ for convenience. (The subscripts t for thermodynamic properties denote stagnation values; the superscript * denotes critical condition for $M = 1$.)

$$\frac{p}{p^*} = \frac{1+\gamma}{1+\gamma M^2} \tag{7.10}$$

$$\frac{T}{T^*} = \frac{(1+\gamma)^2 M^2}{(1+\gamma M^2)^2} \tag{7.11}$$

$$\frac{V}{V^*} = \frac{\rho^*}{\rho} = \frac{p^*/RT^*}{p/RT} = \frac{(1+\gamma)M^2}{1+\gamma M^2} \tag{7.12}$$

$$\frac{p_t}{p_t^*} = \left(\frac{1+\gamma}{1+\gamma M^2}\right)\left[\left(\frac{2}{\gamma+1}\right)\left(1+\frac{\gamma-1}{2}M^2\right)\right]^{\gamma/(\gamma-1)} \tag{7.13}$$

$$\frac{T_t}{T_t^*} = \frac{2(1+\gamma)M^2}{(1+\gamma M^2)^2}\left(1+\frac{\gamma-1}{2}M^2\right) \tag{7.14}$$

$\gamma = 1.4$

Mach Number	p/p^*	T/T^*	$V/V^*=\rho^*/\rho$	p_t/p_t^*	T_t/T_t^*
0	2.4	0	0	1.26787629	0
0.02	2.39865675	0.00230142	0.00095946	1.26752152	0.001918
0.04	2.39463602	0.00917485	0.00383142	1.26646001	0.00764816
0.06	2.38796466	0.02052855	0.00859667	1.26470013	0.01711944
0.08	2.37868696	0.03621217	0.0152236	1.26225566	0.03021544
0.1	2.36686391	0.05602045	0.02366864	1.2591456	0.04677707
0.12	2.35257215	0.07969818	0.03387704	1.25539382	0.06660642
0.14	2.33590283	0.10694626	0.0457837	1.25102873	0.08947124
0.16	2.31696015	0.13742859	0.05931418	1.24608281	0.11511019
0.18	2.2958598	0.1707795	0.07438586	1.24059214	0.14323846
0.2	2.27272727	0.20661157	0.09090909	1.23459588	0.17355372
0.22	2.24769611	0.24452347	0.10878849	1.22813574	0.20574205
0.24	2.22090613	0.28410762	0.12792419	1.22125541	0.23948379
0.26	2.19250164	0.32495749	0.14821311	1.21400003	0.2744591
0.28	2.16262976	0.36667425	0.16955017	1.20641566	0.31035308
0.3	2.13143872	0.40887279	0.19182948	1.19854878	0.34686042
0.32	2.09907641	0.45118687	0.21494542	1.1904458	0.38368931
0.34	2.06568891	0.49327337	0.23879364	1.18215267	0.42056487
0.36	2.03141928	0.53481569	0.26327194	1.17371444	0.45723176
0.38	1.99640647	0.57552624	0.28828109	1.16517497	0.4934562
0.4	1.96078431	0.61514802	0.31372549	1.15657661	0.5290273
0.42	1.92468082	0.6534555	0.3395137	1.14795997	0.56375784
0.44	1.88821752	0.69025474	0.36555891	1.13936373	0.59748451
0.46	1.85150898	0.72538289	0.3917793	1.1308245	0.63006758
0.48	1.81466247	0.75870717	0.41809823	1.1223767	0.66139033
0.5	1.77777778	0.79012346	0.44444444	1.1140525	0.69135802
0.52	1.74094708	0.81955447	0.47075209	1.10588181	0.71989665
0.54	1.70425496	0.84694781	0.49696075	1.09789224	0.74695151
0.56	1.66777852	0.87227376	0.52301534	1.0901092	0.77248564
0.58	1.63158753	0.895523	0.54886605	1.08255586	0.79647816
0.6	1.59574468	0.91670439	0.57446809	1.07525332	0.81892259
0.62	1.56030582	0.93584265	0.59978156	1.06822062	0.8398252
0.64	1.52532032	0.95297621	0.6247712	1.06147487	0.85920335
0.66	1.49083139	0.96815507	0.64940615	1.05503135	0.87708395
0.68	1.45687646	0.98143892	0.67365967	1.04890365	0.89350199
0.7	1.42348754	0.99289523	0.6975089	1.04310374	0.90849913
0.72	1.39069164	1.00259764	0.72093454	1.03764211	0.92212247
0.74	1.35851107	1.01062446	0.74392066	1.0325279	0.93442337
0.76	1.32696391	1.01705726	0.76645435	1.027769	0.94545643
0.78	1.29606428	1.02197975	0.78852551	1.02337216	0.95527854
0.8	1.26582278	1.02547669	0.81012658	1.01934312	0.96394809
0.82	1.23624675	1.02763298	0.83125232	1.01568668	0.97152422
0.84	1.20734063	1.02853294	0.85189955	1.01240683	0.97806626
0.86	1.17910624	1.02825961	0.87206697	1.00950681	0.98363314
0.88	1.15154307	1.02689423	0.89175495	1.00698925	0.98828301
0.9	1.12464855	1.02451583	0.91096532	1.00485619	0.99207283
0.92	1.09841828	1.02120082	0.92970123	1.0031092	0.99505808
0.94	1.07284626	1.0170228	0.94796696	1.00174943	0.99729256

(continued)

γ = 1.4

Mach Number	p/p^*	T/T^*	$V/V^*=\rho^*/\rho$	p_t/p_t^*	T_t/T_t^*
0.96	1.04792511	1.01205231	0.96576778	1.00077767	0.99882816
0.98	1.02364623	1.00635668	0.98310984	1.00019444	0.99971472
1	1	1	1	1	1
1.02	0.97697593	0.99304305	1.01644576	1.00019444	0.99972954
1.04	0.95456281	0.98554328	1.03245514	1.00077769	0.99894666
1.06	0.93274881	0.97755486	1.04803656	1.0017496	0.99769249
1.08	0.91152163	0.96912874	1.06319883	1.00310993	0.99600591
1.1	0.8908686	0.9603127	1.077951	1.00485842	0.99392364
1.12	0.87077673	0.95115146	1.09230233	1.00699479	0.99148028
1.14	0.85123287	0.94168678	1.10626224	1.00951882	0.98870834
1.16	0.8322237	0.93195757	1.11984021	1.01243029	0.98563832
1.18	0.81373586	0.92200001	1.13304581	1.01572906	0.98229881
1.2	0.79575597	0.91184769	1.14588859	1.0194151	0.97871652
1.22	0.77827068	0.9015317	1.15837808	1.02348846	0.97491638
1.24	0.76126675	0.89108081	1.17052375	1.02794929	0.97092165
1.26	0.74473103	0.88052155	1.18233498	1.0327979	0.96675396
1.28	0.72865054	0.86987835	1.19382104	1.03803471	0.9624334
1.3	0.71301248	0.85917368	1.20499109	1.04366031	0.95797865
1.32	0.69780424	0.84842816	1.21585411	1.04967542	0.953407
1.34	0.68301346	0.83766065	1.22641896	1.05608095	0.94873445
1.36	0.66862798	0.82688841	1.2366943	1.06287797	0.94397581
1.38	0.65463591	0.81612715	1.24668863	1.0700677	0.93914472
1.4	0.64102564	0.80539119	1.25641026	1.07765156	0.93425378
1.42	0.6277858	0.79469351	1.26586729	1.08563116	0.92931459
1.44	0.6149053	0.78404586	1.27506764	1.09400827	0.92433779
1.46	0.60237335	0.77345885	1.28401903	1.10278485	0.91933319
1.48	0.59017941	0.76294204	1.29272899	1.11196306	0.91430974
1.5	0.57831325	0.75250399	1.30120482	1.12154523	0.90927566
1.52	0.56676491	0.74215237	1.30945364	1.13153389	0.90423845
1.54	0.55552469	0.73189398	1.31748236	1.14193177	0.89920495
1.56	0.54458321	0.72173488	1.32529771	1.15274177	0.8941814
1.58	0.53393134	0.71168038	1.33290619	1.163967	0.88917347
1.6	0.52356021	0.70173515	1.34031414	1.17561073	0.88418629
1.62	0.51346124	0.69190324	1.34752768	1.18767645	0.87922451
1.64	0.50362611	0.68218814	1.35455278	1.20016783	0.87429233
1.66	0.49404674	0.67259285	1.36139519	1.21308872	0.86939352
1.68	0.48471531	0.66311987	1.36806049	1.22644318	0.86453147
1.7	0.47562426	0.65377127	1.3745541	1.24023542	0.85970922
1.72	0.46676624	0.64454876	1.38088125	1.25446987	0.85492947
1.74	0.45813417	0.63545364	1.38704702	1.26915114	0.8501946
1.76	0.44972117	0.62648692	1.39305631	1.28428402	0.84550674
1.78	0.4415206	0.61764928	1.39891386	1.29987347	0.84086773
1.8	0.43352601	0.60894116	1.40462428	1.31592466	0.83627919
1.82	0.42573119	0.60036273	1.410192	1.33244292	0.83174252
1.84	0.41813012	0.59191392	1.41562134	1.34943378	0.8272589
1.86	0.41071697	0.5835945	1.42091645	1.36690294	0.82282935
1.88	0.40348612	0.57540403	1.42608134	1.38485627	0.81845469

(continued)

$\gamma = 1.4$

Mach Number	p/p^*	T/T^*	$V/V^*=\rho^*/\rho$	p_t/p_t^*	T_t/T_t^*
1.9	0.39643211	0.56734189	1.43111992	1.40329985	0.81413561
1.92	0.38954968	0.55940734	1.43603594	1.42223991	0.80987266
1.94	0.38283374	0.5515995	1.44083305	1.44168287	0.80566623
1.96	0.37627935	0.54391735	1.44551475	1.46163532	0.80151661
1.98	0.36988176	0.53635979	1.45008446	1.48210404	0.79742398
2	0.36363636	0.52892562	1.45454545	1.50309598	0.79338843
2.02	0.3575387	0.52161355	1.45890093	1.52461825	0.78940994
2.04	0.35158447	0.51442221	1.46315395	1.54667816	0.78548842
2.06	0.34576951	0.50735019	1.46730749	1.56928319	0.78162371
2.08	0.34008978	0.50039601	1.47136444	1.59244097	0.77781556
2.1	0.3345414	0.49355815	1.47532757	1.61615933	0.7740637
2.12	0.32912059	0.48683504	1.47919958	1.64044628	0.77036776
2.14	0.32382371	0.48022508	1.48298306	1.66530997	0.76672736
2.16	0.31864724	0.47372665	1.48668055	1.69075875	0.76314205
2.18	0.31358776	0.4673381	1.49029446	1.71680114	0.75961134
2.2	0.30864198	0.46105777	1.49382716	1.74344583	0.75613474
2.22	0.3038067	0.45488398	1.49728093	1.77070168	0.75271168
2.24	0.29907884	0.44881504	1.50065797	1.79857772	0.74934159
2.26	0.2944554	0.44284928	1.50396043	1.82708317	0.74602389
2.28	0.28993351	0.43698499	1.50719035	1.85622741	0.74275795
2.3	0.28551035	0.43122049	1.51034975	1.88601999	0.73954313
2.32	0.28118322	0.42555409	1.51344056	1.91647064	0.73637879
2.34	0.27694949	0.41998412	1.51646465	1.94758927	0.73326427
2.36	0.27280663	0.4145089	1.51942383	1.97938595	0.73019888
2.38	0.26875218	0.40912679	1.52231987	2.01187094	0.72718195
2.4	0.26478376	0.40383613	1.52515446	2.04505465	0.7242128
2.42	0.26089906	0.3986353	1.52792924	2.07894769	0.72129071
2.44	0.25709585	0.39352268	1.53064582	2.11356084	0.71841501
2.46	0.25337196	0.38849668	1.53330574	2.14890504	0.71558498
2.48	0.2497253	0.38355571	1.5359105	2.18499142	0.71279994
2.5	0.24615385	0.37869822	1.53846154	2.22183129	0.71005917
2.52	0.24265562	0.37392267	1.54096027	2.25943612	0.70736199
2.54	0.23922873	0.36922754	1.54340805	2.29781757	0.70470769
2.56	0.23587131	0.36461133	1.54580621	2.33698749	0.70209558
2.58	0.23258158	0.36007257	1.54815602	2.37695787	0.69952498
2.6	0.2293578	0.3556098	1.55045872	2.41774092	0.6969952
2.62	0.22619829	0.35122159	1.55271551	2.459349	0.69450556
2.64	0.22310141	0.34690653	1.55492757	2.50179467	0.6920554
2.66	0.22006558	0.34266324	1.55709601	2.54509067	0.68964403
2.68	0.21708927	0.33849035	1.55922195	2.58924991	0.68727081
2.7	0.21417098	0.33438653	1.56130644	2.63428548	0.68493508
2.72	0.21130927	0.33035046	1.56335052	2.68021066	0.68263619
2.74	0.20850274	0.32638085	1.56535518	2.72703893	0.68037352
2.76	0.20575003	0.32247642	1.56732141	2.77478392	0.67814642
2.78	0.20304981	0.31863594	1.56925014	2.82345948	0.67595428
2.8	0.2004008	0.31485817	1.57114228	2.87307962	0.67379649
2.82	0.19780176	0.31114192	1.57299874	2.92365854	0.67167244

(continued)

$\gamma = 1.4$

Mach Number	p/p^*	T/T^*	$V/V^*=\rho^*/\rho$	p_t/p_t^*	T_t/T_t^*
2.84	0.19525148	0.30748601	1.57482037	2.97521066	0.66958154
2.86	0.19274879	0.30388929	1.57660801	3.02775053	0.66752321
2.88	0.19029254	0.30035061	1.57836247	3.08129294	0.66549685
2.9	0.18788163	0.29686887	1.58008455	3.13585286	0.66350192
2.92	0.18551499	0.29344297	1.58177501	3.19144542	0.66153784
2.94	0.18319156	0.29007186	1.5834346	3.24808598	0.65960407
2.96	0.18091034	0.28675448	1.58506404	3.30579008	0.65770007
2.98	0.17867034	0.2834898	1.58666405	3.36457344	0.6558253
3	0.17647059	0.28027682	1.58823529	3.42445199	0.65397924
3.02	0.17431017	0.27711455	1.58977845	3.48544186	0.65216138
3.04	0.17218817	0.27400203	1.59129417	3.54755935	0.65037121
3.06	0.17010371	0.2709383	1.59278307	3.61082099	0.64860824
3.08	0.16805593	0.26792245	1.59424576	3.67524349	0.64687197
3.1	0.166044	0.26495357	1.59568286	3.74084377	0.64516194
3.12	0.16406711	0.26203075	1.59709492	3.80763893	0.64347766
3.14	0.16212448	0.25915315	1.59848251	3.87564629	0.64181868
3.16	0.16021533	0.25631988	1.59984619	3.94488336	0.64018454
3.18	0.15833892	0.25353014	1.60118649	4.01536788	0.63857481
3.2	0.15649452	0.25078308	1.60250391	4.08711775	0.63698904
3.22	0.15468143	0.24807792	1.60379898	4.16015112	0.6354268
3.24	0.15289896	0.24541387	1.60507217	4.23448631	0.63388767
3.26	0.15114645	0.24279016	1.60632397	4.31014188	0.63237125
3.28	0.14942323	0.24020603	1.60755484	4.38713656	0.63087712
3.3	0.14772867	0.23766075	1.60876523	4.46548934	0.62940489
3.32	0.14606216	0.2351536	1.6099556	4.54521937	0.62795417
3.34	0.1444231	0.23268387	1.61112636	4.62634604	0.62652458
3.36	0.1428109	0.23025086	1.61227793	4.70888894	0.62511574
3.38	0.14122499	0.22785391	1.61341072	4.7928679	0.62372728
3.4	0.1396648	0.22549234	1.61452514	4.87830292	0.62235885
3.42	0.13812981	0.22316551	1.61562156	4.96521426	0.62101009
3.44	0.13661949	0.22087278	1.61670037	5.05362237	0.61968066
3.46	0.13513331	0.21861352	1.61776192	5.14354793	0.61837021
3.48	0.13367078	0.21638714	1.61880659	5.23501183	0.61707841
3.5	0.1322314	0.21419302	1.61983471	5.32803518	0.61580493
3.52	0.13081471	0.21203059	1.62084663	5.42263933	0.61454946
3.54	0.12942024	0.20989927	1.62184269	5.51884583	0.61331167
3.56	0.12804753	0.2077985	1.62282319	5.61667647	0.61209127
3.58	0.12669614	0.20572774	1.62378847	5.71615326	0.61088795
3.6	0.12536565	0.20368644	1.62473882	5.81729842	0.6097014
3.62	0.12405563	0.20167408	1.62567455	5.92013442	0.60853135
3.64	0.12276566	0.19969013	1.62659595	6.02468394	0.60737751
3.66	0.12149536	0.19773411	1.62750331	6.13096992	0.60623959
3.68	0.12024434	0.19580551	1.6283969	6.23901548	0.60511733
3.7	0.1190122	0.19390384	1.629277	6.34884402	0.60401046
3.72	0.11779858	0.19202863	1.63014387	6.46047915	0.6029187
3.74	0.11660312	0.19017943	1.63099777	6.5739447	0.60184181
3.76	0.11542546	0.18835576	1.63183896	6.68926478	0.60077953

(continued)

$\gamma = 1.4$

Mach Number	p/p^*	T/T^*	$V/V^* = \rho^*/\rho$	p_t/p_t^*	T_t/T_t^*
3.78	0.11426526	0.18655719	1.63266767	6.80646368	0.59973161
3.8	0.11312217	0.18478328	1.63348416	6.92556597	0.59869782
3.82	0.11199588	0.18303359	1.63428866	7.04659644	0.5976779
3.84	0.11088605	0.18130772	1.63508139	7.16958013	0.59667163
3.86	0.10979238	0.17960525	1.63586258	7.2945423	0.59567877
3.88	0.10871456	0.17792578	1.63663246	7.42150848	0.59469911
3.9	0.10765228	0.1762689	1.63739123	7.55050442	0.59373242
3.92	0.10660526	0.17463425	1.6381391	7.68155612	0.59277849
3.94	0.10557321	0.17302143	1.63887628	7.81468983	0.5918371
3.96	0.10455585	0.17143008	1.63960297	7.94993206	0.59090805
3.98	0.1035529	0.16985983	1.64031936	8.08730953	0.58999112
4	0.1025641	0.16831032	1.64102564	8.22684925	0.58908613
4.02	0.10158919	0.16678121	1.641722	8.36857846	0.58819287
4.04	0.10062792	0.16527216	1.64240863	8.51252465	0.58731115
4.06	0.09968003	0.16378283	1.64308569	8.65871556	0.58644079
4.08	0.09874528	0.16231288	1.64375337	8.80717919	0.58558159
4.1	0.09782343	0.160862	1.64441184	8.95794381	0.58473338
4.12	0.09691425	0.15942988	1.64506125	9.11103792	0.58389598
4.14	0.09601751	0.15801619	1.64570178	9.26649028	0.58306922
4.16	0.095133	0.15662065	1.64633357	9.42432994	0.58225291
4.18	0.09426048	0.15524294	1.6469568	9.58458616	0.5814469
4.2	0.09339975	0.15388278	1.64757161	9.7472885	0.58065101
4.22	0.0925506	0.15253988	1.64817814	9.91246677	0.5798651
4.24	0.09171283	0.15121396	1.64877655	10.080151	0.57908899
4.26	0.09088623	0.14990475	1.64936698	10.2503716	0.57832253
4.28	0.09007062	0.14861197	1.64994956	10.4231592	0.57756557
4.3	0.08926579	0.14733537	1.65052444	10.5985446	0.57681796
4.32	0.08847157	0.14607467	1.65109174	10.7765589	0.57607955
4.34	0.08768776	0.14482963	1.6516516	10.9572335	0.57535019
4.36	0.0869142	0.1436	1.65220414	11.1406002	0.57462975
4.38	0.0861507	0.14238553	1.6527495	11.3266909	0.57391809
4.4	0.0853971	0.14118598	1.65328779	11.5155377	0.57321507
4.42	0.08465322	0.14000111	1.65381913	11.7071733	0.57252055
4.44	0.0839189	0.1388307	1.65434364	11.9016303	0.5718344
4.46	0.08319398	0.13767452	1.65486144	12.0989419	0.57115649
4.48	0.08247831	0.13653233	1.65537264	12.2991412	0.57048671
4.5	0.08177172	0.13540394	1.65587734	12.5022621	0.56982491
4.52	0.08107407	0.13428912	1.65637566	12.7083382	0.56917099
4.54	0.08038521	0.13318765	1.65686771	12.9174039	0.56852481
4.56	0.07970499	0.13209934	1.65735358	13.1294935	0.56788627
4.58	0.07903327	0.13102399	1.65783338	13.3446417	0.56725524
4.6	0.07836991	0.12996138	1.65830721	13.5628837	0.56663162
4.62	0.07771477	0.12891133	1.65877516	13.7842547	0.56601528
4.64	0.07706773	0.12787365	1.65923734	14.0087902	0.56540613
4.66	0.07642864	0.12684815	1.65969383	14.2365262	0.56480405
4.68	0.07579739	0.12583463	1.66014472	14.4674989	0.56420895
4.7	0.07517384	0.12483293	1.66059011	14.7017446	0.5636207

(continued)

$\gamma = 1.4$

Mach Number	p/p^*	T/T^*	$V/V^*=\rho^*/\rho$	p_t/p_t^*	T_t/T_t^*
4.72	0.07455787	0.12384287	1.66103009	14.9393002	0.56303922
4.74	0.07394936	0.12286426	1.66146474	15.1802027	0.5624644
4.76	0.0733482	0.12189694	1.66189414	15.4244894	0.56189614
4.78	0.07275426	0.12094074	1.66231839	15.6721979	0.56133435
4.8	0.07216743	0.11999549	1.66273755	15.9233663	0.56077894
4.82	0.0715876	0.11906104	1.66315172	16.1780327	0.5602298
4.84	0.07101466	0.11813721	1.66356096	16.4362357	0.55968685
4.86	0.0704485	0.11722386	1.66396536	16.6980141	0.55915
4.88	0.06988902	0.11632083	1.66436499	16.9634071	0.55861916
4.9	0.06933611	0.11542797	1.66475992	17.232454	0.55809424
4.92	0.06878967	0.11454513	1.66515024	17.5051946	0.55757516
4.94	0.0682496	0.11367217	1.665536	17.781669	0.55706184
4.96	0.06771581	0.11280894	1.66591728	18.0619175	0.55655418
4.98	0.0671882	0.1119553	1.66629415	18.3459808	0.55605211
5	0.06666667	0.11111111	1.66666667	18.6338998	0.55555556
5.5	0.05536332	0.0927192	1.67474048	27.2113249	0.54472528
6	0.04669261	0.07848718	1.68093385	38.9459449	0.53632909
6.5	0.03990025	0.06726326	1.68578554	54.683031	0.5296982
7	0.03448276	0.05826397	1.68965517	75.4137931	0.52437574
7.5	0.03009404	0.0509429	1.69278997	102.287485	0.52004206
8	0.02649007	0.04491031	1.69536424	136.623525	0.51646858
8.5	0.02349486	0.03988261	1.69750367	179.923629	0.51348863
9	0.02097902	0.03564967	1.6993007	233.88395	0.51097853
9.5	0.0188457	0.03205323	1.7008245	300.407217	0.50884502
10	0.01702128	0.02897239	1.70212766	381.614879	0.50701675
12	0.011846	0.0202072	1.70582428	904.053983	0.50181208
14	0.0087146	0.01488506	1.708061	1896.28727	0.49864962
16	0.0066778	0.0114158	1.70951586	3625.3135	0.49658724
18	0.00527937	0.00903043	1.71051474	6445.22491	0.49516881
20	0.00427807	0.00732077	1.71122995	10809.6488	0.49415196

$\gamma = 5/3$

Mach Number	p/p^*	T/T^*	$V/V^*=\rho^*/\rho$	p_t/p_t^*	T_t/T_t^*
0	2.66666667	0	0	1.29903811	0
0.02	2.66489007	0.00284066	0.00106596	1.29860542	0.00213078
0.04	2.65957447	0.01131734	0.00425532	1.29731135	0.00849253
0.06	2.65076209	0.02529554	0.00954274	1.29516772	0.01899442
0.08	2.63852243	0.04455552	0.01688654	1.292194	0.03348793
0.1	2.62295082	0.06879871	0.02622951	1.28841689	0.05177103
0.12	2.60416667	0.09765625	0.0375	1.28386985	0.07359375
0.14	2.58231117	0.13069929	0.0506133	1.27859245	0.09866489
0.16	2.55754476	0.1674505	0.06547315	1.27262967	0.12665956
0.18	2.53004428	0.20739642	0.08197343	1.26603107	0.15722723
0.2	2.5	0.25	0.1	1.25885001	0.19

(continued)

$\gamma = 5/3$

Mach Number	p/p*	T/T*	V/V*=ρ*/ρ	p_t/p_t*	T_t/T_t*
0.22	2.46761258	0.29471301	0.11943245	1.2511428	0.22460079
0.24	2.43309002	0.3409878	0.14014599	1.24296779	0.26065107
0.26	2.3966447	0.38828803	0.16201318	1.23438461	0.29777809
0.28	2.35849057	0.43609826	0.18490566	1.22545337	0.33562122
0.3	2.31884058	0.48393195	0.20869565	1.21623392	0.37383743
0.32	2.27790433	0.53133805	0.2332574	1.20678522	0.41210579
0.34	2.23588597	0.57790591	0.25846842	1.19716472	0.45013091
0.36	2.19298246	0.6232687	0.28421053	1.18742794	0.48764543
0.38	2.14938205	0.66710536	0.31037077	1.17762796	0.52441152
0.4	2.10526316	0.70914127	0.33684211	1.16781519	0.56022161
0.42	2.06079341	0.74914777	0.36352396	1.15803703	0.59489825
0.44	2.01612903	0.78694069	0.39032258	1.14833776	0.62829344
0.46	1.97141449	0.82237813	0.41715131	1.13875839	0.6602874
0.48	1.92678227	0.85535768	0.44393064	1.12933664	0.69078686
0.5	1.88235294	0.88581315	0.47058824	1.12010692	0.71972318
0.52	1.83823529	0.91371107	0.49705882	1.11110041	0.74705017
0.54	1.79452669	0.93904708	0.52328398	1.10234511	0.77274184
0.56	1.75131349	0.96184222	0.54921191	1.09386598	0.7967901
0.58	1.70867151	0.98213942	0.5747971	1.08568511	0.81920249
0.6	1.66666667	1	0.6	1.0778218	0.84
0.62	1.62535555	1.01550048	0.62478667	1.07029283	0.85921496
0.64	1.58478605	1.02872958	0.64912837	1.06311255	0.8768891
0.66	1.54499807	1.03978549	0.67300116	1.05629311	0.89307176
0.68	1.5060241	1.04877341	0.69638554	1.04984465	0.90781826
0.7	1.46788991	1.05580338	0.71926606	1.04377544	0.92118845
0.72	1.43061516	1.06098841	0.7416309	1.03809208	0.93324541
0.74	1.39421401	1.06444279	0.76347159	1.03279965	0.94405431
0.76	1.35869565	1.06628072	0.78478261	1.02790188	0.95368147
0.78	1.32406488	1.06661512	0.80556107	1.02340132	0.9621935
0.8	1.29032258	1.06555671	0.82580645	1.01929942	0.96965661
0.82	1.25746621	1.06321317	0.84552028	1.01559672	0.97613602
0.84	1.2254902	1.05968858	0.86470588	1.01229294	0.9816955
0.86	1.19438638	1.05508291	0.88336817	1.00938708	0.98639702
0.88	1.16414435	1.04949172	0.90151339	1.00687757	0.99030039
0.9	1.13475177	1.04300589	0.91914894	1.0047623	0.99346311
0.92	1.10619469	1.03571149	0.93628319	1.00303874	0.99594017
0.94	1.07845781	1.02768975	0.95292532	1.00170401	0.99778397
0.96	1.05152471	1.01901701	0.96908517	1.00075493	0.99904427
0.98	1.02537811	1.00976481	0.98477314	1.00018812	0.99976814
1	1	1	1	1	1
1.02	0.97537186	0.98978482	1.01477688	1.00018687	0.99978164
1.04	0.95147479	0.9791771	1.02911513	1.00074495	0.99915231
1.06	0.92828963	0.96823042	1.04302622	1.00167041	0.99814874
1.08	0.9057971	0.95699433	1.05652174	1.0029594	0.99680529
1.1	0.8839779	0.94551448	1.06961326	1.00460809	0.99515399
1.12	0.86281277	0.93383291	1.08231234	1.00661265	0.99322468
1.14	0.84228259	0.92198816	1.09463045	1.00896936	0.99104508

(continued)

γ = 5/3

Mach Number	p/p*	T/T*	V/V*=ρ*/ρ	p_t/p_t*	T_t/T_t*
1.16	0.82236842	0.91001558	1.10657895	1.01167453	0.98864093
1.18	0.8030516	0.89794743	1.11816904	1.01472458	0.98603608
1.2	0.78431373	0.88581315	1.12941176	1.018116	0.9832526
1.22	0.76613676	0.87363949	1.14031795	1.02184542	0.98031087
1.24	0.74850299	0.86145075	1.1508982	1.02590958	0.97722973
1.26	0.73139514	0.84926891	1.16116292	1.03030534	0.97402651
1.28	0.71479628	0.83711382	1.17112223	1.03502968	0.97071718
1.3	0.69868996	0.82500334	1.18078603	1.04007972	0.96731641
1.32	0.68306011	0.81295351	1.19016393	1.04545273	0.96383768
1.34	0.66789113	0.80097867	1.19926532	1.05114609	0.96029333
1.36	0.65316786	0.78909163	1.20809928	1.05715734	0.95669469
1.38	0.63887558	0.77730372	1.21667465	1.06348414	0.95305209
1.4	0.625	0.765625	1.225	1.07012431	0.949375
1.42	0.61152729	0.75406429	1.23308363	1.07707578	0.94567202
1.44	0.59844405	0.74262931	1.24093357	1.08433664	0.94195101
1.46	0.5857373	0.73132677	1.24855762	1.09190509	0.93821911
1.48	0.5733945	0.72016244	1.2559633	1.09977948	0.93448279
1.5	0.56140351	0.70914127	1.26315789	1.10795827	0.93074792
1.52	0.54975261	0.69826742	1.27014843	1.11644006	0.92701982
1.54	0.53843048	0.68754433	1.27694171	1.12522356	0.92330329
1.56	0.52742616	0.67697484	1.2835443	1.13430762	0.91960263
1.58	0.5167291	0.66656119	1.28996254	1.14369118	0.91592173
1.6	0.50632911	0.65630508	1.29620253	1.15337332	0.91226406
1.62	0.49621635	0.64620776	1.30227019	1.16335321	0.90863273
1.64	0.48638132	0.63627004	1.30817121	1.17363013	0.90503051
1.66	0.47681488	0.62649235	1.31391107	1.18420347	0.90145984
1.68	0.46750818	0.61687475	1.31949509	1.19507272	0.89792289
1.7	0.45845272	0.60741702	1.32492837	1.20623747	0.89442156
1.72	0.44964029	0.59811863	1.33021583	1.21769739	0.89095751
1.74	0.44106296	0.58897882	1.33536222	1.22945227	0.88753218
1.76	0.43271311	0.5799966	1.34037213	1.24150195	0.88414681
1.78	0.42458338	0.57117078	1.34524997	1.25384638	0.88080246
1.8	0.41666667	0.5625	1.35	1.26648559	0.8775
1.82	0.40895614	0.55398275	1.35462632	1.2794197	0.87424018
1.84	0.4014452	0.54561737	1.35913288	1.29264887	0.87102358
1.86	0.3941275	0.53740211	1.3635235	1.30617337	0.86785067
1.88	0.3869969	0.52933508	1.36780186	1.31999353	0.86472179
1.9	0.38004751	0.52141435	1.3719715	1.33410975	0.86163721
1.92	0.37327361	0.51363786	1.37603583	1.34852249	0.85859705
1.94	0.36666972	0.50600354	1.37999817	1.36323228	0.85560139
1.96	0.36023055	0.49850925	1.38386167	1.37823971	0.85265022
1.98	0.35395098	0.49115279	1.38762941	1.39354543	0.84974344
2	0.34782609	0.48393195	1.39130435	1.40915015	0.84688091
2.2	0.29411765	0.41868512	1.42352941	1.58183529	0.82062284
2.4	0.25157233	0.36454254	1.4490566	1.78555468	0.79834817
2.6	0.2173913	0.3194707	1.46956522	2.02170011	0.77950851
2.8	0.18957346	0.28175468	1.48625592	2.29192691	0.76355518

(continued)

γ = 5/3

Mach Number	p/p*	T/T*	V/V*=ρ*/ρ	p_t/p_t*	T_t/T_t*
3	0.16666667	0.25	1.5	2.59807621	0.75
3.2	0.14760148	0.22309064	1.51143911	2.94212205	0.73843003
3.4	0.13157895	0.2001385	1.52105263	3.32613496	0.72850416
3.6	0.1179941	0.180437	1.52920354	3.75225662	0.71994361
3.8	0.10638298	0.16342236	1.53617021	4.22268193	0.7125215
4	0.09638554	0.14864276	1.54216867	4.73964614	0.70605313
4.5	0.07673861	0.11924849	1.55395683	6.25059468	0.69313183
5	0.0625	0.09765625	1.5625	8.10261339	0.68359375
5.5	0.05186386	0.08136826	1.56888169	10.3322275	0.67637363
6	0.04371585	0.06879871	1.57377049	12.9762868	0.67078742
6.5	0.03733956	0.05890675	1.57759627	16.0718414	0.66638255
7	0.03225806	0.05098855	1.58064516	19.6560699	0.6628512
7.5	0.02814424	0.04455552	1.58311346	23.7662369	0.6599787
8	0.0247678	0.03926042	1.58513932	28.4396661	0.65761198
8.5	0.02196294	0.03485128	1.58682224	33.7137226	0.65563966
9	0.01960784	0.03114187	1.58823529	39.6258016	0.65397924
9.5	0.01761145	0.02799222	1.58943313	46.2133204	0.65256858
10	0.01590457	0.02529554	1.59045726	53.513713	0.65136023
12	0.01106501	0.01763055	1.593361	90.5930848	0.64792273
14	0.00813835	0.01298162	1.59511699	142.075496	0.64583577
16	0.00623539	0.00995329	1.59625877	210.359806	0.64447548
18	0.00492914	0.00787205	1.59704251	297.845382	0.64354024
20	0.00399401	0.00638084	1.59760359	406.931842	0.64286994

γ = 1.3

Mach Number	p/p*	T/T*	V/V*=ρ*/ρ	p_t/p_t*	T_t/T_t*
0	2.3	0	0	1.25517379	0
0.02	2.29880462	0.0021138	0.00091952	1.25484764	0.0018382
0.04	2.29522593	0.0084289	0.00367236	1.25387164	0.00733124
0.06	2.28928614	0.01886699	0.00824143	1.25225299	0.01641494
0.08	2.2810219	0.03329959	0.01459854	1.25000364	0.02898396
0.1	2.27048371	0.05155096	0.02270484	1.24714002	0.04489416
0.12	2.2577352	0.0734021	0.03251139	1.24368284	0.06396578
0.14	2.24285213	0.09859556	0.0439599	1.23965678	0.08598733
0.16	2.22592134	0.12684098	0.05698359	1.23509015	0.11072004
0.18	2.2070395	0.15782116	0.07150808	1.23001446	0.13790275
0.2	2.18631179	0.19119837	0.08745247	1.22446405	0.16725701
0.22	2.16385052	0.22662086	0.10473037	1.21847558	0.19849228
0.24	2.13977374	0.26372918	0.12325097	1.21208764	0.23131113
0.26	2.11420377	0.30216237	0.14292018	1.20534027	0.26541417
0.28	2.08726586	0.34156362	0.16364164	1.19827451	0.3005047
0.3	2.05908684	0.38158548	0.18531782	1.19093196	0.33629294
0.32	2.02979384	0.42189446	0.20785089	1.18335443	0.37249979

(continued)

$\gamma = 1.3$

Mach Number	p/p*	T/T*	V/V*=ρ*/ρ	p_t/p_t^*	T_t/T_t^*
0.34	1.99951316	0.46217491	0.23114372	1.1755835	0.40886002
0.36	1.96836916	0.50213224	0.25510064	1.16766021	0.44512495
0.38	1.93648335	0.54149534	0.2796282	1.15962477	0.48106446
0.4	1.90397351	0.58001842	0.30463576	1.15151625	0.51646858
0.42	1.87095305	0.61748208	0.33003612	1.14337242	0.5511484
0.44	1.83753036	0.65369385	0.35574588	1.13522948	0.58493663
0.46	1.80380839	0.68848815	0.38168586	1.12712199	0.61768762
0.48	1.76988426	0.72172577	0.40778133	1.11908271	0.64927705
0.5	1.73584906	0.75329299	0.43396226	1.11114253	0.67960128
0.52	1.70178762	0.78310033	0.46016337	1.10333045	0.70857641
0.54	1.66777852	0.81108108	0.48632422	1.09567349	0.73613719
0.56	1.63389407	0.83718964	0.51238918	1.08819679	0.76223569
0.58	1.60020037	0.86139971	0.53830741	1.08092355	0.78683995
0.6	1.56675749	0.88370246	0.5640327	1.07387513	0.80993251
0.62	1.53361961	0.90410461	0.58952338	1.0670711	0.83150894
0.64	1.50083525	0.92262664	0.61474212	1.06052928	0.85157636
0.66	1.46844753	0.9393009	0.63965574	1.05426585	0.87015202
0.68	1.43649445	0.95416994	0.66423504	1.04829544	0.88726189
0.7	1.40500916	0.96728487	0.68845449	1.04263122	0.9029394
0.72	1.37402026	0.97870379	0.7122921	1.03728499	0.91722417
0.74	1.34355212	0.98849044	0.73572914	1.03226728	0.93016091
0.76	1.31362515	0.99671293	0.75874989	1.02758745	0.94179838
0.78	1.28425614	1.00344253	0.78134143	1.0232538	0.95218843
0.8	1.25545852	1.00875269	0.80349345	1.01927364	0.96138518
0.82	1.22724265	1.01271813	0.82519796	1.01565341	0.96944425
0.84	1.19961612	1.01541403	0.84644914	1.01239874	0.97642213
0.86	1.17258397	1.01691536	0.8672431	1.00951458	0.98237561
0.88	1.14614894	1.01729628	0.88757774	1.00700521	0.98736124
0.9	1.12031174	1.0166297	0.90745251	1.0048744	0.99143496
0.92	1.09507123	1.01498679	0.92686829	1.00312541	0.99465175
0.94	1.07042463	1.01243674	0.94582721	1.00176111	0.99706531
0.96	1.04636774	1.00904643	0.96433251	1.000784	0.99872783
0.98	1.02289506	1.00488026	0.98238842	1.00019631	0.99968983
1	1	1	1	1	1
1.02	0.977675	0.99446469	1.01717307	1.00019688	0.99970509
1.04	0.95591169	0.98833056	1.03391408	1.00078859	0.99884983
1.06	0.93470098	0.98165102	1.05023002	1.00177668	0.99747695
1.08	0.91403319	0.97447667	1.06612831	1.00316263	0.99562705
1.1	0.89389817	0.96685527	1.08161679	1.00494792	0.9933387
1.12	0.87428537	0.95883188	1.09670356	1.00713399	0.99064842
1.14	0.8551839	0.95044882	1.111397	1.00972236	0.98759071
1.16	0.83658267	0.94174583	1.12570564	1.01271457	0.98419809
1.18	0.81847039	0.93276009	1.13963816	1.01611227	0.98050118
1.2	0.80083565	0.92352635	1.15320334	1.0199172	0.97652874
1.22	0.78366702	0.91407703	1.16640999	1.02413125	0.97230772
1.24	0.766953	0.9044423	1.17926693	1.02875642	0.96786337
1.26	0.75068214	0.89465019	1.19178297	1.03379488	0.96321929

(continued)

$\gamma = 1.3$

Mach Number	p/p^*	T/T^*	$V/V^*=\rho^*/\rho$	p_t/p_t^*	T_t/T_t^*
1.28	0.73484306	0.88472671	1.20396687	1.03924899	0.95839751
1.3	0.71942446	0.87469593	1.21582734	1.04512128	0.95341856
1.32	0.70441515	0.86458011	1.22737296	1.05141447	0.94830154
1.34	0.6898041	0.85439979	1.23861223	1.05813151	0.9430642
1.36	0.67558041	0.84417389	1.24955353	1.06527554	0.93772304
1.38	0.6617334	0.83391979	1.26020508	1.07284995	0.93229332
1.4	0.64825254	0.82365345	1.27057497	1.08085835	0.92678918
1.42	0.63512752	0.81338949	1.28067114	1.08930459	0.92122371
1.44	0.62234826	0.80314126	1.29050134	1.09819278	0.91560897
1.46	0.60990485	0.79292095	1.30007319	1.10752727	0.90995608
1.48	0.59778767	0.78273964	1.3093941	1.11731267	0.90427529
1.5	0.58598726	0.77260741	1.31847134	1.12755385	0.89857601
1.52	0.57449444	0.76253335	1.32731197	1.13825596	0.89286688
1.54	0.56330025	0.7525257	1.33592288	1.14942441	0.88715582
1.56	0.55239596	0.74259185	1.3443108	1.16106488	0.88145007
1.58	0.54177306	0.73273845	1.35248226	1.17318334	0.87575625
1.6	0.53142329	0.72297143	1.36044362	1.18578604	0.87008039
1.62	0.52133862	0.71329605	1.36820106	1.19887952	0.86442798
1.64	0.51151123	0.70371699	1.37576059	1.21247059	0.85880398
1.66	0.50193354	0.69423835	1.38312805	1.22656638	0.8532129
1.68	0.49259818	0.68486373	1.39030909	1.24117429	0.84765881
1.7	0.483498	0.67559622	1.39730923	1.25630204	0.84214538
1.72	0.47462608	0.66643851	1.40413379	1.27195763	0.83667588
1.74	0.46597567	0.65739286	1.41078794	1.28814937	0.83125326
1.76	0.45754026	0.64846116	1.41727672	1.30488589	0.82588014
1.78	0.44931353	0.63964497	1.42360498	1.32217612	0.82055882
1.8	0.44128933	0.63094553	1.42977744	1.3400293	0.81529135
1.82	0.43346174	0.62236378	1.43579866	1.35845499	0.81007953
1.84	0.42582499	0.61390043	1.44167308	1.37746307	0.80492489
1.86	0.41837351	0.60555591	1.44740499	1.39706373	0.79982877
1.88	0.4111019	0.59733046	1.45299854	1.41726751	0.79479232
1.9	0.40400492	0.58922411	1.45845776	1.43808527	0.78981649
1.92	0.39707751	0.58123671	1.46378653	1.45952817	0.78490205
1.94	0.39031476	0.57336795	1.46898864	1.48160776	0.78004965
1.96	0.38371193	0.56561738	1.47406775	1.50433589	0.77525977
1.98	0.37726441	0.55798439	1.47902738	1.52772477	0.77053277
2	0.37096774	0.55046826	1.48387097	1.55178695	0.76586889
2.2	0.31541415	0.48151266	1.5266045	1.83245772	0.7226877
2.4	0.27097078	0.42292895	1.56079171	2.19700355	0.68551266
2.6	0.23498161	0.37326257	1.58847568	2.66440918	0.65369637
2.8	0.20550393	0.33109783	1.61115082	3.25825213	0.62649467
3	0.18110236	0.29518259	1.62992126	4.00738323	0.60319921
3.2	0.1607043	0.26445694	1.64561207	4.94671565	0.58318505
3.4	0.14349888	0.23804268	1.65884702	6.11812114	0.56592061
3.6	0.12886598	0.21521947	1.67010309	7.57143378	0.55096185
3.8	0.11632612	0.1953987	1.67974914	9.36556261	0.53794111
4	0.10550459	0.17809949	1.68807339	11.5697154	0.526555

(continued)

$\gamma = 1.3$

Mach Number	p/p*	T/T*	V/V*=ρ*/ρ	p_t/p_t*	T_t/T_t*
4.5	0.084172	0.14346976	1.70448307	19.4370944	0.5037036
5	0.06865672	0.11784362	1.71641791	32.0625477	0.48674538
5.5	0.05703658	0.09840843	1.72535648	51.779242	0.47385798
6	0.04811715	0.08334938	1.73221757	81.7942018	0.46385743
6.5	0.04112651	0.07146122	1.73759499	126.417257	0.45595363
7	0.03554869	0.06192175	1.74188563	191.326139	0.44960572
7.5	0.03102867	0.05415628	1.74536256	283.869894	0.44443465
8	0.02731591	0.04775419	1.74821853	413.412737	0.44016904
8.5	0.02422965	0.04241625	1.75059257	591.720416	0.43661079
9	0.02163688	0.03792051	1.75258702	833.391101	0.43361278
9.5	0.01943799	0.03409964	1.75427847	1156.33277	0.43106399
10	0.01755725	0.03082571	1.75572519	1582.28905	0.42887944
12	0.01222104	0.02150695	1.75982997	4919.12069	0.42265842
14	0.0089914	0.01584567	1.76231431	13079.431	0.4188769
16	0.00689035	0.0121541	1.7639305	30834.3282	0.4164102
18	0.00544766	0.00961533	1.76504027	66110.8737	0.41471339
20	0.00441459	0.00779543	1.76583493	131309.281	0.41349686

Bibliography

Ames Research Staff, Equations, Tables, and Charts for Compressible Flow, NACA Report No. 1135, 1953.

Anderson, J. D. Jr. *Hypersonic and High Temperature Gas Dynamics*, McGraw-Hill, New York, 1989.

Anderson, J. D. Jr. *Modern Compressible Flow*. 3rd ed., McGraw-Hill, New York, 2002.

Bertin, J. J., and Cummings, R. M. *Aerodynamics for Engineers*, Prentice Hall, New Jersey, 2009.

Clarke, J. F., and McChesney. *The Dynamics of Real Gases*, Butterworth, Washington, DC, 1964.

Courant, R., and Friedrichs, K. O. *Supersonic Flow and Shock Waves*, Springer Verlag, Reprinted by Interscience Publishers Inc. New York, 1948.

Ferri, A., *Elements of Aerodynamics of Supersonic Flows*. Dover Phoenix Editions, Mineola, NY, 2005.

Fox, R. W., McDonald, A. T., and Prichard, P. J. *Introduction to Fluid Mechanics*, 6th ed., John Wiley & Sons, Hoboken, NJ, 2004.

Glassman, I., and Yetter, R. A. *Combustion*, 4th ed., Academic Press, New York, 2008.

Goldstein, R. J., ed. *Fluid Mechanics Measurements*, Taylor and Francis, Washington, DC, 1996.

Hill, P., and Peterson, C. *Mechanics and Thermodynamics of Propulsion*, 2nd. Ed., Prentice Hall, New Jersey, 1992.

John, J. E., and Keith, T. G. *Gas Dynamics*, 3rd ed., Prentice Hall, New Jersey, 2006.

Kuethe, A. M., and Chow, C.-Y. *Foundations of Aerodynamics*, John Wiley & Sons, New York, 1998.

Liepmann, H. W., and Roshko, A. *Elements of Gasdynamics*, John Wiley & Sons, New York, 1957.

Park, C. *Nonequilibrium Hypersonic Aerothermodynamics*, John Wiley & Sons, New York, 1990.

Pierce, A. D. *Acoustics: An Introduction to Its Physical Principles and Applications*, McGraw-Hill, New York, 1981.

Shapiro, A. H. *The Dynamics and Thermodynamics of Compressible Fluid Flow*, vol.1, Ronald Press, New York, 1953.

Shivamoggi, B. K. *Theoretical Fluid Dynamics*, 2nd ed., John Wiley & Sons, New York, 1998.

Sonntag, R. E., and Van Wylen, G. J. *Introduction to Thermodynamics: Classical and Statistical*, 3rd ed., John Wiley & Sons, New York, 1991.

Sutton, G. P., and Biblarz, O. *Rocket Propulsion Elements*, 7th ed., John Wiley & Sons, New York, 2001.

Thompson, P. A. *Compressible-Fluid Dynamics*, McGraw-Hill, New York, 1972.

Vincenti, W. G., and Kruger, C. H., *Introduction to Physical Gas Dynamics*, Krieger Publishing, Malabar, FL, 1975.

Von Mises, R. *Mathematical Theory of Compressible Flow*, Academic Press, New York, 1958.

Whitham, G. B. *Linear and Nonlinear Waves*, John Wiley & Sons, New York, 1974.

Zeldovich, Ya. B., and Raizer, Yu. P. *Physics of Shock Waves and High Temperature Hydrodynamic Phenomena*, Academic Press, New York, 1966.

Zucrow, M. J., and Hoffman, J. D. *Compressible Flow*, vol.1, John Wiley & Sons, New York, 1976.

Index